History of Solid State Physics

Joginder Singh Galsin

History of Solid State Physics

Springer

Joginder Singh Galsin
Department of Mathematics, Statistics
and Physics
Punjab Agricultural University
Ludhiana, Punjab, India

ISBN 978-981-95-0503-6 ISBN 978-981-95-0504-3 (eBook)
https://doi.org/10.1007/978-981-95-0504-3

This Springer imprint is published by the registered company Springer Nature Singapore Pte Ltd.
The registered company address is: 152 Beach Road, #21-01/04 Gateway East, Singapore 189721, Singapore

With love, respect, and gratitude, I dedicate this book to my beloved late parents whose guidance continues to illuminate my path in the beautiful journey of life.

Preface

History of science is both fascinating and crucial for fostering new ideas. One can ride on the wings of history and witnesses the evolution of different scientific disciplines unfold. Understanding this history prepares us for further systematic investigations in science.

When I was in college in the 1960s, we envisioned a future where television sets would hang on walls like framed art. That futuristic vision has become a reality today, thanks to Plasma, LCD and LED innovations. Furthermore, the advent of augmented reality, virtual reality and artificial intelligence is revolutionising the way we interact with our screens.

I believe it is imperative for science students and aspiring scientists to delve into the historical aspects of the development of science, particularly in the area of their interest. It is a journey through the minds of genius thinkers, their struggles and triumphs, which shaped the discoveries and innovations that make our lives easier today. It reveals the power of human ingenuity, turning yesterday's impossibilities into today's realities. It inspires us to push the boundaries of what is possible.

No one can understand nature fully, as nature is perfect. Our scientific endeavours comprising hypotheses, theories and experiments rely on a series of assumptions, simplifications and approximations, rather than absolute certainty. But that is nothing to be discouraged about. In fact, they serve as stepping stones towards further refinements and explorations of the different aspects of science.

Although the study of various properties of solids commenced much earlier, but Solid State Physics as a subject came into existence only in the middle of the 20th century. Solid State Physics revolutionised the world in ways previously unimaginable. During the entire teaching career which spans for about 35 years I came across only one book entitled, *Out of Crystal Maze-Chapters from History of Solid State Physics* published by Oxford Press in 1992 which describes the history of some aspects of Solid State Physics. Since then lot of developments have taken place with the emergence of new fields like high temperature superconductivity, quantum Hall effect, smart materials and nanoscience etc. which need a systematic historical survey. This fact encouraged me to write the historical development of solid state physics in a systematic manner. Actually, physics of solids is a very vast subject as it covers all the aspects of solids. A large number of scientists have contributed towards the development of Solid State Physics, and it is not possible to mention the contribution of all in one book. I will focus on the key milestones that have shaped the field of conventional Solid State Physics. I have concentrated on the developments I consider most important and the choice is never fully impartial. In the text, the important classical discoveries are mentioned with the name of the scientist and his life span. But there are many recent discoveries/inventions in which this information is not available, so I have given their original references in the book.

This book is written for a wide section of science students: for students with basic knowledge of physics, chemistry and for undergraduate students with the knowledge of elementary solid-state physics. I have tried to incorporate original references of various discoveries which may be helpful to researchers in various fields of Solid State Physics.

I am very grateful to my teacher Prof. K. N. Pathak, Ex-Vice Chancellor of Panjab University, Chandigarh, India, for creating my interest in the subject of Theoretical Solid-State Physics. I am also thankful to my colleagues, students and well-wishers especially Dr. Nathi Singh, Dr. Paramjit Singh, Late Dr. S. D. Raj, Dr. B. Pal who silently urged me to move forward on this project. I am grateful to my daughter Amardeep Kaur Galsin for reading carefully some parts of the book. I am also very grateful to my wife, Surinder Kaur, for her suggestions at various stages of the book and for inspiring me to devote time generously for the completion of this work. I am also thankful to Mr. Akshay Kumar of Somalia Printers, Ludhiana (Punjab), for doing the art work of the book. I would welcome and appreciate comments/suggestions/feedback for the improvement of the book in the near future.

Ludhiana, India

Joginder Singh Galsin

Competing Interests The author has no competing interests to declare that are relevant to the content of this manuscript.

Links, Licenses and Attributions of Photographs of Scientists

Chapter 1

1. **Aristotle**
   https://commons.wikimedia.org/wiki/File:Aristotle_Altemps_Inv8575.jpg?uselang=en#Licensing
   PUBLIC DOMAIN
   Attribution:
   By After Lysippos - Jastrow (2006), Public Domain, https://commons.wikimedia.org/w/index.php?curid=1359807
2. **Archimedes of Suracuse**
   https://en.wikipedia.org/wiki/Archimedes#/media/File:Domenico-Fetti_Archimedes_1620.jpg
   PUBLIC DOMAIN
   Attribution:
   By Domenico Fetti - http://archimedes2.mpiwg-berlin.mpg.de/archimedes_templates/popup.htm, Public Domain, https://commons.wikimedia.org/w/index.php?curid=146592
3. **Nicolaus Copernicus**
   https://en.wikipedia.org/wiki/Nicolaus_Copernicus#/media/File:Nikolaus_Kopernikus_MOT.jpg
   PUBLIC DOMAIN
   Attribution:
   By Unknown author - http://www.frombork.art.pl/Ang10.htm, https://www.welt.de/img/kultur/mobile152954235/1212503297-ci102l-w1024/Kopernikus-Gemaelde-in-Krakau.jpg, https://muzeum.torun.pl/wp-content/uploads/2023/03/Portert-Mikolaja-Kopernika-MOT.jpg, Public Domain, https://commons.wikimedia.org/w/index.php?curid=147956893

Chapter 2

1. **Isaac Newton**
   https://en.wikipedia.org/wiki/Isaac_Newton#/media/File:Portrait_of_Sir_Isaac_Newton,_1689_(brightened).jpg
   PUBLIC DOMAIN
   Attribution:
   By Godfrey Kneller - File:Portrait of Sir Isaac Newton, 1689.jpg from https://exhibitions.lib.cam.ac.uk/linesofthought/artifacts, Public Domain, https://commons.wikimedia.org/w/index.php?curid=132521185
2. **Galileo Galilei**
   https://en.wikipedia.org/wiki/Galileo_Galilei#/media/File:Galileo_Galilei_(1564-1642)_RMG_BHC2700.tiff
   PUBLIC DOMAIN
   Attribution:
   By Justus Sustermans - http://collections.rmg.co.uk/collections/objects/14174, Public Domain, https://commons.wikimedia.org/w/index.php?curid=62614082
3. **Sir William Rowan Hamilton**
   https://en.wikipedia.org/wiki/William_Rowan_Hamilton#/media/File:William_Rowan_Hamilton_portrait_oval_combined.png
   PUBLIC DOMAIN
   Attribution:
   By William_Rowan_Hamilton_portrait_oval.png: UnknownWilliam_Rowan_Hamilton_portrait_oval_2.png: Unknownderivative work: Quibik (talk) - William_Rowan_Hamilton_portrait_oval.pngWilliam_Rowan_Hamilton_portrait_oval_2.png, Public Domain, https://commons.wikimedia.org/w/index.php?curid=11336587
4. **Joseph-Louis Lagrange**
   https://commons.wikimedia.org/wiki/File:Lagrange_crop.jpg
   License: CC0 1.0 Universal – No Copyright
   Attribution:
   By Unknown author - File:Joseph_Louis_Lagrange.jpg, CC0, https://commons.wikimedia.org/w/index.php?curid=155608232

Chapter 3

1. **Christiaan Huygens**
   https://en.wikipedia.org/wiki/Christiaan_Huygens#/media/File:Christiaan_Huygens-painting.jpeg
   PUBLIC DOMAIN
   Attribution:
   By Caspar Netscher - http://ressources2.techno.free.fr/informatique/sites/inventions/inventions.html, Public Domain, https://commons.wikimedia.org/w/index.php?curid=44047

2. **Thomas Young**
   https://en.wikipedia.org/wiki/Thomas_Young_(scientist)#/media/File:Thomas_Young_by_Briggs_cropped.jpg
   License: CC BY-SA 4.0
   Attribution:
   By Henry Perronet Briggs - http://rstb.royalsocietypublishing.org/content/370/1666/20140308, CC BY-SA 4.0, https://commons.wikimedia.org/w/index.php?curid=109315084
3. **James Clerk Maxwell**
   https://en.wikipedia.org/wiki/James_Clerk_Maxwell#/media/File:James-Clerk-Maxwell-1831-1879.jpg
   PUBLIC DOMAIN
   Attribution:
   By Unknown author - https://www.researchgate.net/figure/James-Clerk-Maxwell-1831-1879_fig1_360164096, Public Domain, https://commons.wikimedia.org/w/index.php?curid=129476608

Chapter 4

1. **Max Planck**
   https://en.wikipedia.org/wiki/Max_Planck#/media/File:Max_Planck_by_Hugo_Erfurth_1938cr_-_restoration1.jpg
   PUBLIC DOMAIN
   Attribution:
   By Hugo Erfurth - This file was derived from: Max Planck by Hugo Erfurth 1938cr.jpg. Original source: https://www.dhm.de/lemo/bestand/objekt/max-planck, Public Domain, https://commons.wikimedia.org/w/index.php?curid=153625300
2. **Albert Einstein**
   https://en.wikipedia.org/wiki/Albert_Einstein#/media/File:Albert_Einstein_1947.jpg
   PUBLIC DOMAIN
   Attribution:
   By Photograph by Oren Jack Turner, Princeton, N.J. - This image is available from the United States Library of Congress's Prints and Photographs division under the digital ID cph.3b46036.This tag does not indicate the copyright status of the attached work. A normal copyright tag is still required. See Commons:Licensing., Public Domain, https://commons.wikimedia.org/w/index.php?curid=254353
3. **Werner Heisenberg**
   https://en.wikipedia.org/wiki/Werner_Heisenberg#/media/File:Bundesarchiv_Bild183-R57262,_Werner_Heisenberg.jpg
   License: CC BA-SA 3.0 (photograph can be used by giving credit to the creator)
   Attribution:
   By Bundesarchiv, Bild 183-R57262 / Unknown author / CC-BY-SA 3.0, CC BY-SA 3.0 de, https://commons.wikimedia.org/w/index.php?curid=5436254

4. **Max Born**
https://en.wikipedia.org/wiki/Max_Born#/media/File:Max_Born.jpg
PUBLIC DOMAIN
Attribution:
Public Domain, https://commons.wikimedia.org/w/index.php?curid=137838
5. **Louis de Broglie**
https://en.wikipedia.org/wiki/Louis_de_Broglie#/media/File:Broglie_Big.jpg
PUBLIC DOMAIN
Attribution:
By Unknown author - http://www.physics.umd.edu/courses/Phys420/Spring2002/Parra_Spring2002/HTMPages/whoswho.htm, Public Domain, https://commons.wikimedia.org/w/index.php?curid=622169
6. **Paul Dirac**
https://en.wikipedia.org/wiki/Paul_Dirac#/media/File:Paul_Dirac,_1933.jpg
PUBLIC DOMAIN
Attribution:
By Nobel Foundation - http://nobelprize.org/nobel_prizes/physics/laureates/1933/dirac.html, Public Domain, https://commons.wikimedia.org/w/index.php?curid=624401

Chapter 5

1. **Christian Samuel Weiss**
https://www.wikidata.org/wiki/Q66165#/media/File:Weiss-christian-samuel-1780-1856--bildnisse-beruehmter-mitglieder-der-deutschen-akademie-der-wissenschaften-zu-berlin-berlin-akademie-verlag-1950-abb024.jpg
PUBLIC DOMAIN
Attribution:
https://www.wikidata.org/wiki/Q66165#/media/File:Weiss-christian-samuel-1780-1856--bildnisse-beruehmter-mitglieder-der-deutschen-akademie-der-wissenschaften-zu-berlin-berlin-akademie-verlag-1950-abb024.jpg
2. **Rene-Just Hauy**
https://en.wikipedia.org/wiki/Ren%C3%A9_Just_Ha%C3%BCy#/media/File:Ren%C3%A9_Just_Ha%C3%BCy_2.jpg
PUBLIC DOMAIN
Attribution:
By Ambroise Tardieu (1788–1841) - http://digitalgallery.nypl.org/nypldigital/dgkeysearchdetail.cfm?trg=1&strucID=1762983&imageID=1624797&total=6&num=0&word=Tardieu%2C%20Ambroise&s=3&notword=&d=&c=&f=4&k=0&lWord=&lField=&sScope=&sLevel=&sLabel=&imgs=20&pos=5&e=w, Public Domain, https://commons.wikimedia.org/w/index.php?curid=6739765
3. **Auguste Bravais**
https://en.wikipedia.org/wiki/Auguste_Bravais#/media/File:Bravais2.gif
PUBLIC DOMAIN

Attribution:
Public Domain, https://commons.wikimedia.org/w/index.php?curid=472338

4. **Max von Laue**
https://www.wikidata.org/wiki/Q57067#/media/File:(UAZ)_AB.1.0583_Laue.tif
PUBLIC DOMAIN
Attribution:
By Franz Schmelhaus, Zürich - https://mobile.cmistar.ch/webclients/uzh/#/content/0b0d0fdea3304d3cb344cc589bff6e92, Public Domain, https://commons.wikimedia.org/w/index.php?curid=73562188, https://commons.wikimedia.org/wiki/File:(UAZ)_AB.1.0583_Laue.tif?uselang=en#Licensing

Chapter 7

1. **Walter H. Schottky**
https://en.wikipedia.org/wiki/Walter_H._Schottky#/media/File:Walter_Hermann_Schottky_(1886-1976).jpg
PUBLIC DOMAIN
Attribution:
By Unknown author - [1], Public Domain, https://commons.wikimedia.org/w/index.php?curid=16884280
2. **William Hume-Rothery**
https://eehe.org.uk/wp-content/uploads/2023/09/William-Hume-Rothery-3.jpg
License: CC BY-SA 4.0
Image source: sadeghi via wikimedia (CC BY-SA 4.0)
3. **Jacob Frenkel**
https://www.wikidata.org/wiki/Q951957#/media/File:Yakov_Frenkel.jpg
PUBLIC DOMAIN
Attribution:
By Unknown author - Archives of the museum of Saint-Petersburg Karl May school, Russia. See www.kmay.ru, Public Domain, https://commons.wikimedia.org/w/index.php?curid=12436643

Chapter 8

1. **WILLIAM Thomson (Lord Kelvin)**
https://en.wikipedia.org/wiki/Lord_Kelvin#/media/File:Sir_William_Thomson,_Baron_Kelvin_by_T._&_R._Annan_&_Sons.jpg
PUBLIC DOMAIN
Attribution:
By T. & R. Annan & Sons; restored by Adam Cuerden[1] - National Galleries of Scotland Accession number: PGP 230.1, Public Domain, https://commons.wikimedia.org/w/index.php?curid=141105716

2. **Antoine Lavoisier**
   https://en.wikipedia.org/wiki/Antoine_Lavoisier#/media/File:David_-_Portrait_of_Monsieur_Lavoisier_(cropped)2.jpg
   PUBLIC DOMAIN
   Attribution:
   By Jacques-Louis David - Metropolitan Museum of Art: entry 436106 (accession number: 1977.10), Public Domain, https://commons.wikimedia.org/w/index.php?curid=141652400
3. **Rober William Boyle**
   https://en.wikipedia.org/wiki/Robert_Boyle#/media/File:The_Shannon_Portrait_of_the_Hon_Robert_Boyle.jpg
   PUBLIC DOMAIN
   Attribution:
   By Johann Kerseboom - https://digital.sciencehistory.org/works/3r074v879, Public Domain, https://commons.wikimedia.org/w/index.php?curid=27091904
4. **Nicolas Leonard Sadi Carnot**
   https://en.wikipedia.org/wiki/Nicolas_L%C3%A9onard_Sadi_Carnot#/media/File:Sadi_Carnot.jpeg
   PUBLIC DOMAIN
   Attribution:
   By Louis-Léopold Boilly - https://www.alamy.com/portrait-of-lazare-nicolas-marguerite-comte-carnot-1753-1823-museum-private-collection-image212355036.html, Public Domain, https://commons.wikimedia.org/w/index.php?curid=577483
5. **Satyendra Nath Bose**
   https://en.wikipedia.org/wiki/Satyendra_Nath_Bose#/media/File:SatyenBose1925.jpg
   PUBLIC DOMAIN
   Attribution:
   By Unknown author - Picture in Siliconeer, Public Domain, https://commons.wikimedia.org/w/index.php?curid=2201891

Chapter 9

1. **Henri Becquerel**
   https://en.wikipedia.org/wiki/Henri_Becquerel#/media/File:Paul_Nadar_-_Henri_Becquerel.jpg
   PUBLIC DOMAIN
   Attribution:
   By Paul Nadar / Adam Cuerden - This file comes from Gallica Digital Library and is available under the digital ID btv1b532297273/f1, Public Domain, https://commons.wikimedia.org/w/index.php?curid=120421283

2. **Ernest Rutherford**
   https://en.wikipedia.org/wiki/Ernest_Rutherford#/media/File:Sir_Ernest_Rutherford_LCCN2014716719_-_restoration1.jpg
   PUBLIC DOMAIN
   Attribution:
   By Bain News Service, publisher Restored by: Bammesk - Library of Congress Catalog: https://lccn.loc.gov/2014716719
   Image download: https://cdn.loc.gov/service/pnp/ggbain/36500/36570v.jpg
   Original url: https://www.loc.gov/pictures/item/2014716719/, Public Domain, https://commons.wikimedia.org/w/index.php?curid=112190894
3. **Otto Hahn**
   https://en.wikipedia.org/wiki/Otto_Hahn#/media/File:Otto_Hahn_1970.jpg
   License: CC BY-SA 3.0 nl
   Attribution:
   By Basch, [...] / Opdracht Anefo - [1] Dutch National Archives, The Hague, Fotocollectie Algemeen Nederlands Persbureau (ANEFO), 1945-1989 bekijk toegang 2.24.01.04 Bestanddeelnummer 923-4491, CC BY-SA 3.0 nl, https://commons.wikimedia.org/w/index.php?curid=36966973
4. **Julius Robert Oppenheimer**
   https://en.wikipedia.org/wiki/J._Robert_Oppenheimer#/media/File:Oppenheimer_(cropped).jpg
   PUBLIC DOMAIN
   Attribution:
   By Unknown author - https://catalog.archives.gov/id/558579, Public Domain, https://commons.wikimedia.org/w/index.php?curid=134271687
5. **Enrico Fermi**
   https://en.wikipedia.org/wiki/Enrico_Fermi#/media/File:Enrico_Fermi_1943-49.jpg
   PUBLIC DOMAIN
   Attribution:
   By Department of Energy-Office of Public Affairs, restored by Yann - This tag does not indicate the copyright status of the attached work. Commons: Public Domain, https://commons.wikimedia.org/w/index.php?curid=156854

Chapter 10

1. **Felix Bloch**
   https://en.wikipedia.org/wiki/Felix_Bloch#/media/File:Felix_Bloch,_Stanford_University.jpg
   License: CC BY 3.0
   Attribution:
   By Stanford University / Courtesy Stanford News Service - Stanford News Service, CC BY 3.0, https://commons.wikimedia.org/w/index.php?curid=5683254

2. **Arnold Wilhelm Sommerfeld**
   https://en.wikipedia.org/wiki/Arnold_Sommerfeld#/media/File:A._Sommerfeld_LCCN2014715010.jpg
   PUBLIC DOMAIN
   Attribution:
   By Bain News Service, publisher - Library of Congress Catalog: https://lccn.loc.gov/2014715010
   Image download: https://cdn.loc.gov/service/pnp/ggbain/34800/34858v.jpg
   Original url: https://www.loc.gov/pictures/item/2014715010/, Public Domain, https://commons.wikimedia.org/w/index.php?curid=67763649
3. **Rudolf Ernst Peierls**
   https://en.wikipedia.org/wiki/Rudolf_Peierls#/media/File:Peierls,Rudolf_1966_G%C3%B6ttingen.jpg
   License: CC BY 3.0
   Attribution:
   By GFHund - Own work, CC BY 3.0, https://commons.wikimedia.org/w/index.php?curid=11494196
4. **Frederick Seitz**
   https://en.wikipedia.org/wiki/Frederick_Seitz#/media/File:Dr._Frederick_Seitz_&_Dr._H.A._Kramers_Oak_Ridge_(8365596582).jpg
   PUBLIC DOMAIN
   Attribution:
   By doe-oakridge - Dr. Frederick Seitz & Dr. H.A. Kramers Oak Ridge, Public Domain, https://commons.wikimedia.org/w/index.php?curid=65679711
5. **Walter Kohn**
   https://en.wikipedia.org/wiki/Walter_Kohn#/media/File:Walter_Kohn.jpg
   License: CC BY-SA 3.0
   Attribution:
   By Markus Pössel (User name: Mapos) - Own work, CC BY-SA 3.0, https://commons.wikimedia.org/w/index.php?curid=20120170

Chapter 11

1. **Charles-Augustin de Coulomb**
   https://en.wikipedia.org/wiki/Charles-Augustin_de_Coulomb#/media/File:Charles_de_Coulomb.png
   PUBLIC DOMAIN
   Attribution:
   By After Hippolyte Lecomte - Unknown source, Public Domain, https://commons.wikimedia.org/w/index.php?curid=76393504
2. **Michael Faraday**
   https://en.wikipedia.org/wiki/Michael_Faraday#/media/File:Michael_Faraday_sitting_crop.jpg
   PUBLIC DOMAIN

Attribution:
By Unknown author - [1], Public Domain, https://commons.wikimedia.org/w/index.php?curid=141504840

3. **Georg Simon Ohm**
https://en.wikipedia.org/wiki/Georg_Ohm#/media/File:Georg_Simon_Ohm_(1789-1854).jpg
PUBLIC DOMAIN
Attribution:
By Unknown author - Bayerische Akademie der Wissenschaften https://www.facebook.com/badw.muenchen/photos/a.1504466919851549/2050320341932868/, Public Domain, https://commons.wikimedia.org/w/index.php?curid=154471715
4. **Hans Christian Oersted**
https://en.wikipedia.org/wiki/Hans_Christian_%C3%98rsted#/media/File:Hans_Christian_%C3%98rsted_daguerreotype.jpg
PUBLIC DOMAIN
Attribution:
By Unknown author - digitaltmuseum.se, Public Domain, https://commons.wikimedia.org/w/index.php?curid=11218804
5. **Paul Langevin**
https://en.wikipedia.org/wiki/Paul_Langevin#/media/File:Paul_Langevin_Wellcome2.jpg
License: CC BY 4.0
Attribution:
By Henri Manuel - https://wellcomeimages.org/indexplus/obf_images/f5/f5/fddc87e79e22bf5412905fdc09f9.jpg
Gallery: https://wellcomeimages.org/indexplus/image/V0028151.html, CC BY 4.0, https://commons.wikimedia.org/w/index.php?curid=41301998
6. **Pierre-Ernest Weiss**
https://en.wikipedia.org/wiki/Pierre_Weiss#/media/File:ETH-BIB-Weiss,_Pierre_(1865-1940)-Portrait-Portr_01257.tif_(cropped).jpg
PUBLIC DOMAIN
Attribution:
By Johannes Meiner - This image is from the collection of the ETH-Bibliothek and has been published on Wikimedia Commons as part of a cooperation with Wikimedia CH. Corrections and additional information are welcome., Public Domain, https://commons.wikimedia.org/w/index.php?curid=59562558
7. **Louis Neel**
https://en.wikipedia.org/wiki/Louis_N%C3%A9el#/media/File:Louis_N%C3%A9el_1970b.jpg
PUBLIC DOMAIN
Attribution:
By Farabola - [1], Public Domain, https://commons.wikimedia.org/w/index.php?curid=62219660

8. **Klaus Von Klitzing**
   https://en.wikipedia.org/wiki/Klaus_von_Klitzing#/media/File:Klaus_von_Klitzing_2015.jpg
   License: CC BY 4.0
   Attribution:
   By Pontifical Academy of Sciences - http://www.casinapioiv.va/content/accademia/en/academicians/ordinary/vonklitzing.html, CC BY-SA 4.0, https://commons.wikimedia.org/w/index.php?curid=107740587

Chapter 12

1. **Pieter Zeeman**
   https://en.wikipedia.org/wiki/Pieter_Zeeman#/media/File:Pieter_Zeeman.jpg
   PUBLIC DOMAIN
   Attribution:
   By Unknown author - http://he.wikipedia.org/wiki/Image:Zeeman.jpg, Public Domain, https://commons.wikimedia.org/w/index.php?curid=584960
2. **John Hasbrouck Van Vleck**
   https://en.wikipedia.org/wiki/John_Hasbrouck_Van_Vleck#/media/File:JH_van_Vleck_1974.jpg
   License: CC BY-SA 3.0 nl
   Attribution:
   By JH_van_Vleck_and_HBG_Casimir,_1974.jpg: Unknownderivative work: Materialscientist - This file was derived from: JH van Vleck and HBG Casimir, 1974.jpg:, CC BY-SA 3.0 nl, https://commons.wikimedia.org/w/index.php?curid=20427072
3. **Nicolaas Bloembergen**
   https://nl.wikipedia.org/wiki/Nico_Bloembergen#/media/Bestand:Nico_2006_Tucson_Wiki.png
   License: CC BY-SA 3.0
   Attribution:
   Door Pinoccio op de Nederlandstalige Wikipedia, CC BY-SA 3.0, https://commons.wikimedia.org/w/index.php?curid=4757773

Chapter 13

1. **Hendrik Antoon Lorentz**
   https://www.researchgate.net/publication/309449683/figure/fig1/AS:570806994051077@1513102400889/Hendrik-Antoon-Lorentz-in-1911-Fotoboek-Solv
   License: CC BY 3.0 nl
2. **Peter Debye**
   https://en.wikipedia.org/wiki/Peter_Debye#/media/File:Debye100.jpg
   PUBLIC DOMAIN
   Attribution:
   By Unknown author - http://chem.ch.huji.ac.il/~eugeniik/history/debye.html, Public Domain, https://commons.wikimedia.org/w/index.php?curid=1083819

3. **Rudolf Julius Emanuel Clausius**
   https://en.wikipedia.org/wiki/Rudolf_Clausius#/media/File:Rudolf_Clausius_01.jpg
   PUBLIC DOMAIN
   Attribution:
   By Picture taken by Theo Schafgans (1859–1907), Bonn; heliogravüre by Meisenbach, Riffarth & Co. Berlin. Scanned, image processed and uploaded by Kuebi = Armin Kübelbeck - Zeitschrift für Physikalische Chemie, Band 21, von 1896, Public Domain, https://commons.wikimedia.org/w/index.php?curid=1202617
4. **David Brewster**
   https://en.wikipedia.org/wiki/David_Brewster#/media/File:David_Brewster_1618.jpg
   License: CC BY-SA 4.0
   Attribution:
   By Scottish Scientists - Google, CC BY-SA 4.0, https://commons.wikimedia.org/w/index.php?curid=134636542

Chapter 14

1. **Ernest O Wollan**
   https://en.wikipedia.org/wiki/Ernest_O._Wollan#/media/File:Wollan_and_Shull_1949.jpg
   License: CC BY 3.0
   Attribution:
   By Photo is from US Atomic Energy Commission or Oak Ridge National Laboratory. Article rights are T. E. Mason et al. 2013 (for Oak Ridge National Laboratory). - http://europepmc.org/wicket/bookmarkable/uk.bl.ukpmc.web.utilities.redirect.RedirectPage?figure=fig12/&articles=PMC3526866; From: Acta Crystallogr A. 2013 January 1; 69(Pt 1): 37–44. Published online 2012 December 5. doi: 10.1107/S0108767312036021, CC BY 3.0, https://commons.wikimedia.org/w/index.php?curid=31343101
2. **Clifford G. Shull**
   https://en.wikipedia.org/wiki/Clifford_Shull#/media/File:Clifford_Shull_(cropped).jpg
   PUBLIC DOMAIN
   Attribution:
   By ENERGY.GOV - HD.3F.113, Public Domain, https://commons.wikimedia.org/w/index.php?curid=153296873

Chapter 15

1. **Kamerlingh Onnes**
   https://en.wikipedia.org/wiki/Heike_Kamerlingh_Onnes#/media/File:Kamerlingh_portret.jpg
   Copyrighted Free Use

Attribution:
By Unknown author - http://www.kb.nl/hkc/nobel/onnes/onnes.html. Further source: (cropped to 600×800), Museum Boerhaave, http://www.museumboerhaave.nl/contact/pers2a.html, Copyrighted free use, https://commons.wikimedia.org/w/index.php?curid=879737

2. **Walther Meissner**
https://en.wikipedia.org/wiki/Walther_Meissner#/media/File:Walther_Meissner.jpg
PUBLIC DOMAIN
Attribution:
By Wietek at German Wikipedia(Original text: Wietek) - Self-photographed, Public Domain, https://commons.wikimedia.org/w/index.php?curid=3165158
3. **John Bardeen**
https://en.wikipedia.org/wiki/John_Bardeen#/media/File:Bardeen.jpg
PUBLIC DOMAIN
Attribution:
Public Domain, https://commons.wikimedia.org/w/index.php?curid=6155225
4. **Johannes Georg Bednorz**
https://en.wikipedia.org/wiki/Georg_Bednorz#/media/File:8th_Silesian_Science_Festival,_Johannes_Georg_Bednorz,_KsP_709,_09.12.2024.jpg
License: CC BY-SA 4.0
Attribution:
By Krzysztof Popławski - Own work, CC BY-SA 4.0, https://commons.wikimedia.org/w/index.php?curid=156516437
5. **Karl Alexander Muller**
https://en.wikipedia.org/wiki/K._Alex_M%C3%BCller#/media/File:Karl_Alexander_Mueller_(cropped).jpg
License: CC BY-SA 3.0
Attribution:
By Armin Kübelbeck - own picture --Kuebi 21:37, 2 February 2007 (UTC), CC BY-SA 3.0, https://commons.wikimedia.org/w/index.php?curid=153212809
6. **Ivar Giaever**
https://en.wikipedia.org/wiki/Ivar_Giaever#/media/File:Ivar_Gi%C3%A6ver_3x4.jpg
License: CC BY 2.0
Attribution:
By NTNU Vitenskapsmuseet - Prisutdeling DKNVS., CC BY 2.0, https://commons.wikimedia.org/w/index.php?curid=153297269

Chapter 16

1. **Joseph John Thomson**
https://en.wikipedia.org/wiki/J._J._Thomson#/media/File:J.J_Thomson.jpg
PUBLIC DOMAIN

Attribution:
By Not Mentioned - First World War.com, Public Domain, https://commons.wikimedia.org/w/index.php?curid=2969861

2. **William Bradford Shockley**
https://en.wikipedia.org/wiki/William_Shockley#/media/File:William_Shockley,_Stanford_University.jpg
License: CC BY 3.0
Attribution:
By Chuck Painter / Stanford News Service - Stanford News Service, CC BY 3.0, https://commons.wikimedia.org/w/index.php?curid=5683262
3. **Charles Hard Townes**
https://en.wikipedia.org/wiki/Charles_H._Townes#/media/File:Charles_Townes.jpg
PUBLIC DOMAIN
Attribution:
By Unknown author - [1], Public Domain, https://commons.wikimedia.org/w/index.php?curid=92738267
4. **Zhores Ivanovich Alferov**
https://en.wikipedia.org/wiki/Zhores_Alferov#/media/File:Zhores_Alfyorov_2012.jpg
License: CC BY 4.0
Attribution:
By Council.gov.ru, CC BY 4.0, https://commons.wikimedia.org/w/index.php?curid=66830911

Chapter 17

1. **Alexander Edmond Becquerel**
https://en.wikipedia.org/wiki/Edmond_Becquerel#/media/File:Edmond_Becquerel_by_Nadar.jpg
PUBLIC DOMAIN
Attribution:
By Nadar - This file comes from Gallica Digital Library and is available under the digital ID btv1b53100803v, Public Domain, https://commons.wikimedia.org/w/index.php?curid=41635023
2. **Jonas Ferdinand Gabriel Lippman**
https://en.wikipedia.org/wiki/Gabriel_Lippmann#/media/File:Gabriel_Lippmann2.jpg
PUBLIC DOMAIN
Attribution:
By Nobel Foundation - http://nobelprize.org/nobel_prizes/physics/laureates/1908/lippmann-bio.html, Public Domain, https://commons.wikimedia.org/w/index.php?curid=1265606

Chapter 18

1. **Kim Eric Drexler**
   https://en.wikipedia.org/wiki/K._Eric_Drexler#/media/File:Drexler763x1000.jpg
   License: CC BY-SA 3.0
   Attribution:
   By Eric Drexler - Own work, CC BY-SA 3.0, https://commons.wikimedia.org/w/index.php?curid=32740676
2. **Harold Walter Kroto**
   https://en.wikipedia.org/wiki/Harry_Kroto#/media/File:Harry_Kroto.jpg
   License: CC BY-SA 3.0
   Attribution:
   By Pd1000 at English Wikipedia, CC BY-SA 3.0, https://commons.wikimedia.org/w/index.php?curid=48597530
3. **Gerd Binnig**
   https://en.wikipedia.org/wiki/Gerd_Binnig#/media/File:Gerd_Binnig_at_the_Memorial_Symposium_for_Heinrich_Rohrer_(cropped)_2.jpg
   License: CC BY 2.0
   Attribution:
   By ibm research zurich - https://www.flickr.com/photos/ibm_research_zurich/11432794546, CC BY 2.0, https://commons.wikimedia.org/w/index.php?curid=51007207
4. **Bernard Lucas Feringa**
   https://en.wikipedia.org/wiki/Ben_Feringa#/media/File:Ben_Feringa_-_april_2017_05.png
   License: CC BY-SA 4.0
   Attribution:
   By Richard Broekhuijzen - Own work, CC BY-SA 4.0, https://commons.wikimedia.org/w/index.php?curid=57771868
5. **Alexei Ekimov**
   https://en.wikipedia.org/wiki/Alexey_Ekimov#/media/File:US_Embassy_Sweden_2023_Nobel_Reception_(53390424929)_(cropped).jpg
   License: CC BY 2.0
   Attribution:
   By US Embassy Sweden - 4A0A1305-Enhanced-NR, CC BY 2.0, https://commons.wikimedia.org/w/index.php?curid=143178186

Contents

Abbreviations

ac	Alternating Current
AD	Anno Domini (indication the year after the estimated birth of Jesus Christ)
AEC	Atomic Energy Commission
AECL	Atomic Energy Canada Limited
AFM	Atomic Force Microscope
AM	Amplitude Modulation
APW	Augmented Plane Wave
ARPANET	Advanced Research Projects Agency Network
ARPES	Angle-resolved Photoemission Spectroscopy
AS	Axially Symmetric
BC	Before Christ
BCS	Bardeen-Cooper-Schrieffer
BvK	Born- von Karman
BZ	Brillouin Zone
CDD	Charge Density Distortion
CDW	Charge Density Wave
CGS	Centimeter-Gram-Second
CMOS	Complementary Metal-Oxide-Semiconductor
CNO	Carbon-Nitrogen-Oxygen
CNT	Carbon Nanotube
CPA	Coherent Potential Approximation
CPU	Central Processing Unit
CQD	Carbon Quantum Dot
CRN	Continuous Random Network
CRT	Cathode Ray Tube
CSIRO	Commonwealth Scientific and Industrial Organization
CTR	Computing-Tabulation-Recording
dc	Direct Current
DELTA	Depleted Lean-channel Transistor
DFT	Density Functional Theory

DPSS	Diode-pumped Solid State
EBR	Experimental Breeder Reactor
EEL	Electron Energy Loss
emf	Electromotive Force
EPR	Electron Paramagnetic Resonance
ER	Electrorheological
ESR	Electron Spin Resonance
FEL	Free-Electron Laser
FM	Frequency Modulation
GER	Giant Electrorheological
GSM	Global System of Mobile
GTF	General Tensor Force
IBM	International Business Machines
ICs	Integrated Circuits
IEEE	Institute of Electrical and Electronics Engineers
IMT	International Mobile Telecommunications
IP	Internet Protocol
IR	Infrared
ISTN	Inelastic Scattering of Thermal Neutrons
ITER	International Thermonuclear Experimental Reactor
JET	Joint European Torus
KKR	Korringa-Kohn-Rostoker
LAN	Local Area Network
LAPW	Linear Augmented Plane Wave
LCAO	Linear Combination of Atomic Orbitals
LCD	Liquid Crystal Display
LED	Light Emitting Diode
LEED	Low Energy Electron Diffraction
LSI	Large Scale Integration
MAS	Modified Axially Symmetric
MFM	Magnetic Force Microscope
MKS	Meter-Kilogram-Second
MOS	Metal-Oxide-Semiconductor
MOSFET	Metal-Oxide-Semiconductor Field Effect Transistor
MR	Magnetorheological
MRI	Magnetic Resonance Imaging
MWNT	Multi-walled Nanotube
Nd:YAG	Neodymium-doped Yttrium, Aluminium, Garnet
NMR	Nuclear Magnetic Resonance
NQR	Nuclear Quadrupole Resonance
NTT	Nippon Telegraph and Telephone
OLED	Organic Light Emitting Diode
OPW	Orthogonalized Plane Wave
OSRD	Office of Scientific Research and Development
PAC	Perturbed Angular Correlation

PC	Personal Computer
PCB	Printed Circuit Board
PDP	Plasma Display Panel
PLATO	Programmed Logic for Automatic Teaching Operation
PLED	Polymeric Light Emitting Diode
PLZT	Lead Lanthanum Zirconate Titanate
PMN	Lead Magnesium Niobate
PMN-PT	Lead Magnesium Niobate-Lead Titanate
PVC	Polyvinyl Chloride
QD	Quantum Dot
RCA	Radio Corporation of America
RCM	Random Coil Model
RCP	Random Close Packing
REM	Reflection Electron Microscope
REMs	Rare Earth Metals
RF	Radio Frequency
RPA	Random Phase Approximation
SBZ	Surface Brillouin Zone
SDW	Spin Density Wave
SEM	Scanning Electron Microscope
SGT	Single Gate Technology
SMOLED	Small Molecule Light Emitting Diode
SQUID	Superconducting Quantum Interference Device
STEM	Scanning Transmission Electron Microscope
STM	Scanning Tunnelling Microscope
SWNT	Single-walled Nanotube
TCP	Transmission Control Protocol
TDPAC	Time Differential Perturbed Angular Correlation
TDS	Thermal Diffused Scattering
TEM	Transmission Electron Microscope
TFT	Thin Film Transistor
TFTR	Tokamak Fusion Test Reactor
TMCs	Transition Metal Compounds
TMs	Transition Metals
TN	Twisted Nematic
TOF	Time of Flight
TPE	Thermoplastic Elastomer
TV	Television
UHF	Ultra High Frequency
UNIVAC	Universal Automatic Computer
UV	Ultra Violet
VCA	Virtual Crystal Approximation
VLSI	Very Large Scale Integration
WCDMA	Wide-band Code Division Multiple Access
WiFi	Wireless Fidelity

WiMAX	World-wide Interoperability to Microwave Access
WKB	Wentzel-Kramers-Brillouin
WS	Wigner-Seitz
XPS	X-ray Photoelectron Spectroscopy
ZEEP	Zero Energy Experimental Pile

Part I
Roots of Solid State Physics

Chapter 1
Early Developments in Science

In early human history, people lived in the jungles and focused primarily on securing food. They either hunted animals or foraged for plants to sustain themselves. Life was perilous in this environment, where survival depended on strength and adaptability. As the time went on, humans began to form small groups, ranging from a few individuals to several tens. In these groups, the strongest member often emerged as the leader. The leader's primary responsibility was to protect the group from wild animals and make arrangements for their food by killing animals, with the assistance of other group members of course. Slowly and steadily the small groups grew into tribes and then a number of tribes merged to form small empires. With time, humans started domesticating wild animals for help. They also began cultivating plants, and the concept of agriculture developed. During this time, humans also developed rational thought processes, and different civilizations emerged in different parts of the world. Some people who were deep thinkers and curious observers started investigating the nature around. Later these people were called philosophers. Naturally, in the beginning, the aspects of nature that came under observation were the ones that were most abundant in the surroundings and could be observed daily: like the earth, sun, stars, planets and the light coming from the sun. Therefore, astronomy was the dominant physical science throughout antiquity, but it had never been reduced successfully to a coherent system. The following were the common natural questions that intrigued the thinkers:

1. How did the world come into existence?
2. What was it composed of and how was it organised?
3. Why, when and how did the natural phenomena occur?

The word science comes from a Latin word "scientia" which means "*knowledge*". In science, developments can go on for decades or centuries before it takes the shape of a particular field. The history of the development of science gives us insight into the following:

J. S. Galsin, *History of Solid State Physics*,
https://doi.org/10.1007/978-981-95-0504-3_1

1. It tells us how the present situation came about and who are the people behind it, enabling us to understand the concepts within the subject better.
2. It explains how a possibility became a reality with time.
3. It points towards the missing links in the development of different fields on the subject to inspire future exploration and experimentation.

The quest to know about the surroundings started from the very beginning when humans came into existence. But the power of reasoning started with the development of civilizations such as Sumerians, Egyptians, Babylonians, Greece and Roman civilization etc. In the development of science, the following three factors play a crucial role:

1. **Chance Factor**: Chance often plays an important role in the development of science. In 1895, for instance, Wilhelm Roentgen (1845–1929) was studying electric discharge through the gases at very low pressures when he ended up discovering X-rays by chance. He could not determine even the nature of X-rays.
2. **Necessity**: Necessity fuels the development of science. In the last decade of the nineteenth century, Max Planck (1858–1947) was trying to explain the spectrum of black-body radiations by applying the established theories of the time, particularly the wave theory of electromagnetic radiations, but could not succeed. In 1900, he had to assume that in electromagnetic radiation, energy flows in the form of discrete energy packets ($E = h\nu$) called quanta of energy. He could explain the spectrum of black body radiations in terms of quanta of energy.
3. **Infrastructure and Experimental Facilities**: Another crucial factor is the type of infrastructure and knowledge available at the time of discovery. For instance, the experimental determination of atomic structure of a crystalline solid became possible only after the discovery of X-rays. In 1913, William Lawrence Bragg (1890–1971) and William Henry Bragg (1862–1942) determined, for the first time, the structure of rock salt crystal with the help of X-ray diffraction technique. Further, the magnetic structure determination of a crystalline solid on the atomic scale became possible only after the discovery of neutrons by Sir James Chadwick (1891–1974) in 1932. In 1946, Ernest Omar Wollan (1902–1984) and Clifford G. Shull (1915–2001) applied neutron diffraction technique to study the magnetic structure of a number of crystalline solids.

Most of the early developments of science can be traced back to central Asia and Europe, particularly Greece and Italy. Miletus had been a prosperous and flourishing city of ancient Greece before Athens and Sparta came about, and was probably one of the first centers of scientific activities. The people there were of the opinion that it was possible to understand the world through discussions. Caravans from other civilizations (Mesopotamia and all over the Mediterranean) used to visit Miletus and exchange their ideas. At Miletus, Thales (626/623BC–548/545BC) was regarded as the first philosopher of the Western World who broke the old methods of explaining the World and found a different way of looking for the answers. He was the first to have engaged in mathematics, science and logical reasoning. According to Herodotus

(484BC–425BC), a Greek historian, Thales predicted a solar eclipse and the same was observed on May, 28, 585BC. Herodotus is regarded as the first ever historian in the West. According to Isaac Asimov (1920–1992), an American fiction writer, this solar eclipse is the earliest historical event whose date is known with precision and this event is regarded as the birth of science. The writings of Thales show that load stone or magnetite was known to attract iron filings. It is worth mentioning here that magnetism was known to Chinese way back in 2500BC and to Greeks since 800BC. Chinese were able to make a magnetic compass somewhere around 200BC. Thales set up his own school known as "*Milesean School*" and tutored a number of students and among these Anaximander (610BC–546BC) and Hecataeus (550BC–476BC) were the most prominent. However, the findings and predictions made by Thales were not documented. We came to know about these from the writings of other people.

This revolution in rational thinking inaugurated the first dawn of scientific thought. Anaximander (610BC–546BC) was one of the greatest thinkers of his times. He succeeded Thales and became the master of Milesean School. His work stretched across different fields which are today known as physics, geography, astronomy and biology. He created the first map of the World and contributed immensely to the advancement of geography. In astronomy, he attempted to create a mechanical model for celestial bodies in relation to earth. It was Anaximander who, for the first time, considered Sun as a huge mass and tried to estimate its distance from the Earth. He believed the Earth is flat, probably the top face of a cylindrical object. Further, he believed that the Earth is situated at the centre of the universe and floats in the sky and the sky continues even beneath the Earth. Anaximander speculated and argued that the universe consists of infinite space—"the Boundless". Anaximander's book quoted under the title, "*On the Nature of Things (periphyseos)*" probably contained the information about his World map and celestial model (the original writings are unavailable). Anaximander was the first person to postulate that most of the substances are made of a single constituent, which he called *apeiron* (meaning infinite, unlimited or indefinite). He assumed that apeiron can be compressed and rarefied thus transforming one element into another. According to the available historical documents, Anaximander was the first philosopher who had written down his studies, although only a fragment of his work remains. Miletus was also the place where people, for the first time, made their own laws and had the first parliament (called Panionium) in the history of the world. It would be appropriate to say that most of the scientific and philosophical traditions (Mediterranean and modern) had their roots in the speculations of the thinkers of Miletus in the 600BC era.

Pythagoras (570BC–495BC) was an ancient Greek thinker and disciple of Anaximander. He was born in Samos, a small island near Miletus. It is believed that he was the first person to call himself a *philosopher*. Pythagoras was a genius and was a great philosopher through the Middle Ages. In Samos, he founded a school known as '*semicircle*'. Pythagoras was a charismatic political and religious teacher. He believed that all souls are immortal and after death, a soul is transferred into a new body. His political and religious teachings had a significant impact on the philosophies of Plato and Aristotle and through them, the Western philosophy. Pythagoras made many mathematical discoveries including *Pythagoras theorem*. It states that in a right-angled triangle, the square of the hypotenuse is equal to the sum of the squares of the other two sides. He was a great advocate of numbers and believed that numbers govern forms and ideas. He had said that mathematics is the language best adopted to understand and describe the world. Pythagoreans believed that all things are made of numbers. They attached significance to each number from one to ten. Further, they discovered that the musical notes could be translated into mathematical equations. According to Xenophanes (560BC–478BC) and Herodotus (484BC–425BC), Pythagoras and his contemporary Parmenides (515BC–460BC) of Elea were the first to propose and teach that Earth was spherical and divided the globe into five climatic zones. Eratosthenes (276BC–194BC) was the chief librarian of Library of Alexandria and believed that earth was spherical. Using shadows and sunlight he proved that earth is spherical and calculated the earth's diameter to be 7900 miles. The actual value of earth's diameter is 7918 miles—a stunning accuracy over 2200 years ago. The philosophy of Pythagoras also significantly impacted the great scientists such as Nicolaus Copernicus, Johannes Kepler and Isaac Newton. However, no authentic writings of Pythagoras have survived and almost nothing is known for certain about his life.

The city of Miletus was destroyed in 494BC by the invasion of the Persian Empire. But twenty years later, the Greeks repelled the Persians and Miletus was rebuilt and repopulated to become the centre of thoughts and spirit again. Leucippus (480BC–420BC), one of the thinkers, moved from Miletus to Abdera in 450BC and founded a scientific and philosophical school there. Democritus (460BC–370BC) who hailed from Abdera became his disciple and joined his school. Very few details about Leucippus' life are known and even his historical existence has on occasion been questioned. But Leucippus is considered to be the first philosopher who had proposed and developed a theory of atomism. But Leucippus' reputation was shadowed by Democritus who was a genius. Democritus wrote dozens of works on every field of knowledge. He wrote books on physics, philosophy, ethics, politics, cosmology, religion, origin of human societies and much more, though unfortunately these books have been lost. Democritus is credited for the first development of atomic theory and had postulated an elementary substance called *atom* and supposed that everything was made of atoms. The name atom is derived from a Greek word, "*atomos*" meaning uncuttable or indivisible. Atoms have no weight, no color and no taste but possess definite size and shape. Further, atoms are indivisible and form elementary particles. By collecting atoms with finite dimensions, one obtains finite-sized solids. He said the universe has infinite space in which there are innumerable atoms in motion. They move freely in space, collide with one another and push and pull one another. Similar atoms attract each other and join to form a substance. Therefore, in space there are atoms and vacuum. The infinite variety of substances of which the world is made of, originated solely from the different combinations of atoms: their shape, arrangement and order is important. The most important fact about the hypothesis of atoms is that it was purely based on observation, imagination and speculation. That time his ideas were not accepted but later they were revived in the modern era and became fundamental to the development of modern science. Actually, it was Aristotle who reported the hypothesis of atoms put forward by Democritus. Near about 390BC, Christianity dominated and became the only religion of the empire. Resultantly, all the scientific literature which was not in conformity with Christianity, including that of Democritus, was systematically destroyed. It is unfortunate that we lost the great treasure of intellectual knowledge given by Democritus. We know about his thoughts only through his quotations, references made by other ancient authors and by their summaries of the ideas.

After Miletus, Athens became the centre of philosophical thoughts. Socrates (470BC–399BC) was a Greek philosopher from Athens. He worked on the development of society, ethics and goodness. Socrates is considered to be the first moral philosopher of western ethical tradition of thought, and the founder of Western philosophy. He had a rare quality—he would admit ignorance instead of pretending to know something he did not. Socrates gave three principles of philosophy and asked to work on them. First, discover what the purpose of your life is. In other words, try to search who you are, what is your life's mission and what you are trying to become. Second, care for your soul. And the third and the last is be a good person then no outside forces will harm you. Actually, Socrates wrote nothing and what we know of his ideas and methods comes to us mainly from his contemporaries and disciples: mainly Plato (428/427BC–348/347BC) and Xenophon (340BC–355BC). Plato was Socrates' most famous disciple, and the majority of what most people know about Socrates is known from the book "Plato's Socrates". These works are known as the logoi sokratikoi, or Socratic accounts. Socrates strongly opposed the established ideologies of the government of his time, which was immensely influenced by the church. In 399BC, Socrates went on trial for corrupting the minds of the youth of Athens. As a consequence, he was made to drink poison, bringing an end to his life. Plato (428/427BC–348/347BC) is considered by many to be the most reputed philosopher who ever lived. He is known as the father of idealism in philosophy. In 380BC, Plato founded the *Platonist School of Thought and the Academy*, which is regarded as an ancient form of a university. His academy was the first institution of higher learning in the Western World and attracted many men of outstanding abilities including Aristotle. Plato, throughout his life, helped people reach a state of fulfilment. Plato is also considered as the first political philosopher. The influence of Pythagoreans on him was significant. He was the most innovative writer of his times. He wrote many philosophical texts (at least 25) and some known texts are The Republic, The Symposium, Apology, Crito, Phaedo, and The Laws. Out of these "The Republic" is Plato's best known work which is widely acknowledged as the corner stone of Western philosophy. Plato made Aristotle his successor at the

Academy. Both Plato and Aristotle knew the ideas of Democritus but Plato did not agree with them as is evident from several of Plato's texts.

Marble sculpture of Aristotle (384BC–322BC), Public Domain, URL: https://commons.wikimedia.org/wiki/File:Aristotle_Altemps_Inv8575.jpg?uselang=en#Licensing

Aristotle (384BC–322BC) was one of the greatest intellectual figures and the most outstanding philosopher and scientist of his times. After the death of Plato, the control of the Academy was passed on to Plato's nephew Speusippus (408BC–339BC). Aristotle was disappointed with this, and therefore he left Athens.Philip II of Macedon requested Aristotal to tutor Alexander the Great in 343BC. In 335BC, Aristotle returned back to Athens and founded his own school called *Lyceum* where he spent the rest of his life studying, teaching and writing. He spoke extensively about the ideas of Democritus with respect and admiration. Aristotle's intellectual range was vast, covering most of the sciences and arts. He himself was a biologist (with special interest in zoology) and did lot of work in biology. He made pioneering contributions in the fields of philosophy and science. His most important treatises include Physics, Metaphysics, Nicomachean Ethics, Politics, On the Soul and Poetics. Theophrastus (371BC–287BC) was the successor of Aristotle who independently elaborated Aristotle's metaphysics and psychology and made additions to his study of nature. Most of Aristotle's work is probably not in its original form because later his students are believed to have edited it.

Aristotle's contributions to physical sciences are no less impressive. Aristotle studied nature systematically with keen observation. Aristotle wrote a book entitled "*Physics*" which is credited as the first systematic study of science that we know. The book did not take its name from the discipline. It is the discipline that got its name from Aristotle's book. Aristotelian physics is a broad field and includes a kind of natural science and natural philosophy. It discusses the most general principles of moving objects, both living and non-living, and the motion of celestial and terrestrial objects. The main purpose of this work was to discover principles and causes of change, movement or motion of natural objects and cosmic bodies. He postulated that Earth is in the centre of the universe and the moon, sun and other planets revolve around it in concentric spheres (geocentric model of universe). Aristotle further postulated that heavenly bodies have souls with supernatural intellects, which guide them in their travel through the universe. He posited in his work that terrestrial spheres were made of four elements namely earth, air, fire and water which undergo change and decay and a fifth element '*aether*' is unchangeable. One of the most important contributions of Aristotle is that he developed the concept of logic and introduced it in natural philosophy. That is why Aristotle is generally called the *father of logic philosophy* and is also considered the *first scientist*. Both Islamic philosophy and Christian theology were influenced by the work of Aristotle. Islamic scholars respected Aristotle as 'the First Teacher' and Christians regarded him as a great philosopher. Socrates, Plato and Aristotle are considered to be the greatest philosophers of the Western World. Later, Greek astronomer Aristarchus (301BC–230BC) of Samos suggested that it is Earth that revolves around the Sun, not vice versa, proposing the first heliocentric model of universe. Today even the best of Aristotle's scientific work is only of historical interest. In ancient times, the serious thinkers, the philosophers, explained natural phenomena based on myths involving Gods, especially Greek Gods of ancient Greece. *Natural history* is the qualitative and descriptive investigation, based on predictions and speculations, of the natural phenomena. The process of reasoning emerged with the development of civilizations. From the time of Socrates, the philosophers tried to explain nature with reasoning and they are called Natural philosophers and this study is called *Natural Philosophy*.

Painting of Archimedes (287BC–212BC), Public Domain, URL: https://en.wikipedia.org/wiki/Archimedes#/media/File:Domenico-Fetti_Archimedes_1620.jpg

Archimedes (287BC–212BC), born in Syracuse, Sicily, was a genius mathematician, physicist, engineer, inventor and astronomer. One of the greatest of all times, he is regarded as one of the leading scientists and the great mathematician of ancient times. He developed and perfected a method of integration which allowed him to find areas, volumes and surface areas of many bodies. In a book entitled "*A History of Greek Mathematics II*" by Thomas Little Heath (1861–1940), Michel Chasles (1793–1880) said that Archimedes' method of integration gave birth to the calculus of the infinites which was later conceived and improved by a number of scientists. Using this method Archimedes discovered pi (π) and estimated its value as 3.1605. He rigorously proved a range of geometrical theorems including area of a circle, area of an ellipse and area under a parabola etc. Archimedes showed that volume and surface of a sphere is (2/3)rd of the volume and surface, respectively, of a circumscribed cylinder including its plane faces. He was very proud of this discovery. Archimedes devised a system using exponentials for expressing very large numbers and proved the law of exponents. Archimedes determined the value of square root of 3 with great precision. He was also one of the first to apply mathematics to physical phenomena and founded hydrostatics and statics. He made pioneer contributions to the field of mechanics (which will be described in Chap. 2 on Classical Mechanics) and also did some work on optics (presented in Chap. 3 on Optics). Archimedes was killed during the siege of Syracuse by a Roman soldier despite orders that he should not be harmed. The last words of Archimedes were "Do not disturb my circles", referring to the circles in the mathematical drawing he had made". The works of Archimedes which have survived are: On Plane Equilibriums (two books), On Floating Bodies (two books), On the Sphere and Cylinder (two books), Quadrature of the Parabola, On Spirals, On Conoids and Spheroids, Measurement of a Circle, and The Sandreckoner. Unlike his inventions, Archimedes' mathematical writings were little known in antiquity.

Lucretius (99BC–55BC) was a Latin poet and philosopher who believed in the philosophy of Epicurus (341BC–270BC), pupil of Democritus. Lucretius' only best known work, based on the philosophies of Epicurus and Democritus, is his philosophical poem "*De rerum natura*" in which he described many aspects of their philosophies. This poem is also translated in English under the title '*On the Nature of Things*' and also as '*On the Nature of Universe*'. Very little is known about Lucretius' life but the only certainty is that he was either a friend or a client of Gaius Memmius (99BC–49BC) to whom the poem was addressed and dedicated. Lucretius believed that religion is ignorance and reason is the torch that enlightens. No attention was paid to the works of Lucretius for many centuries. In January 1417, Gian Francesco Poggio Bracciolini (1380–1459), an Italian scholar and humanist, rediscovered one of the text books of Lucretius from the library of German monastery. In this book, he said atoms in solids are in constant motion and interact with each other. He called them *dancing atoms*. The Catholic Church was against the works of Lucretius and in 1516 prohibited his work from being taught in schools and ultimately banned it in 1551. The BC era of scientific research ended with the work of Lucretius. It is pointed out here that the original scientific works of BC era philosophers were either not documented or were lost and got destroyed with time. The existing information of their works is derived from the references and writings of other philosophers. Furthermore, much less is known about the life of some of the BC era philosophers.

Claudius Ptolemy (100AD–170AD)[1] was an astronomer, mathematician, geographer and astrologist living in Alexandria in the Roman Empire. He wrote about a dozen scientific works. Out of these, three each on astronomy, geography and astrology are considered to be his most important contributions. Ptolemy devoted a lot of time to studying astronomy and in 150AD, he wrote a book titled "*Almagest*" which means 'the greatest' in Arabic. It is a 13-volume textbook on astronomy. In 1984, Gerald James Toomer (1934–present) translated and annotated Ptolemy's Almagest. Ptolemy supported Aristotle's views and proposed that Earth is at the centre of the universe and sun, stars and other heavenly bodies are positioned on

[1] From now onwards AD will not be written and it is understood that it is Christian era.

spheres and they daily rotate around the Earth in circular orbits. He further assumed that there is nothing beyond the spheres of sun, stars and other heavenly bodies. It is called *geocentric model* of solar system or geometric system that prevailed for about 1400 years. Ptolemy gave a precise and complex mathematical framework (called Ptolemy theory) capable of predicting movements of planets in the sky with almost complete precision. As a mathematician, his contribution to trigonometry is especially important. Ptolemy created a world map where he drew longitudinal and latitudinal lines for the first time. With the help of these lines, one can find the coordinates of a place thus pin pointing its position. Ptolemy's geographical work was almost unknown in Europe until about 1300AD. But much of the medieval astronomy and geography were built on his ideas. Ptolemy wrote an astrological treatise named, '*Tetrabiblos*' or its Latin equivalent '*Quadripartitum*'. On geography, Ptolemy wrote a book entitled, "*Geographike hyphegesis*" known today simply as Geography. He also contributed towards music theory through his work known as "*Harmonics*". In addition to this, he contributed towards Optics as described in Chap. 3 on Optics.

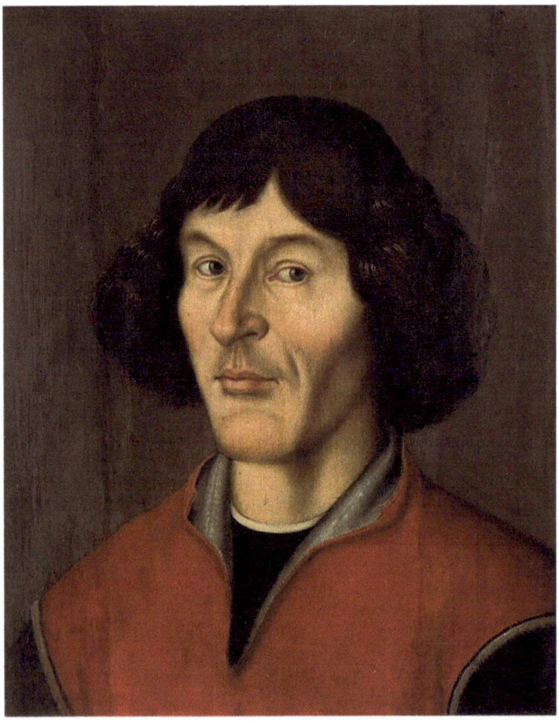

Portrait painting of Nikolas Copernicus (1473-1543), Public Domain, URL: https://en.wikipedia.org/wiki/Nicolaus_Copernicus#/media/File:Nicolaus_Kopernikus_MOT.jpg

It is very hard to believe that after this, scientific activity remained dormant for more than one thousand years. But, in these years science was dominated by religious dogma and the people at the helm of affairs were either religious priests or monks. Not much priority was given to the scientific achievements made in the scientific world. The first sign of a return to scientific thinking came from Christopher Columbus (1451–1506) who took his first adventurous voyage in 1492 assuming earth to be a sphere and ended up discovering America. Columbus' expeditions inaugurated a period of exploration, conquest and colonization that lasted for centuries helping create the modern Western World. Nikolas Copernicus (1473–1543) was a polymath and was active as a mathematician, astronomer and catholic canon. He studied in Krakow University of Poland and was a pupil of Albert Brudzewski (1445–1497) who was Professor of Aristotelian philosophy. Copernicus attended lectures on astronomy while he was in Krakow University, which piqued his interest in learning astronomy. He studied astronomy independently too and acquired good knowledge of philosophy and natural science. Copernicus was aware of the contradictions of Aristotle's and Ptolemy's models of the Universe. Aristotle proposed the system of concentric circles as mentioned earlier while Ptolemy gave the geocentric model assuming eccentric and epicycles. According to the epicycle system a terrestrial object is moving in a circle, which in turn is moving on another larger circle. For example, according to Ptolemy, a planet is moving around the Earth in a circle but its moon is moving in a smaller circle around the planet. The geocentric model could explain the variation in speed and direction of the motion of the moon, planet and sun. Nikolas Copernicus developed a *heliocentric model* of the solar system independently of Aristarchus (310BC–230BC) of Samos, an ancient Greek philosopher, who had formulated such a model, eighteen centuries earlier. Helios is a Greek word meaning 'sun', therefore heliocentric means 'sun at the centre'. In the heliocentric model Copernicus assumed that sun is fixed at the centre and the various planets revolve around it. He further assumed that the sun and stars are at rest and are situated on a sphere and there is nothing beyond the sphere of stars. Heliocentric model contradicted Ptolemy's geocentric model but resolved all the issues arising out of the geocentric model. By 1533, Copernicus completed the work on heliocentric theory in the form of a book "*De revolutionibus orbium coelestium*". In the same year, Copernicus explained his book to the Pope through his secretary and Pope was pleased to know about his revolutionary work. But the Protestants reacted to Copernicus' theory. Copernicus delayed the publication of his work due to the fear of philosophical and religious objections. Finally, in 1543, Copernicus published his book "De Revolutionibus Orbium Coelestium" just before his death and dedicated it to Pope III. This book triggered Copernican revolution and is now widely considered to have marked the beginning of modern astronomy and also of the scientific revolution.

Johannes Kepler (1571–1630), a German mathematician, astronomer and astrologer made significant contributions towards astronomy. At the time of Kepler there was no sharp distinction between astronomy and astrology. They were considered a single pursuit. Kepler was aware of Ptolemy and Copernicus theories of planetary motion and defended the heliocentric theory of Copernicus both from the theoretical and theological perspectives. In 1596, while Kepler was a student, he published his own ideas about astronomy in a book entitled, "*Mysterium Cosmographicum*". Johannes Kepler and Galileo Galilie (1564–1642) were contemporaries and came in contact with each other when, in 1597, Kepler sent his book Mysterium Cosmographicum to Galileo. Galileo wrote back to Kepler, acknowledging the gift. When in 1610, Galileo published his book entitled, "*Sidereus Nuncius*" Kepler supported the findings of Galileo in his open letter. Kepler and Galileo stayed in touch with each other through correspondence but never met personally. In 1600, Kepler shifted to Prague and became an assistant to the astronomer Tycho Brahe (1546–1601). While in Prague, in 1609, Kepler published his second important book entitled, "*Astronomia Nova*" which contained two laws of planetary motion. In 1621, Kepler published improved, expanded and revised edition of his book, 'Mysterium Cosmographicum' under the title, "*Mysterium*". Kepler used the concept of epicycles and eccentric cycles given by Ptolemy in advancing the heliocentric theory of Copernicus. It can be regarded as the first step in moderninzing the heliocentric theory of universe. Kepler described his new astronomy as 'Celestial Physics'. Kepler wanted to cover all the fundamentals of heliocentric astronomy in the form of a book. He worked for a number of years (from 1615 to 1621) and published a book entitled, "*Epitome Astronomiae Copernicanae*" which comprises of seven volumes. In this book, he described all the three laws of planetary motion which are stated below:

1. According to the first law, planets move in elliptical orbits around the sun, sun being at the focus of the ellipse.
2. The second law states that the radius vector of an elliptical orbit sweeps equal areas in equal intervals of time. The constant areal velocity is equivalent to saying that angular momentum is conserved in a central force motion. According to this law when a planet comes near to the sun, it speeds up, but when the planet goes away from the sun, it slows down.
3. The third law states that the squares of time periods of revolution of various planets are proportional to the cube of their major axis.

The above three laws of planetary motion laid the foundations for Newton's theory of universal gravitation. Kepler also did some fundamental work in the field of optics which will be described in Chap. 3 on Optics. He also worked on geometry and showed that an ellipse becomes a parabola when one of its foci is moved towards infinity. When two foci of an ellipse merge into one another the ellipse reduces to a circle. Kepler published a pamphlet entitled, "*Hexagonal Snow*" in which he explained that snow is built of closely-packed spheres with hexagonal symmetry—first thought of symmetry in nature. Kepler conjectured that the hexagonal symmetry is the most efficient arrangement for the packing of atoms if they are assumed to be spheres.

Galileo Galilei (1564–1642), a mathematician, astronomer and physicist, was born in Pisa, Italy. He was highly sophisticated, exceptionally intelligent and overflowing with inventiveness. He made pioneer contributions to both physics and astronomy. Up to the sixteenth century the development of science was based purely on speculation, prediction, observation and experience but Galileo incorporated the component of experimental observation into the study of science. In 1609, Galileo independently made his own telescope with 3X magnification and later its design was improved to get 30X magnification. It was known as a terrestrial telescope. For the first time in the history of astronomy Galileo directed the telescope towards the sky with a view to observe celestial objects. In 1609, Galileo observed mountains on the surface of Earth's moon and estimated their height. In 1610, Galileo made a number of important observations on celestial objects. He observed four satellites (moons) of Jupiter and named these as Medicean stars. Galileo's observation of satellites of Jupiter ignited a revolution in astronomy as it contradicted the Aristotelian cosmology. Initially many astronomers and philosophers refused to believe that Galileo could have discovered such a thing. Galileo observed Saturn with its ring and also observed that Venus exhibits a full set of phases similar to that of moon. In 1610, Galileo published his initial telescopic observations in a small book entitled, "*Sidereus Nuncius* (*Starry Messenger*)". It was his first publication in which he presented observation of Earth, moon, hundreds of stars not visible to the naked eye in the Milky Way and in certain constellations. Before the publication of 'Sidereus Nuncius' the Catholic Church had proclaimed that the Copernican heliocentric system is strictly mathematical and hypothetical in nature. In 1611, Galileo obtained remarkably accurate time periods of satellites of Jupiter and in 1612, he observed Neptune considering it a star, not a planet. In 1613, Galileo observed Sun Spots and reported these in "*Letters on Sun Spots*". He further observed that although the Milky Way appears to be clouds but it actually comprises of densely packed stars. Galileo also made pioneer contributions to mechanics which are described in Chap. 2 on Classical mechanics.

Galileo's observations proved that Copernicus' system is real: Earth and other planets move around the sun and follow some mathematical laws. In 1615, Galileo's work on heliocentric theory was brought into the knowledge of Roman Inquisition and it was conveyed that his work contradicted the Holy Bible. In 1616, Inquisition declared the heliocentric system to be a '*foolish and absurd philosophy*' and ordered Galileo to abandon the heliocentric theory completely. It banned the Copernicus' book, "*De Revolutionibus Orbium Coelestium*" and the other work done on heliocentric system. Galileo remained silent for about a decade until 1623, when the new Pope Urban VII was elected who was his friend and admirer. Thereafter, Galileo felt encouraged to write a book on his work on the heliocentric theory. He got the formal permission of the Inquisition with the condition that the heliocentric theory is not to be highlighted. In 1632, Galileo published his book, "*Dialogue Concerning the Two Chief World Systems*" in the form of questions and answers. Inquisition felt that Galileo's book was advocating heliocentric theory but Galileo denied the charge. Inquisition sentenced Galileo with formal imprisonment which was later commuted to house arrest. Galileo remained under house arrest for the rest of his life. During the house arrest Galileo wrote one of his finest works entitled, "*Two New Sciences*" and these sciences are now called kinematics and strength of materials. Albert Einstein (1879–1955) had high praise for this book. It is remarked here that Johannes Kepler is known as a mathematical astronomer while Galileo is considered an observational astronomer and is called the "*father of observational astronomy*".

Before seventeenth century there were no scientific journals and the method of informing the results of new investigations was through either letters or books. The Royal Society of London was founded in 1662 and started the first journal named

"*Philosophical Transactions of Royal Society*" in 1664. Henry Oldenburg (1619–1677) was its founding editor and publisher. The first issue of Philosophical Transactions was published in 1665 under the name "*Micrographia*" which solely contained the work of Robert Hooke (1635–1703). After this a number of Scientific Societies and Academies started their own journals thus streamlining the method of information of discoveries and inventions. The overall division of science into different branches was a gradual and historical process. In ancient Greece, Aristotle divided the theoretical science into three different branches: physics, mathematics and theology. The division of science into physics, chemistry and biology as we know today was done in eighteenth and nineteenth centuries. Biology was considered a branch of natural history up to the eighteenth century and chemistry was considered a branch of physics up to the nineteenth century. The development of new technologies, growth of scientific literature and the increasing specializations of scientific research forced the scientific community to divide science into the following four basic sciences:

1. Physics: science, which deals with physical properties of matter.
2. Chemistry: science, which deals with chemical properties of matter.
3. Biology: science dealing with plants and animals.
4. Mathematics: an abstract science, which forms the language of science.

Mathematics has been included in the basic sciences as it quantifies the different sciences. At present, there are hundreds of branches of science, which have emerged from the basic sciences. Science in totality can be schematically represented by a tree (*Tree of Science*), where the roots represent the need-based researches done in different directions in the early stages. The basic sciences form the stem of the Tree of Science which got separated from one another with time. Other scientific subjects emerging from the basic sciences form branches, sub-branches and leaves of the tree. The flowers (or fruits) of the Tree of Science represent patents and technologies emerging from the various fields. Figure 1.1 is the schematic representation of the Tree of Science where only some of the main scientific streams are shown.

This book traces the historical developments of the subject of solid state physics. One of the main aims of science is to be able to understand and use the materials around us, and out of these, solids are the most abundant in nature. Solids are a result of densely packed atoms which interact strongly with one another. These interactions are responsible for the various properties of solids such as mechanical, electrical, magnetic, electromagnetic and optical properties etc. The early studies on different properties of solids were done in isolation using the available classical knowledge. Many physicists studied solids in the hope of gaining deeper understanding of the fundamental forces. In the beginning of twentieth century, quantum mechanics, relativity, radioactivity, X-rays and statistical mechanics were just being developed. When quantum mechanics emerged, most of the frontline research workers started working on its development, and as a result the study of the properties of solids got slowed down. During the development of quantum mechanics, it was realized that the same could be used for the theoretical study of various properties of solids from a microscopic point of view. Therefore, from 1920 onwards, the study of the various

Fig. 1.1 Schematic representation of different branches of science in the form of a tree. This representation is by no means complete as every branch of science has split up into a large number of sub-branches and it is not possible to depict all of them in a single tree

properties of solids such as electrical conductivity, photoconductivity, and ferromagnetism etc. using quantum mechanics grew rapidly. Until 1930, solid state physics as a subject did not exist and the work on solids was considered to be a limited part of atomic physics. At that time, a number of schools, universities and institutions in Europe, Britain, United States and the Soviet were involved in the study of physics of solids, and the scientists involved in this field grew to a few hundreds. In 1938, Mervin Kelly (1894–1971) set up a group named "Physics of the Solid State" at Bell Laboratories. To popularize the field of physics of solids, in 1940, Friedrick Seitz (1911–2008) wrote the first book entitled, "*Modern Theory of Solids*". In 1945, Mervin Kelly formed the Solid State Physics Department at Bell Laboratories. At that time, nuclear physics and high energy physics were at the forefronts using big accelerators and in these fields scientists were working in large groups. Some of the nuclear and high energy physicists looked down on the study of solids. They believed that everything outside the nucleus of an atom had already been studied and a physicist of high calibre would not study physics of solids.

In 1945, after the bombing of Hiroshima and Nagasaki, scientists studying physics of solids gradually began to assert their right to be considered at par with nuclear and high energy physicists. They started aspiring for a separate identity and they had a number of meetings with American Physical Society for the same. A separate subject under the name "Solid State Physics" came into existence in June, 1947 with the establishment of the "*Division of Solid State Physics*" within the American Physical Society. For the first time, there could be such a thing as a 'solid-state physicist'—now a generally recognized name and status. Solid state Physics can be defined as a subject that deals, both experimentally and theoretically, with various physical properties of atoms, molecules and solids. Solid state physics comprises of an array of different fields, old and new, forming a novel constellation: prominent fields are the structure of solids, electromagnetism, dielectric properties, lattice vibrations, electronic band structure of solids, semiconductors, superconductivity, nuclear resonance, photoconductivity, nanophysics and so. Therefore, solid state community is diverse in nature (unity in diversity). With the invention of transistor in 1947 the research on semiconducting materials got accelerated. In 1950s, General Electric Company created a "Semiconducting Branch" which was distinct from Metal Physics. After World War II, a large number of communities of Solid State Physicists emerged in different countries of the world. In 1953, Charles Kittel (1916–2019) published a book, "*Introduction to Solid State Physics*" which made the subject of Solid State Physics popular among students.

In 1957, Russia launched a satellite "*Sputnik*" in space and with this the need for strengthening research in basic sciences was strongly felt. The funding in basic research in all fields of physics increased immensely in United States, United Kingdom, Germany, France, Japan and other countries. Department of Defence and Atomic Energy Commission of United States were the biggest patrons. It was realized that material is a crucial factor in making advanced weapons which accelerated research in Solid State Physics. The discovery of new materials was becoming indispensable for everything from microelectronics to jet planes and from nuclear reactors to ballistic missiles. To develop new materials interdisciplinary research was required where physicists, chemists and metallurgist were asked to work together. Cyril Stanley Smith (1903–1992) had already introduced such an interdisciplinary work in his institute for the study of metals in Chicago, United States. From 1950–1960, the number of Solid State Physicists increased immensely and the same was the situation in the funding of Solid State Physics. But at that time there was no journal fully devoted to the subject of Solid State Physics. In 1956, Harvey Brooks (1915–2004) of Harvard University, USA, started "*Journal of Physics and Chemistry of Solids*" published by Pergamon Press, which was fully devoted to the study of solids. At that time, it was not possible to distinguish fully between the physics and chemistry of solids, thus the name of the journal. Russian journal "*Fizika Tverdogo Tela* (Physics of Solids)" came into existence in 1959 and two years later East Germany started the journal "*Physica Status Solidi*" and both these journals were fully devoted to the study of solids. In 1968, *Journal of Physics* (UK) was split into different volumes and one of these was fully dedicated to Solid State Physics. In 1970, "*Physical Review*" was split into two volumes A and B and Physical Review B was fully devoted to Solid State Physics. In 1971, "*Zeitschrift fur Physik*" followed a similar divide. Later more number of research journals fully devoted to Solid State Physics and published from different parts of the world, came into existence. Solid state physics became one of the most important subjects with the discovery and fabrication of what are called solid state devices such as transistors, integrated circuits, lasers, fibre optics and microelectronics. In the greater perspective, it is not wrong to say that the subject of Solid State Physics is panoramic in nature both from intensive and extensive points of views but in the present text we will focus our attention only on the main fields of the subject. Solid State Physics is more like chemistry than any other branch of Physics because it deals with common properties of a huge class of solids. In order to comprehend the physics of solids, one is expected to have the basic knowledge of classical mechanics, quantum mechanics, optics etc. Therefore, we would first like to review the historical development of these subjects in the first part of the book. In the second part, the historical developments of different conventional fields of solid state physics are presented.

Bibliography

Andolfo M (1999) Atomisti antichi: Frammenti e testimonianze [Ancient atomism: fragments and testimonies]. Rusconi, Milan. Fragments and testimonies concerning Democritus has been made by Solomon Luria

Aristotle (1984) On generation and corruption, A1.315b.6. In: The complete works of Aristotle, vol 1. Princeton University Press, Jonathan Barnes

Conche M (1991) Anaximandre, fragments et témoignages: texte grec, traduction, introduction & commentaire. Presses Universitaire de France, Paris

Democritus (2007) Raccolta dei frammenti [Collected fragments]. Interpretation commentary by Solomon, Italian translation. Bompiani, Milan

Drake S (1990) Galileo: pioneer scientist. University of Toronto Press, Toronto, Canada

Heath TL (1931) A history of Greek mathematics II. Oxford University Press, London

Kirk GS, Raven JE, Schofield M (1983) The presocratic philosophers, 2nd edn. Cambridge University Press, Cambridge, UK (chap 3)
Kittel C (1953) Introduction to solid state physics. Wiley, New York
Lucretius (1951) On the nature of the universe (trans: Latham EA). Penguin, Harmondsworth, UK, p 173
Rovelli C (2007) The first scientist: Anaximander and his legacy (trans: Rosenberg ML). Westholme, Yardley, Penn
Rovelli C (2015) Aristotle's physics: a physicist's look. J Am Philos Assoc 1:23–40
Seitz F (1940) Modern theory of solids. McGraw-Hill Book Company, New York
Toomer GJ (1984) Ptolemy's almagest. Springer, New York, Duckworth and London, p 1984

Chapter 2
Classical Mechanics

Mechanics is a branch of physics which deals with the motion of objects or more specifically the relationship between force, matter and motion. Mechanics is one of the very oldest sciences and there are many mechanical inventions of which the exact origin is not known. For example, the wheel is one of the greatest inventions in mechanics but no one knows who invented it. Some of the most important inventions in mechanics are lever, screw, inclined plane etc. with large number of applications. Ancient Greek philosophers were the first to observe the motion of objects. It is considered that the history of mechanics started with Aristotle (384BC–322BC). Aristotle categorized the motion of objects as natural or violent. The motion of celestial objects is a natural one and is uniform, circular, eternal and perfect and goes on without the application of any force. On the other hand, the motion of terrestrial objects is violent and is rectilinear and finite in both distance and time. The terrestrial motion lasts as long as the force is applied and a common example is the up and down motion of objects. According to Aristotle, for an object to move, a force in direct contact with the object is required in case of all types of motion. Aristotle assumed that the velocity of an object is determined by the ratio of force to resistance to the motion, that is

$$\text{volocity} = \frac{\text{Force}}{\text{Resistance}} \tag{2.1}$$

This is nothing but the idea of dragging a load. The assumption of Aristotle that force must be in contact with the body when in motion caused some problems. For example, in case of a projectile thrown at an angle, why does the projectile continue to move when it has left the hand of the thrower? Aristotle studied the motion of falling objects. The question to which he was seeking an answer is: what is the force acting on falling objects? In case of falling objects, Aristotle considered weight of the object as the force and the resistance of air as the resistance. Under this assumption,

J. S. Galsin, *History of Solid State Physics*,
https://doi.org/10.1007/978-981-95-0504-3_2

heavy bodies fall faster than light bodies. Further, with these assumptions, a falling body in vacuum will acquire infinite velocity, which is absurd. In order to avoid this absurdity Aristotle said that vacuum is not possible. Equation (2.1) yields uniform velocity when constant force acts on a body, which is in contradiction with Newton's second law of motion, which says force accelerates a body. Aristotle believed that in vacuum a body at rest will remain at rest and a body in motion will continue to have the same motion. Therefore, Aristotle was the first to approach something similar to the law of inertia. The Aristotelian mechanics is only of historical interest, as many of his predictions were eventually proven wrong.

Archimedes (287BC–212BC) made very important contributions to mechanics. Around 5000BC, people were using lever in a simple balance scale. Even in other civilizations like Egypt and Mesopotamia, lever was being used in different forms. Archimedes in his work, "*On the Equilibrium of Planes*" gave the mathematical proof of the principle of working of a lever. According to Pappus (290–350) of Alexandria, Archimedes remarked, "Give me a place to stand on, and I will move the Earth". In this book he was the first to give the concept of centre of mass (centre of gravity) and gave a method to calculate the centre of gravity of different geometric figures like triangles, trapezium, parallelograms and parabola etc. In his book, "*On the sphere and cylinder*" he gave the method of finding the surface area and volume of a sphere and cylinder. Archimedes is best known for his principle of floating, which has an interesting story behind it. Actually, King Hiero-II of Syracuse gave pure gold to a goldsmith to make a crown for a temple but he had some doubts about its purity. Therefore, the King entrusted the responsibility upon Archimedes for checking the purity of the crown without damaging it. Archimedes worked on this problem and ultimately found that when a body is immersed in a liquid it displaces liquid whose volume is equal to the volume of the body. From this finding, he calculated the density of the crown and found that some silver had been added by the goldsmith. Further Archimedes found the principle of floating which has been described in his book, "*On Floating Bodies*". It states that an object, immersed in a fluid, experiences a buoyant force equal to the weight of the fluid it displaces (principle of buoyancy). Archimedes also developed block and tackle (compound) pulley system to lift heavy loads. Archimedes invented a screw pump which is used for crop irrigation and in sewage treatment plants even today. He also made War machines with which ships of the enemy could be burnt. Other works of Archimedes are '*Measurement of a circle*', '*Quadrature of the parabola*', '*The sand reckoner*', '*The method of mechanical theorems*', '*Archimedes palimpsest*' and '*Ostomachion*'. In 1897, Thomas Little Heath (1861–1940) has described the works of Archimedes very well.

Ibn Bajja (1085–1138), also known as Avempace, was an Islamic philosopher. He contributed towards astronomy, physics, music, philosophy, medicine, botany and poetry. Ibn Bajja gave his own theory of the Milky Way. According to him, Milky Way is the light of many stars which are very close to (or almost touch) each other.

The light from these stars form a continuous image which is a result of refraction. Ibn Bajja proposed that for every force there is always a reaction force. Although he did not specify that these forces be equal but it is still an early version of the third law of motion, which states that for every action there is equal and opposite reaction. Ptolemy (100–170), Nikolas Copernicus (1473–1543) and Johannes Kepler (1571–1630) studied the motion of celestial objects like planets and stars as described earlier. Out of these, Kepler' study was the most significant and complete.

Painting of Galileo Galilei (1564–1642), Public Domain, https://en.wikipedia.org/wiki/Galileo_Galilei#/media/File:Galileo_Galilei_(1564-1642)_RMG_BHC2700.tiff

In 1589, Galileo Galilei (1564–1642) started studying the motion of freely falling bodies. Galileo's disciple Vincenzo Viviani (1622–1703) described that Galileo had performed a number of experiments by dropping balls with different masses from the Leaning Tower of Pisa and found that objects did not fall with velocities proportional to their masses. This was contrary to Aristotle's belief that heavier objects fall faster than the light ones. Galileo found that freely falling objects have accelerated motion and to his surprise the acceleration turned out to be the same for all the objects irrespective of their masses. He measured the acceleration due to gravity, which came out to be 980 cm/s^2 approximately. This was the first mathematical law discovered for falling bodies. But Galileo could not give mathematical form to gravitational force, may be because mathematics was not developed to that extent. *It was for the first time in the history of science that an experiment was performed and it is said that Galileo marked the beginning of experimental science.* In 1602, Galileo started performing experiments with a *simple pendulum* to study its properties. He showed that the time period of a pendulum is independent of its amplitude of vibration (isochronisms). He further found that the square of time period depends linearly on the length of the pendulum.

He was of the opinion that the motion of objects on the surface of Earth should also follow some mathematical laws. He initiated experiments concerning the laws of motion of bodies and is credited with conclusions that anticipated Newton's laws of motion. In 1589, he started his study on the path of a projectile and in 1608 found that the path of a projectile is a parabola. In the first decade of seventeenth century, Galileo for the first time formulated the *law of inertia* from his experiments with balls rolling down an inclined plane and later Rene Descartes (1596–1650) generalized it. Galileo deduced from his experiments that a body in motion would remain in motion unless a force (such as friction) caused it to come to rest (incomplete form of Newton's first law of motion). Galileo worked for several years on the motion of objects and based on his study, wrote a book entitled "*De Motu*". The text of this book was abandoned by him, so it remained incomplete and unpublished during Galileo's life time. In Du Motu, Galileo rejected Aristotle's assumptions of motion proving them to be false. Later Galileo included some selected portions of De Motu in his subsequent book entitled "*Le Meccaniche*" on the motion. In 1890, Favaro published the complete text as a book under the name, "De Motu". In 1632, Galileo put forward the *principle of relativity* in his book entitled, "*Dialogue Concerning the Two Chief World Systems*". He said that the laws of physics are the same in any frame of reference moving with a constant velocity in a straight line. He concluded that there is no absolute rest or absolute motion. Using this principle of relativity, Albert Einstein (1879–1955) put forward the theory of special relativity. In 1638, Galileo designed an experiment to measure the speed of light but got inconclusive results. It can be said that Galileo Galilei was the founder of the actual science of motion. Galileo has been called the "*father of modern science*" particularly, the "*father of modern physics*". Simon Stevinus (1548–1620) studied the equilibrium of objects on an inclined plane. During this study, he deduced the resolution of a force into two components and discovered the law of the parallelogram of forces.

Portrait of Isaac Newton (1642–1727), Public Domain, URL: https://en.wikipedia.org/wiki/Isaac_Newton#/media/File:Portrait_of_Sir_Isaac_Newton,_1689_(brightened).jpg

Isaac Newton (1642–1727) was an English mathematician, physicist, astronomer and theologian. He was a key figure in the scientific revolution. He studied in the Cambridge University, London and, in 1665, obtained a degree as Bachelor of Arts. In 1665, while he was a student, Newton discovered and generalized binomial theorem and started working on the subject of mathematics. In 1665, Plague broke out in London and the Cambridge University was closed temporarily. Newton continued his work at home in Woolsthrope for two years and developed theories in mathematics and other fields. We will here concentrate on his work on mathematics and mechanics only. In 1666, Newton developed mathematical theory of *gravitation* and of *circular motion*. Newton used the Latin word '*gravitas*' meaning weight. Newton found that moon is rotating around the earth in an orbit and its acceleration is directed towards the centre of the earth. His calculations showed its value to be 980 cm/s^2, which is the same as the acceleration of freely falling bodies. Newton argued that it cannot be a coincidence but the force, which caused the moon to move around the earth in an orbit, must be the same which caused the objects to fall freely on to the earth's surface. Further, the planets move around the sun in different elliptical orbits due to the force of attraction exerted by the sun. Therefore, Newton concluded that there exists a universal force, called gravitational force, due to which everybody attracts every other body. According to his law the gravitational force of attraction **F** between two bodies having masses m_1 and m_2, separated by distance r is given by

$$\mathbf{F} = G\frac{m_1 m_2}{r^2}\hat{\mathbf{r}} \tag{2.2}$$

where $\hat{\mathbf{r}}$ is a unit vector in the direction of r and G is the universal gravitational constant. Therefore, the gravitational force acts along the line joining the two bodies. Further, according to Newton, particles interact directly in space through either electrical or gravitational forces: *direct action at a distance.* It is noteworthy that Newton himself was not clear as to how a force acts between two distant bodies that do not touch: he thought that there was some missing link. He was of the view that there must be some mechanism, which transmits force between two distant bodies but could not find that. Newton also knew that his equations do not describe all the forces that exist in nature. The law of gravitation provided a firm foundation for theoretical astronomy whose most important problem was the calculation of perturbations of the planetary orbits because of mutual attraction between the planets. Theoretical astronomy predicted the existence of Neptune and Pluto. Later, in 1846, Johann Gottfried Galle (1812–1910) discovered Neptune and in 1930, Pluto was discovered from the computations of Percival Lowell (1855–1916).

In 1666, Newton started developing *calculus* and presented it in the tract at the end of the year and became a leading mathematician of the World. In 1667, Newton returned to Trinity as a fellow where he continued his research activities. By 1671, Newton completed the development of calculus but unfortunately got involved in a controversy with Gottfried Wilhelm Leibniz (1646–1716) who claimed precedence in the development of calculus. Most of the scientists and historians believe that both Newton and Leibniz developed calculus independently using very different notations. Leibniz's notation is found to be more convenient to use. In 1672, Newton was elected a Fellow of Royal Society (FRS). Towards the end of 1679, Robert Hooke asked Newton to study the trajectory of a body under inverse square law. Newton was already aware of the results of Galileo and Kepler and in the same year started work on celestial motion considering gravitation and its effect on the orbits of planets. He found that planets would move in elliptical orbits under the attractive force inversely proportional to the square of the distance. In 1684, Newton wrote a manuscript entitled, "*De motu corporum in gyrum*" and sent it to Edmond Helley. In this book, Newton put forward the theory of motion of bodies in an orbit and established Kepler's laws of planetary motion. After reading the manuscript Helley communicated it to the Royal Society of London.

Newton contributed a lot towards the motion of bodies on the surface of earth (mechanics) and gave us three laws of motion. The first law of motion states that in vacuum a body continues in its state of rest or of uniform motion unless some external force is applied: it is also called law of inertia and was known at the time of Galileo. Second law of motion states that force **F** is determined by the product of mass m and acceleration **a**, that is, $\mathbf{F} = m\,\mathbf{a}$. In other words, constant force acting on a body produces acceleration in the body. Third law of motion states that every action has equal and opposite reaction. The second and third laws of motion laid the firm foundations of the motion of bodies. In 1687, Newton wrote a book entitled, "*Philosophiae Naturalis Principia Mathematica* (*The Mathematical Principles of Natural Philosophy*)" which is referred to as *Principia.* All the contents of "*De* *motu corporum in gyrum*" also appear in Principia. It is considered to be one of the most important works in the history of science and forms the foundation of classical mechanics. The corrected and improved versions of Principia were published in 1713 and 1726. The Principia mainly deals with the motion of macroscopic bodies in non-resisting and resisting media. It includes the formulation of three laws of motion and universal law of gravitation. In Principia, Newton presented the derivation of Kepler's laws of motion, trajectories of comets and confirmed the heliocentric theory of universe. Newton showed that the motion of both celestial bodies and bodies on earth can be accounted for by the same principles. In 1703, Newton became the President of Royal Society of London and in 1705, he was knighted by Queen Anne.

Newton's approach to mechanics is usually called Newtonian mechanics. It is based on the assumption that **space**, **time** and **particle** are independent and constant physical quantities: that is mass, length and time are independent constant quantities. For example, consider a car moving on the surface of earth with a speed of 60 km/hour: car is an object (particle), surface of earth is space and speed is related with time. One of the major contributions of Newton is the development of calculus. He used the imagination of Democritus and combined it with mathematics. Therefore, one can say that the world of Newton is the world of Democritus rendered mathematical. Newton has admitted that the new science always owes to the science of antiquity. Newtonian mechanics was well established by the end of eighteenth century and became the basis of technology of nineteenth century. Building of engines, hydraulic systems, bridges, trains, and skyscrapers etc. would not have been possible without Newtonian mechanics and it was thought to be the final key to the understanding of the reality of universe. It is appropriate to point out here that Newton has also made significant contributions to chemistry.

An important concept in mechanics is the principle of *virtual work* which has been used from antiquity in one form or the other. In 1715, Johann Bernoulli (1667–1748) derived systematically the principle of virtual work and introduced the explicit concept of infinitesimal displacements. In a mechanical system, the work done by force acting on a particle depends on its displacement. A particle can have large number of possible displacements in moving from point A to point B (see Fig. 2.1) and these are called *virtual displacements*. The work done by a force on a particle moving along a virtual displacement is called virtual work. Out of all these displacements the particle will move along that particular displacement in which the work done is the least and such a displacement is called real displacement. The principle of virtual work states that in equilibrium, the virtual work done by applied forces on a system is zero. Bernoulli solved the problems for both the rigid bodies and fluids. The principle of virtual work as is known today is due to Bernoulli. Jean-Baptiste le Rond D'Alembert (1717–1783) was a French mathematician, physicist, philosopher and music theorist. In 1743, d'Alembert published his most famous work, "*Traite de Dynamique*" in which he developed his own laws of motion. He gave a principle based on the concept of virtual work usually called d'Alembert principle and forms an alternate form of Newton's second law of motion. According to Newton's second law of motion force acting on a particle is given by $F = ma$, where m is the mass of particle and a is the acceleration produced in it. According to D'Alembert, a particle in the equilibrium state is described by the equation

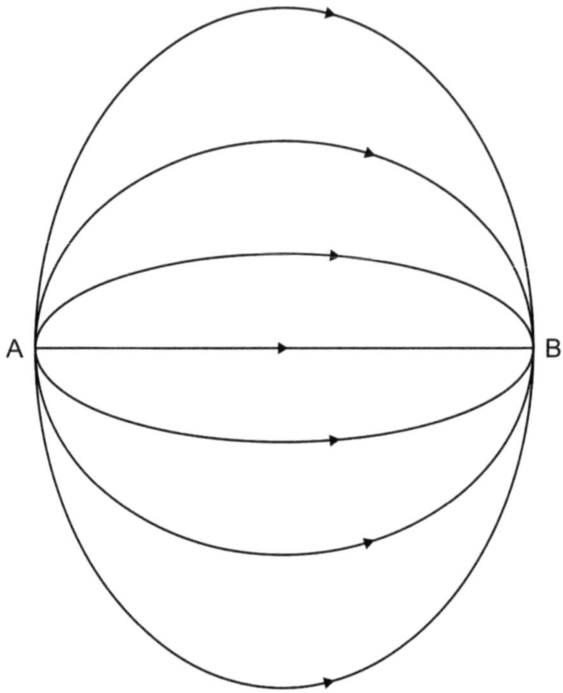

Fig. 2.1 Virtual displacements between points A and B

$$F - ma = 0 \tag{2.3}$$

Here F is the real applied force and -ma is the fictitious force, also called inertial force. Therefore, d'Alembert's principle states that for a system of particles of mass m, the difference of the real force acting on the system and the inertial force (equal to the time derivatives of the momenta of particles or ma) is zero when projected onto any virtual displacement. D'Alembert's idea was to convert a dynamic problem into a static problem by introducing inertial forces.

Leonhard Euler (1707–1783) and D'Alembert were contemporaries working in different fields of mathematics and mechanics. Leonhard Euler was a Swiss mathematician and physicist and one of the founders of pure mathematics. He worked in almost all areas of mathematics like geometry, infinitesimal calculus, trigonometry, algebra, number theory, as well as continuum physics and so on. Euler introduced the concept of a function and represented it for the first time as f(x). He introduced the use of exponential function, logarithm, and purely imaginary number $\iota = \sqrt{-1}$. He gave the current definition of constant 'e' which is the base of natural logarithm and is called Euler's number. Euler introduced different symbols like letter $\sum$ for summation and letter π for the ratio of circumference to the diameter of a circle. He is also known for the development of the summation of infinite power series. In 1736, Euler published a book, "*Mechanica, Vol. I & II*", in which he described mathematics of motion. In this book, Euler gave the concept of rigid body which has a unique place in physics as it greatly simplifies the study of dynamics of a body. In a rigid body, interatomic distances are fixed as a result of which body cannot be deformed. The parameters, which describe the configuration of a rigid body, are reduced to translations and rotations. Euler gave new laws, known as Euler's laws, to explain in a better way the motion of rigid bodies. He explained the laws of Galileo and Newton in terms of points in reference frames. Euler worked on the rigid bodies from 1738 to 1775 and developed the theory for the rigid body dynamics. In 1746, Daniel Bernoulli (1700–1782) and Leonhard Euler gave the proof of conservation of angular momentum. He, for the first time, specified the orientation of a rigid body in terms of three rotations performed in three different frames of references. The angles through which these rotations are performed are called Euler angles. Euler realized that two rotations are equivalent to a single rotation about a different axis (Euler's rotation theorem). Therefore, three rotations corresponding to the Euler angles are equivalent to a single rotation but the evaluation of the axis of rotation is a little bit complicated. Euler also gave equation of motion of a rotating rigid body in the form of a first order differential equation. It related torque with angular momentum of the rotating body.

Joseph-Louis Lagrange

https://www.britannica.com/biography/Joseph-Louis-Lagrange-comte-de-lEmpire

Joseph-Louis Lagrange (1736–1813), CCO 1.0 Universal, URL: https://commons.wikimedia.org/wiki/File:Lagrange_crop.jpg

Another drastically different and more powerful approach used in developing classical mechanics is variational method. Euler is well known for the discovery of calculus of variations. In 1744, Euler discussed the principle of least action and emphasized that the dynamics of a physical system is determined by the variational principle. The variational principle involves the concept of virtual work and allows for the derivation of differential equations of motion for a physical system. In 1744 and then in 1746, Pierre Louis Maupertuis (1698–1759) gave formulation of the principle of least action which is basically a variational principle. According to Maupertuis the principle of least action states that in all natural phenomena the quantity 'action' tends to be minimized. He said it is roughly equivalent to the idea of economy of effort in daily life: economy principle. He gave this principle after a study of two decades and it seems that it is a universal principle applicable to humans and animals in addition to natural phenomena. While in Berlin, Euler published two books on infinitesimal calculus and differential calculus. In 1748, he published, "*Introductio in analysin infinitorum*" and in 1755, "*Institutiones calculi differentialis*". Joseph-Louis Lagrange (1736–1813), an Italian mathematician, physicist and astronomer, made significant contributions in the field of analysis, number theory and in both classical and celestial mechanics. Lagrange was one of the creators of calculus of variations. While working in this field he corresponded with Euler a number of times. He improved Euler's equations by including constraints. In 1754, Lagrange started working on the development of a method for maximizing and minimizing a function. This work led to the formulation of Euler–Lagrange's equations of variational calculus and gave the Lagrange's method of multipliers, considerably simplifying Euler's earlier analysis. He applied this method to the problems of classical mechanics and generalized the results of Maupertuis and Euler.

In 1768, Lagrange reformulated the principle of virtual work by introducing generalized coordinates. It is called d'Alambert-Lagrange's principle and the equations of motion are called Lagrange's equations defined as

$$\frac{d}{dt}\left(\frac{\partial L}{\partial \dot{q}_i}\right) - \left(\frac{\partial L}{\partial q_i}\right) = 0 \tag{2.4}$$

where

$$L = T - V \tag{2.5}$$

T and V are kinetic and potential energies of the system. L is called Lagrangian of the system, which contains all its physical information, and q_i are the generalized coordinates. D'Alambert-Lagrange's principle formed an efficient alternate principle of mechanics by which all problems of equilibrium could be solved. Lagrange presented most comprehensive treatment of the new approach to classical mechanics, called *Lagrangian mechanics,* in his book, "*Mecanique Analytique, Vol. I & II (Analytic mechanics, Vol. I &II)*" published in 1788, almost about 120 years after the publication of Newton's Principia. It formed the basis for the further development of mechanics and mathematical physics in the nineteenth century. Therefore, Lagrange is best known for transforming Newtonian mechanics into Lagrangian mechanics.

It is noteworthy that different systems for measuring fundamental quantities (length, mass and time) were in practice up to the eighteenth century. In England, scientists used FPS system in which length, mass and time were measured in foot, pound and second, respectively. In most of the other countries, CGS system of units was used in which length, mass and time are measured in centimeter, gram and second. In 1790s, a Commission was set up by the Legislative Assembly in Paris for the development of metric system of measurements and Lagrange was the main scientist in it. He played the main role for choosing the MKS system of units in which length, mass and time are measured in meter, kilogram and second. In this system, decimal subdivision could be used, thus simplifying the system of units. In 1799, the MKS system of units was adopted which is also called Standard International (SI) system of units and now is widely used in all countries.

William Roman Hamilton (1805–1865), Public Domain, URL: https://en.wikipedia.org/wiki/William_Rowan_Hamilton#/media/File:William_Rowan_Hamilton_portrait_oval_combined.png

William Rowan Hamilton (1805–1865) was an Irish mathematician, astronomer and physicist. He worked in the fields of optics, Fourier analysis and linear algebra, but his major contribution is in classical mechanics. Hamilton studied most of Newton's Principia by the age of 16 years and learnt a number of languages while studying. John Mortimer Brinkley (1763–1835), the Royal Astronomer of Ireland, gave the following remark about Hamilton when he was just 18 years old: "*This young man, I do not say will be, but is the first mathematician of his age.*" In 1827, Hamilton started work on dynamics and gave the principle of "*Varying Action*" based on the calculus of variations usually called Hamilton's principle. In 1834 and 1835, he published two most important papers on dynamics of solids in Philosophical Transactions of the Royal Irish Academy which presented a new approach for the solution of problems generally called Hamiltonian mechanics. The Lagrange's equations were derived for the instantaneous state of a system with small virtual displacements about the instantaneous state. On the other hand, in 1834, Hamilton considered the motion of a system for a finite interval of time with small virtual variations in motion from the actual motion. He defined the action integral I for the system as

$$I = \int_{t_1}^{t_2} L dt \tag{2.6}$$

According to Hamilton's principle, the motion of the system is such that variation in action integral is zero, that is,

$$\delta I = \delta \int_{t_1}^{t_2} L dt = 0 \tag{2.7}$$

Hamilton derived the expression for a function $\widehat{H}$, called Hamiltonian, which contains whole of the information of the system, and is given by

$$\widehat{H} = \sum_i p_i q_i - L \tag{2.8}$$

where p_i are the canonical momenta. The Hamiltonian for a conservative system reduces to the total energy and is given by

$$\widehat{H} = T + V \tag{2.9}$$

Hamilton also worked on quaternions which made him one of the founders of linear algebra.

In 1835, Gaspard-Gustave de Coriolis (1792–1843), a French mathematician, engineer and scientist, found that a force acts on an object in motion within a frame of reference that rotates with respect to an inertial frame of reference. Such a force is named *Coriolis force* and the deflection of the object due to this force is called Coriolis effect. Newton's laws of motion describe the motion of an object in an inertial frame of reference but when these are transformed to a rotating frame of reference, the Coriolis and centrifugal forces appear. Both the Coriolis and centrifugal forces are proportional to mass of the body. The Coriolis force is linearly proportional to angular velocity while the centrifugal force is proportional to the square of angular velocity. Carl Gustav Jacob Jacobi (1804–1851) made significant contributions to elliptical functions, dynamics, differential equations, determinants and number theory. In 1837, Jacobi applied Hamilton's method to the general problems of classical variational calculus. He made a fundamental contribution to classical mechanics in the form of Hamilton–Jacobi theory and clarified the mathematical issues. In physics, Hamilton–Jacobi equation provides an alternate formulation of classical mechanics and is particularly useful in identifying conserved physical quantities for mechanical systems.

In 1850, James Joseph Sylvester (1814–1897) introduced the concept of matrix for an array of numbers. Arthur Cayley (1821–1895), friend and colleague of Sylvester, quickly saw the significance of the matrix concept and in 1853 published a note giving, for the first time, the inverse of a matrix. He further worked on matrices and developed matrix algebra in "*A memoir on the theory of matrices*" published in 1858. With the introduction of matrices, Euler theorems were rewritten and rotations were represented by orthogonal matrices. With the help of matrix algebra, it was easily proved mathematically that Euler's three rotations are equivalent to one rotation and it became easy to evaluate the axis of equivalent rotation.

Another important aspect of classical mechanics is the theory of elasticity of solids. In about 1493, Leonardo da Vinci (1452–1519) illustrated bending of solids in some of his sketches but he did not develop any mathematics related to it. In 1638, Galileo Galilei (1564–1642), studied for the first time that solids, like wood, resist bending when force is applied, because he treated solids to be inelastic in nature. Later Galileo made the first attempt at developing a theory of beam bending. He fixed one end of a solid rod and applied load to the other end for investigating the breaking loads. He found that the breaking load is independent of the length of the rod but is proportional to its area of cross section: the second property pointed towards the concept of stress. In 1680, Edme Mariotte (1620–1684), a French physicist and priest, published similar discoveries as found by Galileo. In 1660, Robert Hooke (1635–1703) discovered (and published in 1678), that many materials exhibit displacement under the action of applied force. Based on the experimental findings he gave a law, generally called *Hooke's law*, which stated that the displacement produced in solids is linearly proportional to the applied load. At that time, Hooke's law was not in the form as it is today but it gave the first idea of elasticity in solids. Hooke's law resulted in the development of a balance spring, which was used in producing portable time-pieces and watches to give reasonably accurate time. But there arose a dispute between Hooke and Huygens over the credit of this invention which was later resolved by the Royal Society in favour of Robert Hooke. It is worth mentioning here that Hooke's law was the biggest and most famous discovery of its times and formed the basis of mathematical theory of elasticity. In 1684, Gottfried Wilhelm Leibniz (1646–1716) pointed out that the cause of deformation in a solid is the internal tension acting across the surfaces of a body. In 1691, Jakob Bernoulli (1655–1705) observed that the deformation in solids arose due to the extension and compression of its fibres and could be described in terms of force per unit area (stress) as a function of elongation per unit length (strain). This investigation was of great importance from the point of view of theory of elasticity. Charles-Augustin de Coulomb (1736–1806) studied the bending of beams and was the first to relate it to stress and strain. In 1713, Antoine Parent (1666–1716) gave the concept of shear stress acting on elastic bodies. In 1727, Leonhard Euler (1707–1783) proposed a linear relation between stress σ and strain ε, that can be written as

$$\sigma = Y\varepsilon \tag{2.10}$$

where Y is the constant of proportionality and is called modulus of elasticity. In 1744, Euler introduced the concept of strain energy per unit length of a beam and showed that it is proportional to the square of beam's curvature. He regarded the total strain energy analogous to potential energy of a discrete mechanical system. In 1807, Thomas Young (1773–1829), an English physicist, material scientist and engineer, defined linear modulus of elasticity in rods in the same way as Eq. (2.10) and is called Young's modulus of elasticity. He described Young's modulus of elasticity in his "*Course of Lectures on Natural Philosophy and the Mechanical Arts*" written in 1845. Louis Poinsot (1777–1858) was the inventor of geometrical mechanics and showed how a system of forces acting on a rigid body can be resolved into a single force and a couple.

In 1820s, Augustin-Louis Cauchy (1789–1857) generalized the theory of elasticity to three-dimensional solids and represented stress and strain in terms of 3 × 3 matrices. Cauchy expressed the strain components in terms of displacement. He developed elasticity theory for isotropic solids in the linear approximation in which strain is linearly proportional to stress: this was the foundation of classical linear elasticity theory. He further derived the equation of motion for continuum elastic solids. He extended the linear elasticity theory to crystals (anisotropic substances) assuming them to be comprising of lattice points linked with attractive and repulsive forces. Simultaneously, Claude-Louis-Marie Navier (1785–1836) had developed an elasticity theory based on a simple particle model of matter in which particles interacted with their neighbors by a central attractive force. Woldemar Voigt (1850–1919) was a student of Franz Ernst Neumann and obtained Ph.D. degree on the theory of elasticity. He carried out experimental work in crystal physics and measured elastic, piezoelectric and pyroelectric constants. Gabriel Lame (1795–1870) introduced two parameters λ and μ, called Lame's coefficients, in the expression for Helmholtz free energy and hence the stress in an elastic material. The first coefficient λ represents the compressibility of the solid while the second coefficient μ represents the shear modulus. These coefficients established the linear elasticity theory and started the field of *continuum mechanics.*

There were controversies concerning the possible number of elastic constants in the general anisotropic solids. In 1837, George Green (1781–1840) settled the problem and showed that the independent elastic constants are 21 in number. Sieon-Danis Poisson (1781–1840), Augustin-Louis Cauchy (1789–1857), and George

Stokes (1819–1903) proved that the equations of elasticity predicted the existence of two types of deformation waves which could propagate through an isotropic elastic solid: longitudinal and transverse waves. In 1946, R. F. S. Hearmon suggested that elastic constants can be readily measured by passing ultrasonic waves through a crystal. John Douglas Eshelby (1916–1981) worked on the continuum elastic theory of point defects and dislocations in crystals (see Chap. 6).

By the end of nineteenth century three approaches to classical mechanics were available, that is Newtonian mechanics, Lagrangian mechanics and Hamiltonian mechanics. The classical mechanics is largely compatible with rest of the classical physics theories such as classical electrodynamics and thermodynamics etc. But in nineteenth century some difficulties were discovered such as Gibb's paradox in which entropy is not a well-defined quantity. Black-body radiations could also not be explained using classical mechanics. As the experiments reached the atomic level, the classical mechanics failed to explain basic things like the energy levels in solids. In the twentieth century quantum mechanics came in to existence (see Chap. 4) which played an important role in explaining atomic and subatomic systems. The basic concepts of classical mechanics have been used in the development of quantum mechanics. The variational principle, Lagrangian formalism and Hamilton's principle formulated for classical mechanics played an important role in quantum mechanics. Due to the close relation of Hamiltonian with energy spectrum and time evolution of a system, it is of fundamental importance in most formulations of quantum mechanics. The wave-particle duality nature emerged in quantum mechanics: both the particle and wave mechanics formalisms (Schrodinger wave equation) have been developed. It is noteworthy that Hamilton–Jacobi equation is the only formulation in classical mechanics in which motion of a particle can be represented by a wave and the wave equation is similar but not identical with Schrodinger equation in quantum mechanics. For this reason, the Hamilton–Jacobi equation in classical mechanics is considered the closest approach to quantum mechanics.

Bibliography

Aristotle (350BC) On the heavens (de Caelo), Book 13; Aristotle (1939) On the heavens (trans: Gutherie). Harvard University Press, Cambridge

Aristotle (1957) Physics (trans: Wicksteed PH, Conford FM). Harvard University Press, Cambridge

Cauchy AL (1823) Récherches sur l'équilibre et le movement intérieur des corps solides ou fluides, élastiques ou non élastiques. Bulletin de la Societé Philomathique **10**:9–13; Œuvres **2**:300–304

Cauchy AL (1827a) De la pression ou tension dans un corps solide. Exercices de Mathématiques 2:42–56; Œuvres 7:60–78; (1827b) Sur la condensation et la dilatation des corps solides. Exercices de Mathématiques 2:60–69; Œuvres 7:82–93; (1827c) Sur les relations qui existent dans l'état d'équilibre d'un corps solide ou fluide, entre les pressions ou tensions et les forces accélératrices. Exercices de Mathématiques 2:108–111; Œuvres 7:141–145

Cauchy AL (1828a) Sur les équations qui expriment les conditions d'équilibre ou les lois du mouvement intérieur d'un corps solide, élastique ou non élastique. Exercices de Mathématiques 3:160–187, Œuvres 8:195–226; (1828b) Sur l'équilibre et le mouvement d'un système de points matériels sollicités par des forces d'attraction ou de répulsion mutuelle. Exercices

de Mathématiques 3:188–213, Œuvres 8:227–252; (1828c) De la pression ou tension dans un système de points matériels. Exercices de Mathèmatiques 3:214–238, Œuvres 8:254–277
Cauchy AL (1829) Sur l'équilibre et le mouvement intérieur des corps considérés comme des masses continues. Exercices de Mathèmatiques 4, 293–319, Œuvres 9:243–369
Cayley A (1858) A memoir on the theory of matrices. Philos Trans Roy Soc (London) 148:24
Coriolis GG (1835) Sur les équations du mouvement relatif des systèmes de corps. Journal de l'École Royale Polytechnique 15:144–154
Coulomb CA (1784) Recherches théoriques et expérimentales sur la force de torsion et sur l'élasticité des fils de metal. Histoire de l'Académie Royale des Sciences, pp 229–269
Eshelby JD (1956) In: Seitz F, Turnbull D (eds) Solid state physics, vol 3 Academic, New York
Eshelby JD (1957) The determination of the elastic field of an ellipsoidal inclusion, and related problems. Proc Roy Soc (London) A241:376–396
Euler L (1765) Theoria motus corporum solidorum seu rigidorum The Euler Archive, E-289
Hearmon RFS (1946) The elastic constants of anisotropic materials. Rev Mod Phys 18:409
Hamilton WR (1834) On a general method in dynamics; by which the study of the motions of all free systems of attracting or repelling points is reduced to the search and differentiation of one central relation, or characteristic function. Philos Trans Roy Soc part II:247–230; (1835) Second essay on a general method in dynamics. Philos Trans Roy Soc part I:95–144
Hamilton WR (1845) On a theory of quaternions. British Association Report. 1844, published 1845, part 2, p 2; (1847) On quaternions. Proc Roy. Irish Acad 3:1–16
Hamilton WR (1844) On quaternions, or on a new system of imaginaries in algebra. Philos Mag 25:10–13, 241–246, 489–495; (1845) Philos Mag 26:220–224; (1846) Philos Mag 29:26–31, 113–122, 326–328; (1847) Philos Mag 30:458–461; (1847) Philos Mag 31:214–219, 278–293, 511–519; (1848) Philos Mag 32:367–374; (1848) 33:58–60; (1849) Philos Mag 34:294–297, 340–343, 425–439; (1849) Philos Mag 35:133–137, 200–204; (1850) Philos Mag 36:305–306
Heath TL (1897) The works of Archimedes. Cambridge University Press, London
Hooke R (1678) De potentia restitutiva, or of spring (Explaining the power of springing bodies). London
Jacobi CGJ (1837a) Zur theorie der variations-rechnung und der differential-gleichungen. Journal fur die Reine und Angewandte Mathemutik 17:68–82 = Werke 4:39–55; (1837b) Uber die reduction der integration der partiellen differential-gleichungen erster ordnung zwischen irgend einer zahl variabeln auf die integration eines einzigen systems gewohnlicher differential-gleichungen. Journal fur die Reine und Angewandte Mathematik 17:97–162 = Werke 4:57–127
Lame G (1852) Lecons sur la theorie mathematique de l'elasticite des corps solides: Bachelier
Maupertuis PL (1751) Essai de cosmologie: Mens agitat molem Aeneid. Lib. VI, M.DCC. LI
Navier CLM (1827) Mémoire sur les lois de l'équilibre et du mouvement des corps solides élastiques. Mémoires de l'Academie des Sciences de l'Institut de France 7:375–393; (1864) Résumé des leçons données a l'École des ponts et chaussées [...]. Troisième editionavecdes notes et des appendices, par M, Barré de Saint-Venant. Dunod, Paris
Newton I (1686) Philosophiae naturalis principia mathematica. IMPRIMATUR, Pepys, Reg. Soc. Press, Londini; (1687) A treatise of the system of the world. Printed by Fayram at the south entrance under the Royal Exchange, London; (1707) Arithmatica univerfalis. CANTABRIGIAE Tupis Academicis, Londini; (1736) Method of fluxions andinfinite series. Printed by Henry Woodfall, London
Poinsot L (1852) Théorie nouvelle de la rotation des corps, Copy located in the Niels Bohr Library & Archives, American Institute of Physics, in College Park, Maryland
Poisson CLM (1828) Mémoire sur l'équilibre et le mouvement des corps élastiques; (1829) Mémoires de l'Académie des Sciences de l'Institut de France 8:357–570; (1829) Mémoire sur les équations générales de l'équilibre et du mouvement des corps solides élastiques et des fluids. Journal de l'École Polytechnique 13:1–174 (1831)
Stokes GG (1845) On the theories of the internal frictions of fluids in motion, and of the equilibrium and motion of elastic solids. In: Mathematical and physical papers. The University Press, Cambridge, pp 75–129 (1880)

Voigt W (1887) Theoretische studien über die elasticitätsverhältnisse der krystalle, Königliche Gesellschaft der Wissenschaften zu Göttingen; (1900) L'état actuel de nos connoissances sur l'élasticité des cristaux. In: Poincaré GL (ed) Rapports présentés au Congrès international de Physique, rassemblés et publiés par Ch. Gauthier-Villars, Paris, pp 277–347; (1910) Lehrbuch derkristallphysik:mit ausschluss der kristalloptik, vol 34. BG Teubner

Young T (1845) Course of lectures on natural philosophy and the mechanical arts. Taylor and Walton, London, p 106

Chapter 3
Optics

Just like mechanics, optics is also one of the oldest topics in the study of Physics. A number of facts about light such as the concept of ray and rectilinear propagation of light were known in antiquity but remained of unknown origin. Although the phenomenon of reflection was known from the ancient times, it is unclear who first discovered the laws of reflection. Around 300BC, Euclid (325BC–265BC), a Greek mathematician, wrote the first known book on light entitled, "*Optics*" which includes an introductory discussion on geometrical optics. He believed the law of reflection was described around 300BC, stating that light travels in straight lines and reflects from a surface at the same angle at which it is incident on. Euclid was of the opinion that light is emitted in the form of rays from the human eye and vision occurred when these rays struck an object. He is best known for his treatise, "*Elements-13 book compilation*" which established the foundations of geometry known as *Euclidean geometry*. In this treatise, Euclid assembled the works originating from the earlier mathematicians along with his own work. Greeks also knew how to concentrate light with the help of mirrors. Archimedes (287BC–212BC) worked on the reflection of light from mirrors, reportedly with the aim of producing war weapons. He arranged several mirrors in such a way that these collectively acted as a parabolic reflector. He was able to concentrate light with such parabolic reflectors and these were used to burn ships of the enemy during the siege of Syracuse. Most of the original work by Archimedes is lost but a part of it is known through references made by others. Heron of Alexandria, in "*Catoptrica*" mentioned that reflection is one of the works by Archimedes on optics. Diocles (240BC–180BC), a Greek mathematician, studied and described parabolic mirrors in his work "*On Burning Mirrors*". He was the first to prove that the rays of light reflected from a spherical mirror meet at a point causing burning: demonstrating the focal property of spherical mirrors. He believed it is possible to obtain a lens using this property. But the work of Diocles was largely ignored by the Greeks. Although the original work "On Burning Mirrors" is lost, Gerald J. Toomer (1934–present) translated it from an Arabic version into English.

J. S. Galsin, *History of Solid State Physics*,
https://doi.org/10.1007/978-981-95-0504-3_3

Claudius Ptolemy (100–170), the Greek mathematician, physicist and astronomer, wrote a book entitled "*Optica (Optics)*" which dealt with some aspects of light. The original manuscript of Optica is lost but around 1154, Eugenius of Palermo (1130–1202) translated it from Arabic to Latin, which is available in a somewhat crude Latin version. The book was divided into three parts. In the first part, Ptolemy discussed the theory of vision. He believed that the rays of light emerged from the human eye in the form of a cone with its vertex within the eye and the base forming the field of view. Light after suffering reflection from an object, returned to the observer, enabling vision. The size, shape and orientation of the object were determined from the visual angle. In the second part of the book, he described a number of experiments with curved polished iron mirrors and studied reflection from plane, convex and concave mirrors. In the third part, Ptolemy explained how a ray of light changes its direction when it travels from one medium to another—phenomenon of refraction. He attempted to explain the law of refraction concluding that the angle of incidence is proportional to the angle of refraction. He could not derive the exact relationship between the angle of incidence and angle of refraction (Snell's law).

Ibn Sahl (940–1000), a Persian mathematician and physicist studied Ptolemy's work on optics and advanced it significantly. He dealt with parabolic mirrors, ellipsoidal mirrors and convex lenses. In 984, he discovered the law of refraction, though not in the modern form. Ibn Sahl used this law to derive the lens shapes that focus light with no geometric aberration. In 984, he wrote a treatise '*On the Burning Instruments*' in which it is described how mirrors and lenses could be used to focus sunlight to generate intense heat. Ibn al-Haytham (965–1040), also known as Alhazen in Europe, was an Iraqi mathematician, astronomer and physicist. He is the first philosopher who was of the opinion that a hypothesis must be supported by either experiments or mathematical reasoning. Therefore, Al-Haytham was an early pioneer of the scientific method and is sometimes described as the world's "*first true scientist*". Al-Haytham worked in the field of optics from 1011–1021 (about ten years) and wrote a seven-volume treatise entitled "*Kitab al-Manazir*". Volumes 1–3 are devoted to the physiology of the eye and the theory of perception and volumes 4–7 cover physical optics. Al-Haytham proved through experiments that light travels in straight lines. He was the first philosopher to explain that an object is seen when light falling on the object is reflected into the eyes of a person and that vision occurs in the brain. He was the first physicist to give the complete statement of the law of reflection: that is incident ray, reflected ray, and the normal to the surface all lie in the same plane perpendicular to the reflecting plane. He found the focus and focal length of spherical mirrors using the law of reflection—an impressive achievement at the time. He developed mathematical equations to explain the reflection of light from the curved mirrors. Al-Haytham constructed a refractometer for studying in detail the phenomenon of refraction. He proposed the principle of least time for refraction which later became *Fermat's principle*. Al-Haytham invented pin-hole camera which produced an inverted image. He conducted early experiments on the dispersion of light, studied shadows, rainbows and eclipses, and is often referred to as the *father of optics*.

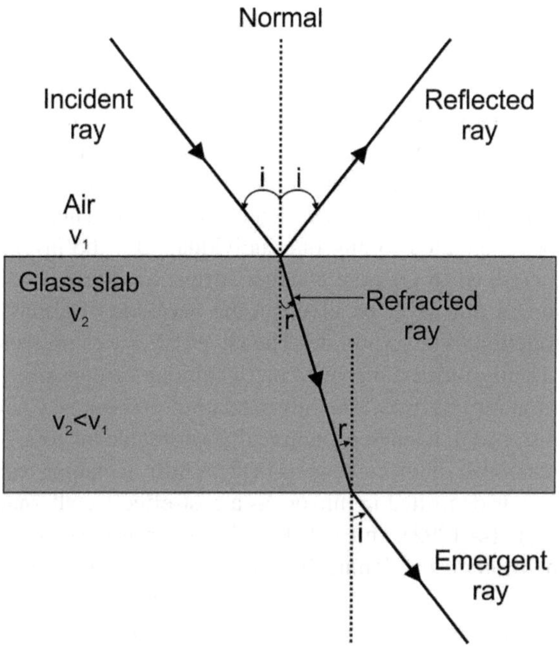

Fig. 3.1 Schematic representation of reflection and refraction of light from a glass slab

The law of refraction is credited to two scientists: Willebrord Snell (1591–1626) and Rene Descarte (1596–1650). It is believed that the modern form of the law of refraction was given in 1621 by Dutch scientist Willebrord Snell but unfortunately his published work was lost. According to Christian Huygens (1629–1695), the law of refraction given by Willebrord Snell is written as

$$\frac{\sin i}{\sin r} = \frac{n_2}{n_1} \tag{3.1}$$

where i (r) is the angle of incidence (refraction) in medium 1 (2) with refractive index $n_1(n_2)$: it is usually called *Snell's law*. In 1637, Rene Descartes published his work "*La Dioptrique*" where he derived the law of refraction from his corpuscular concept of light. Figure 3.1 shows the propagation of light through a glass slab depicting the phenomenon of reflection and refraction beautifully. With the discovery of the laws of reflection and refraction the stage was set for the development of geometrical optics.

The invention of mirror in optics is also of unknown origin. Early natural mirrors used by humans were probably in the form of a pool of still water or still water collected in a vessel. In the Stone Age, the first human-made mirrors were highly smooth and polished stone surfaces, such as obsidian, used in the regions like Turkey, Mesopotamia and Egypt etc. In the Bronze and Copper ages (around 6000BC to

2000BC), mirrors in the form of polished smooth metallic surfaces such as bronze mirrors and copper mirrors, came into existence. But the stone and metallic mirrors had low reflectivity and poor color rendering. With the advent of soda glass and glass blowing technique, the glass mirrors came into existence in the post Christian era. In 1508, glass mirrors were made by coating mercury-tin amalgam on the backside of glass sheets but these mirrors were toxic in nature due to the mercury coating. Later the amalgam layer was covered with a layer of paint. Therefore, the invention of mirror cannot be credited to any one individual. But the invention of modern glass mirrors is credited to German chemist Justus von Liebig (1803–1873) who, in 1835, deposited a thin layer of silver on the backside of a glass sheet through the chemical reduction of silver nitrate. The silver layer was protected by a coat of paint. In 1856, Liebig further improved the technique of depositing silver coating to produce high quality mirrors. The silver coating process was adopted for mass production of mirrors and it led to economically affordable mirrors. Since then there have been a large number of improvements in glass mirror-making technology. Today aluminium coating is deposited in mirrors as a cost-effective alternative to silver.

Roger Bacon (1214–1292), an English philosopher, is considered the first European who was in favour of abandoning the old method of imagination and speculation. He advocated that science is an inductive study of nature and should be supported by experiment. He applied the empirical method of Ibn Al-Haytham to the observations made by Aristotle in his texts and found that the results obtained contradicted the predictions made by Aristotle. Bacon played a crucial role in transforming philosophy into science and philosophers into scientists. Bacon worked on different aspects of science and in 1267, completed his major work "*Opus Majus*", which he sent to Pope Clement IV upon his request. The work included mathematics, optics, alchemy, and astronomy. The most important contribution of Roger Bacon is the application of geometry to optics. In 1250, Bacon invented the first convex lens, studied refraction from lenses and defined focal point and image formation of a point object in a lens. He explored human vision in detail, discussed anatomy of the eye and the brain considering light, distance, position and size regarding direct and reflected vision. Actually the structure of human eye is complex and is made of over 2 million parts working together. It is truly a wonder of nature and even the best cameras of the world cannot replicate its precision. He is also credited to be first European who described in detail the process of making gun powder. Actually, gun powder was invented in China somewhere in 142AD. Bacon also proposed flying machines and motorized ships and carriages. In 1284, Salvino Degli Armati (1258–1317) invented spectacles for far-sightedness (hypermetropia), though this claim is debated. In 1451, Nicholas Krebs (1401–1464), also known as Nicholas of Cusa, invented concave lens and used it to correct nearsightedness (myopia).

After the development of spherical mirrors and lenses, a number of major inventions followed and the most important among these were optical telescopes and microscopes. Telescopes and microscopes were invented at nearly the same time. In 1590, two Dutch spectacle makers Zacharias Janssen (1585–1632) and his father Hans Janssen invented the first compound microscope having objective and eyepiece lenses. The name microscope comes from Greek word 'micron' meaning 'small' and 'skopein' meaning 'to look at'. The first microscope was crude and mainly used to study insects. Galileo (1564–1642), upon hearing about the experiments with microscope, made its improved version, in 1609, by adding a focusing device in it yielding 30 times magnification (30X). Robert Hooke (1635–1703) further enhanced the design and capabilities of microscope significantly and was able to obtain magnification of 50X. He used his microscope to see minute living objects which cannot be seen with naked eye and is credited to be the first scientist to discover microorganism. Robert Hooke is known for his work "*Micrographia*" published in 1665. In Micrographia he coined the term 'cell' which spurred microscopic investigations. Hooke also studied the reflection of light from the surface of thin films and found that light reflected from the top and bottom surface of a thin film produced colored surface and the color depended on the thickness of the film and angle of incidence.

On the other hand, Hans Lippershey (1570–1619), a Dutch lens maker, invented the first telescope in 1608 although a number of other men also claimed the invention. The first telescope could magnify the object only three times (3X). In 1609, Galileo Galilei (1564–1642) constructed his own telescope independently with some improvements and obtained magnification of 30X. The discovery of telescope revolutionized the study of astronomy as described in Chap. 1. Johannes Kepler (1571–1630) did very fundamental work in the field of optics. In 1604, he published a book "*Astronomiae Pars Optica (Optical parts of astronomy)*" that laid the foundation of modern optics. He discovered inverse square law governing intensity of light, described reflection by flat and curved mirrors, explained the principle of *pinhole camera* and parallax and studied optics of human eye. For the first time Kepler described that images are projected inverted and then reversed by the eye lens on to the retina. In 1611, Kepler published another book entitled, "*Dioptrice*" in which he described the basic theory of double convex and double concave lenses and explained how they are combined to form a telescope. He described the concept of real and virtual images in a telescope and the effect of focal length on magnification. In 1611, Kepler immensely improved the design of telescope, called astronomical or Keplerian telescope, to obtain higher magnification than that of Galileo's telescope. With the help of his telescope he discovered that the orbit of Mars is an ellipse. In 1662, Christiaan Huygen (1629–1695) invented the first compound eyepiece (Huygens eyepiece) with which one obtains the erect image of objects. It comprises of two plano-convex lenses, separated by some distance, with plane faces towards the eye. In 1611, Kepler discovered the phenomenon of *total internal reflection*—about ten years before Snell formulated the famous formula for the refraction of light. In 1658, Pierre de Fermat (1601–1665) gave a general principle, called *Fermat's principle*,

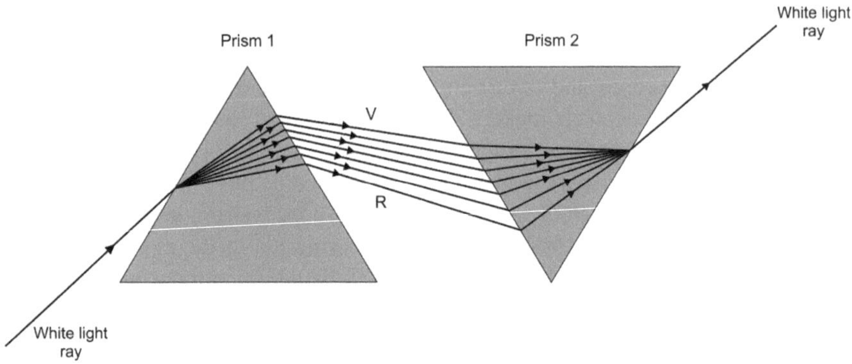

Fig. 3.2 Dispersion of light through a prism

that can be used to determine the path of light it follows from point to point. Fermat's principle states that when a ray of light travels between two points, its follows the path that requires the minimum time interval. In other words, one can say that light travels in straight lines between two points.

Isaac Newton (1642–1727) studied several aspects of optics. In 1666, he found that when white light passes through a glass prism it splits up into seven colors: phenomenon of *dispersion* of light. When light emerging from a glass prism is projected on to a screen the order of colors from top to bottom is violet, indigo, blue, green, yellow orange and red and this order is usually remembered by the acronym VIBGYOR. Rainbow is a beautiful natural example of dispersion of light. Newton placed a second glass prism in the upside-down position in the way of light emerging from the first prism (Fig. 3.2) and found that all the colors merged to form white light again. This experiment proved beyond doubt that white light comprises of seven colors. To further demonstrate this fact Newton made a circular disc with seven segments having different colors of rainbow (Newton's disc). When Newton's disc was rotated the seven colors merged to form white light. He concluded that color is the intrinsic property of light and it does not change its properties on reflection, refraction, scattering and transmission. This is known as Newton's theory of color. In 1666, Isaac Newton observed a new phenomenon when white light was reflected between a spherical surface and an adjoining touching flat surface. He found alternate bright and dark concentric colored circular rings: *Newton's rings.*

The early telescopes and microscopes did not produce a faithful image of an object because of two aberrations: spherical and chromatic aberration. Ibn Al-Haytham discovered the *spherical aberration* and described it in his book, "*Kitab al-Manazir*". In this defect, the light rays passing through the central part of the lens are focused at a point different from the rays passing through the outer part near periphery of the lens resulting in a blurred image. Thus the spherical aberration can be minimized by reducing the aperture of a lens. After the discovery of dispersion, Newton was the first person to discover *chromatic aberration*. In this aberration, the light while passing through a lens suffers dispersion and different colors are focused at different points on the axis of the lens resulting again in a blurred colored image. In 1668, Isaac Newton (1643–1727) made for the first time a reflection telescope using spherical mirrors instead of lenses and it was free from the chromatic aberration. From 1670–1672, Newton gave lectures on optics in Cambridge University. In 1671, during these lectures, the Royal Society showed interest in the reflecting telescope and asked him for a demonstration. This encouraged Newton to publish his findings in his treatise '*Opticks*' in 1704. After the discovery of chromatic aberration scientists started developing achromatic lenses (chromatic aberration-free lenses). Some of the key improvements in the lenses are mentioned below:

- In 1729, Chester Moore Hall (1703–1771) made the first compound achromatic lens.
- In 1733, Chester Moore Hall made the first achromatic telescope using these achromatic lenses.
- In 1757, John Dollond (1706–1761) invented achromatic lens using glass and water lenses. Later he made achromatic lens using lenses made of different quality of glasses and got it patented in 1758.
- In 1782, Jesse Ramsden (1735–1800), a scientific instrument maker and mathematician, invented an improved compound eyepiece (Ramsden eyepiece). It comprises of two plano-convex lenses of the same glass having same focal lengths with curved surfaces facing each other. The two lenses were separated by a distance less than the focal length of the individual lens.
- In 1849, Carl Kellner (1826–1855), a mechanic and self-educated mathematician, designed an eyepiece which was an achromatized version of the Ramsden eyepiece. It comprises of three lenses and produces a fairly good image. It is superior to the Huygens and Ramsden eyepieces.
- In 1860, Georg Simon Plossl (1794–1868), an Austrian optical instrument maker made an improved achromatic microscope objective. His major achievement is the construction of an eyepiece for a telescope, called Plossl eyepiece, which consists of two sets of doublets. The special feature of Plossl eyepiece is that it provides a large field of view.
- In 1866, Ernst Abbe (1840–1905) invented the apochromatic lens, a lens which eliminates both the primary and secondary color distortions. In 1871, he designed the first refractometer but published it in 1874. In 1880, he made a four-element eyepiece: one plano-convex lens and a cemented convex-convex triplet lens. It has a narrow field of view. Abbe was the first to define "Numerical Aperture" for a lens.

A lot of improvements had also been made in the design and lenses of the microscope. In 1853, Carl Zeiss (1816–1888) invented an improved microscope with magnifying power twice that of other microscopes. Carl Zeiss and Ernst Karl Abbe (1840–1905) worked together and in 1869, they introduced *Abbe condenser lens* to further improve the performance of microscopes. The new microscope had newly designed eyepiece and oil-immersion objective lens. In the nineteenth century, Carl Zeiss along with Ernst Karl Abbe revolutionized the quality of lenses and constructed microscopes with magnification of the order of 2000X and with these it was possible to see objects having 200 nm length. Ernst Leitz (1843–1920) developed a mechanism to put multiple lenses on a movable turret at the end of the lens tube, which allowed different magnifications of the object. The discovery of microscope revolutionized the study of biology but its advanced versions have applications in various scientific fields such as physics, chemistry and nanotechnology etc. On the other hand improvements were also made in the design and size of telescopes, in their objective and eyepiece lenses to increase the field of view, magnification and resolution. Space telescopes were proposed as early as 1923, and the project of Hubble telescope was funded by NASA, USA and the European Space Agency. In 1990, Hubble space telescope was launched which revolves around the Earth. It is the first telescope designed to be maintained in space by astronauts. Later a number of improvements were made in it. The project of Webb space telescope was undertaken and funded by NASA, European Space Agency and Canadian Space Agency. In 2022, James Webb space telescope was launched and it revolves around the sun at a distance of 1.5 million kilometres from the Earth, and it forms the World's space science observatory. Hubble and Webb's telescopes are the two advanced versions of telescope and are designed to probe our solar system, origins of our universe, galaxies and other worlds around the stars.

Although the various properties of light were studied and optical instruments were made but still the question arose what actually the light is? In 1613, Rene Descartes (1596–1650) put forward *corpuscular theory* of light. It stated that light is made of small discrete particles called corpuscles, which travel in straight lines with a finite velocity and possess impetus. In 1660, Francesco Maria Grimaldi (1618–1663) observed that when light falls on a coin it bends around its corners and enters into the geometrical shadow of the coin. Similarly, when light passes through a narrow slit, it bends around its sharp edges and enters into the shadow. Grimaldi performed a number of experiments described in his work "*Physicomathesis de lumine, coloribus, et iride, aliisqueannexislibri duo*" published in 1665, and established the fact that light bends very slightly around a sharp corner of an object. Grimaldi was the first person to coin the word *diffraction* for the bending of light at sharp corners and edges. In 1672, Newton supported and more importantly elaborated further the corpuscular theory of light. According to Newton light comprised of a large number of very small, elastic, rigid and massless particles called corpuscles. The corpuscles were assumed to travel through vacuum and through transparent medium in straight lines with very high speed. Newton explained reflection, refraction and dispersion of light with the help of corpuscular theory. But he could not explain why light is partially reflected and partially refracted from the surface of transparent materials like glass and water. Newton also could not explain the phenomenon of diffraction on the corpuscular theory of light. The corpuscular theory yielded higher velocity of light in denser medium than that in the rare medium which is contrary to the experimental fact.

Christiaan Huygens (1629–1695), Public Domain, URL: https://en.wikipedia.org/wiki/Christiaan_Huygens#/media/File:Christiaan_Huygens-painting.jpeg

In 1678, Christiaan Huygens (1629–1695), a Dutch mathematician, physicist and astronomer, proposed *wave theory* of light which he described in his book "*Traite la Lumiere* (*Treatise on Light*)" published in 1690. It is regarded as the first mathematical theory of light. In the wave theory, Huygens assumed that light travels in the form of longitudinal waves with the velocity of light. In 1678 itself, he gave a principle, called *Huygens principle*, for the propagation of light waves with the help of which he could explain reflection, refraction and diffraction of light on the wave theory. But he could not explain the rectilinear propagation of light in a precise manner. The wave theory was initially rejected in favor of Newton's corpuscular theory, might due to the great reputation of Newton. But there were some objections to the wave theory. The first and strong objection was that the propagation of a wave requires a medium. But beyond about 200 km from the surface of earth there is all vacuum but the light is coming from the sun to the surface of earth. To overcome this objection hypothetical medium called '*ether*' was assumed to be present in space. Otto von Guericke (1602–1686) showed experimentally that a sound wave is transmitted only through a medium and not in vacuum. Newton, in his "*Principia*", gave the formula for the velocity v of sound, assuming it a longitudinal wave, and it is given by

$$v = \sqrt{\frac{B_M}{\rho}} \tag{3.2}$$

where B_M is the bulk modulus of elasticity and ρ is the density of the medium. Actually, it is a general formula for the velocity of a wave. To obtain the actual velocity of light from the above formula one has to assign a very high value to B_M and very low value to ρ for medium ether: contradictory properties. No real medium was known to possess such values of B_M and ρ: the reason why ether was called hypothetical medium. The second objection was that the longitudinal waves exhibit only single polarization and therefore could not explain the phenomenon of birefringence. Thirdly, the rectilinear propagation of light could not be explained on the wave theory. Huygens made another groundbreaking advancement in regulating a clock by using pendulum for the first time. By harnessing the steady motion of pendulum, Huygens created a clock that was far more reliable than any before it. His invention significantly improved the accuracy of timekeeping and laid the ground work for the advancements in navigation and scientific research. In 1800, Friedrich Wilhelm Herschel (1738–1822) found that spectrum of light is not limited to the visible region only but extends beyond red (on the higher side of wavelength). In 1801, Johann Wilhelm Rotter (1776–1810) and William Hyde Wollaston (1766–1828) discovered spectrum beyond violet also (on the lower side of wavelength). In 1887, Albert Abraham Michelson (1852–1931) and Edward Williams Morley (1838–1923) preformed an experiment to find the presence of ether but they obtained negative results and contradicted the presence of ether medium.

Thomas Young (1773–1829). CC BY-SA 4.0, URL: https://en.wikipedia.org/wiki/Thomas_Young_(scientist)#/media/File:Thomas_Young_by_Briggs_cropped.jpg

On the other hand, in 1801, Thomas Young (1773–1829) presented his paper titled, "On the theory of light and color" to the Royal Society, London and described *interference* of light waves. He also presented an analogy with sound waves and with water waves, and even developed a demonstration wave-tank to show interference patterns in water. In 1803, he performed the famous experiment known as Young's double slit experiment (see Fig. 3.3). In this experiment, he proved that light coming from two closely spaced narrow slits interfere with each other and produce interference pattern on the screen. In his subsequent paper, titled "Experiments and calculations relative to physical optics" published in 1804, Young supported the wave nature of light. The phenomenon of interference could not be explained on the corpuscular theory but was fully explained by the wave theory establishing its validity. In 1818, Augustin-Jean Fresnel (1788–1827) used Huygens principle in conjunction with his own principle of interference, commonly known as *Huygens-Fresnel principle*, in explaining the phenomenon of diffraction on the wave theory. In addition he included the phase and amplitude of the secondary waves and also an obliquity factor in the principle. These additional assumptions have no physical foundation, but could explain the phenomenon of diffraction (called Fresnel's diffraction). In the same year, Fresnel explained the approximate rectilinear propagation of light using the Huygens-Fresnel

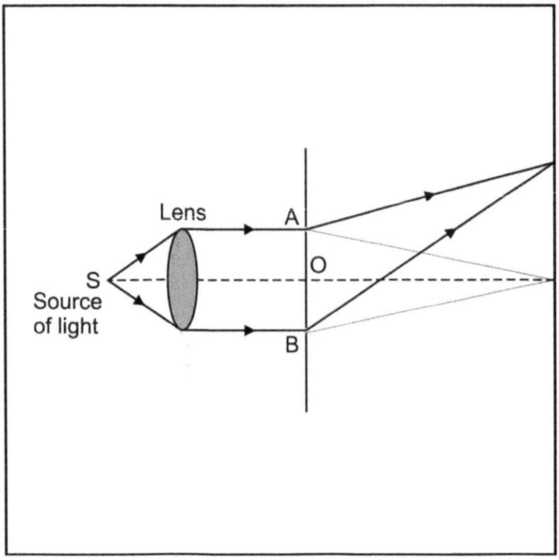

Fig. 3.3 Young's double slit experiment

principle. In 1785, David Rittenhouse (1732–1796) made the first diffraction grating by stretching hairs between two threaded screws. In 1814, Joseph von Fraunhofer (1787–1826) invented the modern spectroscope and in 1821, he made improved form of diffraction grating and studied the phenomenon of diffraction using grating (called Fraunhofer diffraction). He used collimator to have parallel beam of light incident on the grating and diffracted beam was passed through a telescope to see the diffraction pattern. He was the first person to obtain line spectra and measured wavelengths of the spectral lines using the diffraction grating. The advantage of the Fraunhofer diffraction is that it is easy to obtain in the laboratory as compared to the Fresnel diffraction. Both the phenomena of interference and diffraction gave the required impetus to the wave theory of light. In 1947, Dennis Gabor (1900–1979) invented holography which is an application of phenomenon of diffraction and gives three-dimensional images. With the help of electron microscope he obtained the interference pattern of an object. To the interference pattern he gave the name hologram because it contains the whole information required for the reconstruction of the image of the object which may be two-dimensional or three-dimensional. The name hologram was derived from the Greek word '*holos*' meaning the whole. Gabor developed its theory between the years 1946–1951 in a series of papers. In 1960, Theodore Maiman (1927–2007) was the first to invent optical laser which made available coherent light beam. In 1962, Yuri Nikolayevich Denisyuk (1927–2006) of Soviet Union and Emmett Norman Leith (1927–2005) and Juris Apatnieks (1936–present) of USA obtained the first practical three-dimensional hologram of an object using laser.

In 1669, Erasmus Bartholin (1625–1698) got the first clues about polarized light. He found that transparent and colorless crystal of Iceland spar (calcite) produces double image when objects are viewed through the crystal (phenomenon of *birefringence*). Huygens was trying to explain double refraction by assuming that in an Iceland spar crystal, in addition to the spherical primary wave, there exists a secondary ellipsoidal wave. It was in the course of this investigation that, in 1690, Huygens made the fundamental discovery of *polarization.* Actually, the longitudinal nature of light, as assumed by Huygens, was a stumbling block in explaining the polarization of light. In 1808, Etienne-Louis Malus (1775–1812) discovered that light can be polarized by reflection which was published in 1809. In 1810 he gave the theory of double refraction through crystals. In 1811, Sir David Brewster (1781–1868) discovered that reflected light is completely polarized when falls on an optical surface at a particular angle called polarizing angle or *Brewster's angle.* In this case electric vector of light is parallel to the surface of incidence. Brewster gave a relation between the refractive index of the material and the polarizing angle and this relation is named *Brewster law.* In 1816, Brewster invented an optical toy called kaleidoscope. In kaleidoscope a number of images of an object are formed by multiple reflections. The toy is named after Greek words 'kalo' meaning beautiful, 'eidos' meaning form and 'scopos' meaning watcher. In 1817, Brewster got kaleidoscope patented. In 1817, Fresnel showed that light was a transverse wave instead of longitudinal one. It became possible to explain the phenomenon of polarization with the help of transverse wave nature of light. In 1819, Augustin Jean Fresnel (1788–1827) and Dominique Francois Arago (1786–1853) furnished the proof that rays polarized at right angle to each other do not interfere, a discovery which finally put the theory of transverse vibrations on a firm footing. It is worth-mentioning here that Newton could not explain the phenomena of diffraction and polarization with the help of corpuscular theory. On the other hand, Huygens-Fresnel principle became the basis of advancement of physical optics explaining approximately all aspects of propagation of light.

Jean-Baptiste Biot (1774–1862) worked on the polarization of light using corpuscular theory of light. In 1815, he invented first polarimeter and demonstrated that polarized light when passed through an organic crystal could be rotated clockwise or anticlockwise depending on the optical axis of the crystal (*optical activity*). It was further found that the angle of rotation was a direct measure of the concentration of the substance. He said his result could be obtained if light were considered to be made of corpuscles. Later in 1874, Jacobus Henricusvan't Hoff (1852–1911) and Joseph Achille Le Bell (1847–1930) found that optical activity is due to the asymmetrical positioning of the carbon atoms in molecules of the organic crystals: sugar is a common optically active substance. Biot's work on chromatic polarization and rotator polarization greatly advanced the field of optics. Later it was shown that his findings could also be explained using wave theory of light. Biot's work on polarization led to many breakthroughs in the field of optics. For example, computer screens and liquid crystal displays (LCDs) in television use polarized light. Polarized light can be obtained using a number of methods. In 1828, William Nicol (1770–1851) invented a polarizing prism named *Nicol prism*. He bisected a parallelepiped of iceland spar crystal along its shortest diagonal, then cemented the two halves together with Canada balsam to form a Nicol prism. It splits unpolarized polarized light into two plane polarized light components with their plane of polarization perpendicular to each other. Nicol prisms greatly facilitated the study of refraction and polarization, and were later used to investigate molecular structures and optical activity of organic compounds. In 1932, Edwin Herbert Land (1909–1991) discovered a material called Polaroid, which polarizes light through selective absorption by oriented molecules. In 1937 Herbert Land set up Polaroid Corporation and initially he began to use numerous types of Polaroid materials in sunglasses and other optical devices. Once, Herbert Land was enjoying vacation in Santa Fe, New Mexico, with his three year old daughter Jennifer. He took her photograph with a camera, but she asked him why she cannot see her photograph immediately. Herbert Land thought about it and within hours he had an idea of instant camera. In 1947, Land invented *Polariod Land Camera* that produced finished print in 60 s. Land performed experiments on color vision and gave retinex theory of color vision. Polarized light can also be obtained by double refraction through a calcite crystal. It is noteworthy that Huygens is among the greatest scientists of all times and a major figure in scientific revolution. He is regarded as the first theoretical physicist and founder of modern mathematical physics. Elkan Rogers Blout (1919–2006), a close colleague of Edwin Land at Polaroid, wrote: "*What was Land like? Knowing him was a unique experience. He was a true visionary; he saw things differently from other people, which is what led him to the idea of instant photography. He was a brilliant, driven man who did not spare himself and who enjoyed working with equally driven people*".

Although various properties of light like the laws of reflection, refraction etc. were well understood but the velocity of light was not known. Rene Descartes said the velocity of light is infinite but Galileo was of the view that it is finite. But there was no experimental evidence for their speculations because of the absence of sufficient resources and technology. In 1676, Olaus Romer (1644–1701) estimated the velocity of light from the periodic occurrence of eclipses of one of the moons of Jupiter and found it to be finite with value of 2.12×10^{10} cm/s. Armand Hippolyte Louis Fizeau (1819–1896) was the first physicist who in 1849 measured experimentally the velocity of light using a rotating toothed wheel and found it to be 3.1×10^{10} cm/s. In 1862, Jean Bernard Leon Foucault (1819–1868) measured the velocity of light using rotating mirror method and found it to be 2.98×10^{10} cm/s. In 1879, Albert Abraham Michelson (1852–1931) measured the velocity of light when he was in U.S. Navy. In 1881, he devised an interferometer and measured the wavelength of light with great precision. In 1925 and 1926, Albert Abraham Michelson (1852–1931) performed the most accurate measurement of the velocity of light and its value was found to be 2.99796×10^{10} cm/s.

James Clerk Maxwell (1831–1879), Public Domain, URL: https://en.wikipedia.org/wiki/James_Clerk_Maxwell#/media/File:James-Clerk-Maxwell-1831-1879.jpg

James Clerk Maxwell (1831–1879), a Scottish physicist, was specialized in the field of mathematical physics. He is most famous for formulating Maxwell equations for electromagnetic radiations (discussed in detail in Chap. 11). The most important and beautiful success of the Maxwell equations is that light is a transverse electromagnetic radiation which does not require any medium for propagation. All the objections that were levelled against the Huygens wave theory were overcome in the Maxwell theory of electromagnetic radiations. The Maxwell theory also gave explanation about the color of light which is determined by the frequency of light wave. In 1861, he presented the first durable color photograph. Actually, Newton and Maxwell theories contradicted each other. Maxwell equations determined velocity while the Newton's equations of motion involved acceleration. In Newtonian mechanics velocity of a body is measured with respect to another body: relative velocity. Now the question arises with respect to what the speed of light is determined in the Maxwell's theory of light. The theory of light up to Maxwell and before was developed using classical physics now known as classical optics.

Heinrich Hertz (1857–1894), a German physicist was working on the production of radio waves. In 1887, he observed experimentally that when ultraviolet (UV) light falls on two metal electrodes with a voltage applied across them, the light changes the voltage at which sparking takes place—a relation between light and electricity. A year later, Wilhelm Hallwachs (1859–1922) observed that when a negatively charged zinc plate was illuminated with UV light, it lost its negative charge and finally got positively charged. These observations indicated that negatively charged particles are emitted from the metal surface when UV light falls on it thereby confirming the observations of Heinrich Hertz. In 1902, Philipp Lenard (1862–1947), a German physicist, clarified the relation between light and electricity by demonstrating that electrically charged particles are librated from the metal surface when illuminated and these particles are identical to electrons. Maxwell's electromagnetic wave theory is not able to explain the photoelectric effect because it is not possible for a wave to eject particles from a metal. In 1900, Max Planck (1858–1947) explained black body radiations in terms of quanta of electromagnetic radiations. In 1905, Albert Einstein (1879–1955) assigned particle properties to the quanta of light (photons) and explained the photoelectric effect. These facts marked the beginning of quantum optics which is a branch of atomic, molecular and optical physics dealing with how individual photons of light interact with atoms and molecules. The development of quantum mechanics is described in Chap. 4.

The origin of optical spectroscopy goes back to seventeenth century when in 1648, Johannes Marcus Marci von Kronland (1595–1667) discussed about color, rainbow and in general optics in his book entitled "*Thaumantius*". Generally, Sir Isaac Newton (1643–1727) is regarded as the founder (discoverer) of optical spectroscopy, but his work would not have been possible without the discoveries made by scientists before him. In 1666, he built his first spectrometer to examine the sun light. His spectrometer comprised of a small aperture to produce a beam of light, a lens to collimate it, a glass prism to disperse it, and a screen to display the resulting spectrum. This first spectroscope was nearly in modern form. With the spectrometer Newton studied dispersion of white light and introduced the word "spectrum" to describe the collection and order of different colors obtained on the screen. Newton's analysis of light was the beginning of the science of spectroscopy. Newton published his experiments and theoretical explanation of dispersion of light in his book "*Opticks*". In 1802, William Hyde Wollaston (1766–1828) improved the Newton's model of spectrometer by including a lens to focus sun's spectrum on the screen. He analyzed the sun's light and found black lines in its spectrum. He believed that these black lines were natural boundaries between the colors but could not characterize them. Joseph Ritter von Fraunhofer (1787–1826) conducted experiments with dispersive spectrometers by using diffraction grating in place of glass prism thus enabling the spectroscopy to become a more precise and quantitative technique. In 1814, he analyzed the sun's spectrum and independently found that a large number of black lines (more than five hundred) crossed the spectrum: named *Fraunhofer lines* in his honor. He studied systematically the sun's spectrum and measured the wavelengths of the black lines. Since the discovery of Fraunhofer lines, spectroscopy is playing a significant role in physics, chemistry, and astronomy. In 1820, Sir David Brewster (1781–1868) explained that these black lines in the spectrum arise due to the absorption of light in the sun's atmosphere. Therefore, the absorption spectroscopy commenced with the discovery of the Fraunhofer lines in the sun light. Charles Wheatstone (1802–1875) studied the emission spectra of a number of metals and, in 1835, he found that these can be distinguished from the bright lines appearing in their spectrum. In 1849, Jean Bernard Leon Foucault (1819–1868) experimentally demonstrated that the absorption and emission lines appearing at the same wavelength are due to the same material. In 1859 and 1860, Gustav Robert Kirchhoff (1824–1887) and Robert Wilhelm Eberhard Bunsen (1811–1899) explained the origin of the Fraunhofer lines in their two classic papers. They also established the general laws of emission and absorption of light.

It is very easy to understand the principle of absorption spectroscopy in terms of quantized states of an atom or molecules. Suppose a beam of light is incident on a sample material. Those photons are absorbed whose energy matches with the energy gap between the quantum states of the atoms or molecules of the material thus exciting the atoms or molecules to higher states. It produces absorption lines in the form of black lines in the spectrum. The rest of the photons are transmitted unaffected. The absorption spectrum exhibits maximum intensities at the wavelengths where absorption is the strongest. But the transmission spectrum will have its maximum intensities at wavelengths where the absorption is the weakest because most of the light is transmitted through the material. If the excited molecules make transition from the excited state to the original state light is emitted and the emission spectrum exhibits colored lines at the positions where the absorption spectrum showed black lines. After the invention of optical spectroscopy, a number of other spectroscopies such as infrared, ultraviolet, laser and X-ray spectroscopies etc. were invented, which played a vital role in determining the structure and composition of various crystalline materials. In 1949, Peter Berners Fellgett (1922–2008) invented Fourier transform infrared (FTIR) spectroscopy as a part of his Ph.D. thesis.

In 1955, Sir Alan Walsh (1916–1998) published his classic paper on atomic absorption spectroscopy and established its basic principles. In the same year Cornelis Theodorus Joseph Alkemade (1923–1989) and Joannes Marius Wilhelm Milatz (1910–2000) independently published similar findings on the absorption flame photometer. Walsh constructed atomic spectral lamps that form the key component of atomic absorption spectrometers. Atomic absorption spectroscopy has potential applications in chemical analysis and is immensely used in determining even very small amounts of elements present in a sample material.

Nano-optics is the study of the interaction of light with nanometer scale objects such as carbon nanotubes and metallic nanoparticles etc. It often involves dielectric structures such as nanoantennas or metallic components that can transport and focus light via surface plasmon and polaritons. Normal optical components, like lenses and microscopes cannot focus light to nanometer scale because of the diffraction limit (Rayleigh criterion). But it is possible to squeeze light into a nanometer scale using other techniques like surface plasmons, localized surface plasmons around nanoscale metal objects, nanoapertures and nanoscale sharp tips used in near-field scanning optical microscopy. Nano-optics will be described in brief in Chap. 18.

Bibliography

Abbe E (1873) Beiträge zur theorie des mikroskops und der mikroskopischen wahrnehmung [Contributions to the theory of the microscope and of microscopic perception]. Archiv für Mikroskopische Anatomie (in German). Bonn, Germany: Verlag von Max Cohen & Sohn 9:413–468

Abbe E (1880) Crisp F (ed) Conditions on aplanatism of systems of lenses. J Roy Microsc Soc (London) UK: Williams & Norgate 3:509–515; (1880) Some remarks on the apertometer. J Roy Microsc Soc (London) UK: Williams & Norgate 3:20–31; (1880) Conditions of aplanatism for wide-angled pencils. J Roy Microsc Soc 2 (London) UK: Williams & Norgate 1:831

Alkemade C (1954) A contribution to the development and understanding of flame photometry. Proefschrift Universiteit Utrecht; (1979) Fundamentals of analytical flame spectroscopy. Met Roland Herrmann

Arago DFJ, Fresnel A (1819) Mémoire sur l'action que les rayons de lumière polarisée exercent les uns sur les autres. Annales de Chimie et de Physique 10:288–305; reprinted in Fresnel A (1866–70) 1:509–522; translated as On the action of rays of polarized light upon each other, in Crew, 1900, pp 145–155

Bacon R (1897) In: Bridges JH (ed) The 'Opus Majus' of Roger Bacon (in Latin and English). Horace Hart for the Clarendon Press, Oxford.

Bartholin E (1669) Experimenta crystalli islandici disdiaclastici quibus mira & insolita refractio detegitur. Daniel Paulli, Copenhagen (Hafniæ), Denmark [English translation: (1959) Experiments with the double refracting Iceland crystal which led to the discovery of a marvelous and strange refraction, by Werner Brandt. Westtown, PA]

Biot JB (1815) Phenomene de polarisation successive, observés dans des fluids homogenes (Phenomenon of successive polarization, observed in homogeneous fluids). Bulletin des Sciences, par la Société Philomatique de Paris 190–192

Descartes R (1637) Discourse on method, optics, geometry, and meteorology (The first translated edition in English was published in 1979 and the latest third edition was published in 1998 by Hackett Publishing Company Inc.)

Fellgett PB (1951) Theory of infra-red sensitivities and its application to investigations of stellar radiation in the near infra-red. PhD thesis

Fizeau AHL (1849) Comptes Rendus Hebdomadaires de séances de l'Academie de Sciences (Paris), vol 29, pp 90–92)

Foucault L (1849) Lumière électrique [Electric light], L'Institut. J Universel Sci (in French) 17:44–46; (1849) Lumière électrique [Electric light], Société Philomatique de Paris. Extraits des Procès-Verbaux de Séances (in French) 13:16–20

Fraunhofer J (1814–1815) Bestimmung des brechungs- und des farben-zerstreuungs – vermögens verschiedener glasarten, in bezug auf die vervollkommnung achromatischer fernröhre (Determination of the refractive and color-dispersing power of different types of glass, in relation to the improvement of achromatic telescopes). Denkschriften der Königlichen Akademie der Wissenschaften zu München (Memoirs of the Royal Academy of Sciences in Munich) 5:193–226; see especially pp 202–205 and the plate following page 226

Frauhofer J (1821) Neue modifikation des lichtes durch gegenseitige einwirkung und beugung der strahlen, und gesetze derselben (New modification of light by the mutual influence and the diffraction of light rays, and the laws thereof). Denkschriften der Königlichen Akademie der Wissenschaften zu München (Memoirs of the Royal Academy of Science in Munich) 8:3–76; (1823) Kurzer bericht von den resultaten neuerer versuche über die gesetze des lichtes, und die theorie derselben (Short account of the results of new experiments on the laws of light, and the theory thereof). Ann Phys 74:337–378

Fresnel A (1818) Mémoire sur la diffraction de la lumière (deposited 1818, "crowned" 1819). In: Oeuvres complètes (Paris: Imprimerie impériale, 1866–70) 1:247–363

Grimaldi FM (1665) Physico mathesis de lumine, coloribus, et iride, aliisque annexis libri duo (Bologna "Bonomia"). Vittorio Bonati, Italy, pp 1–11 (in Latin)

Hallwachs W (1888) Uber den einfluss des lichtes auf electrostatisch geladene korper. Ann Phys 269:301–312

Hertz HR (1887) Uber einen einfluss des ultravioletten lichtes auf die electrische entladung. Ann Phys 267:983–1000

Hooke R (1965) Micrographia: or, Some physiological descriptions of minute bodies made by magnifying glasses (first edition). John Martyn and James, Allestry, Printers to Royal Society, London

Huygens C (1690) Traité de la lumière. Leyden (reprinted and published in 1962)

Kepler J (1604) Ad Vitellionem Paralipomena, Quibus Astronomiae Pars Optica Traditur. Claudium Marnium and Haeredes Joannis Aubrii, Frankfurt

Kirchhoff G, Bunsen R (1860) Chemical analysis by observation of spectra. Ann Phys Chem 110:161–189; (1860) Chemical analysis by spectrum-observations. Philos Mag 20:88–109

Land EH (1959) Experiments in color vision. Sci Am 200:286–298; (1977) The retinex theory of color vision. Sci Am 237:108–128

Lenard P (1902) Uber die lichtelektrische wirkung. Ann Phys 313:149–198

Maiman TH (1960) Stimulated optical radiation in ruby. Nature 187:493–494; (1960) Optical and microwave-optical experiments in ruby. Phys Rev Lett 4:564–566; (1961) Stimulated optical emission in fluorescent solids I: theoretical considerations. Phys Rev 123:1145–1150

Maiman TH (1961) Ruby laser systems. U.S. Patent 3,353,115

Maiman TH, Hoskins RH, D'Haenens IJ, Asawa CK, Evtuhov V (1961) Stimulated optical emission in fluorescent solids II: spectroscopy and stimulated emission in ruby. Phys Rev 123:1151–1157

Newton I (1687) Philosophiae naturalis principia mathematica. IMPRIMATUR, S. Pepys, Reg. Soc. Preses, Londini; (1704) Opticks. Sam Smith and Benj. Walford, Printers to Royal Society, London (After this a number of revised editions appeared)

Rittenhouse D (1786) An optical problem, proposed by Mr. Hopkinson, and solved by Mr. Rittenhouse. Trans Amer Philos Soc 2:201–206; Cope TD (1932) The Rittenhouse diffraction grating. David Rittenhouse Papers MSS. SMS Coll 11:377–382

Romer O (1676) Démonstration touchant le mouvement de la lumière trouvé par M. Roemer de l'Académie des sciences [Demonstration concerning the movement of light found by Mr. Romer of the Academy of Sciences]. Le Journal des Sçavans (in French) 233–236

Toomer GJ (1976) Diocles on burning mirrors. Sources in the history of mathematics and the physical sciences, vol 1. Springer, Berlin, Heidelberg, New York

Toomer GJ (1984) Ptolemy's Almagest. Springer, New York, Duckworth, London

Walsh A (1955) The application of atomic absorption spectra to chemical analysis. Spectrochim Acta 7:108–117 (erratum, *ibid.*, p 252); (1961) The application of atomic absorption spectra to chemical analysis. Adv Spectrosc 2:1–22; (1966) Some recent advances in atomic absorption spectroscopy. Jl. N Z Inst Chem 30:7–21

Wheatstone C (1836) On the prismatic decomposition of electrical light, Report of the Fifth Meeting of the British Association for the Advancement of Science; Held at Dublin in 1835; Notices and abstracts of communications to the British Association for the Advancement of Science, at the Dublin Meeting, Aug 1835, Murray, London, England, pp 11–12

Wollaston WH (1802) A method of examining refractive and dispersive powers, by prismatic reflection. Philos Trans Roy Soc 92:365–380

Young T (1804) Bakerian Lecture: experiments and calculations relative to physical optics. Philos Trans Roy Soc 94:1–2

Chapter 4
Quantum Mechanics and Relativity Theory

In the nineteenth century, great advances were made in the concepts of macrophysics such as the laws of mechanics, thermodynamics and electromagnetism. During the last decades of the nineteenth century, a number of important discoveries were made and most important of these were the discoveries of electrons, photoelectric effect, X-rays and radioactivity. Towards the end of the nineteenth century, the electromagnetic theory of James Clerk Maxwell (1831–1879) was fully established and it managed to explain most of the physical phenomena. Many scientists believed that the work of science was nearly at an end and only a few minute corrections and more precise measurements are needed to be added. In 1897, Lord Kelvin said:

> There is nothing new to be discovered in physics now. All that remains is more and more precise measurement.

In 1903, Albert Michelson (1852–1931) said:

> The more important fundamental laws and facts of physical science have all been discovered..... our future discoveries must be looked for in the sixth place of decimals.

But three phenomena, namely blackbody radiation spectrum, photoelectric effect and bright line spectra of elements, remained unexplained. In fact, science was moving from the world of macrophysics to the world of microphysics. In the twentieth century, two very powerful subjects that is quantum physics and relativity theory came into existence and physics was going to enter the quantum age. Solid State Physics as a subject came into existence only in the middle of the twentieth century and most of its properties are explained using quantum mechanics. Therefore, before discussing various properties of solids, let us first review briefly the development of quantum mechanics and relativity theory.

The radiations emitted by a hot body, at room temperature or above, are mostly invisible infrared radiations and the body, which emits these radiations, appears black. In 1859, Gustav Robert Kirchhoff (1824–1887) worked on the emission of thermal radiation and gave the law of thermal radiation according to which emissivity is equal to absorptivity in thermal equilibrium. In 1862, Kirchhoff measured thermal radiation and coined the word blackbody radiation for it. The schematic representation of

J. S. Galsin, *History of Solid State Physics*,
https://doi.org/10.1007/978-981-95-0504-3_4

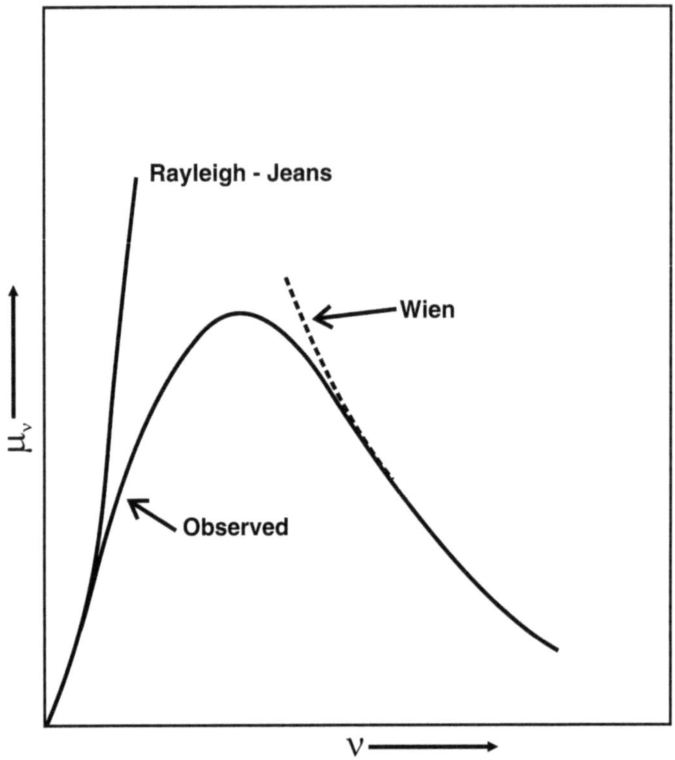

Fig. 4.1 Schematic representation of blackbody radiation spectrum

blackbody radiation spectrum is given in Fig. 4.1. In 1879, Josef Stefan (1835–1893) formulated the law of blackbody radiation (Stefan's law) by examining the experimental measurements of blackbody radiation. Later, in 1884, Ludwig Boltzmann (1844–1906), a student of Josef Stefan, derived the law from the theoretical considerations. According to *Stefan-Boltzmann law*, heat energy radiated per unit area per unit time E_T is directly proportional to the fourth power of its absolute temperature, that is

$$E_T = \sigma T^4 \tag{4.1}$$

where σ is the Stefan's constant and T is the temperature in absolute degree. In 1893, Wilhelm Wien (1864–1928) formulated a theory that gave good results only at high frequencies and also predicted the frequency at which peak of the blackbody radiation spectrum occurred at low temperatures—*Wien displacement law.* In 1900, Lord Rayleigh (1842–1919) and James Jeans (1877–1946) gave a law for blackbody radiation from the classical concepts and it was found that the Rayleigh-Jeans law showed correct behavior at low frequencies only. The classical theory predicted an infinite intensity in the ultraviolet region and beyond and it was dubbed as the *ultraviolet catastrophe*. In 1894, Max Planck (1858–1947) started working on the problem of blackbody radiations. He was of the view that with the help of thermodynamics, the two incomplete descriptions—of Rayleigh-Jeans and of Wien, could be combined to describe the shape of the whole blackbody spectrum. From 1894–1900, he tried the problem using different approaches based on the Maxwell theory of electromagnetic radiations but could not succeed. In 1877, Ludwig Boltzmann (1844–1906) had discussed in one of his research papers that the energy states of a physical system could be discrete. Ultimately, in 1900, Max Planck assumed that the electromagnetic energy is not flowing in a continuous fashion but in the form of packets of energy called quanta of energy. The energy of a quantum of electromagnetic radiation energy was assumed to be given by

$$E = h\nu \tag{4.2}$$

Max Planck (1858–1947), Public Domain, URL: https://en.wikipedia.org/wiki/Max_Planck#/media/File:Max_Planck_by_Hogo_Erfurth_1938cr_-_restoration1.jpg

where h is called Planck constant and ν is the frequency of radiation. Planck derived the equation for the blackbody radiation spectrum by combining thermodynamics and electrodynamics both of which are classical theories. Planck's blackbody radiation law was first proposed in 1900 in a meeting of German Physical Society called Deutsche Physikalische Gesellschaft (DPG) and subsequently published in 1901.

The assumption of quantization of energy was incompatible with classical mechanics. Therefore in the beginning Planck considered it "a purely formal assumption" and did not give much attention to it (Max Planck, 1963). At that time Planck did not understand even the physical meaning and importance of energy quanta. Moreover, Planck's relation did not obey the law of equipartition of energy and not all modes had the same energy. Actually, nobody knows what Planck had in mind when he took the revolutionary step of assuming quantized energy. Later Planck said that this was the most intensive period of work of his entire life. Many attempts failed, until at last he was left with only one alternate that is the assumption of discrete energy. In 1931, Max Planck wrote a letter to his friend Robert Williams Wood (1868–1955) and said that "*the postulate of energy quanta was an act of desperation 'done because' a theoretical explanation had to be supplied at all costs, whatever the price*". Planck himself did not know that the assumption of energy quanta is going to give birth to a new powerful field later called quantum mechanics. That is why Max Plank is called the *father of quantum mechanics*. The emergence of quantum mechanics therefore, is partly a chance, partly a necessity and to a large extent, possibility of experimental verification. The concepts of thermodynamics and quantization of energy made a breakthrough in microscopic models which eventually explained a large number of properties that had previously been handled only phenomenologically. Surprisingly, in spite of the important and revolutionary aspects of the blackbody formula, it did not draw much attention immediately after the quantization of energy.

Albert Einstein (1879–1955), Public Domain, URL: https://en.wikipedia.org/wiki/Albert_Einstein#/media/File:Albert_Einstein_1947jpg

Albert Einstein (1879–1955) was a German physicist who wanted to become an engineer. After attending school, he, in 1896, enrolled at the Swiss Federal Polytechnic in Zurich to be trained as a teacher in mathematics and physics. In 1901, he got the diploma but could not get the position of a teacher neither he could not find any research position. Einstein joined as technical assistant in a patent's office in Bern and got sufficient time there to study and work independently. In 1905, at the age of 25, Einstein published five papers, out of which three revolutionized the approach to physics. The first paper is on *Brownian movement*, the second on *photoelectric effect* and the third on theory of *special relativity*. All the three papers were worthy of a Nobel Prize. In 1803, John Dalton (1766–1844), an English chemist, physicist and meteorologist, proposed atomic theory of matter. According to him an element (matter) is made of very small indivisible particles called atoms which can exist independently and are electrically neutral. Although the idea of atoms was not new but Dalton's contribution was to consider the relative sizes and characters of the atoms. The proposal of atomic theory by John Dalton remained entirely hypothetical for almost a century because many scientists disagreed with him. Einstein's paper on Brownian movement proved beyond doubt the atomic nature of the materials: the atoms/molecules in a liquid have zigzag motion called Brownian movement. He further proved that although the motion is random but it obeys the statistical law proposed by Maxwell and Boltzmann. Albert Einstein submitted the work concerning the Brownian movement and size of molecules for his doctoral thesis to the University of Zurich and received a Ph.D. degree in 1906. In the second paper, Albert Einstein was reminded of Newton's corpuscular theory of light to explain the photoelectric effect. Combining corpuscular theory with the quantization of light, Albert Einstein assumed that light consists of quanta of energy which behave as particles, later named photons, and these possess energy as integral multiple of $h\nu$. Einstein argued that when photons fall on the surface of a photosensitive material, electrons are ejected out (see Fig. 4.2). Based on this he gave the explanation of the laws of photoelectric effect which were earlier studied experimentally by Philipp Lenard (1862–1947) in 1899. *Therefore, the year 1905 is actually the true birth of quantum theory.* It seemed we were going back to the particle nature of light, a concept given by Newton more than 200 years ago. Initially Max Planck rejected Einstein's assumption of light quanta as particles (photons) because he was unwilling to discard completely the Maxwell theory of electrodynamics. Max Planck had once said that Einstein is a brilliant physicist but he may be excused for his idea of photons. After a number of years, it was proved beyond doubt that photons do exist, establishing the particle nature of light. Therefore, not for the first time, Einstein was ahead of his contemporaries in dealing with light radiations. The photoelectric effect proved to have a large number of applications. Therefore Einstein was awarded 1921 Nobel Prize in Physics for his explanation of the photoelectric effect. At this stage, there arose some doubt about the nature of light because it behaved as electromagnetic waves in explaining interference, diffraction and polarization but as photons (particles) in explaining the photoelectric effect. It seemed that light has dual character: both wave and particle nature.

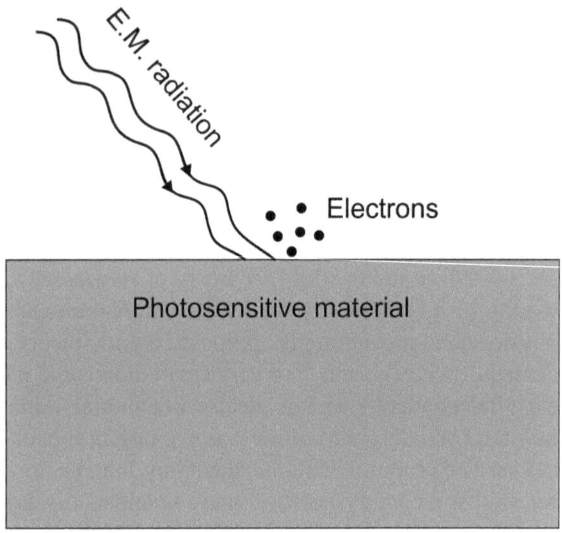

Fig. 4.2 Photoelectric effect

Before the formulation of the theory of special relativity, a number of related developments had already taken place. In 1887, the Michelson-Morley experiment proved that the speed of light is the same for all observers regardless of their motion. Therefore, it was necessary to have a simultaneous transformation of the local coordinates and time. In the same year, Woldemar Voigt (1850–1919) had formulated a transformation between an inertial frame at rest and a frame moving with velocity v in the x-direction just similar to the Lorentz transformation. But Voigt himself said that this transformation was developed in relation to a particular problem and was not a general transformation. In 1889, George Francis FitzGerald (1851–1901), a Danish physicist, had proposed the concept of length contraction which was independently developed, in 1892, by Hendrik Antoon Lorentz (1853–1928) to explain the negative outcome of Michelson-Morley experiment. Hendrik Lorentz, a Dutch theoretical physicist, was a genius who made contributions to many fields. In 1892, Lorentz pointed out that the motion of some charged particles is responsible for the flow of current in a conductor. In 1899 and again in 1904, Hendrik Lorentz added time dilation to his transformations and published what Jules Henri Poincare (1854–1912), in 1905, named Lorentz transformations. In 1904, Hendrik Lorentz gave covariant formalism of electrodynamics in which electrodynamic phenomena in different frames of reference are described by identical equations with well-defined transformation properties. Lorentz is the man who set the stage for Albert Einstein for the discovery of theory of special relativity. According to biography of Lorentz published by the Nobel foundation, "*It may well be said that Hendrik Lorentz was regarded by all theoretical physicists as the world's leading spirit, who completed what was left unfinished by his predecessors and prepared the ground for the fruitful reception of the new ideas based on quantum theory*". When in 1905, Albert Einstein formulated the theory of special relativity, he was the first to remove completely the adhoc character from the length contraction hypothesis and proved that it could be explained using his theory. This changed the notion of space, time and simultaneity. Einstein was trying to make the Maxwell theory and discoveries of Galileo and Newton consistent. In the Newtonian mechanics, present, past and future are well defined. Further, in the Newtonian mechanics, when an event occurs, it is simultaneously observed (principle of simultaneity is followed). But Einstein argued that between the past and future of an event there exists an intermediate zone called extended present; a zone which is neither past nor future. Actually, this is the concept introduced by Einstein in the theory of special relativity. The duration of the intermediate zone is very small and depends where the event takes place relative to the observer (Fig. 4.3): greater the distance between the event and the observer, longer the duration of the extended present. For example, if the distance between event and observer is a few meters, the intermediate zone is of a few nanoseconds (near simultaneity). If the event occurs on the moon and the observer is on the earth, the intermediate zone is of a few seconds. If the event occurs on Mars (with the observer on earth), then the intermediate zone is quarter of an hour. Einstein said that this time is inevitable and cannot be reduced. He had explained that "absolute simultaneity" does not exist. Suppose I am here on the earth and you are at Mars, I ask a question and you reply as soon as you have heard it, your reply reaches me after a quarter of an hour, and this time is neither the past nor the future but the extended present. Because of the extended present it is not possible to hold a smooth conversation between two persons, one on Earth and the other on Mars. In physical problems, the time at which an event occurs (retarded time) and the time at which it is observed (observation time) are different. One can take an example from physics. Suppose alpha particles are emitted from a radioactive source and these are observed at the position of the detector placed at some distance from the radioactive material, the time at which an alpha particle is actually emitted by the radioactive source is different from the time at which it is recorded by the detector: observation time is a little after the retarded time. Therefore, space and time are not absolute but relative to both the observer and the object being observed: the faster one moves the more pronounced these effects become. Therefore, one should call it "*spacetime*" not space and time separately as in the Newtonian mechanics. Due to the merger of space and time the restructuring of Newtonian mechanics is required as follows:

1. As space and time fuse together in a single concept of **spacetime**, the electric and magnetic fields also fuse together in a single field called electromagnetic field. The covariant formalism showed very beautifully that electric and magnetic fields form different elements of the same force tensor and therefore had the same origin.
2. In Newtonian mechanics, mass and energy are separately conserved. But Einstein realized that mass and energy are the two forms of the same entity (two faces of the same coin) just as electric and magnetic fields are the two faces of the same field. In spacetime concept, mass may be transformed into energy and vice versa: mass-energy equivalence. Therefore, sum of the mass and energy is conserved, not separately, and a process must exist for such a transformation.

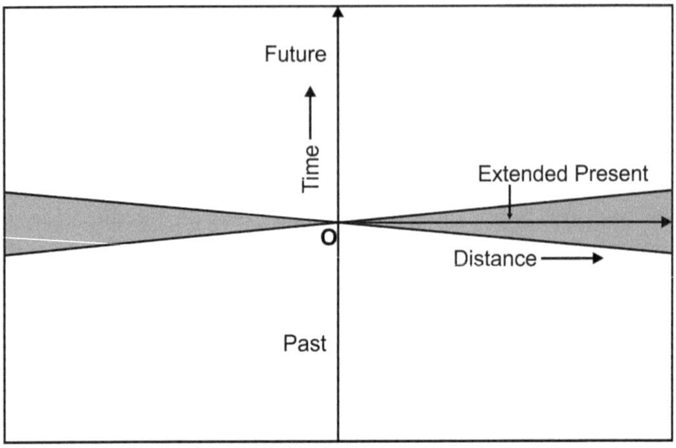

Fig. 4.3 Figure shows the extended present as a function of distance from the observer situated at point 0

In 1905, Einstein published a fourth paper in the month of September entitled, "*Does the inertia of a body depend upon its energy content*" in which he gave the famous relation between mass m and energy E as

$$E = mc^2 \tag{4.3}$$

where c is the velocity of light. The energy obtained from one gram of mass is enormous: it is equivalent to energy of million bombs exploded at the same time, it is enough to illuminate a city and power the industries of a country for months.

At that time Max Planck was the editor of Annalen der Physik research journal. It is believed that Max Planck received all the revolutionary papers of Albert Einstein and accepted them immediately realising the greatness of the work. Einstein's papers attracted little attention at that time. After the publication of these papers Einstein applied for the post of a university lecturer, but his application was rejected. He then applied for the job of a high-school teacher and this application was rejected as well. So he went back to the job of clerk in the patent office. Later Max Planck invited Einstein for the post of a full professor in Berlin and worked as colleagues for about twenty years.

If the space and time merge together to form spacetime, it is natural that spacetime must have four dimensions. In 1908, Hermann Minkowski (1864–1909) gave the concept of four-dimensional space, a combination of three-dimensional Euclidean space and time (Minkowski spacetime), which is closely related with Einstein theory of special relativity. Minkowski concluded that space and time should be treated equally. According to the Newtonian mechanics, length, mass and time are independent constant quantities. But in relativity theory, length, mass and time change with the velocity of the particle as space and time are linked with each other. According to theory of special relativity, length, mass and time vary with velocity as

$$L = L_0\sqrt{1 - \frac{v^2}{c^2}} \tag{4.4}$$

$$m = \frac{m_0}{\sqrt{1 - \frac{v^2}{c^2}}} \tag{4.5}$$

$$t = \frac{t_0}{\sqrt{1 - \frac{v^2}{c^2}}} \tag{4.6}$$

where L_0, m_0, t_0 are the length, mass and time of an object at rest while L, m, t are the corresponding quantities when object is moving with velocity v. Here c is the speed of light. It is evident from Eqs. (4.4) to (4.6) that with the increase in velocity, length of the object decreases, its mass increases and the time dilates. As the velocity of an object approaches velocity of light, length approaches zero, mass approaches the infinite value (two contradicting properties) and the time stands still. One can interpret this result by saying that no object can acquire the velocity of light. In other words, the maximum velocity of an object is the velocity of light. It is well known that a photon travels with the velocity of light. It means time has no meaning for a photon. Let us assume that a photon, after leaving a distant star, arrives at the earth after spending thousands of years of time as measured by a clock on earth. But as the photon is concerned, it takes no time to reach the earth. Therefore, the theory of special relativity shook the well-established basis of the Newtonian mechanics. It is because in relativity, one speaks about *spacetime* and *mass* instead of space, time and mass separately. Because of this reason, in the beginning, a large section of physicists did not believe in the theory of relativity. As of now the concept of spacetime given by Einstein is well understood and tested in laboratories.

One day, Einstein came to the realisation that the theory of special relativity describes the motion of bodies well, but one thing is missing—gravity. What would happen when light and, in general, bodies in motion encounter gravity? Using Faraday's concept of field in terms of the lines of force (see Chap. 11), Einstein thought that the force of attraction between the earth and the sun and between the earth and falling bodies must also be due to a field: gravitational field. It is well known that the motion of freely falling bodies does not depend on mass or the composition of mass. Therefore, Albert Einstein thought that the gravitation might be due to certain properties of space (medium) in which objects are falling. Einstein worked on this problem for about 10 years and ultimately in 1915, formulated a theory of *general relativity* which is altogether different from the theory of special relativity. In the theory of general relativity, Einstein imagined that space and time were woven together into something he called the fabric of the universe. He assumed that the space is curved and not flat (as assumed earlier in Newtonian mechanics). He further assumed that the local curvature in space is produced by the presence of mass, and the degree of curvature is determined by the mass distribution in the universe. Consequently, objects moving through the curved space do not travel in straight lines but follow the path of least resistance along the contour of the curved space. In a nut shell, in the theory of general relativity, Einstein had replaced Newton's gravitational force by the curved surface: a theory with no forces. Einstein derived a set of mathematical equations describing the relationship between the curvature of space and the distribution of mass in the universe. Einstein found that "matter determines the curvature of space and space determines the motion of matter": a new way to describe gravitation without any forces. Einstein's theory of general relativity assumed that the universe is fixed. Vesto Melvin Slipher (1875–1969), an astronomer at the Lowell Observatory of Arizona, was a contemporary of Einstein. Slipher from his spectroscopic measurements of distant stars observed finite Doppler shift. The phenomenon of Doppler shift also applies to light. In case of receding galaxies, light moving away from us shifts towards the red end of the spectrum, called red shift. On the other hand, in case of approaching galaxies the approaching light shifts towards the blue end, called blue shift. In 1913, Slipher measured the blue shift of Andromeda which is moving towards our solar system. In 1915, Slipher published one of his classic papers in which he measured the red shift of a number of galaxies and its results were presented to American Astronomical Society in its meeting held on Aug. 1914 meeting. In 1917, Slipher noted that other galaxies are moving with respect to us but he missed the main point of his data which was that the universe is expanding. Although Vesto Slipher was the first to note the motion of galaxies, his findings were overlooked by the scientific community.

Einstein could solve these mathematical equations only approximately and was of the view that their exact solution might not be possible. In the same year (1915), Karl Schwarzchild (1873–1916) gave an exact solution of the Einstein equations for a spherical object since heavenly bodies like sun and stars are nearly spherical. According to Schwarzchild's solution, there exists a critical radius of a spherical object at which the curvature of space becomes so large that matter cannot escape. In other words, at the critical radius the gravitational force would become so large that the object would collapse indefinitely and nothing would be able to resist the self-gravity caused by the extreme curvature of the surface. Ultimately, the whole object would be compressed to a single point at the center of the object: a singularity. Later this singularity was identified as a *black hole*. Albert Einstein's theory of general relativity was not just a ground breaking thought but a transformative idea that challenged the conventional understanding of the universe. In 1922, Albert Einstein gave a lecture on his theory of general relativity at the University of Berlin that drew a crowd of leading physicists eager to hear Einstein's insights into the nature of space and gravity. Einstein's ideas were so ahead of their time, it seemed that they belong to some other world. Einstein's 1922 lecture was not just an academic event but laid the foundation of modern physics.

In 1922, Alexander Friedmann (1888–1925) gave another solution of the Einstein equations according to which the universe is unstable and a slight perturbation can cause expansion or contraction of the universe. But Einstein did not accept the idea of an expanding universe, rather he was in favour of a static universe. Edwin Hubble (1889–1953) was a good sportsman and equally good scholar. In 1924, Hubble published his landmark paper "*Cepheids in spiral nebulae*" showing that the universe is not just our own galaxy, the Milky Way, but contains lots of independent galaxies—*Island Universes*. Many of these galaxies are bigger than the Milky Way and much more distant. In 1929, Hubble showed experimentally, with the help of Hooker telescope at Mount Wilson that galaxies in the sky are moving away from us in all directions. He observed a relation between red shift and distance and showed that the recessional velocity v of a galaxy increases linearly with its distance d from the Earth: that is $v \propto d$ and the constant of proportionality is called Hubble constant. This is known as Hubble's law. Hubble wrote a popular book entitled, "*The realm of the nebulae*" and in it he described his own work. But he was not aware of Einstein's theory of general relativity and could not explain his results theoretically. Actually the first idea of an expanding universe was given by Georges Lemaitre (1894–1966), a priest and theoretical physicist, in 1927, which was supported experimentally by Hubble. In 1931, Georges Lemaitre used Friedmann-type solution and gave his own model of the universe. Lemaitre said that the universe at one time existed in a very small space, called *primeval atom* or *cosmic egg*, before an expansion took place several billion years ago and that the universe is still expanding. It was an idea that very neatly anticipated the modern concept of the Big Bang, but it did not get attention in cosmology. It is because of this reason Lemaitre is called *the father of big bang*. Julius Robert Oppenheimer (1904–1967) worked on the stability of massive neutron stars. In 1939, Oppenheimer (1904–1967) and George Volkoff (1914–2000) and Tolman (1881–1948) gave a limit to the mass of stars, called Tolman-Oppenheimer-Volkoff limit, beyond which stars would not remain stable as neutron stars but would undergo gravitational collapse. In the same year, Oppenheimer and his student Hartland Snyder (1913–1962) put forward a third solution of the Einstein equations. They said stars may eventually burn out and then collapse under the gravitational contraction. When critical radius is achieved, catastrophic collapse would take place. Their solution predicted the formation of neutron stars in case of small stars. Walter Baade (1893–1960) and Fritz Zwicky (1898–1974) were working on neutron stars. They presented their work at Stanford University and then published it in 1934. They coined the term *supernova* for the first time and hypothesized that supernovae were the transition of normal stars into neutron stars. They further connected supernova explosions to the production of a new phenomenon now called cosmic rays. In 1938, Zwicky proposed that a neutron star may shrink enough and finally it would become so dense that even light cannot escape its immense gravitational pull. Unfortunately his ideas were not noticed by the scientists at that time. Fred Hoyle (1915–2001), an English cosmologist, published two research papers, one in 1946 and the second in 1954, on the synthesis of elements heavier than helium. He realized that if a star collapsed, a huge amount of heat is produced—about 100 million degrees or more, forming heavier elements in a process known as nuclear synthesis. He further realized that at such temperatures iron can become much more abundant than other heavy elements, thus explaining natural high abundance of iron on earth. In 1949, Hoyle coined, for the first time, the word 'Big Bang' on BBC radio broadcast for Georges Lemaitre's model of expanding universe, but was a strong critic of it. Hoyle put forward a *steady state theory* of the universe according to which the universe is constantly expanding creating new matter: in other words, when galaxies get farther apart, new ones develop between them to fill the empty space. John Archibald Wheeler (1911–2008) was one of the last collaborators of Einstein and did lot of work on general relativity. In 1965, using relativity theory he confirmed the gravitational collapse of massive stars and the formation of neutron stars in case of small stars: a supernova is triggered and most likely a neutron star is formed. He is best known using for the first time the term *black hole* for objects with gravitational collapse. Wheeler also worked with Niels Bohr in explaining the basic principles behind nuclear fission. He was the driving force in developing both the atomic and hydrogen bombs. One can say that Einstein's equations contain information about the degree of bending of starlight, the information about the singularity of space, the existence of gravitational waves and the description about the formation of neutron stars, black holes and even the prediction of expansion of the universe.

In 1946, George Gamow (1904–1968) advanced the idea of expanding universe of Lemaitre in connection with the study of origin of elements and suggested that some cosmic background radiations are left over from the Big Bang. In 1964, Arno Allan Penzias (1933–present) and Robert Woodrow Wilson (1936–present) discovered some background noise in the form of radiation at their hissing antenna which made the experimental work impossible. The background radiations came from all directions of the sky, day, night and in all seasons. They could neither identify the background radiations nor find their origin. But Penzias and Wilson were unaware of the work of Gamow. After discussion with Robert Henry Dicke (1916–1997) and co-workers, in 1965, they came to the conclusion that these are cosmic radiations which helped establish the Big Bang theory of cosmology as these were considered as remnants of the Big Bang.

According to the Big Bang, an explosion occurred and the universe was created from nothingness. It is worth mentioning here that it is not an explosion in the traditional sense rather it was the expansion of space itself from an extremely hot, dense and small state. But the question is where did this explosion occur and where is its epicentre? It has been proved experimentally that every point in the universe moves away from every other point making the concept of centre of creation a mere illusion. Big Bang is not a localized explosion in the existing space but was the expansion of the space itself. The galaxies are moving away with a velocity proportional to its distance from the Earth (Hubble's law). The galaxies beyond a distance of roughly

14.5 billion years (critical threshold) move with velocity greater than the velocity of light. Therefore, even if we move with a velocity of light (which is not possible) we would never catch them, so practically they are lost to us forever. According to the horizon of our sight we can see galaxies up to a distance of about 46.1 billion years and beyond this the cosmos is forever hidden. Our own galaxy, the Milky Way, is about 100,000 (one lac) light years wide and comprises of more than 2 billion stars. We want to point out here that the universe is infinite, without any boundary, and is simply expanding. The whole universe comprises of two trillion galaxies and everything we see—stars, planets, galaxies etc.—makes up just 5% of it. About 95% of the universe is beyond our reach and is made up of mysterious unseen forces. Therefore the universe we see is just the tip of the cosmos iceberg.

We have already discussed that the gravitational contraction of massive stars leads to the formation of very dense small mass (singularity in the solution of Schwarzchild) which results in a very strong bending of space around it. The extreme bending of space causes the light to bend back into the star itself as a result of which light emitted by the star does not reach us, forming a black hole. Further, anything can fall into the black hole but nothing can get out of it—not even light or any other radiation. For a black hole the past is outside and the future is inside but the surface of black hole is like present. Therefore, travelling to the past is difficult, but travelling to the future is easy. It would be clearer from the following example. There is a boundary of black hole, called *horizon*, at which the velocity required by an object (say a rocket) to escape exceeds the velocity of light. So according to relativity, for the rocket near the horizon the time slows down. If the rocket stays at a fixed position near the horizon for an hour and then moves away, in the mean time, centuries have passed outside. Stephen Hawking (1942–2018) researched about the black holes in great detail. He combined the theory of general relativity and quantum mechanics to produce quantum gravity. In 1973, James Maxwell Bardeen (1939–present), Brandon Carter (1942–present) and Stephen Hawking proposed four laws of black hole mechanics and found that these laws were identical with the laws of thermodynamics. Hawking concluded that black holes were real just like other bodies in the universe. In 1974, Hawking said that black holes are not really black but they are hot and have finite temperature and finite entropy. He showed that just like hot bodies black holes emit radiation, named *Hawking radiation*, and may continue emitting until all of their energy is exhausted. Finally, the black holes explode and evaporate into pure radiation. In 1974, two astronomers namely Russell Alan Hulse (1950-present) and Joseph Hooton Taylor (1941–present) discovered a new type of binary pulsar 21,000 light years away from the earth and for the first time obtained the indirect evidence of the existence of gravitational waves. On September 14, 2015, Laser Interferometer Gravitational-Wave Observatory (LIGO) detected gravitational waves generated by two colliding black holes 1.3 billion light years away and the results were published in 2016 by Abbott and co-workers (discovered 100 years after gravitational waves were proposed by Einstein). Rainer Weiss (1932–present) and Barry C. Barish (1936–present) are best known for inventing LIGO detector. Kip S. Thorne (1940–present), a theoretical physicist, provided the theoretical support to LIGO. It was found that gravitational waves are in the form of ripples in the curvature of spacetime caused by the motion of massive objects. Rainer Weiss, Barry C. Barish and Kip S. Thorne shared the 2017 Nobel Prize in Physics for their work and detection of the gravitational waves. In the later years of his life, Einstein attempted to develop a unified field theory by generalizing his geometric theory of gravitation to include electromagnetism but could not complete this task during his life time.

There are different theories about the origin of our universe and the first one is the Big Bang theory. Stephen Hawking believed that there was nothing before Big Bang and it represents the beginning of time. Some scientists believe that our universe has originated from the collision of two parallel universes. Roger Penrose (1931–present) has pointed out that our universe has originated from one of the multiple Big Bangs which happened. But actually we know very little about the beginning of the universe and all we have are opinions, theories and hypotheses.

On the other hand quantum mechanics was also being developed at the same time. The classical theories failed to explain the temperature dependence of specific heat of solids. In 1906, Albert Einstein proposed a quantum theory of specific heat assuming atoms as quantum oscillators. It was able to explain the temperature dependence of specific heat at medium and high temperatures but failed to explain its temperature dependence at very low temperatures. In 1912, Peter Debye (1884–1966) improved upon the Einstein theory and explained the temperature variation of specific heat of solids at very low temperatures (details of the theories of specific heat are given in Chap. 8 on Heat and Thermodynamics). The reproduction of correct temperature variation of specific heat of solids provided strong support to the concept of quantization of energy.

Elements exhibit three types of spectra: continuous spectrum, band spectrum and line spectrum. In a continuous spectrum, all sorts of frequencies are emitted, in a band spectrum only a band of frequencies are emitted but in a line spectrum only particular frequencies are emitted. Most of the elements exhibit line spectrum and Fig. 4.4 shows the emission and absorption line spectra of hydrogen. The line spectra of elements were unexplained till the end of nineteenth century. Because the line spectrum is related to the structure of the atom therefore, it was thought that the study of line spectra may reveal something fundamental about the atomic structure. In the beginning, hydrogen, the simplest atom, was chosen for the study of line spectrum. In 1862, Anders Jonas Angstrom (1814–1874), a Swedish physicist, measured experimentally four prominent lines of hydrogen atom in the visible region of the spectrum. In 1885, Johann Jakob Balmer (1825–1898), a Swiss mathematician and school teacher, fitted the frequencies of lines of visible spectrum of hydrogen atom in the following formula

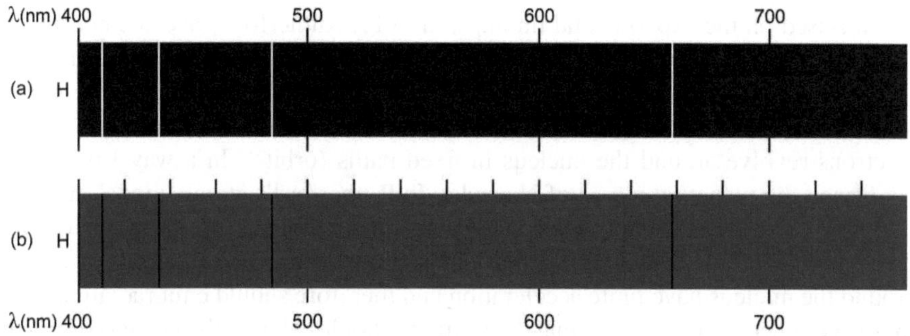

Fig. 4.4 (**a**) Emission and (**b**) absorption line spectra of hydrogen

$$\nu = R\left(\frac{1}{n_f^2} - \frac{1}{n_i^2}\right) \tag{4.7}$$

Here n_f and n_i are initial and final numbers and R is a constant. Using this equation, Balmer could reproduce four lines of visible spectrum of hydrogen with $n_f = 2$ and $n_i = 3, 4, 5, 6$. Other lines in the ultraviolet region were confirmed later: this series is called *Balmer series*. It is interesting to note that the above formula was given before the discovery of electron. On 30th April, 1897, Joseph John Thomson (1856–1940), a British physicist, discovered electron while performing experiment on electric discharge through vacuum tubes at low pressures. The announcement of the discovery of electron took place during the evening lecture at the Royal Institution in London. The discovery of electron disproved the 2000-year-old perception of the atom as a homogeneous smallest particle and further proved that an atom has a complex structure in fact. The electron was the first subatomic particle to be identified: the smallest and fastest particle. Shortly after the discovery of electron, a number of models describing the structure of atom were proposed and a few of them are worth mentioning. In 1900, William Thomson (Lord Kelvin) (1824–1907) proposed a model for an atom. According to Thomson model, an atom is a sphere in which positive charge is uniformly distributed and electrons are embedded in it in such a way that the most stable electrostatic arrangement having minimum energy is achieved and the atom as a whole is electrically neutral. Thomson model of atom is also known as *plum pudding* model and it was strongly supported by J. J. Thomson. But this model failed to explain how a positive charge holds the negatively charged electrons in an atom. Hantaro Nagaoka (1865–1950) rejected Thomson model of atom on the ground that opposite charges are impenetrable. In 1904, Nagaoka proposed *planetary model* of atom. In this model, electrons were assumed to revolve around a positive charge centre just like the rings of Saturn. Ernest Rutherford (1871–1937) was studying the scattering of alpha particles from a gold foil and found that scattering is non-uniform and most of the scattering of alpha particles was taking place from the centre of an atom. Based on the experimental findings, in 1911, Rutherford gave another model for an atom. According to the Rutherford model, the positive charge and most of the mass of an atom is concentrated in an extremely small volume at the centre of an atom: he called this region of the atom a nucleus. He further assumed that the electrons revolve around the nucleus in fixed paths (orbits). In a way Rutherford confirmed the planetary model of Nagaoka. In Rutherford's atomic model, most of the space of an atom is empty with a very small and dense nucleus at the centre. But this model could not explain the stability of an atom because the electrons revolving around the nucleus have finite acceleration and therefore should emit radiations and loose energy continuously. Ultimately electrons should fall into the nucleus. Further, the Rutherford model of atom did not account for different colors of the materials which depend on the frequencies emitted by them. The discovery of proton is also credited to Ernest Rutherford who, in 1909, proved that the nucleus of hydrogen atom (proton) is present in the nuclei of all atoms. Ernest Rutherford did a lot of work on disintegration of elements and radioactive substances (see Chap. 9) and was awarded Nobel Prize in Chemistry in 1908. However, he regarded himself a physicist, as chemistry was considered an inferior branch of science as compared to physics during those days.

Experiments have shown that electrons revolve in fixed orbits around the nucleus of an atom and therefore an atom behaves like a small solar system. In 1913, Danish physicist Niels Henrik David Bohr (1885–1962), a student of Ernest Rutherford, proposed a model called *Bohr model* for an atom. In Sept. 1913, Niels Bohr presented his work on atomic model in the 83rd Annual Meeting of British Association for Advancement of Science. The audience comprised of most eminent atomic physicists of the day who listened to Bohr's work on atom model, but his work was met with criticism. In his atomic model, Bohr used the ideas of both quantum and classical physics and therefore it was considered a *hodgepodge model of atom.* Further, it allowed many more spectral lines than actually observed in the light from different atoms. In 1914, there was an outbreak of World War I and the progress in this direction came to a standstill. In 1916, Albert Einstein introduced the concept of probability into the atomic theory which gave some boost to the work of Bohr. Actually Einstein's concept of probability gave the first clue to the strangeness of the quantum world but its significance was not noticed at that time. After the First World War and up to 1926, a number of refinements were made in the Bohr's atomic model. Bohr assumed that electrons revolve around the nucleus in a series of circular orbits with certain "quantized energies (frequencies)". The energy of an orbit depends on its radius and increases with the increase in radius. The orbits of electrons are labelled by an integer n called principal quantum number. The stationary orbits (states) are attained at distances for which the angular momentum p_φ of the revolving electron is an integral multiple of $\hbar$, that is

$$p_\varphi = m_e v\, r = n\, \hbar \tag{4.8}$$

where n = 1, 2, 3, …, $\hbar = h/2\pi$, m_e is the mass of an electron and r is the radial distance of the electron moving with velocity v. In these stationary orbits electrons do not radiate energy contrary to classical electromagnetism. In an atom the energy levels can be compared to the steps on a staircase: you can be on one step or another, but you cannot stand in the spaces between the steps. Figure 4.5 shows the structure of Na atom. In Bohr's model the energy of an orbit is given by

$$E_n = -\frac{2\pi^2 Z^2 m_e e^4}{h^2}\frac{1}{n^2} \tag{4.9}$$

where Z is the atomic number of element, e and is the charge of an electron. Bohr further assumed that the electrons can jump from one electron orbit to another thereby absorbing or emitting electromagnetic radiations with frequency ν determined by the energy difference between the electron orbits and is given by

$$E_1 - E_2 = h\nu = \frac{2\pi^2 Z^2 m_e e^4}{h^2}\left(\frac{1}{n_2^2} - \frac{1}{n_1^2}\right) \tag{4.10}$$

Here E_1 is the energy of the initial orbit (state) and E_2 is the energy of the final orbit (state) to which the electron makes a transition. For $n_2 = 1$, one obtains *Lyman series*, for $n_2 = 2$, one obtains *Balmer series* and for $n_2 = 3$, the *Paschen series* (see Fig. 4.6). Niels Bohr made the first advancement towards the development of quantum mechanics while studying the structure of an atom. He could explain the line spectra of light elements such as hydrogen and helium but could not explain the spectra of heavy elements satisfactorily. Therefore, Bohr established the connection between the states of atoms and the spectral series and gave a method of calculating energies of the states of an atom. Bohr's model of atom gave new direction to research in quantum theory for the next decade. Another important conclusion of Bohr model is that it proved the quantization of energy of electrons in atoms. The quantization of the light energy and electron energy shows that quantization is wide spread in nature. Bohr's model, for the first time, established the fact that most of the space in an atom is empty. Niels Bohr was awarded the 1922 Nobel Prize in Physics for establishing the basic structure of atoms.

In 1913, Johannes Stark (1874–1957) observed that the spectral lines of atoms or molecules split up in the presence of external electric field just as in the case of magnetic field (Zeeman effect, see Chap. 12). It is named Stark effect after his name. Johannes Stark was awarded 1919 Nobel Prize in Physics. An Italian physicist Antonino Lo Surdo (1880–1949) independently discovered the same effect in the same year accidently. But these two discoveries arose in different contexts. Stark had been working within the context of the emerging quantum theory while Lo Surdo had been working within the context of the classical theory. The discovery of the Stark effect contributed significantly towards the development of quantum

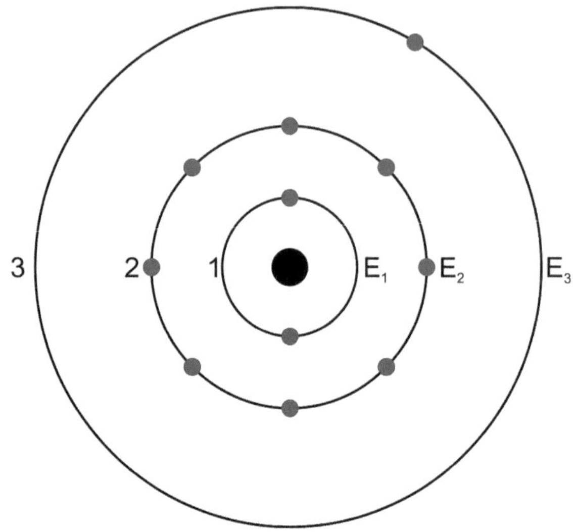

Fig. 4.5 Schematic diagram of an atom of sodium metal

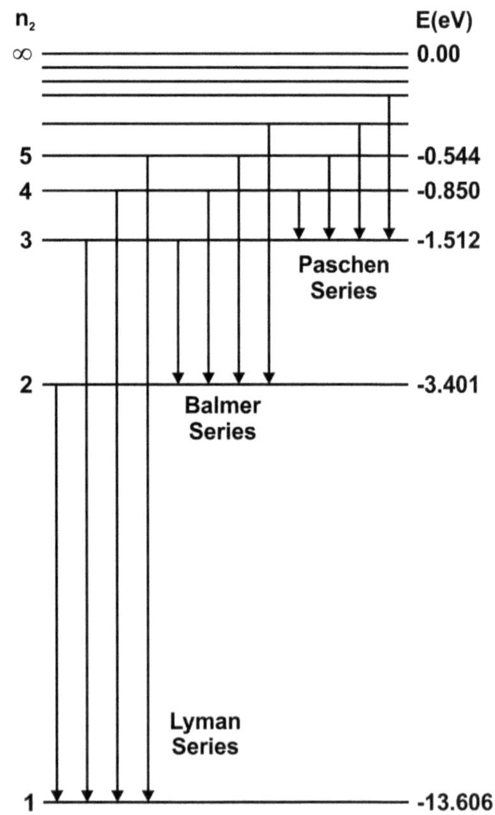

Fig. 4.6 Lyman, Balmer and Paschen series obtained from the spectrum of hydrogen atom. The principal quantum number n_2 and the energy of spectral lines are shown on the left and right sides of the diagram, respectively

theory. We want to point out here that Bohr model could not explain the Zeeman and Stark effects.

Arnold Sommerfeld (1868–1951) was a German theoretical physicist who contributed significantly towards the development of quantum mechanics. He completed his Ph.D. in 1891 at the age of 22 years and in 1906, became the director of the Institute of Theoretical Physics at the University of Munich where he succeeded Ludwig Boltzmann. In 1915, he gave azimuthal quantum number ℓ (second quantum number) and magnetic quantum number m_ℓ (third quantum number) for the motion of an electron in a central field to explain the spectra of elements. In 1916, he also introduced fine structure constant to quantify the tiny gap between two lines in the spectrum of colors emitted by certain atoms. Sommerfeld pioneered development in atomic and quantum physics and educated and mentored a number of students of the new era of theoretical physics. Among his students were Werner Heisenberg, Wolfgang Pauli, Peter Debye and Hans Bethe who received the Nobel Prizes in Physics. It is pointed out that Sommerfeld was nominated for the Nobel Prize a record breaking 84 times but unfortunately could not get it. For the first time Sommerfeld was nominated for the Nobel Prize in 1917 by Max von Laue (1879–1960). In 1922, Otto Stern (1888–1969) and Walther Gerlach (1889–1979) performed an experiment in which neutral silver atoms were passed through a magnetic field applied in a direction normal to their motion. They obtained discrete points on the screen rather than the continuous distribution. For the first time this experiment showed that angular momentum is quantized in all atomic scale systems. But they could not provide correct explanation of their observation. Wolfgang Pauli (1900–1958) was studying the emission spectrum of alkali metals. To explain their spectra Pauli, in 1924, proposed a new degree of freedom with two possible values to resolve discrepancies in the observed spectra and that obtained from quantum mechanics. This degree of freedom was called the fourth quantum number. In 1925, Pauli put forward the exclusion principle which was perhaps his most important contribution. Pauli Exclusion Principle states that no two electrons that have all the four quantum numbers same, can occupy the same quantum state. Pauli's exclusion principle explained the arrangement of electrons in atoms and with its help scientists could explain why elements explain differently chemically. Later in the same year, that is 1925, George Uhlenbeck (1900–1988) and Samuel Goudsmit (1902–1978) from their calculations identified Pauli's new degree of freedom as electron spin. In 1927, Pauli identified the quantized angular momentum observed in Stern and Gerlach experiment as the spin angular momentum of electrons. In the same year Pauli formulated theory of spin using quantum mechanics invented by Schrodinger and Heisenberg and found that electrons have two possible spin angular momentum values along any axis: $\hbar/2$ and $-\hbar/2$ (usually written as 1/2 and $-$ 1/2 in units of $\hbar$). Spin is regarded as the intrinsic property because it always has the same value: spin angular momentum is also called intrinsic angular momentum. In 1927, Pauli introduced 2×2 spin matrices usually called Pauli spin matrices. Spin is purely a relativistic and quantum mechanical property and electrons are half-spin particles.

As a physicist Pauli was known as a perfectionist. He did not perform just his own work but also extended the work of his colleagues as a result he came to be known in physics community as the "conscience of physics". Pauli got the 1945 Nobel Prize in Physics for his discovery that the four quantum numbers are required to define the energy state of an electron. In general, particles are divided into two categories. Particles having integral spin are called bosons while the particles with half-integral spin are called fermions. Bosons obey the Bose–Einstein distribution, fermions obey the Fermi–Dirac distribution and the classical particles having no spin obey the Maxwell–Boltzmann distribution.

In 1923, Arthur Holly Compton (1892–1962) found that when a quantum of X-rays (photon) with wavelength λ (energy E) is scattered by a free electron, having mass m_e, the scattered quantum (photon) had longer wavelength λ' (less energy E') than the incoming X-rays photon and the balance of energy had been transferred to the electron which suffers recoil. The change in wavelength of the photon is given by the relation

$$\lambda' - \lambda = \frac{h}{m_e c}(1 - \cos\theta) \tag{4.11}$$

Here θ is the angle of scattering (see Fig. 4.7), c is the velocity of light. $h/m_e c = \lambda_{comp}$ is called the Compton wavelength and has the value 2.43×10^{-10} cm. This phenomenon is known as Compton scattering or Compton effect. The importance of Compton effect is that it gave direct experimental proof of the particle nature of electromagnetic radiation for the first time, that is the proof of reality of photons. Further, the Compton effect justified the explanation of photoelectric effect given by Einstein. It was a sensational discovery at the time because the idea that electromagnetic radiation had both the wave and particle nature was not easily accepted.

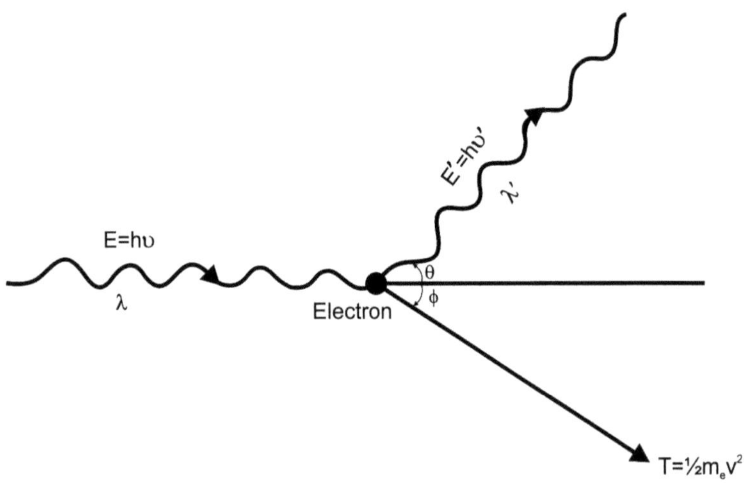

Fig. 4.7 Compton scattering

Compton was awarded 1927 Nobel Prize in Physics for the discovery of Compton effect. Actually it was Gilbert Newton Lewis (1875–1946), an American chemist, who introduced the term *photon* for the quantum of light in 1926 and it was adopted in literature after the 5th Solvay conference held in 1927 under the name "Electrons and Photons.

Louis de Broglie (1892–1987), Public Domain, URL: https://en.wikipedia.org/wiki/Louis_de_Broglie#/media/File:Broglie_Big.jpg

Louis de Broglie (1892–1987) was inclined for a career in humanities and received his first degree in history. After joining the university, he became interested in mathematics and physics, particularly in quantum mechanics. It was de Broglie who thought if light waves behave like particles (photons) then why should electrons not behave like waves? De Broglie worked with Paul Langevin and completed his Ph.D thesis in 1924, which was published in 1925 in Annales de Physique. In his Ph.D. thesis, de Broglie made a ground-breaking contribution to quantum mechanics. He put forward a hypothesis according to which any moving particle (say an electron) has a wave associated with it: wave-particle duality nature. According to de Broglie, the wavelength λ of a wave associated with a particle having mass m and moving with velocity v is given by

$$\lambda = \frac{h}{p} = \frac{h}{mv} \tag{4.12}$$

where p is the momentum of the particle. The de Broglie wave is also called matter wave as it is associated with matter. De Broglie first wrote to Albert Einstein about his idea of wave-particle duality but it came as a shock to Einstein. But de Broglie urgently needed the opinion of Einstein over the concept of wave-particle duality. Einstein immediately realized that matter waves must be real. Originally de Broglie proved the wave-particle duality in case of electrons and later this relation was found to hold for all type of material particles. From the relation it is evident that only very small particles, for example subatomic particles, will exhibit wave phenomenon to an appreciable degree. In 1927, George Paget Thomson (1892–1975) and his research student Alexander Reid conducted the first experiment on electron diffraction. They passed a beam of electrons through a thin film of celluloid and tried to obtain pictures of electron diffraction. Their results were encouraging, but certainly not conclusive. Unfortunately, Reid died soon after in a motorcycle accident. In the same year of 1927, Clinton Joseph Davisson (1881–1958) and Lester Halbert Germer (1896–1971) passed the electron beam through a crystal of nickel and observed the diffraction pattern. Davisson and Germer are credited with the experimental verification of the wave-particle duality nature of matter. In 1937, Joseph Davisson and George Thomson shared the 1937 Nobel Prize in Physics for proving that waves are associated with electrons. It is interesting to note that Joseph John Thomson received the 1906 Nobel Prize in Physics for proving that electrons are particles while he saw his son George Thomson getting the 1937 Nobel Prize in Physics for proving that electrons are waves—actually both father and son were correct in their times. The confirmation of the de Broglie hypothesis also explained the long standing unanswered question of the wave-particle duality of light. The wave-particle duality forms one of the pillars of quantum mechanics and for this de Broglie won the 1929 Nobel Prize in Physics. Yet, the question remained—under what circumstances light behave as a wave or as a particle. It was established that during the propagation, light behaves as a wave but when light interacts with matter it exhibits particle nature. The de Broglie hypothesis created a new field in quantum mechanics called wave mechanics and it united the physics of energy (wave) with matter (particle). The complete break from classical mechanics comes with the establishment of the fact that all ‘particles’ and all ‘waves’ are in fact a mixture of wave and particle.

Erwin Schrodinger (1887–1961) made the most important contribution towards the development of quantum mechanics. He studied in the University of Vienna and received a Ph.D. there. He is best known for *Schrodinger equation*, which he developed in a series of four papers. In 1926, Schrodinger published his first paper on wave mechanics and gave a time-independent equation for evaluating stationary wave functions and energy eigenvalues of a particle system: named time-independent Schrodinger equation. Second paper was published immediately after the first paper in which he solved some physical problems of quantum harmonic oscillator, rigid rotator and diatomic molecule. So Schrodinger gave a new approach towards quantum theory by giving different reasoning. Schrodinger used de Broglie matter waves and sought the wave solution of a differential equation describing atomic electrons. In the third paper of the series, he introduced time-dependent Schrodinger equation and the complex wave function. The time evolution of wave function made the system complex in nature and there is switching from the real to complex systems. Schrodinger theory gave complete description of hydrogen spectrum and reproduced the Balmer's formula in addition to the splitting of spectral lines in the presence of electric and magnetic fields. The Schrodinger equation marked the beginning of wave mechanics of solids and is considered to be the most important achievement of the twentieth century, which created a revolution in quantum mechanics related to physics and chemistry. To understand the philosophy behind the Schrodinger wave mechanics let us consider the total energy E of an electron which is the sum of kinetic energy T and potential energy $V(\mathbf{r})$ [Newtonian mechanics] written as

$$E = T + V(\mathbf{r}) = \frac{p^2}{2m} + V(\mathbf{r}) \tag{4.13}$$

In quantum mechanics, the observables E, $\mathbf{p}$ and $V(\mathbf{r})$ are replaced by their operators that is, $\mathbf{p} \rightarrow -\iota\hbar\nabla$ and $E \rightarrow \iota\hbar\partial/\partial t$. The operators are just mathematical expressions and these as such have no meaning. But when these operators act on some function, say $\Phi(\mathbf{r}, t)$, they yield observable quantities. The function $\Phi(\mathbf{r}, t)$ is the matter wave associated with the electron as defined by de Broglie and is called wave function (a scalar quantity). The wave function describes the state of a quantum system which evolves with time. So, Eq. (4.13) in quantum mechanics becomes

$$\left[-\frac{\hbar^2}{2\,m_e}\nabla^2 + V(\mathbf{r})\right]\Phi(\mathbf{r},\, t) = \iota\,\hbar\,\frac{\partial\Phi(\mathbf{r},\, t)}{\partial t} \tag{4.14}$$

Equation (4.14) is nothing but the time-dependent Schrodinger equation for an electron. From Eqs. (4.13) and (4.14) it is evident that the Schrodinger equation is nothing but the operator form of classical equation for total energy. The time-independent Schrodinger equation, from Eq. (4.14), becomes

$$\left[-\frac{\hbar^2}{2\,m_e}\nabla^2 + V(\mathbf{r})\right]\Phi(\mathbf{r}) = E\,\Phi(\mathbf{r}) \tag{4.15}$$

The wave function of an electron $\Phi(\mathbf{r})$ is shown in Fig. 4.8. The Schrodinger equation describes the wave mechanics of the system under consideration. During the above simple transformation from Eqs. (4.13) to (4.15), energy eigenvalues E and hence momenta $\mathbf{p}$ of the system are quantized. It is called *first quantization* in quantum mechanics. From Eq. (4.15) the energy of a system is given by

$$E = \frac{\langle\Phi(\mathbf{r})| - \left(\hbar^2/2\,m_e\right)\nabla^2 + V(\mathbf{r})\,|\Phi(\mathbf{r})\rangle}{\langle\Phi(\mathbf{r})|\,\Phi(\mathbf{r})\rangle} \tag{4.16}$$

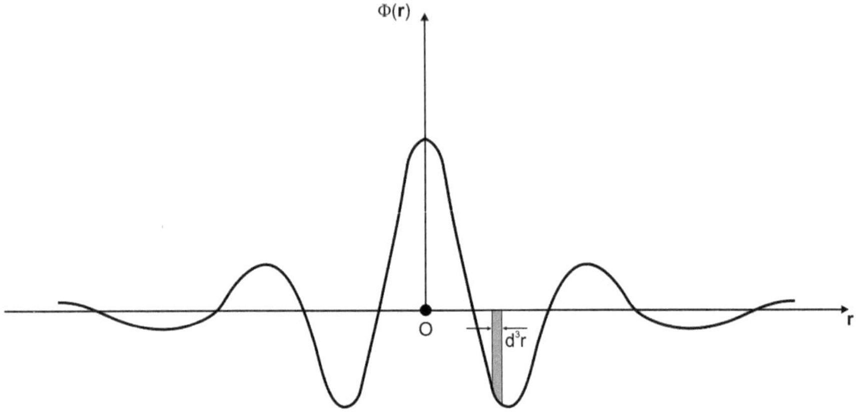

Fig. 4.8 Schematic representation of the wave function for a particle situated at the origin 0

In July 1926, Schrodinger gave a lecture on Schrodinger equation in Munich at Sommerfeld's weekly colloquium and in the end asked if there are any questions. Heisenberg asked if he can explain quantized processes such as photoelectric effect and black body radiation on the continuum wave function. At that time Schrodinger realized that he cannot find the solution to all such problems. In the beginning Schrodinger interpreted wave function as a wave which gives the position of the electron. Then he changed his mind and said it was the density of the electron charge. But Schrodinger could not give proper interpretation of the wave function. Actually, the wave function as such has no direct physical meaning but it stores whole of the available information about the system. In August 1926, Max Born (1882–1970) provided the physical significance of the wave function and interpreted it in terms of probability. According to him wave function gives the probability amplitude for an electron in a state. He normalized the wave function to unity as

$$\int |\Phi(\mathbf{r})|^2 \, d^3r = 1 \tag{4.17}$$

Then modulus square of the normalized wave function gives the probability density $P(\mathbf{r}, t)$ per unit volume at position $\mathbf{r}$ and time t as

$$P(\mathbf{r}, t) = |\Phi(\mathbf{r}, t)|^2 \tag{4.18}$$

The probability in a small volume d^3r around the position vector $\mathbf{r}$ at time t is proportional to $|\Phi(\mathbf{r}, t)|^2 d^3r$ (see Fig. 4.8). Therefore, quantum mechanics gives the probability that a certain quantum state exists and there are no exact answers as in classical physics. Max Born was awarded 1954 Nobel Prize in Physics for fundamental contribution to quantum mechanics especially for his interpretation of wave function in terms of probability. Schrodinger also worked on unified field theory, physics of dielectrics, color theory, electrodynamics and cosmology.

Max Born (1882–1970), Public Domain, URL: https://en.wikipedia.org/wiki/Max_Born#/media/File:Max_Born.jpg

Max Born (1882–1970) was a genius and studied under Joseph Larmor (1857–1942) and Joseph John Thomson (1856–1940). He attended lectures on mathematics given by Oskar Klein (1894–1977), David Hilbert (1862–1943), Hermann Minkowski (1864–1909) and Carl David Runge (1856–1927). He also studied astronomy under David Schwarzchild (1873–1916) and physics under Woldemar Voigt (1850–1919). Among his collaborators were many who later became well-known physicists, such as Werner Heisenberg, Pascual Jordan, Enrico Fermi, Paul Dirac, Friedrich Hund, Egil Hylleraas, Victor Weisskopf, Julius Robert Oppenheimer, Joseph Mayer and Maria Goeppert-Mayer. Max Born was having expertise in matrix algebra but at that time matrices were considered to belong to the world of pure mathematics. In 1921, Born used matrix approach for his work on lattice theory and published a book on dynamical theory of crystal lattices with Kun Huang (1919–2005). Werner Heisenberg (1901–1976) became assistant of Max Born and in 1925 Heisenberg wrote a paper entitled, "*Über quantentheoretische umdeutung kinematischer und mechanischer beziehungen*" and gave it to Max Born (1882–1970) for his expert comments. After reading the paper, an idea immediately came to Born's mind that this work could be translated in the matrix form. Both Born and Heisenberg along with Pascual Jordan (1902–1980) formulated matrix mechanics representation of quantum mechanics. They worked in detail on different aspects of quantum mechanics. The development of quantum mechanics in the matrix notation puzzled physicists all over the world because both mathematicians and physicists were not very familiar with matrices and their noncommutative property seemed a strange thing at that time. Therefore Heisenberg matrix formulation of quantum mechanics was not immediately appreciated by most of the physicists.

Werner Heisenburg (1901–1976), CC BA-SA 3.0, URL: https://en.wikipedia.org/wiki/Werner_Heisenberg#/media/File:Bundesarchiv_Bild183-R57262,_Werner_Heisenberg.jpg

It is well known that if the position and momentum of a particle are known one can study all its properties. But in quantum mechanics the simultaneous determination of position and momentum of a particle is not possible. In 1927, Werner Heisenberg gave the uncertainty principle which is named after him. Let Δx and Δp be the uncertainties in position and momentum, then according to Heisenberg uncertainty principle these are related as follows:

$$\Delta x \, \Delta p \geq \hbar/2 \tag{4.19}$$

According to Eq. (4.19) one cannot determine simultaneously the accurate values of both position and momentum of a particle. Similarly, the Heisenberg uncertainty principle between energy E and time t is given by

$$\Delta E \, \Delta t \geq \hbar/2 \tag{4.20}$$

In general one can say that simultaneous determination of two canonically conjugate quantities P and Q is subject to an uncertainty given by

$$\Delta Q \, \Delta P \geq \hbar/2 \tag{4.21}$$

Heisenberg was awarded the 1932 Nobel Prize in Physics for the creation of quantum mechanics. Heisenberg uncertainty principle is a cardinal rule of quantum mechanics which tells us that in principle it is impossible to measure simultaneously a certain pair of properties. It is important to appreciate, however, that it does not represent any deficiency in the experiments used to measure properties of a particle. Rather it measures the amount of overlap of the pair of properties. For example, the position of a particle is very much a particle property (particles can be located precisely) while the waves have no precise location but they do have momentum. So according to the uncertainty principle, the more you know about the wave aspect of a particle, the less you know about its position. Niels Bohr said that both the particle and wave physics are equally valid and constitute complementary descriptions of the same reality—neither of the two descriptions is complete in itself. Max Born (1882–1970) also made important contributions to solid state physics using quantum theory. In January, 1933, the Nazi party came to power in Germany and Max Born, who was Jewish, was suspended from his Professorship at the University of Gottingen. He migrated to United Kingdom and worked there up to 1952. He again shifted back to West Germany and worked there till death.

Paul Adrien Maurice Dirac (1902–1984), Public Domain, URL: https://en.wikipedia.org/wiki/Paul_Dirac#/media/File:Paul_Dirac,_1933.jpg

Paul Adrien Maurice Dirac (1902–1984), a British physicist, made fundamental contributions to quantum mechanics and developed the subject of quantum mechanics nearly fully. Dirac developed quantum algebra involving addition and multiplication of quantum variables and included matrix mechanics. In 1926, Dirac published his work in four papers that formed a part of his Ph.D. thesis entitled "Quantum Mechanics". After completing his Ph.D., Dirac became a lecturer in St. John's College in Cambridge but soon after he left for Copenhagen to study under Professor Niels Bohr. Dirac discovered that the equations of quantum mechanics resemble with the mathematical structure of equations in classical physics. He further showed that classical mechanics becomes a special case of quantum mechanics. In 1928, Dirac formulated the relativistic equation for an electron named after him which is described briefly in the end of this chapter. In 1930, Dirac wrote a book "*Principles of Quantum Mechanics*" which soon became the standard text book on the subject. In 1931, Dirac proposed that the existence of a single magnetic monopole in the universe would suffice to explain the quantization of electrical charge. But magnetic monopole is still not discovered although scientists are performing experiments for its observation. Dirac shared the 1933 Nobel Prize in Physics with Schrodinger for the formulation of quantum mechanics and the wave theory for a particle system. In 1939, Dirac introduced the *bra-ket notation* in the third edition of his book. He also made significant contribution to the reconciliation of general relativity and quantum mechanics. During the period of one year that is from June, 1925 to June, 1926, three independent and distinct developments in quantum mechanics were published: Schrodinger wave equation, Heisenberg matrix mechanics and Dirac's version of quantum mechanics. It is pointed out that Paul Dirac and Werner Heisenberg had different approaches to scientific methodology and theory. Dirac favoured the mathematical and axiomatic approach, while Heisenberg preferred a physical and intuitive approach. Despite their disagreements, they respected each other's work. Dirac proved mathematically that the three different approaches were exactly equivalent to one another. But Dirac's approach was the most complete since his quantum algebra included both matrix mechanics and wave mechanics as special cases. The equivalence of the three approaches strengthened the confidence of physicists in quantum mechanics. As a general remark, Werner Heisenberg, Wolfgang Pauli and Paul Dirac, the three stalwarts, were good friends throughout their lives and in their free time used to discuss things other than physics too. In 1927, Arnold Sommerfeld gave quantum free-electron model assuming that the free electrons obey the Fermi–Dirac distribution: *Sommerfeld model of metals*. He gave a very crisp and simple mathematics for the evaluation of density of electron states in the parabolic band approximation. This model served as an important step towards the development of electron energy band theory in solids.

One can define an atom according to quantum mechanical principles. Heisenberg uncertainty principle says that the exact position of an electron cannot be defined but one can talk about the probability of an electron at a particular position. As a result an electron does not move around the nucleus in a discrete orbit instead yield an electron cloud. Cloud itself is just a zone of statistical probability beyond which the electron cannot exist. Schrodinger gave quantum mechanical model of atom according to which an atom consists of a nucleus with finite electron cloud around it. In brief, different atomic models like Thomson model, planetary model, Rutherford model and Bohr model were proposed but the latest is the Schrodinger's quantum mechanical model of atom.

The discovery of neutrons is important to complete the basic structure of an atom. Sir James Chadwick (1891–1974) was a student of Ernest Rutherford. Rutherford and Chadwick studied transmutation of elements by bombarding them with alpha particles and identified proton in the nucleus of hydrogen atom in 1919. Chadwick worked on radioactivity and related problems with Rutherford and in 1930, co-authored with Rutherford and Charles Drummond Ellis (1895–1980) a book entitled, "*Radiations from Radioactive Substances*". Rutherford realized that there was another problem with his model of atom, which is how protons with positive charge could reside together inside the nucleus. Moreover, in heavier elements, the atomic mass is greater than the atomic number (equal to the number of protons). Therefore, Rutherford was of the view that there must be some type of neutral particles, which he called neutrons, inside the nucleus to stabilize it. In one of the experiments, Chadwick bombarded beryllium with alpha particles and found that some unknown radiations were emitted. In 1932, Chadwick came to the conclusion that this radiation consisted of particles whose mass was approximately equal to that of the protons but were electrically neutral: establishing the existence of neutrons inside the nucleus—a fundamental discovery in nuclear science. James Chadwick spent eleven intensive years in discovering neutrons. He was awarded 1935 Nobel Prize in Physics for the discovery of neutrons. This discovery led to a new model of atomic nucleus being composed of protons and neutrons. The nucleus is so small that even now it was not understandable how the protons and neutrons reside in it in spite of the strong repelling electrostatic force between the protons. To overcome this problem, in 1935, Japanese physicist Hideki Yukawa (1907–1981), lecturer at Osaka University, introduced at the age of 28 years the concept of a new strong but very short-ranged attractive nuclear force between nucleons (between protons and neutrons). He said that each nucleon is surrounded by its own very short-ranged field comprising of particles called π-mesons (pions). Nuclear force arises due to the interaction of the fields of the nucleons. The strong nuclear force among nucleons is described by the exchange of pions which causes transfer of momentum and energy between them. Yukawa proposed that the potential energy U(r) between two nucleons decays exponentially with distance r between them and is given by

$$U(r) = \frac{C^2}{r} e^{-r/R} \tag{4.22}$$

C^2 is the strength of the potential and R is the range of nuclear force. It is named *Yukawa potential* in his honor. In 1946, Cesare Mansueto Giulio Lattes (1924–2005) and co-workers observed pions experimentally in the cosmic radiations and reported them in early 1947. One can say that the protons give an atom its identity and electrons its personality. Neutrons do not influence the identity of an atom but contribute to its mass. The addition or subtraction of a neutron (neutrons) gives rise to an isotope (isotopes).

It is worth mentioning that Albert Einstein and de Broglie were realists and believed that a particle has definite position and momentum at all times. On the other hand, Heisenberg was antirealist arguing that the direct knowledge of what is ‘real’ was beyond the scope of science. In 1920s, Copenhagen interpretation of quantum mechanics developed by the European physicists was in centre stage. Werner Heisenberg in his book entitled “*Physics and philosophy*” gave a brief history of quantum mechanics and discussed about the Copenhagen interpretation of quantum mechanics. The collection of ideas like uncertainty, complementarity, probability and the disturbance of the system by an observer are together referred to as Copenhagen interpretation of quantum mechanics. One of the key ingredients is Max Born’s probabilistic interpretation of the wave function. In 1927, Niels Bohr first presented this concept in a conference held in Tomo, Italy and declared that theory of quantum mechanics is ready to solve problems involving atoms and molecules.

Ervin Schrodinger was not convinced with Max Born's probabilistic interpretation of the wave function and felt that his wave equation was being misunderstood or misused. In 1935, after ten years of probability interpretation, Schrodinger proposed a thought experiment called *Schrodinger's cat paradox*. In this experiment Schrodinger considered a closed box with a live cat in it. Within the box he put a radioactive source, Geiger-Muller counter, a hammer and a sealed glass bottle containing deadly poison such as poisonous fumes or cyanide. The whole system is set in such a way that when a radioactive decay takes place the Geiger-Muller counter registers it and the hammer falls on the bottle breaking it thereby releasing the poisonous material. The poisonous material then kills the cat. It is well-known that radioactive decay occurs entirely by chance and is unpredictable except in a statistical sense. Therefore, Schrodinger assumed that there is 50% probability that an atom of radioactive material undergoes decay in one hour. Under such circumstances there is equal probability that cat is alive or dead in one hour. According to quantum mechanics, after one hour, the cat is neither wholly dead nor wholly alive but a mixture of the two possible states: the principle of superposition of states. But it is ridiculous because the cat is either alive or dead. Therefore, Schrodinger said the probability interpretation of the wave function is not correct. We have no way of knowing the exact situation until we open the box and make an observation. When observation is made definiteness comes, that is either the cat is alive or dead. So according to quantum mechanics when observation is made the superposition of wave functions collapses to a single wave function making the cat either definitely alive or dead. Even today while teaching the concept of probability and superposition of wave functions, reference is made of Schrodinger cat paradox.

There had been a debate between Albert Einstein and the scientists developing quantum mechanics particularly the Copenhagen group about the completeness of quantum mechanics. Out of this debate originated another paradox called EPR paradox. Albert Einstein believed in the principle of locality according to which two systems are independent of each other that is measurement made on one system can produce no change in the second system. In 1935, Einstein (1879–1955), Podolsky (1896–1966), and Rosen (1909–1995) put forward EPR paradox after the names of the scientists even though it does not really describe a paradox at all. They considered a system containing two particles A and B with opposite spins in a singlet state yielding zero total spin. Let the two particles be allowed to move away from one another to a large distance. In this situation if the particle A is having up spin, the particle B must have down spin. However, in quantum mechanics, according to Copenhagen interpretation, the spin of particle A has no definite value until it is measured. The measurement of particle A instantaneously produces effect on particle B which collapses its spin wave function into the opposite direction or down spin. This effect demands instantaneous action-at-a-distance which is faster than the velocity of light and that is not acceptable according to relativity. Therefore, Einstein said according to EPR paradox quantum mechanics is incomplete or there are missing links in quantum mechanics.

In 1957, Hugh Everett (1930–1982) gave relative state formulation of quantum mechanics in his Ph.D. thesis. In the same year John Archibald Wheeler (1911–2008) published a short paper on Everett's theory which was almost identical to the final version of Ph.D. thesis of Everett. Later in 1973, Bryce Seligman DeWitt (1923–2004) of University of North Carolina developed further the Everett's theory and gave it the name many-world theory of quantum mechanics under which name it became popular. As an illustration of the many-world interpretation, consider the thought experiment of Schrodinger cat. In this experiment, there is a live cat and there is a dead cat and according to Everett both occurred but they belong to different worlds. According to many-world interpretation the whole universe got split into two identical worlds. In one world the atom decayed and the cat died while in the other the atom did not decay and the cat survived. In 1959, Everett visited Copenhagen and discussed his relative state formulation of quantum mechanics with Niels Bohr but he out rightly rejected Everett' approach and defended the Copenhagen interpretation. But Bryce DeWitt (1923–2004) established that the many-world interpretation of quantum mechanics gave exactly the same predictions as the Copenhagen interpretation. Rather it improves the Copenhagen interpretation by removing the paradoxes of EPR and double-slit experiment etc.

Niels Bohr (1885–1962) said that quantum mechanics works on the principle of nonlocality according to which the two particles at a large distance and the observer are not independent. In other words, there is no separation between the observer and the particle being observed. But the big question was whether the remarkable property of nonlocality could ever be experimentally tested. John S. Bell (1928–1990) was examining the principle of nonlocality in quantum mechanics. Assuming the Einstein principle of locality to be true, in 1964, he developed an inequality principle generally called Bell's inequality. Bell said if the experiments showed that the inequality is violated it will show that the assumption of locality is false and prove that nature is nonlocal. John Clauser and coworkers (1969, 1972) and Alain Aspect group (1981, 1982) performed experiments which confirmed the violation of Bell's inequality. This supported the principle of nonlocality and allowed communication faster than the velocity of light, even instantaneously.

One of the main shortcomings of quantum mechanics is that only a few simple systems can be solved exactly, for example, simple harmonic oscillator, hydrogen and helium systems etc. Due to this deficiency approximate methods were developed which provided approximate solutions to a large number of quantum systems. One such method is perturbation theory whose basic principle is to study dynamical systems which form small perturbations of simple exactly solvable systems. In 1918, Niels Bohr developed a perturbation method in quantum mechanics. In 1921, Max Born and E. Brody developed the perturbation method systematically. In 1926, Erwin Schrodinger (1902–1984) gave the standard perturbation method which was the first one to be adopted to solve problems in quantum mechanics in an approximate manner. In this method Hamiltonian and the wave function were assumed to have the form

$$\mathrm{H} = \mathrm{H}_0 + \lambda \mathrm{H}_1 \tag{4.23}$$

$$\Phi = \Phi_0 + \lambda\Phi_1 + \lambda^2\Phi_2 + \cdots \tag{4.24}$$

Here H_0 is the Hamiltonian of a simpler exactly solvable system and H_1 is a small perturbation in the Hamiltonian. Similarly, Φ_0 is the wave function of simpler exactly solvable system and Φ_1 and Φ_2 etc. are the first and second order perturbations in the wave function and λ is a constant. Using Eqs. (4.23) and (4.24), Schrodinger equation is solved up to first order, second order or to some higher order to get results close to the experimental results. There is one problem with this method that is up to how many orders of perturbation the problem should be solved to get reasonable results. In 1926, Gregor Wentzel (1898–1978), Hendrik Kramers (1894–1952) and Leon Brilloiuin (1889–1969) independently developed perturbation method for evaluating energy eigenvalues and eigenfunctions and is known as Wentzel-Kramers-Brillouin (WKB) method. In 1927, Paul Dirac (1902–1984) developed quantum perturbation theory for time-dependent perturbations in the context of emission and absorption of radiations from radioactive elements. According to it the probability of transition between two states per unit time is proportional to the square of matrix elements between two states. On account of the importance of this equation, Fermi called it Golden Rule No. 2 (Fermi Golden Rule No. 2). Fermi's Golden Rule No. 1 is of similar form and considers the probability of indirect transitions per unit time. In order to overcome the convergence problem, in 1932, Leon Brillouin and Eugene Wigner (1902–1995) developed a quantum mechanical perturbation technique known as Brillouin-Wigner formula. The various perturbation techniques resulted in a large number of applications ranging from Zeeman effect to the hyperfine splitting in the hydrogen atom.

Variational method is another way of finding approximate solution for the ground state and some excited states of a system and its basis is the *variational principle*. In this method wave function is approximated by a simple analytical function defined as

$$\Phi(r) = u(r, c_1, c_2, \ldots) \tag{4.25}$$

where $c_1, c_2, \ldots$ are unknown parameters. These parameters are determined mostly by trial and error method in such a way that the integral corresponding to the Schrodinger equation [Eq. (4.16)] attains a minimum. In 1927, George W. Kellner used this method to calculate the ground state of helium atom. Egil Andersen Hylleraas (1898–1965) mastered this method and, in 1928, obtained a very precise value for the ground state energy of helium atom.

The electronic properties in a many-body system like a metal are quite difficult to treat mathematically. In 1927, Max Born (1882–1970) and Julius Robert Oppenheimer (1904–1967) proposed a simplifying approximation according to which the motion of electrons and ions in a metal can be separated and hence the electronic and ionic properties can be treated separately. It is called *Born–Oppenheimer approximation* and sometimes called "*adiabatic approximation*". In a metal, it is very difficult to evaluate the exact crystal potential. In 1928, Douglas Rayner Hartree (1897–1958) estimated crystal potential self-consistently in the one-electron approximation and developed a *self-consistent method* for evaluating energy eigenvalues in crystalline solids. In 1930, Vladimir Fock (1898–1974) extended the Hartree self-consistent method by including symmetry in the wave function: *Hartree–Fock method*. Both these methods make use of the variational technique to derive the equation of motion.

Paul Dirac was the first physicist to develop the subject of quantum field theory in 1927 when he was studying the emission and absorption of electromagnetic radiation. He coined the word quantum electrodynamics for the first time and it formed a part of quantum field theory. In Eqs. (4.14) and (4.15) the number of particles in a system is conserved. But there are a number of physical phenomena in which particles are either created or destroyed such as the Compton effect, electron–positron pair production and annihilation (see Figs. 4.7 and 4.9). In other words, the number of particles in a system may not be constant. If one examines the Eqs. (4.14) and (4.15) carefully one finds that the wave function, which represents the field of a system, is a scalar quantity. In order to quantize the Schrodinger equation fully one has to quantize the wave function also which will quantize the field of the system under consideration. The subject which deals with the quantization of the field is called quantum field theory and its framework allows the combination of classical field theory and quantum mechanics with the theory of special relativity. In quantum field theory, one considers either interaction of two fields or the interaction of particles with a field such as scattering of photons by electrons (Compton effect) or annihilation and creation of electron–positron pairs. A wave field is specified by its amplitudes at all points in space which are a function of time. Figure 4.10 depicts schematically a field at a particular time. The field therefore has an infinite number of degrees of freedom and is analogous to a system that consists of infinite number of particles. In order to explain the above-mentioned physical processes the wave function has to be expressed in terms of creation and annihilation operators. In this process, the field of the system is quantized and it is called *second quantization* in quantum mechanics. The second quantization openes up the subject of quantum field theory, which deals theoretically with the quanta of different fields such as photons, phonons, magnons etc. In the quantum field theory, it is more appropriate to work in the occupation number representation usually called N-representation. In this representation the total number of particles at time t, that is N(t), in a system is given by

$$\mathrm{N(t)} = \int \Phi^{+}(\mathbf{r}, \mathrm{t})\Phi(\mathbf{r}, \mathrm{t})\mathrm{d}^{3}\mathrm{r} \tag{4.26}$$

Here $\Phi^{+}(\mathbf{r}, t)$ is the complex conjugate of the wave function $\Phi(\mathbf{r}, t)$ which is an operator. Hence $\Phi^{+}(\mathbf{r}, t)\Phi(\mathbf{r}, t)$ represent particle density per unit volume at position $\mathbf{r}$ and at time t.

A relativistic form of Schrodinger wave equation was proposed by different scientists. In 1926, Oskar Klein (1894–1977) gave the simple relativistic wave equation for the integral spin particles (bosons). It is nothing but the relativistic Schrodinger equation and describes the scalar fields. A few months later Walter Gordon (1893–1939), independently gave relativistic equation, named Klein–Gordon equation. On

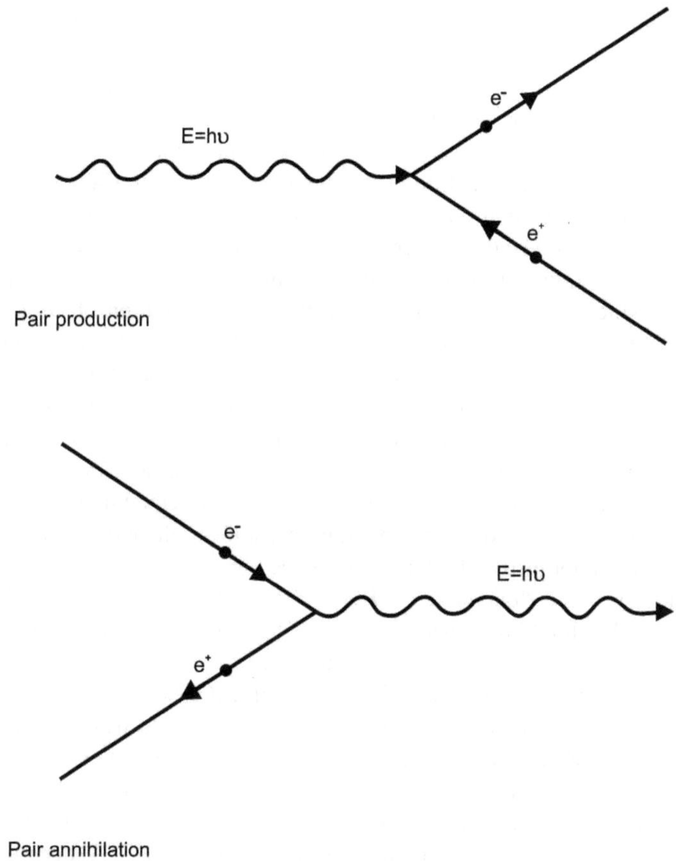

Fig. 4.9 Electron–positron pair production and annihilation

Fig. 4.10 Schematic representation of a field at a particular time

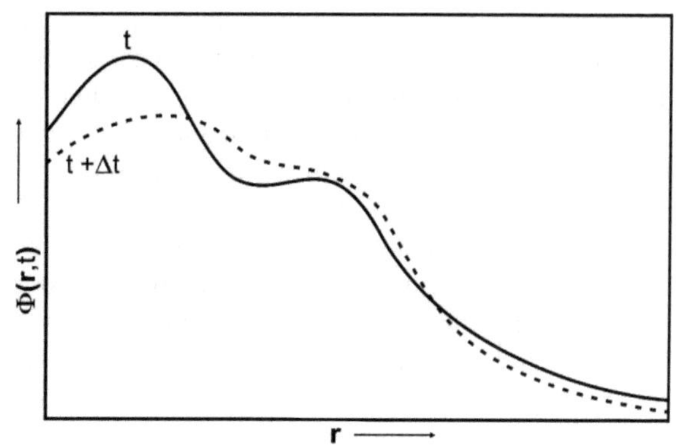

the other hand, in 1928, Paul Dirac proposed a Lorentz invariant relativistic electron equation. The Dirac's equation in the four-vector space is represented by a 4×4 matrix equation. The four-component wave function is neither a scalar nor a vector or tensor: it was called spinor. In fact, Dirac's equation turned out to be a theory of electron spin. It unexpectedly provided the explanation for the doublet splitting of spectral lines that had baffled theorists throughout the decade. Gordon showed that electric current in Dirac's theory comprises of two contributions. First component of electric current is the conduction electron current arising from the actual motion of electrons and the second component of current arises from the polarization or magnetization of the solid which is due to the spin of electrons. The solution of Dirac equation yielded both positive and negative energy states, shown in Fig. 4.11, but the negative energy states are not allowed from classical point of view. Therefore, Dirac assumed that all the negative energy states are filled while the states with positive energy are empty. The transition of an electron from a negative energy state to the positive energy state is allowed. Dirac interpreted the presence of hole in negative energy states as a particle with positive mass and positive charge equal in magnitude to that of the electron. Dirac called it a proton, the only known elementary particle at that time but he could not justify the mass difference between a proton and an electron. Later Dirac in his book "*Directions in Physics*" admitted that it was quite wrong. He expressed that he should have had the courage to predict a new particle and should have asked the experimenters to find the same. Therefore, Dirac's relativistic electron equation predicted theoretically for the first time the existence of *antimatter* but could not establish the existence of *positron* (antiparticle of electron). It is noteworthy that Paul Dirac worked alone throughout his entire career of 44 years in physics. As already mentioned, Carl David Anderson (1905–1991), an American physicist, was studying cosmic rays using a cloud chamber. As a particle enters a cloud chamber it leaves a track like the condensation trail of an aeroplane. In 1932, Anderson observed that some particles as they enter the cloud chamber produce tracks just like those of electrons. These tracks were bent by the magnetic field by the same amount as the tracks of electrons but in the opposite direction. Therefore, he concluded that they could only be particles with positive charge equal in magnitude to that of electron and mass also equal to that of electron (named positrons) and thus established experimentally the existence of antimatter. Carl Anderson was awarded 1936 Nobel Prize in Physics for the discovery of positrons. Dirac's relativistic equation was the first theory to account fully for the theory of special relativity in the context of quantum mechanics. It gave the idea that any particle, in principle, could be created from energy provided it is accompanied by the production of its antiparticle counterpart. On the other hand whenever a particle meets its antiparticle both disappear liberating energy in the form of gamma ray. The discovery of antimatter changed the concept of fundamental particle entirely. Today we have antiparticles of almost every subatomic particle. Actually quantum field theory is a very wide subject which deals with all sorts of physical fields and gave birth to many subfields. The quantum field theory which deals with the electromagnetic field is called *quantum electrodynamics. Quantum chromodynamics* deals with the quantization of strong nuclear force and describes interactions of quarks and gluons which are fundamental constituents of

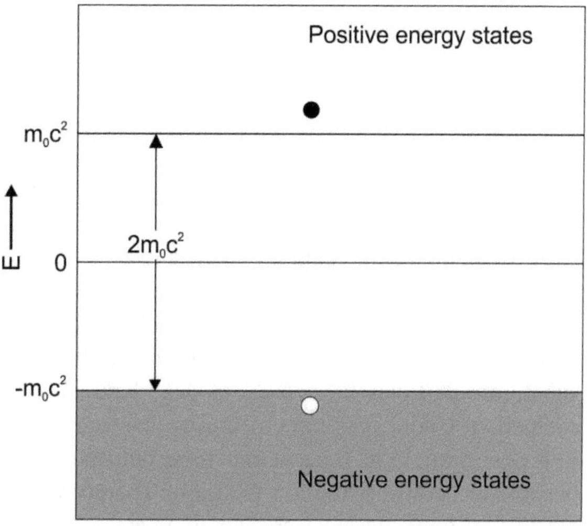

Fig. 4.11 Energy diagram showing positive and negative energy states during the creation of electron–positron pair. An electron from the negative energy state makes a transition to the positive energy state thereby leaving a hole in the negative energy state. The black dot in the positive energy state represents an electron while the white hole in the negative energy state represents a positron

matter. *Quantum flavordynamics* describes weak nuclear force which is responsible for radioactive decay. Similarly the quantum field theory which describes the quantization of gravitational field in curved spacetime is called *quantum gravity*. Here I want to point out that sometimes theories predict some physical effects such as antimatter, black holes, gravitational waves etc. before observation through experiments: power of theories. Usually physical phenomena are first observed experimentally and theories are put forward to explain them later. But actually, theories and experiments go side by side.

The perturbation theory when applied to the quantum fields became quite complicated and went out of hand. In 1948, Richard Feynman (1918–1988) gave a diagrammatic technique by observing that many terms are repeated in a regular fashion: *Feynman diagrammatic technique*. The various terms in the perturbation theory were represented by dots, lines, squiggles and other similar marks and every symbol represents a term, denominator, an integral and so on. In his subsequent papers up to 1951 he developed mathematics for this technique. Thus, complex integrals could be written in terms of simple diagrams, with absolutely no ambiguity as to what they mean—known as Feynman diagrams. The reverse method is also used quite often: the scattering processes can be visualized in terms of diagrams which contribute towards a physical process and one can immediately write the mathematical expression for that physical process. Later it was found that the Feynman diagrammatic technique is broadly applicable to all perturbative series.

Various scientists had different views about quantum mechanics. Einstein observed that '*quantum theory is very worthy of regard*', but he personally did not like it. He further said "*God does not play dice*". On the other hand Robert Feynman expressed that things on a small scale behave nothing like things on a large scale. Richard Feynman said, "*I think I can safely say that nobody understands quantum mechanics*". He also made a statement, "*If you think you understand quantum mechanics, then you don't*". Someone asked John Archibald Wheeler (1911–2008), a well-known figure in relativistic cosmology and quantum theory, about the validity of quantum mechanics. He said, "*I must say, in every day context, quantum theory is unshakeable, unchallengeable and untreatable – it is battle tested*".

In conclusion, we remark that some roots of quantum mechanics extended far back to the nineteenth century such as photoelectric effect, blackbody radiation and line spectra of elements. Quantum mechanics deals with atomic or subatomic particles which cannot be seen with our naked eye. Therefore, one cannot follow the path of a particle as a function of time as in the classical mechanics. Furthermore, subatomic particles like electrons are identical and cannot be distinguished from each other. Under such circumstances one cannot determine with certainty the position of a particle at a particular time but one can talk about the probability of a particle to be at a particular position. Therefore, in studying atomic or subatomic particle systems classical physics does not work but quantum mechanics comes to our rescue. Atomic electrons move with velocities comparable with the velocity of light. Therefore, quantum mechanics in conjunction with relativity, which gives relativistic quantum mechanics, is most suitable for studying atomic systems. It has been found that Newtonian mechanics is a special case of relativistic quantum mechanics and is obtained when mass is large and velocity of a particle is much less than the velocity of light. Further, quantum mechanics agrees with classical mechanics when the difference between the energies of quantized levels becomes vanishingly small. This is usually called *correspondence principle*. In the end, one can say that the development of nonrelativistic and relativistic quantum mechanics has set the stage for the study of various properties of solids. Scientists are trying to develop theory of everything with the help of quantum mechanics and general relativity.

Bibliography

Abbott BP, Abbott R, Abbott TD, Abernathy MR, Acernese F, Ackley K, Adams C, Adams T, Addesso P et al (LIGO Scientific Corporation and Virgo Collaboration) (2016) Phys Rev Lett 116:061102; (2016) Properties of the binary black hole merger GW150914. Phys Rev 116:241102; (2016) Improved analysis of GW150914 using a fully spin-precessing waveform model. Phys Rev 6:041014

Anderson CD (1932) The apparent existence of easily deflectable positives. Science 76:238–239; (1933) The positive electron. Phys Rev 43:491–494

Aspect A, Grangier P, Roger G (1981) Experimental tests of realistic local theories via Bell's theorem. Phys Rev Lett 47:460; (1982) Experimental realization of Einstein-Podolsky-Rosen-Bohm Gedankenexperiment: a new violation of Bell inequalities. Phys Rev Lett 49:91; Aspect A, Dalibard J, Roger G (1982) Experimental test of Bell's inequalities using time-varying analyzers. Phys Rev Lett 49:1804

Baade W, Zwicky F (1934) On super-novae. Proc Natl Acad Sci 20:254–259; (1934) Cosmic rays from super-novae. Proc Natl Acad Sci 20:259–263

Bardeen JM, Carter B, Hawking SW (1973) The four laws of black hole mechanics. Coommun Math Phys 31:161–170
Bell JS (1964) On the Einstein Podolsky Rosen paradox. Phys Phys Fizika 1:195
Bohr N (1913) On the constitution of atoms and molecules. Part II. Systems containing only a single nucleus. Philos Mag 26:476–502; (1913) On the constitution of atoms and molecules. Part III. Systems containing several nuclei. Philos Mag 26:857–875
Bohr N (1918) On the quantum theory of line spectra. Part I. On the general theory; Part II. On the hydrogen spectrum. Kgl Danske Vid Selskab Skr (Naturvidensk og mathem afd) (8) 4:1
Boltzmann L (1877) On the relationship between the second fundamental theorem of the mechanical theory of heat and probability calculations regarding the conditions for thermal equilibrium (trans: Sharp K, Matschinsky F). Sitzungberichte der Kaiserlichen Akademie der Wissenschaften. Mathematisch-Naturwissen Classe. Part II, LXXVI 76:373–435. Vienna [(1909) Wissenschaftliche Abhandlungen, vol II, reprint 42. Barth, Leipzig, pp 164–223]
Boltzmann L (1884) Ableitung des Stefan'schen gesetzes, betreffend die abhängigkeit der wärmestrahlung von der temperatur aus der electromagnetischen lichtttheorie [Derivation of Stefan's law, concerning the dependency of heat radiation on temperature, from the electromagnetic theory of light]. Ann Phys Chem (in German) 258:291–294
Born M (1922) Einstein's theory of relativity. Dutton and Co., New York
Born M (1926) Zur Quantenmechanik der Stoßvorgänge (Vorläufige Mitteilung). Z Phys 37:863–867; (1926) Quantenmechanik der Stoßvorgänge. Z Phys 38:803–827; Born M, Heisenberg W, Jordan P (1926) Zur Quantenmechanik. II. Z Phys 35:557–615
Born M, Brody E (1921) Uber die spezifische warme fester korper bei hohen temperature. Z Phys 6:132–139; Born M, Brody E (1921) Über die schwingungen eines mechanischen systems mit endlicher amplitude und ihre quantelung. Z Phys 6:140–152; Epstein PS (1922) Die storungsrechnung im dienste der quantentheorie. I. Eine methode der storungsrechnung. Z Phys 8:211–228; Epstein PS (1922) Die storungsrechnung im dienste der quantentheorie. II. Die numerische durchfiihrung der methode. Z Phys 8:305–320
Born M, Heisenberg W, Jordan P (1925) Zur Quantenmechanik II. Z Phys 35:557–615
Born M, Huang K (1954) Dynamical theory of crystal lattices. Oxford University Press, London
Born M, Jordan P (1925) Zur quantenmechanik. Z Phys 34:858–888
Born M, Oppenheimer JR (1927) Zur quantentheorie der molekeln [On the quantum theory of molecules]. Ann Phys (in German) 389:457–484
Brillouin L (1926) La mécanique ondulatoire de Schrödinger: une méthode générale de resolution par approximations successive. Comptes Rendus de l'Académie des Sciences 183:24–26
Chadwick J (1932) Possible existence of a neutron. Nature 129:312; (1932) The existence of a neutron. Proc Roy Soc (London) A136: 692–708; (1933) Bakerian lecture. The neutron. Proc Roy Soc (London) A142:1–26
Clauser J, Horne M, Shimony A, Holt R (1969) Proposed experiment to test local hidden-variable theories. Phys Rev Lett 23:880; Freedman SJ, Clauser JF (1972) Experimental test of local hidden-variable theories. Phys Rev Lett 28:938
Compton AH (1923) A quantum theory of the scattering of X-rays by light elements. Phys Rev 21:483–502
Davisson C, Germer LH (1927a) Diffraction of electrons by a crystal of nickel. Phys Rev 30:705–740; (1927b) The scattering of electrons by a single crystal of nickel. Nature 119:558; (1928) Reflection of electrons by a crystal of nickel. Proc Natl Acad Sci USA 14:317–322
De Broglie LV (1924) Ph.D. Thises entitled "Recherches sur la théorie des quanta (Research on the Theory of the Quanta)"; (1924) A tentative theory of light quanta. Philos Mag 47:446–458; (1925) Research on the theory of quanta. Ann Phys 10:22–128
Debye P (1912) Zur theorie der spezifischen waerme. Ann Phys 39:789–883
Dicke RH, Peebles PJE, Roll PG, Wilkinson DT (1965) Cosmic black-body radiation. Astrophys J 142:414–419
Dirac PAM (1927) The quantum theory of emission and absorption of radiation. Proc Roy Soc 114A:243

Dirac PAM (1928) The quantum theory of the electron. Proc Roy Soc (London) A 117:610–624
Dirac PAM (1930) Principles of quantum mechanics. Oxford University Press, London
Dirac PAM (1931) Quantised singularities in the electromagnetic field. Proc Roy Soc (London) A 133:60–72
Dirac PAM (1930) A theory of electrons and protons. Proc Roy Soc (London) A133:360–365 (The physical picture of Dirac's hole first appeared in it)
Dirac PAM (1939) A new notation for quantum mechanics. Proc Camb Philos Soc 35:416
Dirac PAM (1978) Directions in physics. Wiley, New York and London
Einstein A (1905a) Über einen die erzeugung und verwandlung des lichtes betreffenden heuristischen gesichtspunkt [On a heuristic viewpoint concerning the production and transformation of light]. Ann Phys (in German) Leipzig, Germany 17(series 322):132–148; (1905b) Über die von der molekularkinetischen theorie der wärme geforderte bewegung von in ruhenden flüssigkeiten suspendierten teilchen [On the motion—required by the molecular kinetic theory of heat—of small particles suspended in a stationary liquid]. Ann Phys (in German) 17(series 322):549–560; (1905c) Zur elektrodynamik bewegter körper [On the electrodynamics of moving bodies]. Ann Phys (in German) 17(series 322):891–921; (1905d) Drude PKL (ed) Ist die trägheit eines körpers von seinem energieinhalt abhängig? [Does the inertia of a body depend upon its energy content?]. Ann Phys Vierte Folge (in German) Leipzig, Germany: Verlag von Johann Ambrosius Barth 18 (all series: 323) (13):639–641; (1915) Die feldgleichungen der gravitation [The field equations ofgravitation]. Sitzungsberichte 1915 (in German). Berlin, Germany; (1915) Königlich Preussische Akademie der Wissenschaften, pp 844–847—via ECHO, CulturalHeritage Online, Max Planck Institute for the History of Science
Einstein A (1907) Theorie der strahlung und die theorie der spezifischen warme. Ann Phys 4:180–190
Einstein A (1916) Die grundlage der allgemeinn relativitatstheore [The foundation of general theory of relativity]. Ann Phys 49:769–782
Einstein A (1917) Kosmologische Betrachtungen zur allgemeinen Relativitätstheorie. Sitz König Preuss Akad 142–152 [Available in English translation as 'Cosmological considerations on the general theory of relativity' in (Lorentz et al. 1952):175–188]
Einstein A, Podolsky B, Rosen N (1935) Can quantum-mechanical description of physical reality be considered complete? Phys Rev 47:777–780
Everett H (1957a) On the foundations of quantum mechanics (Ph.D. thesis), Princeton University, USA; (1957b) Relative state formulation of quantum mechanics. Rev Mod Phys 29:454–462; Wheeler JA (1957) Assesment of relative state formulation of quantum theory. Rev Mod Phys 29:463–465
Everett H (1973) The theory of the universal wavefunction. In: DeWitt B, Niel Graham R (eds) The many-worlds interpretation of quantum mechanics. Princeton Series in Physics, Princeton University Press, pp 3–140. [The original and most comprehensive paper on many-worlds theory. Investigates and recasts the foundations of quantum theory in information theoretic terms, before moving on to consider the nature of interactions, observation, entropy, irreversible processes, classical objects etc.]
Fermi E (1926) Zur quantelung des idealen einatomigen gases. Z Phys 36:902–912; Dirac PAM (1926) On the theory of quantum mechanics. Proc Roy Soc A112:661–677
Feynman RP (1948) A relativistic cut-off for classical electrodynamics. Phys Rev 74:939; (1949) The theory of positrons. Phys Rev 76:749; (1950) Mathematical formulation of the quantum theory of electromagnetic interaction. Phys Rev 80:440; (1951) An operator calculus having applications in quantum electrodynamics. Phys Rev 84:108
FitzGerald GF (1889) The ether and the Earth's atmosphere. Science 13:390
Friedman A (1922) Über die krümmung des raumes. Z Phys 10:377–386; (1924) Über die möglichkeit einer welt mit konstanter negativer krümmung des raumes. Z Phys 21:326–332 [English translation in: Friedmann A (1999) On the possibility of a world with constant negative curvature of space. Gen Relativ Gravit 31:2001–2008]; (1999) On the curvature of space. Gen Relativ Gravit 31:1991–2000

Fock V (1930) Naherungsmethode zur losung des quantenmechanischen mehrkorperproblems. Z Phys 61:126–148

Gamow G (1946) Expanding universe and the origin of elememts. Phys Rev 70:572

Gerlach W, Stern O (1922) Das magnetische moment des silberatoms. Z Phys 9:353–355

Goudsmit S, Uhlenbeck GE (1926) Over het roteerende electron en de structuur der spectra. Physica 6:273–290

Hartree DR (1928) The wave mechanics of an atom with a non-Coulomb central field. Part I. Theory and methods. Proc Cambr Philos Soc 24:89–110; (1928) The wave mechanics of an atom with a non-Coulomb central field. Part II. Some results and discussion. Proc Cambr Philos Soc 24:111–132; (1928) The wave mechanics of an atom with a non-Coulomb central field. Part III. Term values and intensities in series in optical spectra. Proc Cambr Philos Soc 24:426–437

Hawking SW (1974) Black hole explosions? Nature 248:30–31; (1975) Particle creation by black holes. Commun Math Phys 43:199–220

Heisenberg W (1925) Über quantentheoretische umdeutung kinematischer und mechanischer beziehungen. Z Phys 33:879–893

Heisenberg W (1927) Über den anschaulichen inhalt der quantentheoretischen kinematik und mechanik. Z Phys (in German) 43:172–198

Heisenberg W (1959) Physics and philosophy. Harper and Row Hoyle F (1946) The synthesis of the elements from hydrogen. Monthly Notices of the Royal Astronomical Society 106:343–383; (1954) On nuclear reactions occurring in very hot stars. I. The synthesis of elements from carbon to nickel. Astrophys J Suppl Ser 1:121–146; Burbidge EM, Burbidge GR, Fowler WA, Hoyle F (1957) Synthesis of the elements in stars. Rev Mod Phys Am Phys Soc (APS) 29:547–650

Hoyle F (1981) The universe: past and present reflections. Eng Sci 45:8–12; (1984) The intelligent universe. Holt, Rinehart, and Winston

Hubble EP (1926) Extragalactic nebulae. Astrophys J 64:321–369; (1929) A relation between distance and radial velocity among extra-galactic nebulae. PNAS 15:168–173; (1929) A spiral nebula as a stellar system. Messier 31. Astrophys J 69:103

Hylleraas EA (1928) Über den grundzustand des Heliumatoms. Z Phys 48:469–494; (1929) Neue berechnung der energie des Heliums im grundzustande, sowie des tiefsten terms von Ortho-Helium. Z Phys 54:347–366; (1930) Über den grundterm der zweielektronenprobleme von H^-, He, Li^+, Be^{++} usw. Z Phys 65:209–225

Kellner GW (1927) Die ionisierungsspannung des heliums nach der Schrödingerschen theorie. Z Phys 44:91–109

Kirchhoff G (1859) Ueber die Fraunhofer'schen linien (On Fraunhofer's lines). Monatsbericht der Königlichen Preussische Akademie der Wissenschaften zu Berlin (Monthly report of the Royal Prussian Academy of Sciences in Berlin) 662–665; (1859) Ueber das sonnenspektrum (On the sun's spectrum). Verhandlungen des naturhistorisch-medizinischen Vereins zu Heidelberg (Proceedings of the Natural History/Medical Association in Heidelberg) 1:251–255; (1860) Ueber die Fraunhofer'schen linien. Ann Phys 185:148–150

Kramers HA (1926) Wellenmechanik und halbzahlige quantisierung. Z Phys 39:828–840

Lattes CMG, Muirhead H, Occhialini GPS, Powell CF (1947) Nature 159:694; Lattes CMG, Occhialini GPS, Powell CF (1947) Nature 160:453 and 486

Lemaître G (1927) Un Univers homogène de masse constante et de rayon croissant rendant compte de la vitesse radiale des nébuleuses extra-galactiques (A homogeneous Universe of constant mass and growing radius accounting for the radial velocity of extragalactic nebulae). Annales de la Société Scientifique de Bruxelles 47(A47):49–59; (1931a) Expansion of the universe. The expanding universe. Month Not Roy Astron Soc 91:490–501; (1931b) The beginning of the world from the point of view of quantum theory. Nature 127:706; (1934) Evolution of the expanding universe. Proc Natl Acad Sci USA 20:12–17

Lewis GN (1926) The conservation of photons. Nature 118:874–875

Lorentz HA (1892a) The relative motion of the earth and the aether. Zittingsverlag Akad. V. Wet. 1:74–79; (1892b) La théorie electromagnétique de Maxwell et son application aux corps mouvants. Archives Néerlandaises des Sciences Exactes et Naturelles 25:363–552

Lorentz HA (1899) Simplified theory of electrical and optical phenomena in moving systems. Proc Roy Netherlands Acad Arts Sci 1:427–442; (1904) Electromagnetic phenomena in a system moving with any velocity smaller than that of light. Proc Roy. Netherlands Acad. Arts Sci. 6:809–831

Michelson AA, Morley EW (1887) On the relative motion of the earth and the luminiferous ether. Am J Sci 34:333–345

Minkowski H (1908) Die grundgleichungen fur die elektromagnetischen vorgange in bewegten korpern. Nachrichten der Gesellschaft der Wissenschaften zu Göttingen, Mathematisch-Physikalische Klasse, pp 53–111

Pauli W (1925) Uber die zusammenhang des abschlusses der electronengruppen im atom mit der komplexstruktur der spektren. Z Phys 31:765–783; (1927) Uber gasentartung und paramagnetismus. Z Phys 41:81–102; (1927) Zur quantenmechanik des magnetischen elektrons. Z Phys 43:601–623

Penzias AA, Wilson RW (1965) A measurement of excess antenna temperature at 4080 Mc/s. Astrophys J 142:419–421

Planck M (1901) Ueber das gesetz der energieverteilung im normalspectrum (On the law of distribution of energy in the normal spectrum). Ann Phys 309:553–563; (1906) Vorlesungen uber die theorie der warmestrahlung. J. A. Barth, Leipzig

Planck M (1963) The philosophy of physics. W. W. Norton, New York (original edition published in 1936)

Rovelli C (2017) The journey to quantum gravity. Riverhead Books, New York

Rutherford E, Chadwick J, Ellis CD (1930) Radiation from radioactive substances. Cambridge Library Collections-Physical Sciences

Rutherford E (1911) The scattering of α and β particles by matter and the structure of the atom, The London, Edinburgh, and Dublin. Philos Mag J Sci 21:669–688; (1914) The structure of the atom. Philos Mag 27:488–498

Schrödinger E (1926a). Zur Einsteinschen Gastheorie. Physikalische Zeitschrift 27:95–101; (1926b) Quantisierung als eigenwertproblem (Erste Mitteilung). Annalen der Physik 79:361–376; (1926c) Quantisierung als eigenwertproblem (Zweite Mitteilung). Annalen der Physik 79:489–527; (1926d) Quantisierung als eigenwertproblem (Dritte Mitteilung). Annalen der Physik 80:437–490; (1926e) Quantisierung als eigenwertproblem (Vierte Mitteilung). Annalen der Physik 81:109–139

Schrödinger E (1935) Die gegenwärtige situation in der quantenmechanik (The present situation in quantum mechanics). Naturwissenschaften 23:807–812

Schwarzchild K (1916) Über das gravitationsfeld eines massenpunktes nach der Einstein'schen theorie. Reimer, Berlin, p 189 (Sitzungsberichte der Königlich- Preussischen Akademie der Wissenschaften; 1916); (1916) Über das gravitationsfeld einer kugel aus inkompressibler flüssigkeit. Reimer, Berlin, pp 424–434 (Sitzungsberichte der Königlich-Preussischen Akademie der Wissenschaften; 1916)

Slipher VM (1913) The radial velocity of the Andromeda Nebula. Lowell Observatory Bulletin, No. 58(2) and No. 56; (1914) Spectrographic observations of Nebulae. Publ Am Astron Soc 3:98(A); (1915) Popular Astron 23:21–24; (1917) Nebulae. Proc Amer Philos Soc 56:403–410; (1918) Spectrographic observations of star clusters. Publ Am Astron Soc 3:331(A); (1918) Popular Astron 26(8)

Sommerfeld A (1927) Elekronentheorie der metalle und des volta-effektes nach der fermischen statistik. Atti del Congresso Internazionale dei Fisici Como-Pavia-Roma II:449–473 [reprinted in Sommerfeld A (1968) Collected works, vol 2. Braunschweig, Vieweg]

Stark J (1914) Weitere resultate uber den effekt des elektrischen feldes auf spektrallinien. Sci Nat 2:145–148

Stefan J (1879) Uber die beziehung zwischen der warmestrahlung und der temperature [On the relation between heat radiation and temperature]. Sitzungsberichteder Kaiserlichen Akademie der Wissenschaften: Mathematisch- Naturwissenschaftliche Classe [Proceedings of the Imperial

Philosophical Academy (of Vienna): Mathematical and Scientific Class] (in German) 79:391–428
Stern O, Gerlach W (1922) Das magnetische moment des silberatoms. Z Phys 9:353–355
Surdo AL (1914) Sul fenomeno analogo a quello di Zeeman nel campo elettrico, Atti della R. Accademia dei Lincei. Rendiconti della Classe di Scienze Fisiche, Matematiche e Naturali, Serie V XXII(12) (1913):664–666 [ripubblicato pure ne: (1914). Il Nuovo Cimento 7:335–337]; (1914) Su l'analogo elettrico del fenomeno di Zeeman: effetto longitudinale, Atti della R. Accademia dei Lincei. Rendiconti della Classe di Scienze Fisiche, Matematiche e Naturali, Serie V, XXIII(4):82–84; (1914) Su l'analogo elettrico del fenomeno di Zeeman: le varie righe della serie di Balmer presentano diverse forme di scomposizione, Atti della R. Accademia dei Lincei. Rendiconti della Classe di Scienze Fisiche, Matematiche e Naturali, Serie V XXIII(4):143–144
Thomson GP (1927) Diffraction of cathode rays by a thin film. Nature 119:890
Thomson JJ (1897a) Cathode rays. Electrician 39:104; (1897) On the cathode rays. Proc Camb Philos Soc 9:243; (1897b) Cathode rays. Philos Mag 44:293
Uhlenbeck GE, Goudsmit S (1925) Naturwissenschaften 47:953; (1926) Nature 117:264
Voigt W (1887) Ueber das Doppler'sche Princip (On the principle of Doppler). Göttinger Nachrichten 7:41–51 [Reprinted with additional comments by Voigt in Physikalische Zeitschrift XVI:381–386 (1915)]; (1887) Theorie des Lichts für bewegte Medien. Göttinger Nachrichten 8:177–238
Wentzel G (1926) Eine verallgemeinerung der quantenbedingungen für die zwecke der wellenmechanik. Z Phys 38:518–529
Wien W (1896) Über die energievertheilung im emissionsspectrum eines schwarzen körpers. Ann Phys Chem 294:662–669
Yukawa H (1935) On the interaction of elementary particles. Proc Phys Math Soc Jpn 17:48
Zwicky F (1938) On collapsed neutron stars. Astrophys J 88:522–525

Part II
Some Fields of Solid State Physics

Chapter 5
Structure of Crystalline Solids

Matter exists in three states: solid liquid and gas although there is also a fourth state called plasma state. Solids are perhaps the first substances which had been utilized by mankind since the inception of the universe. In ancient times, humans considered solid as a lump of material and it never came into their minds that a solid can possess some kind of internal symmetric structure. In 1591, Thomas Harriot (1560–1621), an English astronomer, was the first scientist to bring attention to the natural symmetry found in snowflakes. In 1611, Johannes Kepler (1571–1630) published a small pamphlet entitled, "*Hexagonal Snow*" in which he described that those snowflakes are made of closely spaced spheres with hexagonal symmetry (Fig. 5.1). If the atoms are considered to be spheres, their close packing will exhibit hexagonal symmetry as shown in Fig. 5.1. Kepler extended this concept of symmetry to atomic arrangement which later began to be known as *Kepler's conjecture*: a statement about the most efficient arrangement of packing of spheres.

Crystal is a Greek word meaning "clear ice" and in ancient times this term applied only to transparent materials such as quartz. The shape of a crystal is formed by its characteristic faces which, in early times, were of prime interest. The science of crystallography originated in the "Middle Ages" and initially was considered to be a part of mineralogy as it described the macroscopic structure of crystals. Nicolaus Steno (Niels Stenson) (1638–1686) was a Danish scientist who was pioneer in both anatomy and geology. In 1669, he made the first ever observation on crystals which is described in his book, "*De Solido intra Solidum Naturaliter Contento*". He gave the law of constant angles known as *Steno's law* and it is the first law of crystallography. This law states that the angles between the corresponding faces are the same for all the specimens of the same crystal. In 1678, Christiaan Huygens (1629–1695) explained the cleavage phases and anisotropic hardness in crystals and index of refraction for $CaCO_3$ using a structural model consisting of rhombohedral packing of spheres. Jean-Baptiste L. Rome de l'isle (1736–1790), a French mineralogist, is considered one of the creators of modern crystallography. In 1772, he authored a book on crystallography, "*Essai de Cristallographie*". Its second edition, an improved one, was published under the title "*Cristallographie*", in 1783, and is regarded as his major

J. S. Galsin, *History of Solid State Physics*,
https://doi.org/10.1007/978-981-95-0504-3_5

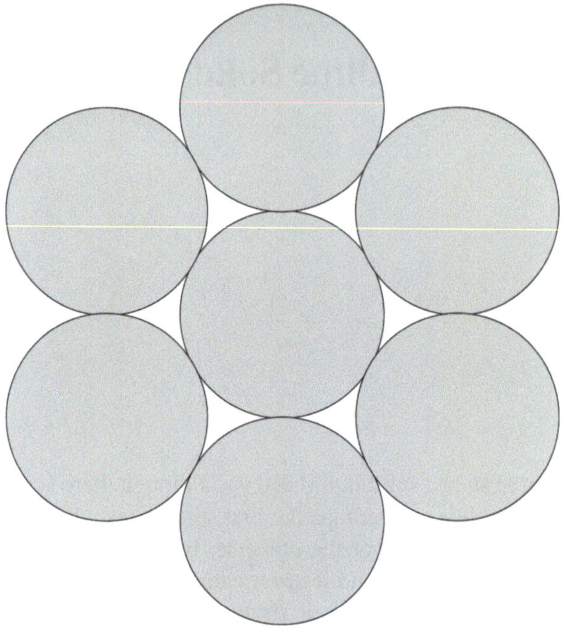

Fig. 5.1 Close-packing of spheres with hexagonal symmetry

work. Rome de l'isle confirmed and extended Steno's findings and formulated the *Law of Consistency of Interfacial Angles.*

Christian Samuel Weiss (1780–1856), Public Domain, URL: https://www.wikidata.org/wiki/Q66165#/media/File:Weiss-christian-samuel-1780-1856--bildnisse-beruehmter-mitglieder-der-deutschen-akademie-der-wissenschaften-zu-berlin-berlin-akademie-verlag-1950-abb024.jpg

Christian Samuel Weiss (1780–1856), a German mineralogist and mathematician, stressed the significance of crystallographic directions in crystals and assumed these as the possible basis for the classification of crystals. Weiss also introduced a scheme for the characterization of crystals. He gave the basic law of crystallography named after him as "*Weiss zones law*[1]". Weiss is credited for creating parameters of modern crystallography and was instrumental in making it a branch of mathematical physics. Franz Ernst Neumann (1798–1895), a student of Samuel Weiss, extended the work on the law of zones (law of rational intercepts). Fredrich Mohs (1773–1839), a German geologist and mineralogist, was also the student of Weiss. He studied the hardness of minerals and in 1812 introduced a scale of hardness called *Mohs scale*. This is a qualitative ordinal scale ranging from 1 to 10. One of the characteristics he used to identify minerals was their hardness. Mohs hardness is a rough measure of resistance of a smooth surface to scratching or abrasion expressed in terms of Mohs scale. Mohs considered diamond as hardest material and rated it as 10 and talc is soft material assigned value 1 on the scale. The rest of the minerals are adjusted in between depending on their hardness. Therefore, Mohs introduced an independent classification (other than that of Weiss) of crystals based on their hardness.

Sketch of Rene-Just Hauy (1743–1822), Public Domain, URL: https://en.wikipedia.org/wiki/Ren%C3%A9_Just_Ha%C3%BCy#/media/File:Ren%C3%A9_Just_Ha%C3%BCy_2.jpg

[1] Weiss zone law states that a lattice direction with indices [uvw] is contained in a lattice plane with Miller indices (hkl) if the following condition is satisfied $hu + kv + lw = 0$.

Rene-Just Hauy (1743–1822), a French priest and mineralogist, had done the most fundamental work on crystal structure. He found that crystals are made of smallest building blocks called "*integrant molecules*" which possess particular shape and size. He assumed that an integrant molecule cannot be broken further without destroying its physical and chemical nature: later an integrant molecule was identified as a unit cell. Hauy also introduced the idea of triple periodicity in crystals. This idea was fundamental to later developments in the field of crystal lattices. He argued that a crystal structure is made up of orderly arrangement of these integrant molecules in successive layers according to the geometrical laws of crystallization. But he could not determine the internal structure of the integrant molecules. Hauy and other mineralogists of his time could observe crystal shapes and cleavage planes and measured interfacial angles with an instrument called *goniometer*. By 1792, Hauy had identified a number of parallelepipeds as possible primitive crystal forms. Further, Hauy gave mathematical theory of his work which became a classic work in the field of crystallography. Hauy's work on the theory of structure led to the culminating point where crystallography splits from mineralogy and became an independent. In 1801, Rene-Just Hauy wrote a book entitled, "*Traite de Mineralogie*" in four volumes and summarized his work on crystallography. He is regarded as the "*Father of Modern Crystallography*". Hauy is also known for his observations on pyroelectricity. He detected pyroelectricity in calamine, an oxide of zinc, as early as 1785. He studied pyroelectricity in a number of minerals including tourmaline and related it to crystalline structure. Hauy also published a book on electricity and magnetism.

In 1803, John Dalton (1766–1844), an English chemist, physicist and meteorologist, proposed atomic theory of matter. According to him an element (matter) is made of very small particles which can exist independently and are electrically neutral: called chemical atoms. Dalton assumed that atoms are hard spheres which cannot be subdivided, created or destroyed. Atoms of an element are all identical but atoms of different elements are different in shape and size and combine in simple whole-number ratios to form chemical compounds. Although the idea of an atom was developed by Democritus (Chap. 1) but Dalton's contribution was to consider the relative sizes and character of these atoms and how they fit together. But at that time the concept of atoms was unable to explain the different colors of various elements and materials. The proposal of atomic theory by John Dalton remained entirely hypothetical for about a century and many scientists disagreed with him till, in 1905, Einstein's paper on Brownian motion proved beyond doubt the existence of atoms.

In 1824, Ludwig August Seeber (1793–1855), a German physicist, said chemical atoms have spherical shape and these are associated with the building blocks of the solids. Seeber gave the law of forces according to which the chemical atoms are bound by forces that are attractive at large distances and repulsive at very small distances. Although Seeber's law of forces was purely empirical in nature but it is qualitatively the same as that predicted by all the modern theories of crystal structure. Johann Friedrich Christian Hessel (1796–1872), a German mineralogist, was aware of Hauy's law of rational intercepts. Further, it was known that only two-, three-, four-, and six-fold rotations could occur in crystals. Based on these facts Hessel, in 1830, proved that by combining the various morphological forms one obtains 32 kinds of crystal symmetries in Euclidean space. But at that time not all of the 32 possible symmetries had actually been observed in real crystals. Unfortunately, Hessel's work went unnoticed but came into light only when it was republished in 1897 as a part of collection of papers on crystallography in "*Oswald's Classics of Exact Sciences*". Due to the delay in the republication of Christian Hessel's work, similar findings had been reported by the French scientist Auguste Bravais (1811–1863) in 1849 and by the Russian crystallographer Alex V. Gadolin (1828–1892) in 1867.

William Hallowes Miller (1801–1880), a British mineralogist, laid the foundation of modern crystallography. It is a well-known fact in geometry that three noncollinear points can specify a plane uniquely. In 1839, William Hallowes Miller gave the law of rational numbers according to which a plane can be specified by three integers usually called *Miller indices* and represented by (hkl) where h, k and l are integers positive, negative or zero. The method of calculating Miller indices had been described in his "*A Treatise on Crystallography*" written in 1839. Figure 5.2 shows a plane with $\mathbf{a}_1$, $\mathbf{a}_2$, and $\mathbf{a}_3$ as the intercepts of the plane on x-, y-, and z-Cartesian axes, respectively. These intercepts are expressed in terms of lattice constant, and then Miller indices are the reciprocals of the intercepts reduced to the smallest integers. The set of equivalent planes corresponding to the Miller indices h, k, and l are written as {hkl}. The direction with Miller indices h, k, and l is written as [hkl]. In hexagonal systems

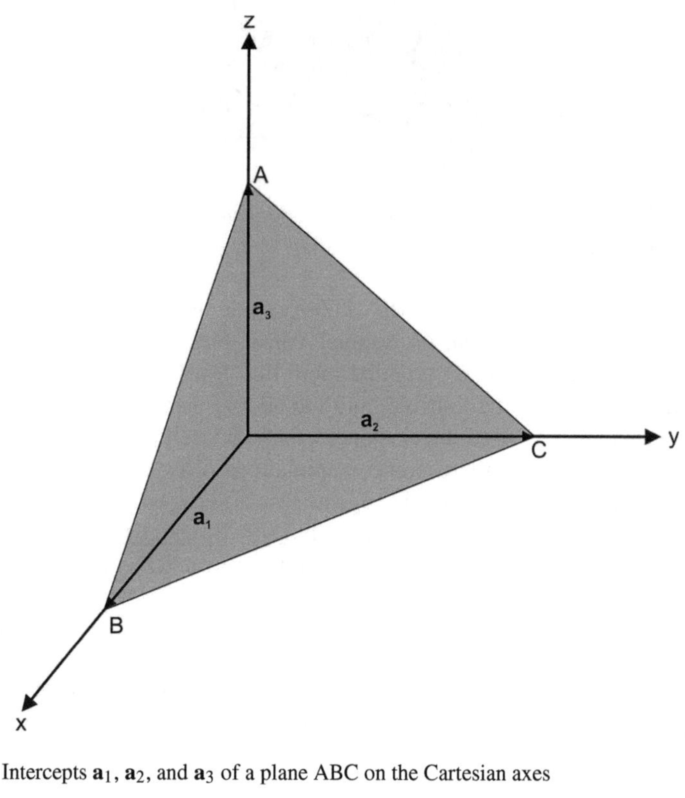

Fig. 5.2 Intercepts $\mathbf{a}_1$, $\mathbf{a}_2$, and $\mathbf{a}_3$ of a plane ABC on the Cartesian axes

four crystallographic axes are defined and a plane has $\mathbf{a}_1$, $\mathbf{a}_2$, $\mathbf{a}_3$, and $\mathbf{a}_4$ as intercepts on them. Bravais had proposed a similar scheme for evaluating the Miller indices in crystals with hexagonal symmetry usually called the Bravais-Miller indices. The planes in this scheme are written as (hkil) where

$$i = -(h + k) \tag{5.1}$$

Auguste Bravais (1811–1863), Public Domain, URL: https://en.wikipedia.org/wiki/Auguste_Bravais#/media/File:Bravais2.gif

Mortiz Ludwig Frankenheim (1801–1869) became interested in crystallography after being inspired by his teacher Samuel Weiss. He was the first to define 32 point groups of symmetry and classified them into four crystal systems: they are regular one, the twofold, the fourfold, and the sixfold systems. From his observations, Frankenheim conceived the existence of 15 lattices in crystals. Frankenheim conducted one of the first microscopic examinations of crystals in polarized light using the Nicol prism as the polarizer. Auguste Bravais (1811–1863), a French physicist, did remarkable work on crystallography. In 1848, Bravais demonstrated that only 14 types of unit cells are compatible with the orderly arrangement of atoms found in crystals. The periodic repetition of these unit cells generates 14 lattices usually called *Bravais lattices*, not the 15 lattices as conceived by Frankenheim. Each lattice point has one or more atoms associated with it in a real crystal. There are seven systems of crystals and the shape of the unit cell determines to which of the seven crystal systems the particular crystal belongs. In 1850, Bravais published a memoir about crystallography. By 1890 it was well established that there are 14 Bravais lattices (comprising of 7 lattice systems) and 32 crystallographic point groups. Space group is made from the linear combination of a Bravais lattice and point groups compatible with the lattice and it describes the symmetry of the crystal. In 1933, Eugene Paul Wigner (1902–1995) and Federick Seitz (1911–2008) defined an alternate primitive cell called *Wigner–Seitz (WS) cell* with the periodic repetition of which whole of the crystal lattice can be generated. The importance of WS cell is that atom is situated at the centre of the WS cell which is nearly spherically symmetric about the atom.

Leonhard Sohncke (1842–1897) was working under Franz Neumann and became aware of the work of Bravais. He classified 65 space groups in which chiral crystal structures are formed, called Sohncke groups. Sohncke groups are the three-dimensional space groups containing only translations, rotations and roto-translations operations. In 1879, Sohncke published his results in his major work, "*Die Entwicklung einer Theorie der Krystallstruktur*". He also made models from cigar boxes to demonstrate his derived space groups. Evgraf Fedorov (1853–1919), a Russian mathematician, crystallographer and mineralogist, studied the structure of crystals for years together. After 10 years of study, in 1891, Fedorov gave total number of possible space groups in crystals but there were some deficiencies (omission and duplicacy). Arthur Mortiz Schoenflies (1853–1928), a German mathematician, also enumerated independently, in the same year, the number of space groups in crystals but again with some omissions and duplicacy. Later, in 1892, Fedorov and Schoenflies jointly corrected the list of space groups and showed that there are 230 space groups in all. Sometimes the space groups are also known as Fedorov groups. It was found that among the 230 space groups found by Fedorov, 65 are the Sohncke groups. These space groups now serve as the mathematical basis for the structure analysis of the various crystalline solids. There are a number of notations used for the representation of the space groups but the notations of Schoenflies and Fedorov are most commonly used. Woldemar Voigt (1850–1919) did pioneering work in crystal physics and in 1910 published a book on crystal physics entitled "*Lehrbuch der Kristallphysik*". Voigt's book is considered to be the bible for workers in crystal physics.

It is worth mentioning here that the nineteenth century ideas concerning the atomic nature of crystals were based on the very indirect evidence. The science of crystallography relied on the overall shape of the crystal to classify its structure. Although the ideas concerning the atomic ordering in crystals developed in line with the work of Hauy and Seeber, but there was no evidence to link the microscopic and macroscopic structure of crystals. Before the experimental determination of crystal structure is discussed one must get familiar with the concept of Fourier transform. In 1807, Jean-Baptiste Joseph Fourier (1768–1830) was studying the heat conduction through solids and in its analysis he used an infinite series now called Fourier series or Fourier transform. Fourier transform is a mathematical transform that decomposes a function, depending on space and time, into functions depending on temporal frequency. Leon Nicolas Brillouin (1889–1969), a French physicist and founder of modern solid state physics, in 1962, wrote a book named "*Fourier Analysis and Sampling Method in Three Dimensions*". In this book, he introduced the concept of reciprocal space (which is nothing but the Fourier space) made up of wave vectors **K** which satisfies the relation

$$e^{\iota \mathbf{G}_p \cdot \mathbf{R}_n} = 1 \tag{5.2}$$

$\mathbf{R}_n$ is the lattice vector in the crystal space and $\mathbf{G}_p$ is the lattice vector in the reciprocal space. From Eq. (5.2) one can derive reciprocal lattice vectors and hence can construct reciprocal lattice in the reciprocal space (momentum space). In 1930, he constructed Brillouin zone which is nothing but the WS cell in the reciprocal space: a uniquely defined primitive cell in the reciprocal space. The significance of the Brillouin zone is that Bragg's reflection condition is satisfied on its entire surface. The reciprocal lattice plays a fundamental role in most of the analytic studies of periodic crystal structures. He gave Brillouin function while studying paramagnetism in the quantum approach.

Max von Laue (1879–1960), Public Domain, URL: https://www.wikidata.org/wiki/Q57067#/File:(UAZ)_AB.1.0583_Laue.tif

In the last decade of the nineteenth century, Wilhelm Conard Rontgen (1845–1923) was performing experiments on the electrical discharge through the cathode ray tube. On Nov. 8, 1895, by chance he found that some invisible radiation is emitted which made the nearby fluorescent screen to glow. He repeated the experiment by shielding the fluorescent screen but obtained the same results. He concluded that a new invisible radiation is emitted but could not determine the nature of radiation. Rontgen named the invisible radiation X-rays as 'X' is used for something unknown in mathematics. On Dec. 22, 1895, Rontgen created the history by capturing the first X-ray image of his wife Anna Bertha's hand. It clearly showed the bones of her hand and the wedding ring astonishing the scientific world. When Anna Bertha saw the image, she reportedly remarked, "*I have seen my death*". Between 1895 and 1897, he published three research papers on X-rays. Later a large number of applications of X-rays were found particularly in medical science. In 1901, Wilhelm Rontgen was awarded the first Nobel Prize in Physics. The wavelength of X-rays is of the order of 1°A which is comparable to the lattice spacing in crystals. Max von Laue (1879–1960) was of the view that X-rays passing through a crystal could suffer diffraction just like the visible light passing through an optical grating. In 1912, Max von Laue along with his students Walter Friedrich (1883–1968) and Paul Knipping (1883–1935) invented X-ray diffraction method to study the structure of crystalline solids. Their experiment proved beyond doubt that X-rays are electromagnetic radiations and thus settled the controversy that had lasted for about 17 years since the discovery of X-rays. Laue's method used a continuous beam of X-rays to produce diffraction pattern. William Lawrence Bragg (1890–1971) carefully studied the diffraction pattern obtained by Laue and came to the conclusion that different Laue diffraction spots arise due to the scattering of X-rays from different planes of the crystalline solid. In 1912, William Lawrence Bragg at the age of 22 years, formulated the law for constructive X-ray reflection defined as

$$2\mathrm{d}\sin\theta = \mathrm{n}\lambda \tag{5.3}$$

where d is the distance between two adjacent parallel planes from which X-ray reflection is considered, θ is the grazing angle (see Fig. 5.3), λ is the wave length of X-rays and n is an integer. In 1913, William Lawrence Bragg (1890–1971) and William Henry Bragg (1862–1942), son and father, studied the structure of rock salt crystal and diamond with the help of X-rays for the first time. In 1912–1913, they invented the first X-ray spectrometer and founded a new science of X-ray crystallography (X-ray spectroscopy) for the determination of crystal structure. X-rays actually interact with the spatial distribution of valence electrons of atoms or molecules and therefore X-ray diffraction technique can be used in determining the structure of crystalline solids. By measuring the intensities of diffracted beams from the single crystal of a solid, one can construct a three-dimensional picture of the density of electrons within the crystal which yields the mean positions of the atoms in the reciprocal space and hence the reciprocal lattice. The mathematical method of Fourier transform combined with the available chemical data, transforms the image into the crystal space and hence determines the positions of atoms in the crystal space. With the help of X-ray

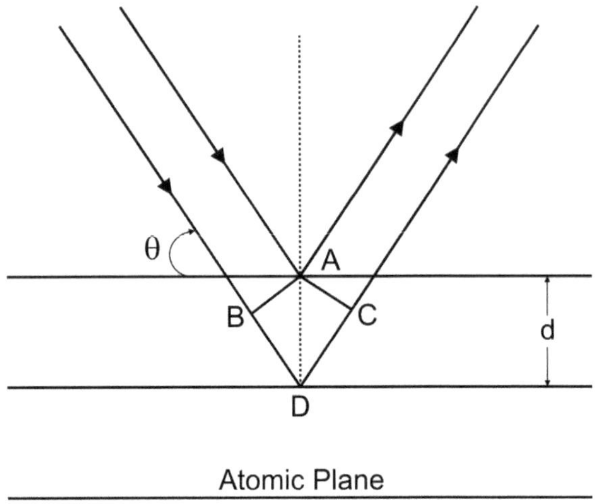

Fig. 5.3 Reflection of a light beam incident at the grazing angle θ from the two adjacent planes of a crystalline solid

diffraction, one can find chemical bond length, crystallographic disorder and lot of other information about the crystal.

In 1916, Peter Joseph William Debye (1884–1966) and Paul Scherrer (1890–1969) invented another X-ray diffraction method in which X-rays were diffracted from a powdered crystalline solid. Actually, a powdered crystalline solid consists of very small crystals oriented randomly in all possible directions. This method yields the X-ray diffraction pattern in the form of a series of concentric circles. It is usually called *powder method* or *Debye–Scherrer method* and allows the determination of unit cell with great precision.

The main thrust of the work of Paul Peter Ewald (1888–1985), a German crystallographer and physicist, was on X-ray crystallography and developed dynamical theory of X-ray diffraction. In 1921, he gave Ewald's construction which describes graphically the Bragg's law of reflection. Louis-Cesar-vector Maurice de Broglie (1875–1960) made many advances in the study of X-ray diffraction and spectroscopy. The *rotating crystal method* of X-ray diffraction is based on the experimental work of Maurice de Broglie and was first employed and developed in *Kaiser-Wilhehm Institut fur Faserstoffchemie* in Berlin. Note that Maurice de Broglie (1875–1960) and Louis de Broglie (1892–1987) were brothers. In 1926, John Desmond Bernal (1901–1971) extended and improved the rotating crystal method in the Royal Institution in London. Bernal developed X-ray spectro-goniometer which is an instrument that measures an angle and allows the crystal to be rotated to a precise angular position: the term goniometer is derived from the Greek words '*gonia*' means angle and '*metron*' means measure. The main advantage of rotating crystal method over the powder method is the possibility to separate different reflections which have the same diffraction angle and to control the direction and range of the rotation. It is noteworthy that rotating crystal method is applicable to only single crystals, the powder method is applicable to finely divided crystalline powder whereas the Laue method is applicable to both single crystals and polycrystalline solids. Crystal structure of a large number of crystalline solids was studied using different methods of X-ray diffraction. In the twentieth century, X-ray spectroscopy became a very useful technique in studying the chemical structure of crystalline solids. Strong X-rays are suited for shallow depths or thin specimens.

According to de Broglie hypothesis, electrons have dual character: both particle and wave nature. The wavelength λ of an electron moving with velocity v is given by

$$\lambda = \frac{h}{m_e v} = \frac{h}{\sqrt{2m_e E}} \tag{5.4}$$

where m_e is the mass of the electron and E is its energy. In 1927, Clinton Joseph Davisson (1881–1958) and Lester Halbert Germer (1896–1971) established the wave nature of electrons. They allowed a low energy electron beam to fall on a nickel target and the intensity of backscattered electrons showed the diffraction pattern consistent with the X-ray diffraction theory. At that time, adequate vacuum techniques were not available therefore low energy electron diffraction (LEED) did not become a popular technique until 1960s. The incident electrons interact with the crystalline solid through the electrostatic forces and being light particles are not expected to penetrate the solid to larger depths. Therefore, LEED is mostly used to study the structure of surfaces or thin films. Electrons with energy in the range of 50–300 eV have wavelength of the order of 1A and are therefore suitable for the study of electron diffraction from crystalline solids. An electron beam with a small energy spread is allowed to fall on the surface of a crystal and the back-scattered electrons form a diffraction pattern on the screen. Most of the incident electrons are scattered from the first atomic layer on the surface and much less number of electrons suffer scattering from a few atomic layers just below the surface. Experiments are usually performed with the use of transmission electron microscope (TEM) or scanning electron microscope (SEM).

Neutrons falling on a crystalline solid are scattered by the atomic nuclei through the strong nuclear forces. Neutrons have high penetration depth, therefore are suited for bulk samples. Further, neutrons possess finite magnetic moment and are scattered by magnetic fields. Therefore, neutron diffraction technique can be used for determining the magnetic structure of a crystal. Neutron diffraction is similar to the X-ray diffraction but due to their different scattering properties, neutrons and X-rays provide complementary information. In December 1944, Ernest Omar Wollan (1902–1984) and Lyle Benjamin Borst (1912–2002), American physicists, using the graphite reactor at Oak Ridge, carried out the first neutron diffraction experiments on gypsum and sodium chloride. In 1946, Clifford G. Shull (1915–2001), an American physicist, joined Ernest Omar Wollan and they together established the basic principles of the neutron diffraction technique and applied them successfully to find the arrangements of atomic magnetic moments in many different materials. On the other hand, Bertram Neville Brockhouse (1918–2003), a Canadian physicist, developed inelastic neutron scattering technique at Chalk River Facility of Atomic Energy of

Canada Limited (AECL). He devised triple-axis spectrometer for neutron diffraction study from crystals. The first results from a prototype triple-axis spectrometer were published in 1955 but the first true neutron triple-axis spectrometer was built in 1956. Neutron diffraction studies yield information about both the chemical and magnetic structures of magnetic crystalline solids. On the other hand, X-ray diffraction studies yield information only about the chemical structure of crystalline solids. Clifford G. Shull and Bertram Neville Brockhouse shared the 1994 Nobel Prize in Physics for their discoveries in neutron diffraction technique (actually Ernest Omar Wollan, pioneer in neutron diffraction studies, died in 1984 before the Nobel Prize was conferred).

Another entirely different method of studying the structure of crystalline solids is with the help of special microscopes. In 1932, Max Knoll (1897–1969), German engineer, and Ernst Ruska (1906–1988), German physicist, developed an *electron microscope* for the first time which was later improved immensely. The basic principle of electron microscope is that electrons interact with the atoms near the surface of a crystalline solid to produce image of the surface. Different types of electron microscopes such as transmission electron microscope (TEM), reflection electron microscope (REM), and scanning transmission electron microscope (STEM) were invented to study the surface profiles of crystalline solids (see Chap. 18 for details of microscopes). Electron microscopes provide much better resolution ($\approx$ 50 pm) than that obtained in optical microscopes. There are number of disadvantages of electron microscopes. Among these the main disadvantages are that ultrahigh vacuum conditions are required and the electron beam may damage some delicate samples like living cells etc. One major limitation of TEM is the need of extremely thin sections of the specimens, typically about 100 nm. A number of probe-type microscopes have also been developed to study the structure of the surface of crystalline solids directly in the crystal space. Gerd Binnig (1947–present) and coworkers developed scanning tunnelling microscope (STM) in 1981, and atomic force microscope (AFM) in 1985. In 1987, Y. Martin, and H. K. Wickramasinghe developed magnetic force microscope (MFM) for the study of surface profiles of magnetic materials (for details see Chap. 18). STM works on the principle of quantum tunnelling while AFM works on the electrostatic force between atoms. With STM one can study the atomic structure of only conducting solids such as metals and semiconductors but with AFM one can study the structure of any type of crystalline solid that is conductors, semiconductors and insulators. Further, in AFM one does not require vacuum conditions rather it works in air. With the help of STM one can manipulate the atoms and can form *quantum corrals*. Therefore, one can create a layer of atoms on the surface of a substrate with STM. The main feature of these microscopes is that these enable us to see the image of individual atoms in the crystal space as their resolution is on the atomic scale. Ernst Ruska, Gerd Binnig and Heinrich Rohrer were awarded 1986 Nobel Prize in Physics for inventing electron microscope and scanning tunnelling microscope.

In 1958, Frederick C. Frank (1911–1998) and John S. Kasper (1915–2005) proposed that approximate icosahedral symmetry is the basis for complicated arrangement of atoms in many complex alloy structures and so proposed Frank-Kasper phases. The most common members of a Frank-Kasper phases family are A15 phases and Laves phases etc. In 1981, Hagen Kleinert (1941–present) and Kazumi Maki (1936–2008) pointed out the possibility of existence of nonperiodic phases in solids. In 1982, Dan Shechtman (1941–present) and co-workers were trying to produce metallic glass by the rapid cooling of molten metallic alloy $Al_{0.86}Mn_{0.14}$ but to their surprise the X-ray pattern showed three-fold and five-fold rotational symmetry. But three-fold and five-fold rotational symmetries are prohibited in a bulk crystalline solid because with these symmetries space cannot be filled completely or uniformly. Shechtman claimed the discovery of a new quasicrystallline phase which was opposed by many physicists particularly the two-time Nobel Laureate Linus Pauling (1901–1994). Therefore, the publication of these results were delayed and ultimately published in 1984. In one of his remarks Pauling said "*there is no such thing as quasicrystals only quasiscientists*". In 1984, Dov Levine (1958-present) and Paul Joseph Steinhardt (1952–present) showed that three-fold and five-fold symmetries are there in the quasi-periodic filling of space which has a true icosahedral symmetry. They called this type of crystal a '*quasicrystal*'. They said that quasicrystals are quasiperiodic structures having an arrangement described by a sum of two or more periodic functions whose periods have ratio equal to an irrational number. The majority of quasicrystals are found to have the icosahedral symmetry, however, octagonal and decagonal quasicrystals have also been found. Ultimately it was established that Shechtman and coworkers discovered new phase called *icosahedral phase* which opened up a new field of quasicrystals. Shechtman was awarded 2011 Nobel Prize in Chemistry for the discovery of quasicrystals.

So far different structures of solids have been discussed but the question arises as to why different solids crystallize in different structures? The second question is what are the different types of crystalline solids? The answer to these questions can be found from the properties of their constituent atoms: atomic electronic structure (shell structure), size and chemical bonding of atoms. When a solid in the molten state is allowed to cool slowly, the atoms settle down in the equilibrium positions to form a periodic structure. Therefore, a solid acquires that crystal structure which yields minimum energy. Theoretically cohesive energy tells us about the equilibrium structure of a solid with the lowest energy. *Cohesive energy* of a solid is the energy required to separate the atoms in the form of a gas. On the other hand, *binding energy* is the energy of formation of a solid from its atoms and thus is equal in magnitude to the cohesive energy. The cohesive energy itself does not have much significance and is not easy to measure experimentally. The cohesive forces can be classified into homopolar, heteropolar, metallic and van der Waals forces and these determine the type of a crystalline solid. Ionic solids exhibit heteropolar binding. Metallic and covalent bindings are homopolar in nature. Many of the solids exhibit mixed binding that is mixture of homopolar and heteropolar bindings. In inert gas crystals the atoms are electrically neutral and possess nearly spherical shape. The only force of attraction that exists in them is van der Waals interaction, given by Johannes Dederik van der Waals (1837–1923), which varies as $1/r^6$. The van der Waals attraction between two atoms arises due to the mutual polarization of the atoms. As the atoms come closer their wave functions start overlapping and experience overlap repulsive interaction in addition which varies as $1/r^{12}$. In 1924, Sir John Edward Lennard–Jones (1894–1954) proposed a potential for liquid argon (inert gas crystal) called Lennard–Jones potential which works fairly well in explaining the properties of inert gas crystals.

Around 1830, Michael Faraday's experiment on electrolysis showed that certain substances when dissolved in water would conduct electricity. He thought that the passage of electricity caused the substances to break into charged particles. In 1884, Svante August Arrhenius (1859–1927) argued that an ion is an atom carrying positive or negative charge. He proposed that a compound like sodium chloride broke up into sodium and chloride ions when dissolved in water and it is these ions that conduct electricity. So, sodium chloride is made of sodium and chloride ions which are bound together by electrostatic forces forming an ionic solid. In fact, the idea of ionic bonding developed gradually over the years. In 1897, Joseph John Thomson (1856–1940) discovered electrons and was of the view that electrons were somehow involved in bonding. In sodium chloride crystal, one electron of sodium atom is permanently transferred to chlorine atom forming positively charged sodium Na^{+1} ion and negatively charged chloride Cl^{-1} ion and these are bound through electrostatic attraction. The structure of sodium chloride is formed by arranging different ions in such a way as to obtain minimum energy. Ionic bond is formed between two atoms with sharply different electronegativities and is not directional. Second type of strong bond is covalent bond. In 1874, Jacobus Henricus van't Hoff (1852–1911) and Joseph Achille Le Bel (1847–1930) discovered the tetrahedral chemical bonds of carbon atoms in organic compounds. In 1916, Gilbert Newton Lewis (1875–1946) suggested that a chemical bond is formed by a pair of two electrons shared by two atoms. In 1919, Irving Langmuir (1881–1957) used the word covalence but in 1939 he used the term covalent bond for this type of bonding. Covalent bond is formed between atoms of similar electronegativities and is highly directional. In 1927, Walter Heinrich Heitler (1904–1981) and Fritz Wolfgang London (1900–1954) gave the first successful quantum mechanical explanation of covalent bonding.

In metals the chemical bonding is entirely different from the ionic and covalent bonding. In 1900, Paul Karl Ludwig Drude (1863–1906) first proposed the concept of metallic bonding. According to him a metal consists of a sea of conduction electrons with ions embedded at the lattice positions. Ions are screened by the conduction electrons and these screened ions interact with one another yielding what is called metallic bonding. Metallic bond is not at all directional rather it is isotropic in nature (homopolar bond). In 1930, John Slater (1900–1976) used Heitler-London theory to calculate cohesive forces in Na metal but this method is complicated. In 1933, Eugene Paul Wigner (1902–1995) and Federick Seitz (1911–2008) gave a method most suited for the metallic bonding but at that time it was not the most accurate method. In the same year of 1933, Hugh Stott Taylor (1890–1974) and coworkers had calculated the binding energies of Na, Cu and H arranged in various configurations. Along with the chemical bonding, atomic size and valence also play an important role in determining the crystal structure.

The naturally occurring crystalline solids possess definite finite size and in a finite crystalline solid there is a transition from the bulk crystal to vacuum at the surface. In an ideal surface there is a translational symmetry in its plane. But due to discontinuity in atomic structure perpendicular to the surface it is expected that a few atomic layers near the surface are strained: there are changes in force constants and lattice parameter near the surface. From the early days the method of *cleavage* is used to create crystalline surfaces and was first used in geology and mineralogy. Cleavage means there is tendency of crystals, minerals and rocks to split along a definite weak crystalline plane in two or more parts thereby creating smooth crystalline surfaces. The relatively weak planes are a result of regular periodic arrangement of atoms in a crystal. The classic examples of cleavage are cutting of diamonds and making layers of mica that seem like pages of a book. In 1968, Alfred Yi Cho (1937–present) and John R. Arthur Jr. (1931–present) invented *molecular beam epitaxy method* for creating crystalline surfaces. In 1970s, they further developed this method to enable the controlled deposition of single atomic layer. The term epitaxy comes from the Greek letters, *epi* means above and *taxis* means in ordered manner. Epitaxy refers to the deposition of a layer of atoms on a crystalline substrate. In the epitaxial deposition, there are some restrictions on the structures and lattice parameters of the substrate and the deposited layer. As has been mentioned earlier the structure of the surface is determined by LEED and direct imaging methods using TEM, STM, AFM etc.

Bibliography

Binnig G, Rohrer H (1982) Scanning tunneling microscope. U. S. Patent No. 4343993A issued on 10 Aug 1982; Binnig G, Rohrer H, Gerber C, Weibel E (1983) 7 × 7 reconstruction on Si(111) resolved in real space. Phys Rev Lett 50:120; Binnig G, Quate CF, Gerber C (1986) Atomic force microscope. Phys Rev Lett 56:930

Bragg WL (1912) The specular reflection of X-rays. Nature 90:410

Bragg WH, Bragg WL (1913) The structure of the diamond. Proc Roy Soc London A89:277–291

Bravais A (1850) Memoire sur les systèmes formés par des points distributes regulierement sur un plan ou dans l'espace. Journal de l'Ecole Polytechnique 33:1–128; (1851) Etudes crystallographiques. Journal de l'Ecole Polytechnique 20:101–276

Brockhouse BN (1955) Energy distribution of neutrons scattered by paramagnetic substances. Phys Rev 99:601

Brockhouse BN, Stewart AT (1955) Scattering of neutrons by phonons in an aluminum single crystal. Phys Rev 100:756; (1958) Normal modes of aluminum by neutron spectrometry. Rev Mod Phys 30:236

Cho AY, Arthur JR (1975) In: Somorjai G, McCaldin J (ed) Progress in solid state chemistry, vol 10. Pergammon Press, New York, pp 157–191

Debye P, Scherrer P (1916) Interferenzen an regellos orientierten teilchen im Röntgenlicht. Nachrichten von der Königl. Gesellschaft der Wissenschaften zu Göttingen, Mathematisch-Physikalische Klasse 29:1–15

Davisson CJ, Germer LH (1927) Diffraction of electrons by a crystal of nickel. Phys Rev 30:705; (1927) The scattering of electrons by a single crystal of nickel. Nature 119:558–560

De l'isle JL R (1783) Cristallographie. Tome Troisieme, A Paris, De L'Imprimerie de Monsieur

Fourier JBJ (1807) Théorie de la propagation de la chaleur dans les solides (Manuscript submitted to the Institute of France), extracted from: (1811) Mémoires del'Académieroyale des sciences del'Institute de France 4:185–556

Frank FC, Kasper JS (1958) Complex alloy structures regarded as sphere packings. I. Definitions and basic principles. Acta Crystallogr 11:184–190
Freedman SJ, Clauser JF (1972) Experimental test of local hidden-variable theories. Phys Rev Lett 28:938–941
Friedrich W, Knipping P, Laue M (1912) In sitzungsberichte der math. Phys. Klasse (Kgl.) Bayerische Akademie der Wissenschaften 303; (1912) Interferenz-erscheinungen bei Rontgenstrahlen, Bayer. Akad. d. Wiss., Sitzungsber., p 314; Friedrich W, Knipping P (1912) Sitz.ber. Bayer. Akad. Wiss., pp 311–322
Hauy RJ (1801) Traite de Mineralogie. A PARIS, Chez Louis, Libraire, Rue de Savoye
Heitler WH, London FW (1927) Wechselwirkung neutraler atome und homöopolare bindung nach der quantenmechanik. Z Phys 44:455–472
Jones JE (1924) On the determination of molecular fields. I. From the variation of the viscosity of a gas with temperature. Proc Roy Soc Math Phys Eng Sci 106A:441–462; (1924) On the determination of molecular fields. II. From the equation of state of a gas. Proc Roy Soc Math Phys Eng Sci 106A:463–477; Lennard-Jones JE (1929) The electronic structure of some diatomic molecules. Trans Faraday Soc 25:668–686; Lennard-Jones JE (1931) Math Proc Camb Philos Soc 27:469–480
Knoll M, Ruska E (1932) Beitrag zur geometrischen elektronenoptik I. Ann Phys 404:607–640; (1932) Beitrag zur geometrischen elektronenoptik II. Ann Phys 404:641–661; (1932) Das elektronenmikroskop. Z Phys 78:318–339
Levine D, Steinhardt PJ (1984) Quasicrystals: a new class of ordered structures. Phys Rev Lett 53:2477–2480
Miller WH (1839) A treatise on crystallography. Pit Press for J. and J. J. Deighton
Röntgen W (1895) Ueber eine neue Art von Strahlen. Vorläufige Mitteilung. In: Aus den Sitzungsberichten der Würzburger Physik.-medic. Gesellschaft Würzburg, pp 137–147; (1896) Eine neue Art von Strahlen. 2. Mitteilung. In: Aus den Sitzungsberichten der Würzburger Physik.-medic. Gesellschaft Würzburg, pp 11–17; (1897) Weitere Beobachtungen über die Eigenschaften der X-Strahlen. In: Mathematische und Naturwissenschaftliche Mitteilungen aus den Sitzungsberichten der Königlich Preußischen Akademie der Wissenschaften zu Berlin, pp 392–406
Shechtman D, Blech I, Gratias D, Cahn JW (1984) Metallic phase with long range orientational order and no translational symmetry. Phys Rev Lett 53:1951–1953
Shull CG, Wollan EO (1948) X-ray, electron, and neutron diffraction. Science 108:69–75
Slater JC (1930) Cohesion in monovalent metals. Phys Rev 35:509
Sohncke L (1879) Die Entwicklung einer Theorie der Krystallstruktur [The development of a theory of crystal structure]. Leipzig, Germany
Steno N (1669) De solido intra solidum naturaliter contento dissertationis prodromus. Florence, Insegna della Stella
Taylor HS, Eyring H, Sherman A (1933) Binding energies in the growth of crystal nuclei from metallic atoms. J Chem Phys 1:68–76
Voigt W (1910) Lehrbuch der Kristallphysik (Reprinted in 2018 by Forgotten Books)
Wigner EP, Seitz F (1933) On the constitution of metallic sodium. Phys Rev 43:804; (1934) On the constitution of metallic sodium. II. Phys Rev 46:509
Wollan EO, Borst LB (1945) Physics section III monthly report for the period ending December 31, 1944, Oak Ridge, Tennessee, Metallurgical Project Report No M-CP-2222
Wollan EO, Shull CG (1948) Neutron diffraction and associated studies. Nucleonics 3:17–31; (1948) Neutron diffraction and associated studies. Nucleonics 3:8

Chapter 6
Structure of Amorphous Solids

Amorphous solid is a relatively new and general term applicable to any solid having non-periodic distribution of atoms. In fact, in amorphous solids the atoms are not completely randomly distributed but exhibit short range order in the form of regular coordination number. Originally amorphous solids were termed collectively as glasses, the only known amorphous solids in early times. Later a number of other amorphous materials such as polymers, pure amorphous elements and other inorganic amorphous solids were found. Mainly four techniques are used for synthesizing amorphous solids depending on the range of quenching rates and these are slow cooling under restricted environments, moderate quenching, rapid "splat quenching" and condensation from the gas phase.

A solid in the molten state when cooled slowly, its atoms or molecules get sufficient time to settle in the lowest energy state yielding crystalline solids. On the other hand, when the molten form of a solid is cooled much faster, the atoms or molecules do not get sufficient time to settle in their equilibrium positions rather the random distribution of atoms or molecules is frozen into the amorphous solid state. Therefore, the amorphous state is not the equilibrium state of a solid but still it is highly stable. In 1969, David Turnbull (1915–2007) gave the following statement in one of his review papers, "*Nearly all materials can, if cooled fast enough and far enough, be prepared as amorphous solids*". The temperature at which amorphous solid/glass is formed is called glass transition temperature. The liquid to glass transition occurs in a narrow range of temperature around the glass transition temperature rather than the sharp melting point observed in crystalline solids (see Fig. 6.1).

As regards the structure, amorphous solids can be classified into three categories:

1. In 1932, William Houlder Zachariasen (1906–1979) proposed *continuous random network* (CRN) model, the first model of glass structure, for inorganic covalent glasses particularly the oxide glasses. In the CRN model, a network is formed by the atomic sites which are connected by lines representing the bonds between the atoms. The network exhibits a small spread in the bond lengths but

J. S. Galsin, *History of Solid State Physics*,
https://doi.org/10.1007/978-981-95-0504-3_6

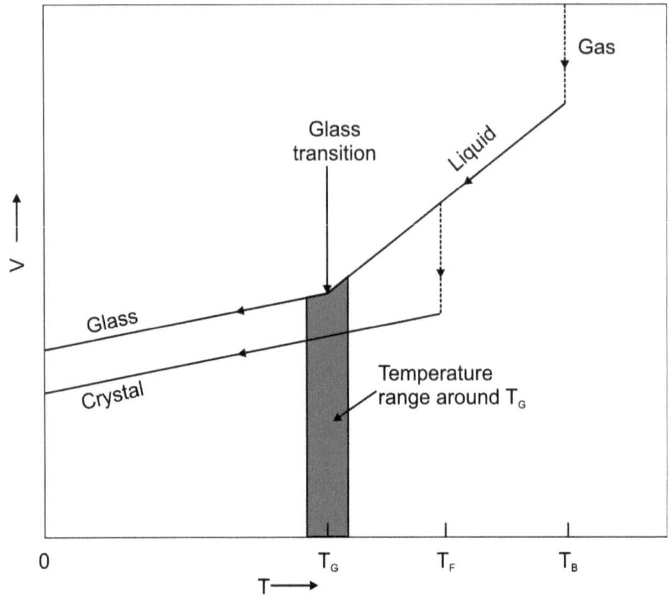

Fig. 6.1 Variation of volume V as a function of temperature T as the material is cooled from the gaseous state to the solid state. Crystal is formed by slow cooling while glass is formed by rapid cooling. T_G, T_F and T_B represent glass temperature, freezing temperature and boiling temperature

a significant spread in the bond angles. The CRN model could explain satisfactorily the structure of oxide glasses and also of amorphous silicon. To illustrate the structure of CRN consider honeycomb lattice of carbon atoms and decorated honeycomb lattice of a binary compound shown in Fig. 6.2. Here the bond length and bond angles are constant. Figure 6.3 shows the CRN model corresponding to honeycomb and decorated honeycomb lattices. In the CRN model, one can construct the WS cell but its shape and volume show significant spread.

2. Second is *random close packing* (RCP) model in which the atoms are distributed randomly on the microscopic scale and form close packing structure: randomness and close packing are two important ingredients. One most suitable example of RCP model is peas in a pot (Fig. 6.4). RCP provides most satisfactory model for the structure of amorphous metals and simple metallic glasses. In RCP model the packing fraction is found to be 0.637 which is less than the close packing fraction in crystalline solids having value 0.74. In 1965, John Desmond Bernal (1901–1971) showed experimentally the existence of RCP structure usually called "*Bernal structure*".
3. Third is *random coil model* (RCM) which is most suited for the structure of polymers. In polymers, the subunits are long chain molecules which are flexible and get tangled very easily. The molecular chains often contain bulky side group of atoms which interfere with the process of close packing. Therefore, polymers

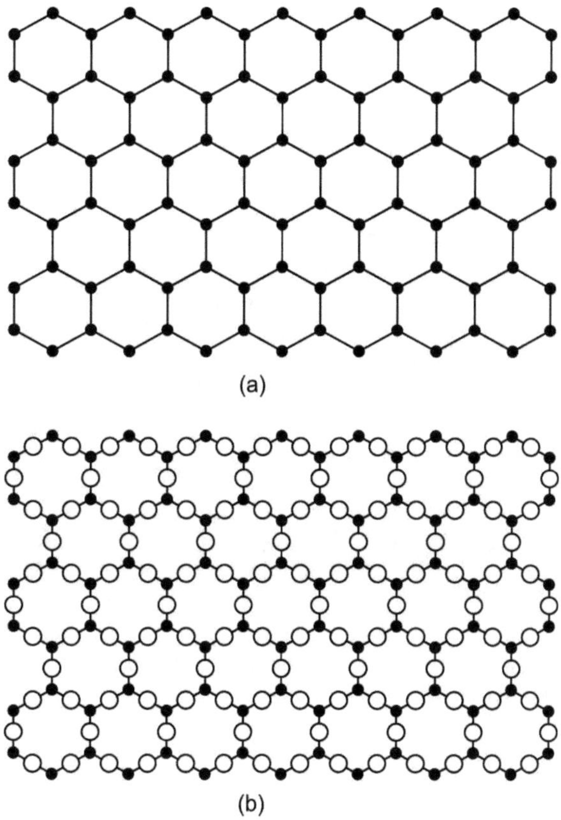

Fig. 6.2 Two-dimensional **a** honeycomb lattice of carbon atoms and **b** decorated honeycomb lattice of a binary compound

have rather loosely packed structures and hence favor the formation of amorphous structures.

Little is known about the first attempts to make glass. It is generally believed that the first manmade glass surfaced around 4000 BC in Mesopotamia, Syria and Egypt. One of the stories goes that some people were cooking food on a beach when sand beneath the fire melted which on cooling hardened into glass. Some people argued that cooking fire cannot reach the melting temperature of sand and this story seemed to be concocted. Around 2000 BC beads were the earliest known glass objects created as a by product of metal-working (slag). Naturally occurring glass, especially the volcanic glass obsidian, had been used by stone-age societies across the globe for making sharp-edged cutting tools. Due to the limited availability of glass, the glass products remained a luxury in early times and these were extensively traded only in certain regions. By 1500 BC glass began to be produced extensively in West Asia, Crete, Egypt and Mycenaean Greek. The first glass making manual was discovered in the library of King Assurbanipal (668–627 BC) of Assyrian. Scientists believe that the ability of glass making developed over a long period of time. By mixing metallic oxides to the glass melt, glass makers were able to produce colored ingots of glass,

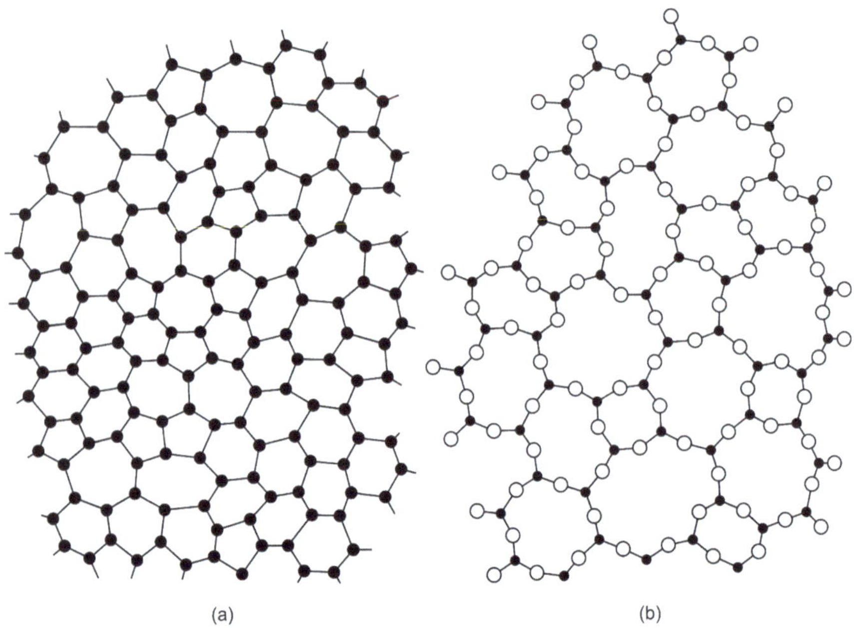

Fig. 6.3 Schematic representation of CRN model of **a** honeycomb lattice and **b** decorated honeycomb lattice

Fig. 6.4 Peas in a pot

the popular colors being royal blue, turquoise blue and red etc. The ingots of glass then used to be sent to workshop where these were remelted and casted into vessels and beads in moulds. During the 100 BC, glass blowing technique was discovered revolutionizing the glass industry. With the glass blowing technique the possibilities of shaping glass into the desired forms was endless. Glass could be blown into a mould or shaped in a completely free-form. In the eighteenth century glass cutting came into fashion. As this technique was perfected, great richness of effect of glass surface became possible: the surface of glass can be deeply cut to reflect light.

The CRN model is most suitable for representing the structure of amorphous semiconducting materials. In 1870, Carl Schulz-Sellack (1844–1879) for the first time performed experiment on glasses without oxide. Wood (1902) and Meier (1910) studied for the first time optical properties of vitreous Se. In 1932, Wilhelm Blitz (1877–1943) and Werner Fischer (1898–1947) at the University of Hannover, Germany, proposed the name *chalcogens* to the VIth group elements O, S, Se, Te, Po. Here O and S are nonmetals, Se and Te are metalloids and Po is a metal. The word chalcogen has been derived from the Greek word Chalcos (ore formers). The glasses formed by mixing the chalcogens S, Se, Te with Ga, In, Si, Ge, Sn, As, Sb are called *chalcogenide glasses*. These are covalently-bonded glasses and their structure can be well represented by the CRN model: amorphous semiconductors. Chalcogenide glasses are molecular structures and can be regarded as glassy semiconductors. Similar to other glasses, chalcogenide glasses exhibit the glass transition temperature. In 1950, Frerichs (1901–1982) performed the study on chalcogenide glass As_2S_3 and also investigated the development of Se glasses and binary compounds with sulfur. Fraser (1953) and Dewulf (1954) also performed studies on chalcogenide glasses. Goriunova and Kolomiets (1955) revealed the various semiconducting properties of chalcogenide glasses. It has been found that mainly S-based chalcogenide glasses are strong. Glass forming ability decreases with the increasing molar weight of constituent elements. Therefore, the strength of glasses decreases as one goes from S to Te through Se. In 1968, Stanford R. Ovshinsky (1922–2012) studied electronic properties of chalcogenide glasses particularly the switching effects. Mott (1969) proposed a rule in which coordination number z in chalcogenide glasses is given by

$$z = 8 - n \tag{6.1}$$

where n is the total number of s- and p-electrons in the outermost shell of atoms. The "*eight minus n rule*" explains the behavior of amorphous semiconductors. Actually, it was proposed to account for the observation that the electrical properties of chalcogenides do not exhibit great sensitivity to impurities and composition changes of a few percent. In 1979, Nevill Francis Mott and Edward A. Davis wrote a classical book entitled, "*Electronic processes in noncrystalline materials*" in which they explained the electronic processes in amorphous solids.

In chalcogenide glasses, there exists a short-range order (at the most up to second nearest neighbors). The CRN is considered to be the most suitable model for chalcogenide glasses which are covalently-bonded semiconductors. The structure of S in the solid, liquid and gaseous states is complex exhibiting S_2, S_3, S_6, S_8, S_9, S_{10}, S_{11}, S_{12}, S_{18}, S_{20} structures but out of these the most stable structures is S_8, a ring of eight atoms: zero dimensionality. It is found that Se possesses one-dimensional structure in the form of long helical chain structure. On the other hand chalcogenide glasses As_2S_3, As_2Se_3, GeS_2, $GeSe_2$ possess two-dimensional layered structures. When the number of components in chalcogenide glasses is increased this tends to widen the glass forming region. In general, the structure of chalcogenide glasses depends on the system, concentration and temperature. Even today the knowledge of structure of different chalcogenide glasses is incomplete, which is a hindrance in the progress of chalcogenide glass science. In 1976, Walter Eric Spear (1921–2008) and Peter George Lecombor (1941–1992) doped the chalcogenide glasses with metals and this widened the glass forming region. The added metallic atoms were named additives or modifiers.

The arsenic-based chalcogenides are the most studied glasses as these are most stable. These glasses possess special properties such as high refractive index, photosensitivity, wide transparency range, low optical losses and these properties make them suitable for optical fiber. Chalcogenides have large transparency in the infrared (IR) region, low thermal conductivity, and high polarizability which make them very attractive. The first report on their application in optical fiber was published by Kapany and Simms (1965). Chalcogenide glasses can be used for making optical lenses and optical fibers. Thin films of chalcogenides are used as optical sensors or phase-change memories. The electronic properties of chalcogenide glasses are not completely understood. For example, "Why holes are more mobile than electrons in chalcogenide glasses?". At present, there is no satisfactory answer to this question. The study of electronic properties of chalcogenide glasses had been carried out throughout the second half of the twentieth century and beyond.

The discovery of amorphous Si is relatively recent and Polk (1971) proposed that its structure is given by the CRN model. In this model, amorphous Si has constant bond length but there is significant spread in bond angles. Further, amorphous Si has no dangling bonds. Bell and Dean (1972) used the CRN model to explain the amorphous structure of SiO_2. In amorphous SiO_2, the Si-O-Si bonds exhibit spread in bond angles but the O-Si-O bonds have constant bond length and bond angles. The experiments indicated that CRN model could not explain the structure of melts and glass containing modified cations. Therefore, Greaves and coworkers (1981) proposed a modified random network model to explain the structure of these glasses.

Metallic glasses constitute an important component of amorphous solids. Metallic glass is either an amorphous metal or an amorphous metallic alloy. Unlike ordinary glasses, metallic glasses possess reasonably high value of electrical conductivity and some of these may also display superconductivity at low temperatures. In 1956, Werner Buckel (1920–2003) and Rudolf Hilsch (1903–1972) discovered superconductivity experimentally in amorphous metal thin films. It was found that the superconducting transition temperature was higher in the amorphous state than in the crystalline state of the metal. In 1960, Klement (Jr.), Willens, and Duwez made the first gold-based metallic glass with composition $Au_{0.75}Si_{0.25}$. Metallic glasses are formed by extremely rapid cooling of the material in the molten state. The metallic glasses can be classified in to two categories:

1. Nonferromagnetic metallic glasses which are composed of elements Cu, Au, Ca, Mg, Pd, Pt, Ln, Zr, and Ti etc.
2. Ferromagnetic metallic glasses which are composed of ferromagnetic elements Fe, Co, Ni such as $Fe_{0.8}B_{0.2}$, $Fe_{0.7}P_{0.2}C_{0.1}$, $Co_{0.8}Fe_{0.1}B_{0.1}$. Ferromagnetic glasses exhibit high saturation magnetization and low loss properties. Another useful

property is that these are magnetically soft but mechanically hard. Because of these properties ferromagnetic glasses are used in the magnetic cores of power transformers.

For the formation of metallic glasses the atomic radii of the component elements has to be significantly different (over 12%) to achieve the high packing density and low free volume. Ichikawa (1973) compared the electron diffraction patterns of crystalline and amorphous Fe metal and found that the diffraction patterns were similar but comparatively diffused diffraction pattern was obtained in the amorphous state. Waseda and Masumoto (1975) obtained the X-ray diffraction pattern of crystalline and amorphous forms of ferromagnetic material $Fe_{0.8}P_{0.13}C_{0.07}$ and found the same trend as was observed by Ichikawa. Cargill (1975) made the comparison between experiment and structure models in case of metallic glasses and found that RCP model describes satisfactorily their structure. Initially metallic glasses were formed by rapidly cooling the melt to avoid crystallization. The rapid cooling process had the limitation that glasses could be produced only in the form of ribbons, foils and wires with thickness less than one hundred micrometers. Libbermann and Graham (1976) developed a new method for manufacturing thin ribbons of amorphous metal, suitable for commercialization, on a super-cooled fast-spinning wheel. In 1990s, bulk metallic glasses, with several cm thickness (maximum thickness depending on the alloy), were synthesized and in 1992, commercial amorphous alloy called Vitreloy was made. In 2004, bulk amorphous steel (glassy steel) was successfully produced. Later improved transformers with core of ferromagnetic glasses were designed for minimum energy loss.

A notable feature of metallic glasses is that some of these exhibit anomalous behavior as regards the electrical resistivity. First, the electrical resistivity of amorphous form of a metal is much greater than that of the same solid in the crystalline form. Second, Mooij (1973) found that the electrical resistivity of a number of amorphous metals decreases with the increase in temperature, a behavior opposite to that found in crystalline metals. It is called *Mooij rule*.

Polymer science is a relatively new discipline which deals with different types of plastics. There are basically two types of polymers: natural and synthetic polymers. Natural polymers such as natural rubber, amber, wool, silk, cellulose, hemp, shellac etc. had been used as essential components of commodities since the early days of mankind. On the other hand, synthetic polymers have a long list including polyethylene (commonly known as polythene), polypropylene, polystyrene, synthetic rubber, Bakelite, synthetic fibers, adhesive sealants etc.: all of these materials today have become very common. It has been found that carbon atoms form the backbone of the polymers. Many other materials do exist in amorphous structure like amorphous silicon and familiar materials like silicones. Silicones are synthetic polymers which comprises of chains of alternate silicon and oxygen atoms with some organic groups attached to the silicon atoms. Such compounds are resistant to chemical attacks, insensitive to temperature changes and are used to make rubber, plastics, polishes and lubricants etc.

In 1830s, Henri Braconnot (1780–1855) created an inflammable material named xyloidine which is considered the first polymer. Henri Braconnot and Christian Friedrich Schonbein (1799–1868) and coworkers developed derivatives of natural cellulose producing new synthetic materials such as celluloid and cellulose acetate. In 1833, Jons Jakob Berzelius (1779–1848) coined the word "*Polymer*" though he did little work on polymers. Polymer is a combination of two Greek words: *poly* means many and *mer* means part. Sometimes natural polymers are modified by adding in them some other suitable element. The ways in which polymers can be modified include oxidation, cross-linking and end capping. In 1840s, Friedrich Wilhelm Ludersdorf (1801–1886) and Nathaniel Hayward (1808–1865) independently discovered that stickiness of rubber can be prevented by adding sulfur to it. In 1844, Charles Goodyear (1800–1860) vulcanized natural rubber with sulfur and received its US patent. But Thomas Hancock (1786–1865) received the UK patent for the same process of vulcanization a year before. The discovery of vulcanization of rubber allowed rubber to withstand heat and cold thus revolutionizing the rubber industry. The vulcanized rubber represents the first successful commercial product of polymer science. In 1856, Alexander Parkes (1813–1890) patented a material named Parkesine, celluloid based on nitrocellulose treated with a variety of solvents. This material was exhibited, in 1862, in London International Exhibition. In 1869, John Wesley Hyatt (1837–1920) invented the first synthetic polymer celluloid by adding camphor to nitrocellulose (a modified natural polymer) when he was asked by some New York firm to provide a substitute for ivory to produce billiard balls. Hyatt is mainly known for discovering the process of polymerization and simplifying the production of celluloid. Polymerization is the process of combining many small molecules, known as monomers, into a covalently bonded chain or network. By 1900, celluloid was already in use for movie films. In 1884, Hilaire de Chardonnet (1839–1924) installed first artificial fiber plant to produce artificial silk. This plant used regenerated cellulose or viscose rayon as a substitute for silk. But the artificial silk was highly inflammable.

In 1907, Leo Hendrik Baekeland (1863–1944) developed phenolic resin as the first true polymer named *Bakelite*: a thermoset plastic. But its official announcement was made in 1909 at a meeting of the New York section of American Chemical Society. The discovery of Bakelite marked the beginning of the age of plastics. Hendrik Baekeland is called "*the father of plastic industry*". After 1930, polymer science entered a "golden age" during which new types of polymers were discovered such as star polymers, graft polymers, copolymers, block copolymers, branched polymers and elastomers etc. During this period, more efficient polymerization techniques were invented such as step-growth polymerization, chain polymerization and plasma polymerization. The improved techniques for polymer characterization were invented and advanced theoretical understanding of polymers was achieved. In 1835, Justus von Liebig (1803–1873) and Henri Victor Regnault (1810–1878) reported the existence of vinyl chloride but its ability to polymerize was not fully known at that time. In 1872, Eugen Baumann (1846–1896) synthesized polyvinyl chloride (PVC) after extensive experimentation which appeared as a white solid. In 1926, Waldo

Semon (1898–1999) invented plasticizers which are substances that prevent crystallization of polymers by keeping the chains separate from one another. Waldo Semon was the first to plasticize PVC. The discovery of plasticized PVC was responsible for the commercial success of PVC and it was the second best-selling plastic. In 1938, Roy J. Plunkett (1910–1994) accidently invented Teflon unlike most of the other polymer products. He found that Teflon is heat resistant, chemically inert and have low surface friction so that most of the substances would not adhere to it. Today there are so many plastic materials it is not possible to mention all.

Another important class of polymers is elastomers and this term is often used interchangeably with rubber. Elastomers exhibit a large and reversible extensibility (5–700%) at room temperature. Elastomers comprise of very long atomic chains with wide-meshed cross-linking between them. The atomic chains may have many bends in them and at room temperature these are in a state of constant motion. The elasticity is derived from the ability of the long chains to reconfigure them to distribute an applied stress. The covalent cross-linkages ensure that the elastomer will return to its original configuration when the stress is removed. Without the cross-linkages the applied stress would result in a permanent deformation. Rubber is the first elastomer found in nature. In 1516, Pietro Martire d'Anghiera (1457–1526) made the first mention of rubber in his book, "*De Orbe Nuovo dcades cum legatione babylonica*". The source of natural rubber was the large forest tree "*Haveea Brasiliensis*" which occurs in Malaysia, Indonesia, Thailand, Ceylon, India, Cambodia, Vietnam, Africa, and Brazil etc: southern equatorial region. In 1932, a staff member Arnold Miller Collins (1899–1982) working under Wallace Hume Carothers (1896–1937) isolated chloroprene, a liquid which polymerized to produce a solid material that resembled rubber. This was the first synthetic rubber made in U.S.A. and today it is known as Neoprene. In the same year 1932, Jullian Werner Hill (1904–1996), another stall member of Carothers, produced polyester called first synthetic silk. In 1935, Gerard Berchet (1902–1990), working under the direction of Carothers, created what is known as nylon. Elastomers are usually thermosets (requiring vulcanization) but also thermoplastics. The history of thermoplastic materials began with the invention of plasticization of PVC. This invention created interest in flexible plastics or thermoplastic elastomers (TPEs). In 1960s TPEs were introduced commercially and soon they gained popularity because of their unique ability to blend the properties of rubber and plastics. In 1980, Lord Todd (1907–1997) said I am inclined to think that the development of polymerization is perhaps the biggest thing that chemistry has done, where it has had the biggest effect on everyday life.

Although the synthesis of polymers began in the first half of the nineteenth century but the nature of these species was not known. In 1861, Thomas Graham (1805–1869) proposed that cellulose and other polymers were colloidal aggregates of small molecules connected by unknown intermolecular forces. In 1920, Hermann Staudinger (1881–1965) published his work in a book, "*Uber Polymerization*" and he was the first to propose that polymers consisted of long chains of atoms, usually called long-chain molecules, held together by covalent bonds. The long-chain molecules are composed of many repeating subunits bonded by covalent bonds. The large molecular mass, relative to the small molecule compounds, produced unique physical properties including toughness, high elasticity, viscoelasticity, and a tendency for amorphous and semicrystalline structures. In 1922, Herman Mark (1895–1992) performed X-ray diffraction studies on polymers which provided direct evidence of long chains of repeating molecular units. High molecular weights of polymers were also confirmed by membrane osmometry and by Staudinger's measurements of viscosity in solutions done in 1933. Staudinger's work took a decade to gain wide acceptance in the scientific community due to lack of experimental evidence for his hypothesis. Staudinger was awarded 1953 Nobel Prize in Chemistry for his work on polymers. Today a large number of polymers are being produced as these are being used in our daily life. Being nondestructible the problem of their disposal causes environmental pollution on a large scale.

Bibliography

Bell RJ, Dean P (1972) The structure of vitreous silica: validity of the random network theory. Philos Mag 25:1381–1398

Bernal JD (1965) In: Hughel TJ (ed) Liquids: structure, properties, solid interactions. Elsevier, Amsterdam, p 25

Buckel W, Hilsch R (1956) Supraleitung und elektrischer widerstand neuartiger zinn-wismut-legierungen. Z Phys 146:27–38

Cargill GS (1975) Structure of metallic alloy glasses, Solid State Phys 30:227–320; (1975) Dense random packing of hard spheres as a structural model for noncrystalline metallic solids. J Appl Phys 41:2248–2250

Dewulf G (1954) Verres transparents dans l'Infrarouge" (Infrared-transmitting glasses). Rev Opt 33:513–518

Fraser WA (1953) A new triaxial system of infrared glasses. J Opt Soc Am 43A:823

Frerichs R (1953) New optical glasses with good transparency in the infrared. J Opt Soc Am 43:1153–1157

Goriunova NA, Kolomiets BT (1955) Properties and structure of ternary semiconducting systems, II. Electrical properties and structure of materials in systems of thallium, antimony and arsenic selenides. Zh Tekh Fiz 25:2069

Greaves GN, Fontaine A, Lagarde P, Raous D, Gurman SJ (1981) Local structure of silicate glasses. Nature 293:611–616

Hansen M (1958) Constitution of binary alloys, 2nd edn. McGraw-Hill, New York

Ichikawa T (1973) Electron diffraction study of the local atomic arrangement in amorphous gallium films. Phys Status Solidi A 19:707

Kapany NS, Simms RJ (1965) Recent developments in infrared fiber optics. Infrared Phys 5:69–76

Klement W, Willens RH, Duwez PL (1960) Non-crystalline structure in solidified gold-silicon alloys. Nature 187:869–870

Liebermann H, Graham C (1976) Production of amorphous alloy ribbons and effect of apparatus parameters on ribbon dimensions. IEEE Trans Magn 12:921–923

Meier W (1910) Untersuchungen über dispersion und absorption bei metallen für das sichtbare und ultraviolette spektrum. Ann Phys 31:1017–1049

Mooij JH (1973) Electrical conduction in concentrated disordered transition metal alloys. Physica Status Solidi 17(a):521–530

Mott NF (1969) Conduction in non-crystalline materials, III. Localized states in a pseudogap and near extremities of conduction and valence bands. Philos Mag 19:835–852

Mott NF, Davis EA (1979) Electronic processes in noncrystalline materials. Clarendon Press, Oxford

Ovshinsky SR (1968) Reversible electrical switching phenomena in disordered structures. Phys Rev Lett 21:1450; (1970) An introduction to ovonic research. J Non-Cryst Solids 2:99–106

Plunkett RJ, Evans WL (1938) The mechanism of carbohydrate oxidation. XXIV. The action of aldehydo-d-glucose and of aldehydo-d-galactose in alkaline solutions. J Am Chem Soc 60:2847–2852

Polk DE (1971) Structural model for amorphous silicon and germanium. J Non-Cryst Solids 5:365–376

Shultz-Sellack C (1870) Diathermansie einer reihe von stoffen für wärme sehr geringer brechbarkeit. Ann Phys 139:182–187

Spear WE, LeComber PG (1975) Substitutional doping of amorphous silicon. Solid State Commun 17:1193–1196

Staudinger H (1920) Über polymerisation. Ber Dtsch Chem Ges 53:1073–1085

Todd L (1980) Chem Eng News Arch 58:28–33

Turnbull D (1969) Under what conditions can a glass be formed? Contemp Phys 10:473–488

Waseda Y, Masumoto T (1975) Structure of amorphous Fe_{80}-P_{13}-C_7 alloy by X-ray diffraction. Z Phys B22:121–126

Wood RW (1902) LXIX. The absorption, dispersion, and surface-colour of selenium. Philos Mag 3:607–622

Zachariasen WH (1932) The atomic arrangement in glass. J Am Chem Soc 54:3841–3851

Chapter 7
Lattice Defects in Crystalline Solids

Naturally occurring elements are not in the pure form rather they are found to have different types of impurities and lattice defects. Even industrially prepared crystalline solids contain some impurities and defects which affect the various properties such as structure, and mechanical, electrical, magnetic, optical and thermal properties of crystalline solids. The lattice defects can be broadly classified into three categories: point defects (zero-dimensional defects), line defects (one-dimensional defects) and planar defects (two-dimensional defects). The line and planar defects are collectively called dislocations.

7.1 Point Defects

Point defects are lattice defects confined to an atomic site in a lattice or are of the order of atomic size. There are different types of point defects in a crystalline solid depending on their positions.

1. A crystalline solid consists of closely-packed atoms at the lattice positions and voids in between because without voids a solid will be a continuous material. The centers of the voids are called interstitial positions. Consider a pure crystalline solid with A type of atoms on the lattice. Let B atoms of a different element are added to this perfect crystalline solid. There are two possible positions which B atoms can occupy. First B atoms may knock out A atoms and occupy their positions, then the positions of B atoms are called substitutional positions and B atoms form *substitutional impurities* (Fig. 7.1). Secondly the B atoms may settle themselves at the interstitial positions resulting in *interstitial impurities* (mostly atoms of light elements form interstitial impurities, Fig. 7.1). Substitutional and interstitial impurities are called extrinsic impurities as these are added externally. In a perfect crystalline solid, if an atom is removed from the lattice site and is taken out of the crystal or to the surface, the defect so produced is called a

J. S. Galsin, *History of Solid State Physics*,
https://doi.org/10.1007/978-981-95-0504-3_7

vacancy: an intrinsic substitutional defect. In fact, vacancies are produced at a finite temperature and can diffuse into the crystal. When the concentration of vacancies is large, some of the vacancies may form clusters of two, three or more, usually called divacancies, trivacancies etc. In 1935, Walter Hans Schottky (1886–1976) discovered that in an ionic crystal, vacancies always occur in pairs and each pair consists of one negatively and one positively charged vacancies and each pair of vacancies is called *Schottky defect*. Schottky was awarded the Royal Society's Hughes medal in 1936 for the discovery of this defect. Schottky defect is shown by ionic substances in which the cations and anions are almost of similar sizes, for example NaCl, KCl, and CsCl. An interstitial impurity always produces local expansion of the lattice around itself while a vacancy produces local contraction of the lattice. But a substitutional impurity may produce local expansion or contraction of the lattice around itself depending on the size and nature of the impurity.

Walter Hans Schottky (1886–1976), Public Domain, URL: https://en.wikipedia.org/wiki/Walter_H._Schottky#/media/File:Walter_Hermann_Schottky_(1886-1976).jpg

2. There are other types of point defects such as *Frenkel defects* and *color centers* in ionic crystals and these are also of much interest. In 1926, Jacov Frenkel (1894–1952) found that a lattice defect is produced as a result of migration of

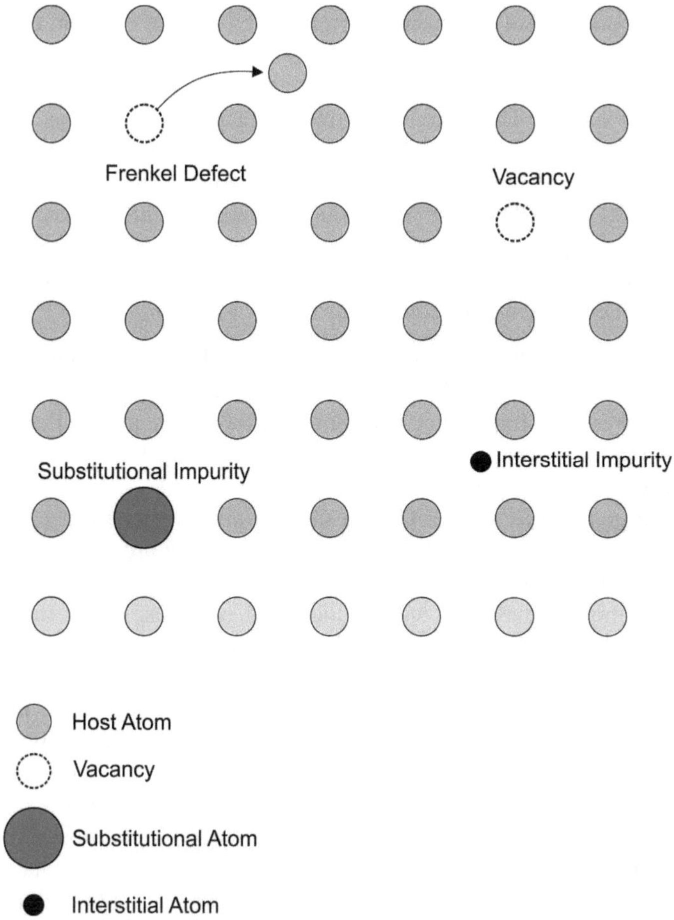

Fig. 7.1 Schematic representation of point defects in a crystalline solid

a lattice atom (ion) to the nearby interstitial position: the vacancy and atom (ion) collectively is called Frenkel defect (Fig. 7.1). The vacancy and interstitial ion (atom) so produced, under favorable environments, may move further away from each other ultimately yielding free vacancy and free interstitial atom. Color centers are created in ionic crystals when irradiated with X-rays or ultraviolet rays and these give deep color to the crystal. The simplest color center is the *F-center*: letter F is taken from Farbe, a German word, which means color. Color centers can also be created by adding foreign metallic atoms to an ionic solid, therefore, color center may be intrinsic or extrinsic in nature. Although color centers were discovered much earlier but the name color center was given for the first time, in 1930, by Rudolf Hilsch (1903–1972) and Robert Wichard Pohl (1884–1976).

A large number of solid substances are made by mixing different elements and among these alloys are worth mentioning. Alloy is a metallic solid obtained by mixing homogeneously, at the atomic level, a metal with one or more elements which may or may not be metals. The main aim of making alloys is to enhance the strength of various properties of metals such as mechanical strength, toughness, hardness, ductility, electrical and thermal conductivities etc. No date and year can be fixed for the discovery of alloys but these have been known since olden times. In antiquity, man was living in jungles using tools and weapons made from stones (Stone Age began around 2.6 million years ago to 4000BC). With the development of human civilizations, it entered into different ages.

1. Copper Age (4000BC–3000BC): Near about 3400BC, the Egyptians learned the process of extracting metals. The first metal found was copper although silver and gold were also found later. Copper being the hardest among these, was used for making tools, weapons and implements. Gold and silver were quite soft and were used more or less for decorative purposes.
2. Bronze Age (3000BC–2000BC): For the first time Sumerians accidently made an alloy by mixing copper and arsenic called arsenical bronze. Later the bronze alloy was made from copper and tin which was much harder than copper and therefore more suitable for making tools and weapons. About the same time brass, which is an alloy of copper and zinc, was also invented.
3. Iron Age (2000BC–500BC): The Iron Age was a long period of very slow development. Iron ore was found in this Age and its smelting began in Egypt. A mixture of iron oxide and impurities (iron ore) was heated to about 1500 °C and the molten iron was drawn off to one side while the slag (waste) on the other side. In 1600BC, Persians produced carbon steel but it was produced by hammering iron in the presence of carbon resulting in the nonuniform distribution of carbon in iron: a crude method. In 1200BC, China produced pig iron which is a very hard alloy of carbon and iron. In around 1100BC, humans came to know that the hardness of steel can be increased by heat treatment. But Europe was lagging behind and made pig iron in the middle ages (500AD–1500AD). In 1709, Abraham Darby (1677–1717) succeeded in smelting iron with coke. And in 1740, Benjamin Huntsman (1704–1776) melted blister steel in a crucible and mixed in it the carbon content homogeneously, thus creating the first process for mass production of steel that is most suitable for making tools. After this, steel became the world's most used material. In the second half of eighteenth century it was realized that carbon is an essential element in alloys for hardening. At that time in Europe, the process of tempering and hardening of iron-based alloys did not meet the required criterion. On the other hand, the craftsmen of Japan, Mesopotamia and India made swords of high quality steel. In 1856, Henry Bessemer (1813–1898) developed the process of steel-making by flowing hot air through the molten pig iron to reduce the carbon content. Even towards the end of nineteenth century the atomic structure of steel was not known: some scientists considered the grain structure while others assumed amorphous form of iron.

Depending on the number of elements (components), alloys are classified as binary (two components), ternary (three components), quarternary (four components) and quinary (five components) alloys etc. Each of these alloys can further be classified with regard to the nature of impurity and their properties. The classification of alloys can be understood by considering a binary alloy. The mixing of an element B (called impurity or solute) with a perfect crystalline metallic solid A (called host or solvent) such that the atoms of element B share various crystal sites of element A, produces a binary alloy. Generally, the concentration of impurity atoms is smaller than that of the host metal atoms. The binary alloy is generally denoted by $A_{1-C}B_C$ where C is the concentration of the impurity. When two or more elements are mixed together, some amount of heat called heat of mixing is required and there may be a change in molar volume of the alloy as compared to that of the host metal. If the heat of mixing is zero, there is no change in molar volume and the alloy so formed is called an *ideal alloy*. Alloys can be classified in a number of ways using different criteria.

William Hume-Rothery (1899–1968), CC BY-SA 4.0, URL: https://eehe.org.uk/wp-content/uploads/2023/09/William-Hume-Rothery-3.jpg

1. **Nature of the impurity**: If the impurity atoms occupy substitutional positions, the alloy formed is called *substitutional alloy*. On the other hand, if the impurity

atoms occupy interstitial positions the alloy formed is called *interstitial alloy*. Bronze was the first substitutional alloy formed by humans somewhere around 2500BC. In the Middle East, people began alloying copper with zinc to form brass. Ancient civilizations mixed elements basically to enhance the properties, most probably hardness and toughness of the material. Around 200BC, arrowheads from the Chinese Qin dynasty were often constructed with a hard bronze-head. In 1926, William Hume-Rothery (1899–1968) gave the following rules for the formation of dilute binary substitutional alloys $A_{1-C}B_C$:

(a) Both the elements in a substitutional alloy should have the same structure.
(b) The radii of the two types of atoms should be approximately the same (within about 15%).
(c) Besides the geometrical factors, the other factors such as valence should be the same.
(d) The two components of a binary alloy should have comparable values of electronegativity.

Small-sized atoms such as hydrogen, boron, nitrogen etc. form interstitial alloys as these atoms cannot knock out the bigger host atoms, but can easily accommodate themselves in interstitial positions in a crystalline solid. In 1930 and 1931, G. Hagg gave the idea of interstitial alloys (compounds) in transition metals and are called Hagg's phases. Hagg gave rules for the formation of interstitial alloys in terms of the radii of impurity atoms R_I and host atoms R_H. According to him if the ratio R_I/R_H is less than 0.59, interstitial alloys with simple structure such as fcc, hcp, hexagonal structure etc. are formed. But for R_I/R_H greater than 0.59, interstitial alloys with complex structure are formed.

2. **Structure**: If the structure of alloy is the same as that of the host metal, it is called *primary alloy*. On the other hand, if the structure is quite different from either of the two components, it is called *intermediate alloy or intermetallic compound*. The term intermetallic compound is used for systems that have long range atomic order and well-defined stoichiometry.
3. **Concentration**: If the concentration of the impurity is small, it is called *dilute alloy*. In a dilute alloy the structure is the same as that of the host metal. For large concentration of the impurity one obtains *concentrated alloy*. With the increase in impurity concentration, the structure of alloy gets distorted and ultimately it may acquire another crystal structure: most probably the structure of the element of impurity.
4. **Solubility**: If the two components of an alloy are soluble at all proportions, it is called *completely soluble alloy*, otherwise it is called *restricted alloy*. Most of the primary alloys exhibit restricted solubility. CuNi y is the common example of completely soluble alloy.
5. **Order**: In 1919, Gustav Heinrich Johann Apallon Tammann (1861–1938) suggested that under certain circumstances the atoms of an alloy may exhibit order in their positions on the host lattice forming an *ordered alloy*. In 1925, C. H. Johansson (1864–1943) and J. O. Linde showed experimentally that in

$Au_{1-C}Cu_C$ alloy there appeared new lines in the X-ray spectrum which is characteristic of the lattice formation in the Au atoms or Cu atoms. For example, in a binary alloy $A_{1-C}B_C$ with equal number of A and B atoms ($C = 0.5$), the B atoms show preference for being first nearest neighbors of A atoms and vice versa forming a superlattice: thus, the alloy may exhibit partial or complete order. One can define degree of order which decreases with the increase in temperature and ultimately disappears above certain temperature just like Curie temperature in ferromagnetic alloys. As expected the appearance of order in an alloy decreases its resistivity as compared to random distribution of atoms. If impurity atoms are randomly distributed on the lattice sites of a host metal, which is the case at very low concentrations of impurity, it is called *random alloy* and this resembles the solid solution.

6. **Physical Property**: Another criterion for the classification of alloys is with respect to a particular property exhibited by them, for example magnetic and nonmagnetic alloys and conducting and nonconducting alloys. In 1901, Carl Ludwig David Friedrich Heusler (1866–1947) discovered ferromagnetic intermetallics, known as Heusler phases, in which the constituent elements are nonmagnetic. In 1896, Charles Edouard Guillaume (1861–1938) discovered invar behavior in FeNi alloys. In the invar alloy, there is no thermal expansion, puzzling and illusive behavior for physicists. Richard M. Bozorth (1896–1981) developed a large number of soft and hard magnetic alloys including invar alloys. In 1951, Bozorth gave compilation work on ferromagnetism and ferromagnetic alloys.

The term *phase* refers to the region of space occupied by a physically homogeneous solid. The phase of an alloy changes with temperature, pressure and composition which was studied as early as in the nineteenth century. Before the twentieth century, only three elementary transitions were known: phase transition due to change in symmetry or order, sharp freezing and boiling points, and latent heat. In 1869, Thomas Andrews (1813–1885), an Irish chemist and physicist, investigated experimentally the continuous change of phase of carbon dioxide into liquid state above a critical point. He is considered to be the first scientist who gave the concept of critical point in physical systems. In 1871, Ludwig Boltzmann (1844–1906) introduced a prototype model of an ensemble that represented a polyatomic gas. In 1873, Johannes Diderik van der Waals (1837–1923) was studying the intermolecular forces and during this study he gave an equation of motion named *van der Waal's equation* which formed the part of his doctoral thesis. Van der Waal's equation is nothing but the extension of the molecular kinetic theory for ideal gases and it implied that liquid and gas phases of a real gas can coexist below certain temperature. The van der Waal's equation provided qualitative theoretical description of Andrews' experiment. The formulation of this equation was the first major step toward the development of the modern theory of phase transitions and critical phenomena. In 1901, Kamerlingh Onnes (1853–1926) generalized the van der Waal's equation by introducing a systematic 'virial expansion'. The calculation of the equation of state of an imperfect gas yielded a power series in particle density. The lowest term yielded equation of motion of an ideal gas and the second virial coefficient gave the van der Waal's equation. Josiah Willard Gibbs (1839–1903) did extensive work in the fields of thermodynamics and kinetic theory of gases. In 1874, James Clerk Maxwell (1831–1879), based on the work of Gibbs, constructed a three-dimensional plot of variables volume, entropy and energy for a given pressure and temperature. He constructed a phase diagram of a fictitious substance in which volume is greater in the solid phase than that in the liquid phase. Gibbs (1839–1903) published two papers one in 1875 and the other in 1878 in which he gave the famous *phase rule*, a thermodynamical relation between the variables of a solid in thermodynamic equilibrium, and is defined as

$$F_d = R_C - P_F + 2 \tag{7.1}$$

where F_d gives degree of freedom which is the number of independent variables like temperature, pressure and composition etc. R_C is the number of components in the solid and P_F is the number of phases that coexist at equilibrium. With the help of Gibbs phase rule, one can draw the *equilibrium phase diagram* which is a graph that gives information about the number of phases present at equilibrium and their relative amounts as a function of temperature, pressure and composition. In one component system that is in pure elements, one can obtain the triple point where all the three phases (solid, liquid and vapor) coexist (see Fig. 7.2). The two component alloys (binary alloys) exist in different forms like isomorphous alloys, eutectic alloys, eutectoid alloys, and peritectic alloys etc. and these exhibit different equilibrium phase diagrams. The phase diagrams of higher component systems are quite complex in nature. Gibb's theory must be viewed as the first general theory of phase transitions. In 1879, James Clerk Maxwell (1831–1879) defined an ensemble as copies of systems of the same energy and called his approach statistical one. Boltzmann and Maxwell regarded the ensemble theory as a much more general approach than the kinetic theory of gases. In 1880s, Ludwig Boltzmann (1844–1906) made use of ensemble theory to establish thermodynamical relations. In 1896, William Chandler Roberts-Austin (1843–1902) published the first true constitutional diagram of carbon steel. In 1903, Charles Thomas Heycock (1858–1931) and Francis Henry Neville (1847–1915) worked on the composition and constitution of alloys and produced superb constitutional diagram for bronze. This opened the way to subsequent great flood of constitutional work in the field of alloys and in 1958 M. Hansen gave nice compilation of constitution of binary alloys.

In 1919, Gustav Tammann (1861–1938) first put forward the hypothesis that under certain circumstances the atoms of an alloy segregate into regular positions. In 1925, Johansson and Linde, using X-ray analysis, proved the occurrence of order in solid solutions. By 1930s, the subjects of quantum mechanics and quantum field theory were developed which gave impetus to the study of phase transitions. In 1932, Felix Bloch (1905–1983) introduced the concept of order parameter in his postdoctoral thesis to describe spatially varying ferromagnetic states. In 1933, Paul Ehrenfest (1880–1933) developed first classification of general type of transition between the phases of matter based on the discontinuity in derivatives of free energy function and these were observed in liquid helium, superconductors and ferromagnetics. In 1935,

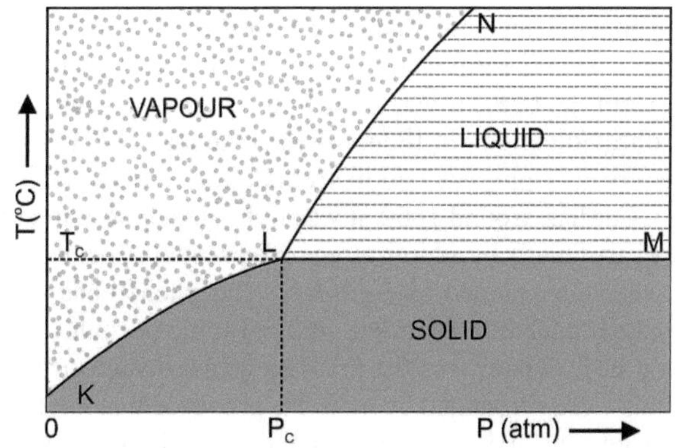

Fig. 7.2 Temperature–Pressure (T–P) equilibrium phase diagram for a pure metal

Lev Landau (1908–1968) in his first paper on phase transitions introduced the general concept of order parameter (varying from 0 to 1) to describe the degree of order in a solid. In his second paper, published in 1935, Landau asserted that the onset of order characterizes a large class of phase transitions and under such a situation free energy can be expanded in the form of a polynomial in order parameter. In 1937, Lev Landau published two classic papers on phase transitions. In the first paper, he dealt with changes of symmetry in phase transitions and in the second paper the famous Landau-Peierls theorem was proved according to which long range order is possible if the order parameter varies in all the three Cartesian directions.

Pure aluminium metal is light and the weakest of all metals. Alfred Wilm (1869–1937) was searching for a method to harden aluminum-based alloys for use in machine-gun cartridge cases. In 1906, Wilm made aluminum (Al) based alloy by adding copper (Cu), magnesium (Mg) and manganese (Mn) usually called *duralumin.* He found that the alloy formed after cooling the melt was quite soft. But to his surprise the alloy hardened with the passage of time. Therefore, these alloys were named *age-hardening alloys* and naturally duralumin was the first age-hardening alloy. Wilm applied for its patent which he obtained in 1909 and then in 1911, he published the paper on age-hardening alloy. Because duralumin was a light and very hard alloy, therefore immediately after its discovery it was extensively used in making aircrafts. In 1919, Paul Dyer Merica (1889–1957) and coworkers worked on the principle of age hardening and found that there occurred with time the precipitation of some elements in duralumin, therefore these are also called precipitation-hardening alloys. Wilm's discovery of age-hardening alloys is one of the very important events in technology. It is usually said that he did not invent only an alloy, but discovered a new universe. This is a rare occurrence experienced only by a few researchers in all of science. Later numerous age-hardening alloys with different elements and compositions were found.

Another very important type of point defects is color centers in crystalline ionic solids. Color centers are studied mostly in alkali halides because they are cheap, easily available, possess simple crystal structure and are transparent to light in contrast to the opacity of metals and semiconductors. Further, it was found that NaCl and KCl in the natural rock salt form were sometimes blue in color instead of colorless and transparent form. In 1830, it was known that fluorspar could be discolored by irradiating with ultraviolet light. In 1834, Michael Faraday (1791–1867) was the first to discover ionic crystals and thus laid the foundation of both *electrochemistry* and *solid state ionics*. Further, around 1860, it was found that coloration of alkali halides could be produced artificially by adding atoms of a particular metal. The discoloration and coloration were later named as *F-centers*. The phenomena of coloration and discoloration created much interest in the study of ionic crystals.

In 1901, Louis E. O. Visser was the first to suggest that electrons were released when ionic crystals were stimulated by light. He proposed that impurities in a crystal are split up into electrons and impurity ions and their vibrations produce phosphorescence. In 1903, Henry Friedrich Wilhelm Siedentopf (1872–1940) and Richard Adolf Zsigmondy (1865–1929) invented ultramicroscope. With the help of ultramicroscope, Siedentopf studied the coloration of alkali halides and after careful examination proposed that particles of atomic dimensions were responsible for it. In 1904, Eduard Antonvon Lenard (1862–1947) and his coworkers gave a photoelectric theory of luminescence in which they assumed that an impurity together with the surrounding host atoms would form a microcrystalline region called *emission center* and such centers are associated with all the impurities in the crystal. Light stimulation would cause electrons to emerge from the impurity atoms and their return would cause phosphorescence. Thus, Lenard assumed that this process takes place well inside the emission center. Before 1914, Carl Tubandt (1878–1942) showed that there is finite ionic conduction even in very pure silver and copper halides: called pure ionic conduction. But it was not understandable how conduction could take place in an insulator. In 1916, Abram Fedorovich Ioffe (1880–1960) developed the first theory of ionic conduction in solids based on his experimental findings. Ioffe assumed that ions get shifted to interstitial positions which are responsible for the transport phenomenon. In 1922, Georg von Hevesy (1885–1966) gave the hypothesis of '*loosened lattice*' at finite temperature in an insulating crystal. According to it, thermal energy causes loosening of ionic lattice positions which should make possible the exchange of ion positions giving rise to finite electrical conduction. It has already been mentioned that in 1926 Jacob Ilyich Frenkel (1894–1952) discovered Frenkel defect which consists of ion at the interstitial position and a vacancy at the lattice position which are bound. The interstitial ion can migrate in the crystal causing finite ionic conduction. Frenkel assumed that only a small fraction of ions is mobile. Frenkel theory for the first time led to the concept of point defects at the atomic level in ionic conduction in solids and it gave support to the Ioffe's theory of ionic conduction.

Jacob Frenkel (1894–1952), Public Domain, URL: https://www.wikidata.org/wiki/Q951957#/media/File:Yakov_Frenkel.jpg

Around 1920, Robert Wichard Pohl (1884–1976) started studying the electrical and optical properties of the ionic crystals. Pohl was very much impressed by his university Professor George Hermann Quincke (1834–1924) who used to say that "*Theories come and go: fact remains*". Pohl made this saying as the guiding principle in his life. Pohl set up the first physical institute in Gottingen, Germany that is regarded as the first great institute of solid state physics. Pohl's institute dominated in the experimental studies of color centers in ionic crystals but lacked interest in theoretical aspect. Unfortunately, the work done at Pohl's institute was not noticed for many years by scientist in the rest of the world. In the beginning Robert Pohl and Bernhard Gudden (1892–1945) did not believe at all that free electrons were being produced by light stimulation. But later in 1920, they observed that free electrons are produced even in pure ionic solids by applying sufficiently high external fields. In 1920, Pohl and Gudden published a number of papers and one of the remarkable discoveries is the measurement of electrical conductivity of diamond in which they observed slow variation in the presence of illumination. They proposed that the variation in conductivity could be due to the Lenard's emission centers (microcrystalline regions).

In 1922, Pohl and Gudden proposed a primary process for the internal photoelectric effect in insulators and indicated that it could be applicable to semiconductors also. In 1923, Karl Przibram (1878–1973) advanced the hypothesis of atomic centers [see Przibram (1953)].

In 1921, Wilhelm Conrad Rontgen (1845–1923) found that when rock salt was irradiated with X-rays it got discolored and its conductivity increased by a huge amount (about 40,000 times). It was observed that photoelectric activation could be produced not only by X-rays but also by artificial coloration caused by metal ions. Rontgen proposed that electrical conduction caused by coloration was due to the movement of electrons. Gudden and Pohl showed that both the natural and artificially prepared NaCl crystals of unknown degree of purity exhibit the discoloration absorption band at the same position. This suggests that neutralized Na atoms are centers of this absorption and this fact provided the thorough experimental evidence for atomic centers.

In 1930, Rudolf Hilsch (1903–1972) and Robert Pohl discovered an absorption band in silver halide in the visible region and named it *color center* for the first time. These color centers were identified as neutral silver atoms (in general alkali atoms) brought about by wandering electrons but these silver atoms must be bonded with the lattice. Alkali halides and rock salt when irradiated by X-rays yielded color centers of the order of 10^{15} per cubic centimeter. In 1930, Alexander Smakula (1900–1993) gave a formula for the calculation of number of discoloration centers, known as *Smakula formula.* Therefore quantification of the physics of color centers commenced with the Smakula formula. In 1931, Erich Mollwo (1909–1993) put forward a relation between the frequency of absorption maximum ν and the lattice constant 'a' given by

$$\nu a^2 = \text{constant} \tag{7.2}$$

It is known as *Mollwo formula.* According to Eq. (7.2), the frequency of absorption maximum decreases with the increase in temperature because the lattice constant increases with the increase in temperature. In 1933, Mollwo confirmed that the color centers are intrinsic atomic defects in the crystal but the position of electrons is not known. In 1933, Rudolf Hilsch and Robert Pohl discovered a new absorption band in the ultraviolet region which corresponded to new color centers called *U-centers.* Later it was confirmed that U-centers are nothing but impurity defects: atomic defects similar to the F-band. Therefore, U-centers were supposed to be electrons in the crystal lattice vacancies. Further, it was found that if U-centers are irradiated these decompose to form F-centers.

In 1937, Bristol conference was held which was mainly focused on lattice defects especially color centers. In this conference, different aspects of color centers were presented and discussed and some important decisions were taken. In the conference, the model of color center was finalized according to which a color center consists of an electron trapped in the neighborhood of a point where the negative ion is missing. In this model, the vacancy center is the point of lowest energy and the wave function of the electron extends up to six nearest neighbor ions. In the 1937 Bristol conference, Jan Hendrikde Boer (1899–1971) discussed about the color centers and argued that F-center is an electron bound to a negative ion vacancy (vacant halogen site) and F′-center consists of two electrons occupying the same negative ion vacancy. In 1937, Nevill Francis Mott (1905–1996) gave a simplified model for F-centers in the continuum elasticity theory. He assumed the crystalline solid as a continuous material with finite dielectric function. F-center was assumed to be an electron trapped in a spherically symmetric cavity (halogen ion vacancy) made in the continuous material. In this approximation, the problem of color center is reduced to the solution of Schrodinger equation in a potential well. Later this method was improved by a number of researchers. The other color centers invented were H-centers, M-centers, N-centers, R-centers, and V-centers etc. Frederick Seitz (1911–2008) suggested that V-bands were due to holes trapped by positive ion vacancies and M-center could be an L-shaped triple vacancy. The basic work for intrinsic atomic defects was completed before 1940 that is before World War II, and it became an integral part of the physics of lattice imperfections. In 1940, Nevill Francis Mott and Ronald Wilfred Gurney (1898–1953) published a classic book on ionic crystals entitled, "*Electronic processes in ionic crystals*" which intensified the research on alkali halides—the book was revised in 1948. In 1946, Frederick Seitz wrote first review article on color centers which became a standard reference for the subject.

In 1945, Evgenii Konstantinovich Zavoisky (1907–1976), performed the first *electron spin resonance* experiment. In 1949, Clyde A. Hutchison (1913–2005) measured for the first time the electron spin resonance absorption spectra of F-centers. Since then the electron spin resonance has become the most powerful technique in studying color centers. In 1957, Theodore Grant Castner (1930–2017) and Werner Kanzig (1922–2002) discovered a new type of color center while studying the electron spin resonance spectra of KCl irradiated with X-rays at liquid helium temperature. They showed that the new color center could most appropriately be described as a self-trapped hole and it came to be known as V_k-center. The electron spin resonance studies confirmed beyond doubt the validity of the Bristol model of F-center. Further, it was found that the wave function of a trapped electron in KCl extended over a range of several atomic spacing.

In 1946, Solomon Isaakovich Pekar (1917–1985), Soviet physicist, published two papers on self-trapped electrons in ionic crystals. In the study of F-centers, in 1950, Pekar used the continuum approximation with a difference where the lattice was replaced by a system of linear oscillators having the same frequency. He viewed the F-center as a trapped charge distribution whose field polarized the lattice around and influenced its vibrational states. It is named as a '*polaron*', a word coined for the first time, which is an elaboration of Landau's earlier idea. In 1956, Clifford Coover Klick (1918–2006) and George Alber Russel (2021–2016) developed a phenomenological model in the localized approximation, an approximation complementary to Pekar's continuum approach. In this model, strongly localized electron in the vacancy is assumed to interact with the vibrations of six surrounding atoms. They assumed that the lattice vibrations with which the electron interacts are different in the ground and excited states. This model was used to describe the relation between absorption and emission energies, the bandwidths and their dependence on temperature. In

1949, T. Muto and in 1950, T. Inui and Y. Uemura assumed that electron in the F-center is well localized and interacts only with its nearest neighbors. It is called molecular approximation because F-center resembles closely with a molecule. The linear combination of atomic orbital (LCAO) approximation developed for molecules could therefore be applied to the F-center electron. This method was applied to study various color centers.

In 1952, William H. Duerig and Jordan J. Markham irradiated alkali halide with X-rays at liquid helium temperature and found a new large absorption band on the ultraviolet side of the F-center. They named it H-band which is believed to be due to one of the trapped hole centers. Duerig and Markham announced the discovery of a new color center named H-center: H for helium and for Hopkins. The H-center is complementary to the F-center as the two annihilate each other if they are brought together reconstituting a perfect lattice structure. Werner Kanzig performed the electron spin resonance study for years together and showed that V_K-center consists of two normal helogen ions that had trapped a hole or in other words had lost an electron. In 1954, Frederick Seitz wrote second review article on color centers giving a comprehensive detail of the development of color centers in ionic solids. It is pointed out that Seitz's guesses about the models of color centers missed the target although sometimes narrowly. In 1960, Heinz Pick (1912–1983) gave final models for M-, R-, and N-centers and according to him these were double, triple, and quadrupole aggregates of F-centers, respectively. Thus, by the early 1960s, the structure of all the essential color centers was established. It is pointed out here that many hypothesis, models and theories of color centers were put forward and many of these were discarded with time. But theorists played an important role in interpreting the experimental results, and integrating color centers into solid state physics. Someone has rightly said that experimental and theoretical physics cannot be separated from each other rather they go side by side for the development of a subject.

7.2 Plastic Deformation and Dislocations

Metals are known for their strength, malleability and ductility and therefore can be deformed permanently. Metals can be shaped into tools, weapons, machines, bridges and building etc. by pressing, rolling or forging. When stress is applied and the solid reaches or crosses the elastic limit, it either deforms plastically or suffers fracture. Plastic deformation is the deformation produced in the shape of a solid which is retained even when stress is removed. The plastic deformation is one of the most desirable and technically important properties of metals. There are two mechanisms of plastic deformation: *slip* on a specific glide plane and *twinning*. In crystalline solids slip is the sliding of one part of a crystal over another part along some particular crystallographic plane called *glide plane* or *slip plane*. It is found that glide planes are normally closely-packed planes or planes of high atomic density. Twinning can be brought about by a combination of slips on many gliding planes simultaneously. Sometimes combination of slip and twinning are also observed with the latter facilitating the former.

With the onset of Industrial Revolution in Europe, the demand for highly reliable and better grade iron, steel and other metals increased many-fold. The art of making high quality metals grew into the field of metallurgy. Henry Clifton Sorby (1826–1908) for the first time used optical microscope for studying the structure of metals and he published in 1858 an important memoir, "*On the microscopical structure of crystals*". He is considered to be one of the founders of the field of metallurgy. Unfortunately, in 1879, a railway bridge named 'Tay Bridge', made of wrought iron, collapsed in high wind with a loss of 78 lives. This disaster contributed towards the establishment of metallurgical engineering: most probably the first subject in engineering faculties. The beginning of modern understanding of mechanical behavior of metals can be dated back to the period of World War I. By this time new techniques came into existence for the study of various properties of solids. X-ray technique was used for studying the structure of solids and the techniques for growing single crystals were developed. Three methods were developed for growing single crystals.

1. In 1918, Jan Czochralsk (1885–1953), in 1923, Percy Williams Bridgman (1882–1961); and in 1924, Obreimov and Schubnikov developed method for growing single crystals from the melt. Czochralski is called the father of crystallization.
2. In 1921, Henry Cort Harold Carpenter (1875–1940) and C. F. Elam developed the method of recrystallization and produced single crystals of brass. In 1925, Anton Eduard van Arkel (1893–1976), Jan Hendrik de Boer (1899–1971) and J. D. Fast produced single crystals of tungsten.
3. In 1930, C. F. Elam described the method for producing single crystals from the vapor phase.

Up to 1934, a large amount of experimental data on plastic deformation of single crystals was generated but no satisfactory theory was put forward to explain it. The available experimental knowledge on plastic deformation of metals was summarized in 1935 by C. F. Elam in her book entitled, "*Distortion of metal crystals*" and in the same year by Erich Schmid (1896–1983) and Walter Boas (1904–1982) in their book entitled, "*Kristallplastizitat*". From the experimental data three puzzling facts came to light.

1. The solids suffered plastic deformation or fracture at stresses much lower (smaller by orders of magnitude) than predicted from the theory of cohesion of metals.
2. The plastically deformed crystals offered increased resistance to further deformation (known as *work hardening*).
3. It was not understandable why a slip band is formed rather than a single slip in a metal.

These puzzles initiated theoretical work on plastic deformation. In 1905, Vito Volterra (1860–1940) developed theory of elastic deformation for the first time and provided everything needed to invent dislocations. In 1920, Alan Arnold Griffith (1893–1963) suggested that a crystal contain very fine cracks and during plastic deformation these cracks propagate. Griffith's theory of crack propagation is considered as one of the direct predecessors of dislocation theory. In 1921, George Beilby (1850–1924) proposed first theory of work hardening in the process of plastic deformation. According to his theory, glide between two planes produces a local liquid region that allows limited movement of the slip and then the liquid region solidifies into an amorphous phase. The hard amorphous layer locks adjacent planes together offering large resistance and thus gives rise to work hardening. In 1925, Richard Becker (1887–1955) suggested a thermally-activated mechanism of plastic deformation which was not widely accepted. In 1928, Ludwig Prandtl (1875–1953) gave a model for work hardening in terms of slip. He assumed that planes of atoms would jump irreversibly from one position of equilibrium to another and such jumps would leave behind defective structure that causes work hardening. In 1928, Geoffrey Ingram Taylor (1886–1975) gave mechanism for slip according to which "increasing deformation leads to an increasing number of cracks". He suggested that slip may propagate from side-to-side of the crystal in a finite time. In 1929, Ulrich Dehlinger (1901–1981) came still closer to the concept of dislocations. He said that imperfection is confined to a small region of the glide plane while in rest of the region atoms are in near perfect alignment. Therefore, he combined the concepts of mosaic structure and localized defect. This model gives rise to internal strains in the solid which are assumed to be the cause of work hardening.

In 1934, Geoffrey Ingram Taylor (1886–1975), Michael Polanye (1891–1976) and Egon Orowan (1902–1989) invented independently the concept of dislocations and proposed that the low stresses can produce plastic deformation which could be explained in terms of the theory of dislocations. Dislocation is an abrupt change in atomic arrangement within the crystal structure and the atomic order is maintained on either side of dislocation but the atoms on one side have moved or slipped. Actually, August Edward Hough Love (1863–1940) introduced the word dislocations in the third edition of his book, "*Treatise on the Mathematical Theory of Elasticity*" published in 1920: dislocation is derived from an Italian word *distorsione*. In 1934, Taylor and coworkers invented *edge dislocation* which can be visualized as being caused by the termination of a plane of atoms in the middle of a crystal (Fig. 7.3). In such a case, the surrounding planes are not straight, instead bent around the edge of the terminating plane so that the crystal structure is perfectly ordered on either side. They explained plastic deformation in terms of edge dislocation but it was not the general concept of dislocations which had to wait a little longer. Geoffrey Taylor gave two mechanisms for a slip. In the first mechanism, he assumed that there is a formation of blocks during deformation and the locking of blocks holds up the slip. In the second mechanism, he assumed that blocks are not formed by deformation but there is a mosaic structure inherent in the crystal. The argument against the locking mechanism is much the same as in the first case. Taylor gave the actual definition of dislocation according to which,

1. Above a certain critical temperature T_D any stress, however small, will cause dislocations to move.

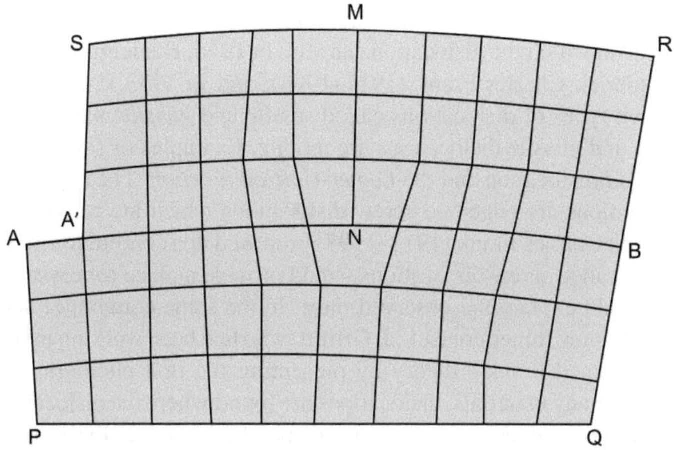

Fig. 7.3 Schematic representation of the strained pattern on the front face PQRS of the crystal where the upper part of the crystal has slipped by AA′. MN represents the straight line edge dislocation [Modified from Cottrell, A. H. (1953). Dislocations and plastic flow in crystals (p. 22), Oxford, New York]

2. At temperatures below T_D, the centers will not move till the shear stress attains some finite value.

Taylor assumed that crystals originally do not contain dislocations. These would be formed quite readily as a result of thermal fluctuations and could move easily under the influence of even small stresses. He further explained that work hardening occurs due to the stress fields arising from the increasing density of dislocations. Egon Orowan in his paper did some work on the temperature dependence of dynamics of dislocations and plastic deformation and found this as a mechanism of glide. Rudolf Peierls (1907–1995) assisted Orowan in understanding the force required to move dislocation. Later Frank Reginald Nunes Nabarro (1916–2006) improved Peierl's estimate of force required to move a dislocation through a perfect lattice and named it Peierls-Nabarro force. In 1934, an International Conference, "*The Solid State of Matter*" was held in London in which Wilhelm Gerard Burgers (1897–1988) supported the mosaic theory and said it works even with improved experimental and theoretical base. Erich Schmid (1896–1983) also reaffirmed the crystallographic nature of slip and opposed to creep.

In 1939, Johannes Martinus Burgers (1895–1981) invented *screw dislocation* and presented its mathematical treatment in a conference entitled "*Internal Strains in Solids*" held in Bristol. He described dislocation in terms of a vector now generally known as *Burgers vector*. A dislocation can be characterized by the magnitude and direction of movement of atoms defined by the Burgers vector. Dislocations can slip in planes containing both the dislocation line and the Burgers vector, the so-called glide plane. For a screw dislocation, the dislocation line and the Burgers vector are parallel, so the dislocation may slip in any plane containing the dislocation. But for an edge dislocation, the dislocation line and the Burgers vector are perpendicular and there is only one plane in which the dislocation can slip. In 1948, Heidenreich and Shockley, in 1949, Frederick Charles Frank (1911–1998), and in 1953 Frank and Nicholas introduced two types of dislocations called *sessile* and *glissile*: sessile dislocations are immobile and glissile dislocations are mobile. Examples of sessile dislocations are the *stair-rod* dislocation and the *Lomer-Cottrell junction*. The two main types of glissile dislocations are edge and screw dislocations which are of interest here. In 1949, Frederick Charles Frank (1911–1998) proposed at a conference in Bristol that spiral features called screw dislocations would provide a place for crystal growth to happen that could explain the observed rates. In the same conference shortly after Frank's presentation, mineralogist L. J. Griffin (who had been working independently of Frank) confirmed Frank's theory by presenting the first photographs of screw dislocations. In many materials, dislocations are found where the dislocation line and Burgers vector are neither perpendicular nor parallel. Such dislocations are called *mixed dislocations*, which comprise of both the screw and edge characters. Burgers published his work just before World War II, and it could not make much impact on the scientific community. With the start of World War II the work on dislocations was also slowed down immensely. Therefore, after many years Burgers work came into the notice of scientists. In 1950, Frederick Charles Frank and William Thornton Read (1896–1972) gave what is known as Frank-Read source of dislocations and it describes the multiplication of dislocations.

Many metallurgists did not believe in dislocations, who considered these as the imagination of solid state physicists working out theories without much supporting experimental evidence. Dislocations have local disordered regions which produce a strain field, and therefore are expected to scatter X-rays in different ways. X-ray diffraction technique has been used to study dislocations using the above fact. But the main drawback of the X-ray technique is that it is slow (needs long exposure of a few hours) and requires a cumbersome apparatus yielding low resolving power of a few microns. In 1947, Heidenreich and Shockley applied electron microscopy to study slip in aluminium crystals which acted as a guideline for the observation of dislocations. Peter Bernhard Hirsch (1925–present) thought of applying improved transmission electron microscopy to observe dislocations within the microstructure of the material. Hirsch and coworkers (1956) observed dislocations, not only static dislocations but moving dislocations also, for the first time with the help of transmission electron microscopy. Bollmann (1956) also independently observed dislocations in stainless steel. Therefore, the debate on the reality of dislocation ended with their experimental observation. Field ion microscopy and atom probe techniques offered methods of producing much higher magnifications (typically 3 million times and above) and permitted the observation of dislocations at an atomic level. At a resolution of the order of atomic size, screw dislocations appear as distinctive spiral features—thus revealing an important mechanism of crystal growth. The main advantage of X-ray method over the transmission electron microscopy is that one can use sheets with large thickness. Nevill Francis Mott (1905–1996) and Frank Reginald Nunes Nabarro (1916–2006) carried out some work on dislocations even during the World War II. In most of the modern theories of slip, it is assumed that slip begins at one end of the crystal and travels across it, thus avoiding the concept of simultaneous slip. Although various aspects of dislocations were studied but the mechanism of formation of dislocations was not known.

7.3 Grain Boundaries

When a material in the molten state is cooled down slowly a large number of nucleation centers are created around which crystallization starts. As a result, a large number of small crystals are formed with random crystallographic orientations usually called crystallites or grains. These grains grow in size till the boundaries of adjacent grains touch one another. The surface at which two grains meet is called grain boundary. Such materials in which grains are oriented randomly are called polycrystalline materials. Therefore, grain boundaries are surface or planer defects. Generally speaking grain boundaries are composed of an array of dislocations and their properties and structure are a function of the misorientation. It has been found through X-ray diffraction studies that most of the crystalline materials exist in the polycrystalline state. Strictly controlled environment is required to grow single crystals, which is quite laborious. *Stacking fault* is another surface defect in crystalline solids which usually occurs parallel to a close-packed plane. A staking fault results from the stacking of one atomic plane out of sequence on another while the lattice on either side of the fault is perfect. It is also possible to describe the production of stacking faults in terms of dislocations. In 1942, D. S. Edward and Henry Lipson (1910–1991), and Arthur James Cochran Wilson (1914–1995) independently observed stacking faults in cobalt through the fcc ⇔ hcp transformation. Later in 1950 and 1952, Charles Sanborn Barrett (1902–1994) observed stacking fault in fcc crystals (CuSi alloy) where both fcc and hcp systems exist which have nearly the same interatomic distances.

Each grain possesses perfect symmetry within but the boundaries of different grains are oriented differently. One can classify grain boundaries into two categories:

1. Small-angle grain boundaries.
2. Large-angle grain boundaries.

In small-angle boundaries, the structures of two adjacent grains are tilted by a few degrees (less than 15°) relative to each other. In 1939, William Lawrence Bragg (1890–1971) and Johannes Martinus Burgers (1895–1981) proposed two similar models of grain boundaries based on two-dimensional array of dislocations. Burgers predicted that if the angle is small the two grains can be joined and during this a series of parallel edge dislocations are formed along the boundary. Lacombe (1947) observed microscopic structure of blocks with dimensions 0.1 μm. The micromosaic structure resides within the macromosaic structure with misorientation between them of 1°–3°. The structure of boundary depends on the relative orientation of the adjacent crystalline blocks. Lacombe and Beaujard (1947) were the first to observe experimentally in aluminium that a low angle tilt boundary is composed of edge dislocations lying one above the other in the boundary (see Fig. 7.4). The angle of tilt will be

$$\theta = \frac{b}{h}$$

In large-angle grain boundaries the angle between two adjacent grain boundaries is more than 15° and the grain boundary structure is nearly that of a liquid. A polycrystalline solid is obtained by the simultaneous formation of a large number of randomly oriented grains. The random orientation of the grains means that the grain boundaries in most solids are large-angle boundaries. The large-angle boundaries are considerably more disordered with large areas of poor fit as compared to low-angle boundaries. Cyril Stanley Smith (1903–1992) developed methods for deriving three-dimensional shapes of the crystalline structures of metals from the two-dimensional microscope images of the grains of the metals. Smith was fascinated by the details of faults and grain boundaries in metals, and in 1952, he developed theoretical models of large-angle grain boundaries. Smith has shown that, because of surface tension,

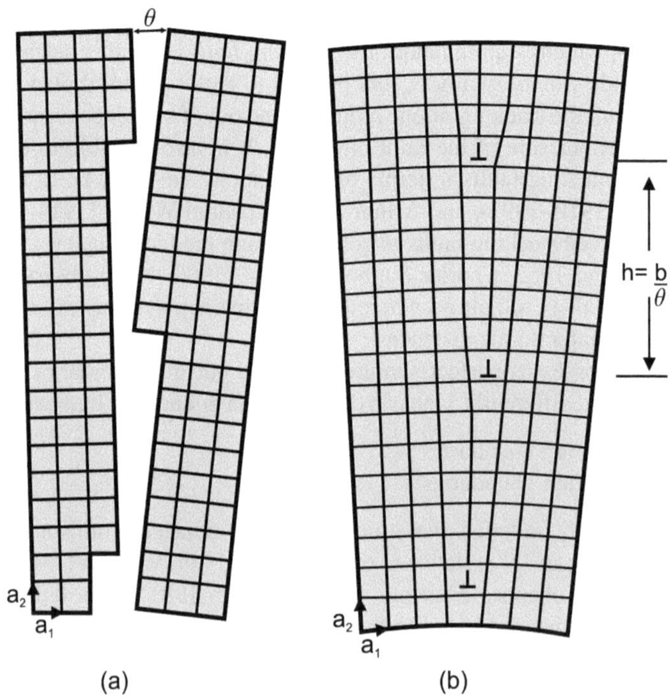

Fig. 7.4 **a** Schematic representation of two crystallites with simple cubic structure inclined to each other by a small angle θ. **b** After joining the crystallites, three positive edge dislocations are formed with spacing h between the two executive dislocations given by $h = b/\theta$ where b is the Burgers vector [Modified from Kubo, R and Nagamiya, T. (1969). Solid State Physics, p. 771, NewYork: McGraw-Hill]

the grain boundaries prefer to form common corners between three grains so that the inter-boundary angles are equal 120°. But it is not possible to fill space completely with polyhedral whose terminal planes form 120° angles with each other.

The grain boundaries have a pronounced influence on mechanical, electrical and other properties of polycrystalline solids. They disrupt the motion of dislocations through a material, so reducing grain size which is a common way to improve mechanical strength, as described by the Hall–Petch relationship [Hall (1951) and Petch (1953)]. It is found that the electrical and thermal conductivities decrease with the formation of grain boundaries. Most grain boundaries are preferred sites for the onset of corrosion and for the precipitation of new phases from a solid.

7.4 Twinning in Crystals

Crystal twinning occurs when two or more adjacent crystals of the same material are mirror image of one another and are oriented so that they share some of the same crystal lattice points in a symmetrical manner. The result is an intergrowth of two separate crystals that are tightly bonded to each other. The surface along which the lattice points are shared in twinned crystals is called *composition surface* or *twin boundary* and if it is a plane, one talks about a *composition plane* or *twin plane*. The shared lattice points give much greater strength to the junction between the crystal segments than that between randomly oriented grains. Therefore, the twinned crystals do not easily break apart. There are three modes of formation of twinned crystals. First is *Growth twins* and they are the result of an interruption in the lattice during the growth due to a possible deformation from a larger substituting ion. Second is *annealing* or *transformation twins* and they are the result of a change in crystal system during cooling as one form becomes unstable and the crystal structure transform into another more stable form. Twinned crystals can also be formed when stress is applied to the crystal and these are called *deformation* or *gliding twins*. Because growth twins are formed during the initial growth of the crystal, they are called *primary twins*, whereas transformation or deformation twins are formed in an existing crystal and are called *secondary twins*. In 1926, Georges Friedel (1865–1933), in 1952, Fritz Laves (1906–1978) and in 1954, Robert Wolfgang Cahn (1924–2007) studied the production of twins and put forward some conditions for their production. In 1945, Charles Crussard (1916–2008) and in 1954, Robert Wolfgang Cahn (1924–2007) studied the growth of twins in zinc single crystals. It was found that large stresses are required to form a nucleus for the initial growth of a twin but smaller stresses are required for its subsequent growth.

The twinned crystals are of two types depending on the nature of the twin plane: *contact twins* or *penetration twins*. Contact twins meet on a single composition plane, often appearing as mirror images across the boundary, for example quartz, gypsum and spinal often show contact twinning. In penetration twins the individual crystals penetrate each other up to several atomic planes in a symmetrical manner, for example pyrite and fluorite often exhibit penetration twinning. The composition surface in penetration twins is usually irregular and extends to the center of the crystal. Contact twinning can arise from either reflection or rotation, whereas penetration twinning is usually produced by rotation. The relationship is defined by a symmetry operation called a *twin operation.* The twin operation is not one of the normal symmetry operations of the untwined crystal structure. Twin laws are symmetry operations that define the orientation between twin crystal segments. These are characteristic of the material as are its crystal face angles. Twin laws include reflection, rotation, and inversion operations. Reflection twinning is described by the Miller indices of the twin plane [i.e. (hkl)] while rotational twinning is described by the direction of the twin axis (i.e. [hkl]). The twin axis will be perpendicular to a lattice plane. Rotational twin laws are almost always twofold rotations, though any other permitted rotation symmetry (threefold, fourfold, or sixfold) is also possible. For rotational twinning the relationship between the twin axis and twin plane falls into one of three types:

1. Parallel twinning, when the twin axis and compositional plane lie parallel to each other,
2. Normal twinning, when the twin plane and compositional plane lie normally, and
3. Complex twinning, a combination of parallel twinning and normal twinning on one compositional plane.

Inversion twinning is typically equivalent to a reflection or rotation symmetry. If several twin crystal parts are aligned by the same twin law they are referred to as *multiple* or *repeated twins.* If these multiple twins are aligned in parallel they are called *polysynthetic twins.* When the multiple twins are not parallel they are *cyclic twins.*

Twinning and slip are two competitive mechanisms for crystal deformation. Each mechanism is dominant in certain crystal systems and under certain conditions. In fcc metals, mostly slip is dominant because the stress required is far less than twinning stress. Twinning can occur by cooperative displacement of atoms along the face of the twin boundary. This displacement of a large number of atoms simultaneously requires significantly higher energy to perform. Therefore, the theoretical stress required to form a twin is quite high. It is believed that twinning is associated with the motion of dislocation on a coordinated scale, in contrast to slip, which is caused by independent glide at several locations in the crystal. Twinning produces a deformation pattern that is more heterogeneous in nature as compared to that of a slip. This deformation produces a local gradient across the material and near intersections between twins and grain boundaries. The deformation gradient can lead to fracture along the boundaries, particularly in bcc transition metals at low temperatures. In 1951, Alan Howard Cottrell (1919–2012) and Bruce Alexander Bilby (1922–2013), in 1954, Robert Wolfgang Cahn, in 1954, E.O. Hall and in 1964, Jacques Friedel (1921–2014) studied various aspects of twinning such as growth of different forms of twins, stability of twins, twinning dislocations, stress relaxation in twins etc.

One can also have low angle *twist boundary* that can be described in terms of at least two set of parallel screw dislocations lying in the boundary. Of course, many nonstructural surface imperfections also exist in materials. For example, when two ferromagnetic domains differ from one another only in the direction of magnetization, the boundary between them is an imperfection that is called a *ferromagnetic domain wall.*

Bibliography

Andrews T (1869) The Bakerian lecture. Philos Trans Roy Soc (London) 159:575

Barrett CS (1950) J Met 123; (1952) Imperfections in nearly perfect crystals. Wiley, New York

Becker R (1925) Elastische nachwirkung und plastizitat. Z Phys 33:185–213

Becker R (1925) Über die plasticität amorpher und kristalliner fester körper. Phys Zs 26:919–925

Blatt FJ (1957) Residual resistivity of copper and silver alloys: dependence on periodic table. Phys Rev 108:285

Bloch F (1932) Zur theorie des austauschproblems und der Remanenzerscheinung de ferromagnetika. Zeitschrift fur Physik 74:295–335

Bollmann W (1956) Interference effects in the electron microscopy of thin crystal foils. Phys Rev 103:1588

Bozorth RM (1951) Ferromagnetism. Van Nostrand, New York

Burgers JM (1939) Proc K Ned Akad Wet 42:293–378

Cahn RW (1954) Twinned crystals. Adv Phys 3:363–445

Carpenter HCH, Elam CF (1921) The production of single crystals of aluminium and their tensile properties. Proc Roy Soc A100:329–353

Cottrell AH, Bilby BA (1951) A mechanism for the growth of deformation twins in crystals. Phil Mag 42:573–581

Crussard MC (1945) Etude des glissements plastiques dans les cristaux d'aluminum. Rev Métallurgie 42(286–294):321–332

Czochralski J (1918) Ein neues verfahren zur messung der kristallisationsgeschwindigkeit der metalle. Z Phys Chem 92:219

Dehlinger U (1929) Zur theorie der rekristallisation reiner metalle. Ann der Physik 2:749–793

Duerig WH, Markham JJ (1952) Color centers in alkali halides at 5 °K. Phys Rev 88:1043

Elam CF (1930) The diffusion of zinc in copper crystals. J Inst Metals 43:217–231

Elam CF (1935) Distortion of metal crystals. Clarendon Press, Oxford

Flinn PA, Maradudin AA (1962) Distortion of crystals by point defects. Ann Phys 18:81–109

Frank FC (1949) Proceedings of Pittsburgh conference, vol 2. ONR Office of Naval Research, p 542

Frank FC, Kasper JS (1958) Complex alloy structures regarded as sphere packings I. Definitions and basic principles. Acta Crystallogr A 11:184–190

Frank FC, Nicholas JF (1953) Stable dislocations in the common crystal lattices. Phil Mag 44:1213–1235

Frank FC, Read WT (1950) Multiplication processes for slow moving dislocations. Phys Rev 79:722

Frenkel J (1926) Über die wärmebewegung in festen und flüssigen körpern. Zeitschrift fur Physik 35:652–669

Friedel J (1964) Dislocations. Pergamon Press, Oxford

Gibson JB, Goland AN, Milgram M, Vineyard GH (1960) Dynamics of radiation damage. Phys Rev 120:1229

Girifalco LA, Weizer VG (1960) Vacancy relaxation in cubic crystals. J Phys Chem Solids 12:260–264

Griffin LJ (1950) Observation of unimolecular growth steps on crystal surfaces. Phil Mag 42:196; (1950) Ph.D. Thesis, London
Griffith AA (1921) The phenomena of rupture and flow in solids. Philos Trans Roy Soc A221:163–198
Guillaume CÉ (1919) The anomaly of the nickel-steels. Proc Phys Soc (London) 32:374; (1896) La Nature 24:234.
Hafner J, Punz G (1983) Static lattice distortions in substitutional alloys: a computer simulation. J Phys F13:1393
Hagg G (1930). Rantgenuntersuchungen uber die hydride von titan, zirkonium, vanadin and tantal. Z Phys Chem B11:433; (1931) Gesetzmassigkeiten im kristallbau bei hydriden, boriden, carbiden und nitriden der übergangselemente [Regularities in the crystal structure with hydrides, borides, carbides and nitrides of the transition elements]. Z Phys Chem B12:33–56
Hall EO (1951) The deformation and ageing of mild steel: III Discussion of results. Proc Phys Soc (London) B64:747–753
Hall EO (1954) Twinning and diffusionless transformations in metals. Butterworths, London
Hansen M (1958) Constitution of binary alloys, 2nd edn. McGraw-Hill, New York
Heidenreich RD, Shockley W (1947) Report of the 1947 Bristol Conference on the strength of solids. The Physical Society, London, p 71
Heidenreich RD, Shockley W (1948) Report of 1948 Bristol conference on the strength of solids. The Physical Society, London, p 57
Heitler WH, London FW (1927) Wechselwirkung neutraler atome und homoopolare bindung nach der quantenmechanik [Interaction between neutral atoms and homopolar binding according to quantum mechanics]. Z Phys 44:455–472
Heusler F (1903) Ueber magnetische manganlegierungen. Verh. Deutsche Physikalische Gesellschaft 5:219–223
Heycock CT, Neville FH (1902) Proc Roy Soc (London) 69:451
Hilsch R, Pohl RW (1930) Zur photochemie der alkali-und siberhelogenidkristalle. Zs Phys 64:606–622; (1930) Einige dispersionsfrequenzen der alkalihelogenidkristalle im schumanngebiet. Zs Phys 59:812–819; (1930) Uber die ausnutzung des latentenbildes bei der photographischen entwicklung. Nachr Gott, 334–337
Hirsch PB, Horne RW, Whelan MJ (1956) Direct observations of the arrangement and motion of dislocations in aluminium. Phil Mag 1:677–684
Hume-Rothery W (1926) Research on the nature, properties and conditions of formation of intermetallic compounds, with special reference to certain compounds of tin. J Inst Metals 35:295–299, 307; (1931) The metallic state. Oxford University, London
Ioffe A, Kirpitchewa (1916) Electric conductivity of pure crystals. J Russ Phys-Chem Soc 48:261–262
Kalia RK, Vashishta P (1990) Advances in statistical physics of liquids and solids (eds: Prakash S, Pathak KN). Wiley Eastern Limited, New Delhi, p 142
Kanzaki H (1957) Point defects in face-centred cubic lattice—I distortion around defects. J Phys Chem Solids 2:24–36
Kanzig W (1955) Electron spin resonance of V1-centers. Phys Rev 99:1890; Castner TG, Kanzig W (1957) The electronic structure of *V*-centers. J Phys Chem Solids 3:178–195
Lacombe P (1947) Sub-boundary and boundary structures in high purity aluminium. Report of 1947 Bristol conference on the strength of solids. Physical Society, London, pp 91–94
Lacombe P, Beaujard L (1947) The application of etch figures on pure aluminum (99.99%) to the study of some micrographic problems. J Inst Met 74:1
Lenard P (1904) Uber sekundare kathodenstrahlung in gasformigen und festen korpern. Ann Phys 15:485–508; Lenard P, Klatt V (1904) Uber die erdalkaliphosphore (Schlussteil). Ann Phys 15:633–672
Levine D, Steinhardt PJ (1984) Quasicrystals: a new class of ordered structures. Phys Rev Lett 53:2477–2480

Martin Y, Wickramasinghe HK (1987) Magnetic imaging by "'force microscopy"' with 1000 Å resolution. Appl Phys Lett 50:1455–1457

Merica PD (1919) Bull Bur Stand 336:101–104; Merica PD, Waltenberg RG, Scott H (1919) Heat treatment of duralumin. Bull Am Inst Min Metall Eng 150:913–949; (1919) Heat treatment of duralumin. Sci Pap Bur Stand 15:271–316 (1919)

Mollwo E (1931) Uber die absorptionspektra photochemisch verfarbter alkalihalogenid-kristalle. Nachr Gott, 97–99; (1933) Uber die farbzentren der alkalihalogenid-kristelle. Zs Phys 85:56–67

Orowan E (1934) Zur kristallplastizität. III. Z Phys 89:634–659

Petch JN (1953) The cleavage strength of polycrystals. J Iron Steel Inst (London) 173:25–28

Pick H (1960) Farbzentren-assoziate in alkalihalogeniden. Z Phys 159:69–76

Polanyi M (1934) Über eine art gitterstörung, die einen kristall plastisch machen könnte. Z Physik 89:660–664

Przibram K (1953) Verfarbung und lumineszenz. Springer, Vienna

Rattan SK, Singh P, Prakash S, Singh J (1993) Strain field due to point defects in metals. Phys Rev B47:599–607; Singh J, Singh P, Rattan SK, Prakash S (1994) Strain field due to substitutional transition-metal impurities in bcc metals: application to dilute vanadium alloys. Phys Rev B49:932–943

Sagalyn PL, Paskin A, Harrison RJ (1961) Contribution of strain to nuclear quadrupole interactions in dilute alloys of copper. Phys Rev 124:428

Schmid E, Boas W (1935) Kristallplastizitat. Springer, Berlin

Schottky W (1935) Uber den mechanismus der ionenbewegung in festen elektrolyten. Zeitschrift fur physikalische chemi B29:335–355

Shechtman D, Blech I, Gratias D, Cahn JW (1984) Metallic phase with long-range orientational order and no translational symmetry. Phys Rev Lett 53:1951–1953

Siedentopf H (1905) Ultramikroskopische untersuchungen uber steinsalzfarbungen. Zs Phys 6:855–866

Singhal SP (1973) Lattice relaxations near a vacancy or interstitial in aluminum. Phys Rev B8:3641

Slater JC (1930) Cohesion in monovalent metals. Phys Rev 35:509

Smakula A (1930) Über erregung und entfärbung lichtelektrisch leitender alkalihalogenide. Zs Phys 59:603–614

Smith CS (1952) Grain shapes and other metallurgical applications of topology. In: Metal interfaces: a seminar on metal interfaces held during the 33rd National Metal Congress and Exposition, Detroit, from October 13–19, 1951; sponsored by the American Society for Metals. American Society for Metals, Cleveland, pp 65–108

Taylor GI (1934) The mechanism of plastic deformation of crystals, Part I: Theoretical. Proc Roy Soc A145:362–387

Taylor HS, Eyring H, Sherman A (1933) Binding energies in the growth of crystal nuclei from metallic atoms. J Chem Phys 1:68–76

Van Arkel AE, De Boer JH (1925) Darstellung von reinem titanium-, zirkonium-, hafnium- und thorium metall. Zeitschrift für anorganische und allgemeine Chemie (in German) 148:345–350

Visser LEO (1901) Essai d'une theorie sur la phosphorescence de longue duree, specialment sur celle des sulfures alcalinoterreux. Recueil des travaux chimiques des Pays-Bas et de la Belgique 20:435–456

Volterra V (1905) Sulle distorsioni dei corpi elastici simmetrici. Rend Accad Lincei, 5e serie, 1er sem 14 (In Italian)

Wigner EP, Seitz F (1933) On the constitution of metallic sodium. I, Phys Rev 43:804; (1934) On the constitution of metallic sodium. II, Phys Rev 46:509

Wilm A (1911) Physikalisch-metallurgische untersuchungen über magnesiumhaltige aluminiumlegierungen. metallurgie: Zeitschrift für de Gesamte Hüttenkunde 8:225–227

Chapter 8
Heat and Thermodynamics of Solids

The sensation of hot and cold had been the intrinsic characteristic of human beings, but they were not able to know how hot one object is with respect to another object. It was known that gases, liquids and solids expand on heating. Empedocles (494 BC–434 BC), a Greek philosopher, wrote two poems, "On Nature" and "On Purification". In the poem "On Nature" he mentioned four elements: Earth, water, fire and air. He formulated classical theory of these four elements and established that all the structures of the world are made of them. Aristotle also reinforced the theory of four elements. Therefore fire was treated as an element in the ancient times which prevailed for many centuries. The concept of temperature arose with the development of science and it is a measure of degree of hotness. Cornelius Drebbel (1572–1633), Robert Fludd (1574–1637), Galileo Galilei (1564–1642), Evangelista Torricelli (1608–1647), Otto von Gueriche (1602–1686) and others were able to measure the relative coldness and hotness of air using rudimentary air thermometers (thermoscopes). The basis of all these was thermal expansion of gases and liquids. But these thermometers were crude, suffering from many deficiencies and therefore, gave different temperatures. In 1667, Johann Joachim Becher (1635–1682) published a book, "*Physica subterranea*" in which he assumed that a combustible substance contained an ignitable substance called *terra pinguis*. Georg Ernst Stahl (1659–1734) was a student of Johann Becher and was very much influenced by his work. Based on the work of Johann Becher, in 1703, Georg Stahl proposed a similar theory called phlogiston theory of combustion in which heat is assumed to be a substance called *phlogiston* instead of terra pinguis. Therefore, it was assumed that all flammable substances contained phlogiston that was released when the substance was burnt.

Denis Papin (1647–1713) worked with Robert Boyle (1627–1691) from 1676 to 1679 and published his work "Continuation of New Experiments" in 1680. In 1679, Papin invented a steam digester, a type of pressure cooker, with safety valve but in 1690 he built a model steam engine with piston and cylinder, a first of its kind. In 1698, Thomas Savery (1650–1715) made the first commercially used steam pump (steam-powered device), which is often referred to as an engine although technically it is not.

J. S. Galsin, *History of Solid State Physics*,
https://doi.org/10.1007/978-981-95-0504-3_8

It was a new invention for raising water and was of great use in draining mines. On July 2, 1698, Savers applied for Patent for the design of his pump which he obtained in the same year. The main disadvantage of Savery's pump was that a large amount of heat was wasted in pumping water. In 1705 (published in 1707), Denis Papin developed an improved steam engine with the help of Gottfried Wilhelm Leibniz (1646–1716) based on the invention of Savery and this engine used steam pressure rather than the atmospheric pressure. Thomas Newcomen (1664–1729) invented the first practical fuel-burning engine for the purpose of lifting water out of mines. Combining the ideas of Denis Papin and Thomas Savery, Newcomen, in 1712, developed a more efficient steam engine with a piston separating the condensing steam from water.

Since 1702, Ole Romer, Danish astronomer, had been making alcohol thermometers because alcohol expands uniformly on heating and contracts on cooling. He invented a temperature scale with two fixed points, one at which water boils and the other at which water freezes, and divided the scale into equal increments. In 1708, Daniel Gabriel Fahrenheit (1686–1736) visited Ole Romer who introduced Fahrenheit with the method of making thermometers and his temperature scale. Romer told Fahrenheit that there was a great demand for accurate thermometers. Fahrenheit got inspired to work on making thermometers and the temperature scale. From 1709 to 1714 Fahrenheit made alcohol thermometers based on the design of Ole Romer. But in 1714, he invented the mercury thermometer which is more accurate and superior to the alcohol-filled thermometer. It consists of a narrow glass tube with a bulb at the bottom containing mercury. The early temperature scales were purely arbitrary but with time these scales were established in a systematic manner. In 1724, Fahrenheit developed a new temperature scale, which bears his name: Fahrenheit scale. He selected values 32 and 90 for the freezing point of water and the normal temperature of the human body and divided the scale into equal parts. Later he adjusted the higher value to 98.6. These were the first reliable thermometers of the time and their popularity led to the widespread adoption of the Fahrenheit scale. It was the first significant step towards the science of heat and the measurement of degree of hotness became possible. In 1730, Rene Antoine Ferchault de Reaumur (1683–1757), a French entomologist, constructed a thermometer containing diluted alcohol (Ethanol). In mercury thermometers, he established a temperature *scale* which bears his name, with zero set at the freezing point of water and its upper mark was at 80° which corresponded to the boiling point of water at normal atmospheric pressure. In 1742, Anders Celcius (1701–1744) proposed a temperature scale for a thermometer in a paper presented to the Royal Society of Sciences of Uppsala. He calibrated the scale with zero representing the boiling point of water while 100° represented the freezing point of water: reverse of the present known Celcius scale. In 1744, coincident with the death of Anders Celcius, the Swedish botanist Carl Linnaeus (1707–1778), a Swedish biologist and physician, reversed the Celcius scale with zero of the scale corresponding to the freezing point of water and 100° corresponding to the boiling point of water. Some gave credit to Jean-Pierre Christin (1683–1755) for independently inverting the Celcius scale. In 1848, William Thomson (1824–1907), also known as Lord Kelvin, proposed an absolute temperature scale that was independent of the properties of a material like ice or human body: called *Kelvin scale.* Kelvin degree is the unit of temperature in the System International (SI) of units and is represented by the symbol K: one Kelvin degree is equal to one centigrade degree. In this scale 0 °K, called absolute zero, corresponds to the lowest attainable temperature in the universe. In 1859, William John Macquorn Rankine (1820–1872) proposed a scale of temperature called *Rankine scale* (an absolute or thermodynamic scale), whose degree is equal to Fahrenheit degree.

William Thomson (1824–1907), Public Domain, URL: https://en.wikipedia.org/wiki/Lord_Kelvin#/media/File:Sir_William_Thomson,_Baron_Kelvin_by_T._&_R._Annan_&_Sons.jpg

In 1754, Jean Andre Deluc (1727–1817) performed an experiment in which ice was melted by supplying heat. He found that the thermometer showed no rise in temperature till whole of the ice melted and the heat supplied was used in melting ice to make water. During the melting process, the heat supplied was hidden (concealed) in ice and is called *latent heat* of ice: latent means hidden. In 1756, Andre Deluc observed that when heat was supplied for evaporating water into vapors the temperature again remained constant and the hidden heat was called latent heat of steam.

Latent heat of ice (steam) is the amount of heat required to change one gram of ice (water) into water (steam) without rise in temperature. Deluc reported these investigations in his book entitled "*Modifications of the atmosphere*" published in 1772. He summarized these experiments in a later book "*New ideas on meteorology*" published in 1787. In 1762, Joseph Black (1728–1799) independently introduced the concepts of *latent heat* and *specific heat*. There arose some doubt about the credit of discovery of latent heat. Black joined University of Glasgow and started giving lectures in 1757 in the subject of chemistry. James Watt was also a Professor in the University of Glasgow. James Watt said that Black must have taught both the laws publically as a Professor of chemistry in the winter of 1757–1758 or at the latest in the winter of 1758–1759. I learned about the latent heat from Black in 1762 or 1763. The theory of latent and specific heat is the most important contribution of Joseph Black which marked the beginning of thermodynamics. He also showed that different substances have different specific heats. In 1772, Johan Carl Wilke (1732–1796) calculated the latent heat of ice and in 1781 he coined the word specific heat in analogy with the term specific gravity. Specific heat of a solid is the amount of heat required to raise the temperature of one gram of the solid by one degree Celsius. Both latent heat and specific heat are the measure of storage of heat by a solid, but do not tell us anything about the mode of flow of heat. In 1780, Antoine Lavoisier (1743–1794) and Pierre Simon Laplace (1749–1827) invented calorimeter in 1782 and used it for the first time to find the latent heat of ice. The use of a calorimeter proved to be a standard method for quantifying heat. In 1783, they published a joint paper, "Memoir on Heat". Lavoisier and Laplace measured experimentally the specific heat of various substances. It can be said that Lavoisier initiated the first quantitative research on heat changes during a chemical reaction using an ice calorimeter. In 1783, Antoine Lavoisier discarded the phlogiston theory of heat in his paper "*Reflexionssur le phlogistique*", and proposed instead *caloric theory* of heat. He assumed that heat is some sort of subtle fluid called caloric which flows from hot to cold bodies. Since heat was assumed to be some sort of a material, therefore, it could neither be created nor destroyed leading to the conservation of heat: the central assumption of the caloric theory. Thus they established its principle of calorimetry that is the heat lost by a hot object equals the heat gained by a cold object. The word caloric came from the Latin word 'calor' meaning heat. In 1786, Deluc reported the facts about the latent and specific heats in his book entitled "*New ideas on meteorology*". Heat energy is measured in unit of calorie which is the heat required to raise the temperature of one gram of water by one degree Kelvin. The term calorie was first introduced by Nicolas Clement (1779–1841) as a unit of heat energy in his lectures on calorimetry delivered during the years 1819–1824: his unit calorie was actually kilocalorie. Definition of calorie was published, in 1824, in the journal *Le Producteur*.

Antoine Lavoisier (1743–1794), Public Domain, URL: https://en.wikipedia.org/wiki/Antoine_Lavoisier#/media/File:David_-_Portrait_of_Monsieur-Lavoisier_(cropped)2.jpg

At the same time, some scientists were trying to make machines using heat energy. In 1765, James Watt (1736–1819), Scottish inventor and engineer, greatly improved the Newcomen's steam engine by adding a separate steam condenser to avoid heating and cooling the cylinder with every stroke. He patented the design of the steam engine with a separate condenser (Patent 913 granted in 1769). Shortly after, James Watt developed a new steam engine that rotated a shaft instead of providing the up and down motion of the pump. James Watt made a compound engine, which connected two or more engines, and two patents were granted for these in 1781 and 1782. He made a number of other improvements to produce a practical power plant. In 1769, Nicolas-Joseph Cugnot (1725–1804), French army captain and inventor, made a three-wheeled steam carriage for roads to carry heavy loads like cannons but it was difficult to manage. In 1770, he made an improved version of steam carriage which was designed to carry a load of four tons with a speed of 8 km/h: the first full-size and self-propelled mechanical land vehicle, the "Fardier a vapeur"—effectively the world's first automobile. The model 1770 of Fardier a vapeur still exists and is on display at the Museum of Arts and Craft in Paris. In 1801, Richard Trevithick (1771–1833) built a steam road locomotive named 'Puffing Devil' and obtained its Patent

2599 on March 24, 1802. Puffing Devil was driven carrying six passengers but it was not able to maintain sufficient steam pressure for a long time. In 1803, Trevithick built another steam road vehicle named, 'London Steam Carriage'. It was driven successfully in London but was uncomfortable for the passengers. Moreover it proved more expensive to run than a horse-driven carriage and therefore was abandoned. Richard Trevithick is credited with the world's first realistic design of the steam locomotive. In 1814, George Stephenson (1781–1848) designed his first locomotive named Blucher in which a lot of improvements were made later. He used a 1.435 m rail gauge sometimes called "Stephenson gauge", which is used by most of the world's railways. In 1829, George Stephenson made the first commercial steam engine "Rocket" which was adopted by railways. In 1830, he built the first intercity railway line in the world "Liverpool and Manchester railway". On September 15, 1830, Rocket was driven successfully from Liverpool to Manchester. George Stephenson is renowned as the "Father of Railways". Prior to the invention of the steam engine, horses were used to lift water out of the mines with the help of the arrangement of pulleys attached to buckets. After the invention of engines, these were utilized for the same task in place of horses. Thus, each engine began to be associated with certain horsepower depending upon how many horses it had replaced. The early engines were crude and slow, converting only 2% of the fuel into useful work.

In 1776, Johann Heinrich Lambert (1728–1777) performed experiments for the study of flow of heat in solids. He took a long metallic rod and heated it at one end. It was found that a part of heat flowed along the length to the other end of the rod and rest of it was radiated to the atmosphere. Constant heat was supplied to the hot end by heating it continuously and it was found that every part of the rod acquired a constant temperature thus establishing the equilibrium state. In the equilibrium state amount of heat radiated by the whole surface of the rod is equal to the amount of heat received by it in a particular time. Benjamin Franklin (1706–1790) was contemporary of Johann Lambert and performed quite different experiments for studying the conduction of heat through metals. In an experiment, he took a rod of metal and coated it with wax. The rod was heated at one end and the distance along the rod up to which the wax melted in a particular time was observed. The same experiment was repeated taking exactly similar rods of different metals and studied relative heat conducting abilities from their comparison. In 1780, Benjamin Franklin visited France and there he suggested Jan Ingen-Housz (1730–1799) to perform these experiments further. Ingen-Housz took exactly similar wires of different metals and coated them with wax. He stretched the wires parallel to each other in a wooden block and dipped one end of each wire in a hot oil bath. He measured the speed of melting of wax along different wires and compared the conductivities from their comparison. Ingen-Housz compared the conductivities of a number of metals such as silver, copper, gold, tin, iron and lead etc. He found that wax on the silver wire melted fastest followed by copper, gold, tin, steel, iron and lead wire. He published his results in 1785. Sir Benjamin Thompson (1753–1814) was a Lieutenant colonel in the British army. He had a scientific bent of mind and was known for drawing designs for warships. He did a lot of experiments on gunnery and explosives which led to his interest in heat. Thompson studied the insulating properties of different materials including fur, wool and feathers. He was in search of a fabric for the soldiers that would better withstand the cold. He found that the insulating property of the materials, which have the ability to trap air inside it, enhanced. In 1785, Thompson moved to Bavaria where he stayed for about eleven years and did a lot of work for the poor. For his sincere efforts, in 1791, he was made an Imperial Count becoming Reichsgraf von Romford. Thereafter he took the name Romford after the town Romford, Hampshire. In 1798, Count Benjamin Thomson OR Count von Rumford observed the production of a large amount of heat during the boring of cast iron cannons. He observed that the specific heats of the material taken away during boring and the remaining material of the cannon were the same. It confirmed that no physical change had taken place in the material machined away. He argued that there is some sort of motion in heated bodies which is responsible for the production of heat. This finding was incompatible with the caloric theory of heat posing the first challenge to it. In 1799, Romford and Joseph Banks, Knight Commander of Bath, helped in establishing the Royal Institution of Great Britain and chose British chemist Sir Humphrey Davy as lecturer.

Up to the eighteenth century a number of physical properties like specific heat, latent heat and flow of heat through a solid were studied but these studies did not shed any light about the mode of flow of heat. In a solid the flow of heat is due to the conduction phenomenon in which heat is transferred from one atom to another. In 1804, Jean-Baptiste Biot (1774–1862) formulated the laws of conduction in solids. Joseph Fourier (1768–1830) was one of the greatest minds of mathematics. Most of his works are so fundamental that they are now applied in almost every area of natural science. He also discovered Greenhouse effect. In 1807, he presented a paper "*On the Propagation of Heat in Solid Bodies*" to the Paris Institute. From 1807 to 1811, he conducted experiments on thermal conduction and formulated a theory, for the first time, to evaluate thermal conductivity of a solid. His methodology had influenced all subsequent work on heat conduction in solids. In 1822, Fourier gave analytical theory of conduction of heat according to which the rate of heat transfer through a solid is proportional to the negative of the temperature gradient and the area at right angles to that gradient. Therefore, the rate of heat transfer per unit area **Q** is given by

$$\mathbf{Q} = -\sigma_{\mathrm{T}} \nabla \mathrm{T} \tag{8.1}$$

where σ_{T} is the coefficient of proportionality and is called thermal conductivity. Equation (8.1) shows that thermal energy transfer is a random process. The energy enters from one end of a specimen and does not proceed directly in a straight path to the other end rather energy diffuses through the specimen suffering frequent collisions. He showed that anharmonic terms are responsible for thermal conductivity. The terms with third, fourth and fifth powers of displacement in the potential energy function describe phonon–phonon coupling which are essential in understanding thermal conductivity. Thermal conductivity has two contributions: first contribution is due to lattice vibrations $\sigma_{\mathrm{T}}^{\ell}$ and second is due to electron motion $\sigma_{\mathrm{T}}^{\mathrm{e}}$. Therefore,

$$\sigma_{\mathrm{T}} = \sigma_{\mathrm{T}}^{\ell} + \sigma_{\mathrm{T}}^{\mathrm{e}} \tag{8.2}$$

and

$$\mathbf{Q} = -\sigma_{T}^{\ell}\nabla T - \sigma_{T}^{e}\nabla T \quad (8.3)$$

In insulators, thermal conductivity is only due to lattice vibrations but in metals both the lattice and electronic contributions are finite. In metals the main contribution to thermal conductivity comes from the electrons and the lattice vibration contribution is only 5–20%.

In 1841, James Prescott Joule (1818–1889) found that heat is generated in a wire carrying current which is proportional to resistance R of the wire and the square of current I, that is $Q = I^2R$: named as *Joule's first law*. He argued that the heating effect is due to the generation of heat, thus challenging the caloric theory of heat. In 1843, Joule proved experimentally that mechanical energy can be converted into heat proving that heat is a form of energy—*Joule's second law*. He gave the mechanical equivalent of heat, a constant. Initially his work was little read by the scientific community. But later using the kinetic theory it was established that with the increase in temperature the average velocity of atoms or molecules increases: showing beyond doubt that heat is a form of energy.

In 1701, Isaac Newton (1643–1727) described his cooling law for the transfer of heat by convection process which is dominant in liquids and gases: the rate of cooling of a warm body is proportional to the temperature difference between the body and its ambient fluid. In 1777, Carl Wilhelm Scheele (1742–1786) found that heat was also transferred by thermal radiation and distinguished it from convection and conduction. In 1791, Pierre Prevost (1751–1839) showed that all bodies radiated heat irrespective of their temperature (cold and hot). In 1804, John Leslie (1766–1832) studied the radiation of heat from different surfaces. He took a hollow cubical box filled with boiling water. One side of the cubical box was highly polished, two sides were made of dull metal and one side painted black. Leslie found that radiation emitted from the black surface was the greatest and negligible from the polished surface. This special type of cube is usually known as *Leslie cube*. The radiation emitted by a black body is known as blackbody radiation. In 1831, Macedonio Melloni (1798–1854) demonstrated that blackbody radiations could be reflected, refracted and polarized just like light. In 1862, James Clerk Maxwell (1831–1879) showed that both light and heat radiations were electromagnetic radiations with different wavelengths. In 1879, Jozef Stefan (1835–1893) observed that total radiant flux from a black body is proportional to the fourth power of temperature in Kelvin degrees. In 1884, this law was derived theoretically by Ludwig Boltzmann (1844–1906) and is called *Stefan-Boltzmann law*. In 1893, Wilhelm Wien (1864–1928) discovered the displacement law which calculates the emission of a blackbody radiation at any temperature—*Wien's displacement law*. In 1896, Wien empirically determined a distribution law of blackbody radiation. But Max Planck, a colleague of Wien, did not believe in empirical laws and proposed a theoretical basis for Wien's law, which became the Wien-Planck law.

Portrait of Robert William Boyle (1627–1691), Public Domain, URL: https://en.wikipedia.org/wiki/Robert_Boyle#/media/File:The_Shannon_Portrait_of_the_Hon_Robert_Boyle.jpg

In 1656, Robert William Boyle (1627–1691) and Robert Hooke (1635–1703) found that in case of air or gas the product of volume V and pressure P is a constant quantity, that is PV = constant. It is usually known as *Boyle's law*. At that time, the concept of thermal motion was not known which came after about two centuries. Therefore, Robert Boyle assumed that air consisted of motionless particles. After the discovery of thermometer, the hotness of a body could be quantified. Jacques Charles (1746–1823) found that volume of a gas is proportional to temperature T in degrees Kelvin: also called law of equal volumes. In 1802, Joseph Louis Gay-Lussac (1778–1850) published this law from the unpublished work of Jacques Charles and named it *Charles law* in his honor. Gay-Lussac worked from 1801–1802 and stated another law according to which if mass and volume of a gas are kept constant, the gas pressure P increases linearly with temperature T, that is P = kT, where k is a constant dependent on mass and volume of the gas: known as *Gay-Lussac's law* or *pressure law*. It is believed that Guillaume Amontons (1663–1705) knew about this law and devised a method to measure change in temperature due to the change in pressure of a gas. He found that pressure is proportional to temperature but his results were at the best semi-quantitative. In 1834, Emile Clapeyron (1799–1864) combined the Boyle's, Charles' and Gay-Lussac's laws and proposed an equation for an ideal gas given by

$$PV = nRT \tag{8.4}$$

Here n is the number of moles of the gas and R is the universal gas constant. It is called *ideal gas equation* and defines the equation of state for an ideal gas.

Portrait of Nicolas Leonard Sadi Carnot (1796–1832), Public Domain, URL: https://en.wikipedia.org/wiki/Nicolas_Léonard_Sadi_Carnot#/media/File:Sadi_Carnot.jpeg

In 1824, Nicolas Leonard Sadi Carnot (1796–1832) published "*Reflections on Motive Power of Fire*" and in it he discussed Carnot heat engine, Carnot theorem and Carnot efficiency. He proposed an ideal heat engine and gave its successful theory in terms of *Carnot cycle* and evaluated its efficiency. Carnot discovered that a steam engine did not work by consuming caloric but it worked on the process of transfer of heat from a hot reservoir to a cold one using a part of heat energy to do work. The greater the temperature difference between the two reservoirs the more is the efficiency of the engine. Unfortunately, Carnot's work attracted little attention during his lifetime. But the concept of an idealized heat engine led to the foundation of a new discipline called *thermodynamics*—a branch of physics which describes the relationship between heat and other forms of energy. The systems consisting of solids, which dealt with heat and the useful work done by it, were studied by thermodynamics. Sadi Carnot is considered to be the *father of thermodynamics.*

Specific heat is classified in two categories: specific heat at constant volume C_V and specific heat at constant pressure C_P: usually C_P is measured experimentally while C_V is evaluated theoretically. Julius Robert von Mayer (1814–1878) derived the relation between C_V and C_P and measured it experimentally. The relation between C_V and C_P is given by

$$C_P - C_V = 3R \tag{8.5}$$

where R is the gas constant. Equation (8.5) is called *Mayer's relation.* Mayer found that chemical reaction produced both heat and work while work could produce definite amount of heat only. Mayer made a statement that in a process at constant pressure, heat used to produce expansion is universally inconvertible with work. In 1842, Julius Mayer was the first person to state the law of conservation of energy but it was not a general statement of the *first law of thermodynamics.* In 1843, James Joule (1818–1889) showed that mechanical energy can be converted into heat energy and vice versa and hence laid the foundation of the theory of conservation of energy which later influenced the first law of thermodynamics. In 1850, Rudolf Clausius (1822–1888) gave the full statement of *first law of thermodynamics*: it states that energy can neither be created nor destroyed but can only be transferred from one form into another. In 1851, Rudolf Clausius published his famous paper "On the Moving Force of Heat" in which he carefully examined the Carnot's cycle and stated the basic ideas of the *second law of thermodynamics.* Clausius's most famous statement of the second law of thermodynamics was published in German in 1854 and in English in 1856. In 1851, William Thomson (1824–1907) gave his statement of second law of thermodynamics as, "It is impossible to convert heat completely into work in a cyclic process". It means some energy is lost in converting the energy of a system into work. In 1854, William Thomson actually coined the word thermodynamics in his paper "On the Dynamical Theory of Heat". In the same year, Rudolf Clausius gave his own statement of second law of thermodynamics according to which "Heat generally cannot flow spontaneously from a material at low temperature to a material at higher temperature". In the year 1854 itself, William John Macquorn Rankine (1820–1872) used a quantity which he named 'thermodynamic function' in a thermodynamic system. In 1865, Rudolf Clausius introduced the concept of *entropy* meaning *content transformative* and is a highlight of nineteenth century physics. He found that thermodynamic function of Rankine was the same as entropy. Entropy is a measure of thermal energy of a system that is unavailable for doing useful work. Because work is obtained from an ordered molecular motion, therefore entropy can be regarded as a measure of molecular disorder or randomness of a system. William Rankine also defined 'curve of nontransmission of heat' which he later called adiabatic curve and the process was called *adiabatic process.* Walther Nernst (1864–1941) worked on thermodynamical systems and in 1905, put forward Nernst theorem which is commonly known as *third law of thermodynamics.* The third law states that entropy of a system at absolute zero is a well-defined constant generally called residual entropy. In crystalline solids, the residual entropy is very close to zero. Entropy approaches zero at absolute zero only when a solid has a unique ground state (a state with minimum energy having only one configuration). In other words, every solid has finite entropy as absolute zero of temperature cannot be approached. In 1931, Ralph H. Fowler (1889–1944) was the first to formulate *zeroth law of thermodynamics* which defined thermal equilibrium between two or more systems and formed the basis for the definition of temperature without reference to entropy, its conjugate variable. Ralph H. Fowler, for the first time, coined the word zeroth law of thermodynamics.

Kinetic theory of gases is of fundamental importance in explaining heat and thermodynamical systems. In 1620, Francis Bacon (1561–1626) described that heat is some form of motion as was considered from ancient times. Mikhail Lomonosov (1711–1765) had written in his writings that heat is due to the motion of very small particles hidden in a substance. He had contributed towards the formulation of kinetic theory of gases. Lomonosov, together with Lavoisier, is regarded as the one who discovered the law of mass conservation in chemical reactions. In '*Hydrodynamica*' published in 1738, Daniel Bernoulli (1700–1782) derived an expression for the pressure exerted by a gas considering collisions of its particles with the walls of a container. It is believed that the principle of superposition was first stated by Bernoulli in 1753. From 1676 to 1689, Gottfried Leibniz (1646–1716) formulated the principle of conservation of kinetic energy when there is no interaction among the particles. On the other hand, Isaac Newton (1643–1727) and Rene Descarte (1596–1650) emphasized the conservation of momentum. Actually, both energy and momentum are conserved in a mechanical system. In 1820, John Herapath (1790–1868) formulated kinetic theory of gases but he made a mistake by associating temperature with momentum. This assumption led him to the incorrect relation that is the product of pressure and volume is proportional to the square of absolute temperature. In 1820, he submitted his paper to the Royal Society, London, which was reviewed by Sir Humphry Davy and rejected. In 1843, John James Waterston (1811–1883) independently formulated the kinetic theory of gases and derived the ideal gas equation. He submitted his paper to Royal Society, London but was rejected by saying that '*The paper is nothing but nonsense*'. Thus, his pioneer work of kinetic theory of gases remained neglected. Based on the work of John Waterston, August Karl Kronig (1822–1879) developed kinetic theory of gases. The real progress in the development of kinetic theory of gases commenced in the middle of nineteenth century. In 1857, Rudolf Clausius (1822–1888), for the first time, stated that heat is a form of energy and is equal to the average kinetic energy of molecules. He gave modern account of kinetic energy of gases in his publication "*On the Nature of Motion called Heat*". In 1870, Clausius also proved the scalar virial theorem. In 1859, James Clerk Maxwell (1831–1879) derived a distribution law for the molecular velocities and hence formulated a new branch of thermodynamics called "*Statistical Mechanics*". With the help of statistical mechanics, a large number of particles in equilibrium can be analyzed and the only thing one needs to know is the average values of temperature T, pressure

P and volume V. In 1872, Ludwig Boltzmann (1844–1906) generalized the distribution for gases in an external field, which is usually known as *Maxwell–Boltzmann distribution.* Later, in 1877, he derived the distribution again under the framework of statistical thermodynamics.

The greatest achievements of Ludwig Boltzmann were the development of statistical mechanics and the statistical explanation of the second law of thermodynamics. He developed the kinetic energy of gases and associated kinetic energy of particles to their degrees of freedom. In 1872, Ludwig Boltzmann formulated a general equation for studying transport properties of particles named as *Boltzmann transport equation* in his honor. Both the classical and quantum mechanical versions of the Boltzmann transport equation were derived. It provided an alternate statistical approach for studying the transport properties of solids such as electrical and thermal conductivities etc. and described the statistical behavior of a thermodynamical system not in the equilibrium state. From 1872–1875, Boltzmann derived the most important relation between entropy S and molecular motion given by

$$S = k_B \ln \Omega \tag{8.6}$$

where k_B is the Boltzmann constant (universal constant) and Ω in the number of possible microstates. Equation (8.6) is a fundamental equation that established the relation between thermodynamical and statistical approaches. It is also viewed as the relation between macroscopic and microscopic quantities of a thermodynamical system. A macrostate is experimentally observable state, while a microstate can be instantaneous and cannot be observed experimentally.

In 1876, Josiah Willard Gibbs (1839–1903) published a paper titled, "*On the Equilibrium of Heterogeneous Substances*" in which he gave the concept of *free energy* and derived free energy equation: here we talk about the internal energy of a system. The Gibbs free energy is a thermodynamical potential that can be used to calculate maximum reversible work which may be performed by a thermodynamical system at constant temperature and pressure. Gibbs showed that thermodynamics is not only applicable to heat and energy but also valid for chemical reactions at the atomic level. Gibbs' Equilibrium has been called "the Principia of thermodynamics" which is applicable to all types of substances such as gases, gas mixtures, surfaces and solids etc. In 1882, in a lecture titled "*On the Thermodynamics of Chemical Processes*", Hermann von Helmholtz (1821–1894) proposed free energy in a solid that measures the useful work obtained from a closed thermodynamical system at constant temperature (isothermal). Gibbs free energy is a more general form of the free energy of a substance and it reduces to the Helmholtz free energy at constant pressure and volume. Gibbs also introduced the concept of *enthalpy*, represented by symbol H, calling it heat function for constant pressure. Actually, the word enthalpy was coined by Heike Kamerlingh Onnes (1853–1926) many years later in 1909. Until 1920, the symbol H was used, somewhat, inconsistently for heat. In 1922, Alfred A.

Porter (1863–1939) fixed the letter H strictly for enthalpy or heat content at constant pressure.

Another important aspect is thermal expansion of solids, liquids and gases on heating. John Dalton (1766–1844) in 1801, Joseph Louis Gay-Lussac (1778–1850) in 1802, Heinrich Gustav Magnus (1802–1870) in 1841–1844 and Henri Victor Regnault (1810–1878) again in 1842, independently, showed that thermal expansion of ideal gases is practically alike and therefore, represents at least more than the property of a single material. In a solid, at a finite temperature, atoms vibrate about their mean positions. The temperature and kinetic energy of atoms or molecules increase when a solid is heated. With the increase in temperature, atoms or molecules vibrate faster with greater amplitude which results in the increase in average separation between atoms and thus the thermal expansion. In other words, it can be said that the vibration of atoms is the origin of thermal expansion. Therefore, in a solid, atoms themselves do not expand, but the average volume occupied by an atom increases. The relative increase in the size of a solid when heated is therefore quite small. In 1896, Charles Edouard Guillaume (1861–1938) was the first to discover thermal expansion in solids. Actually, he measured very small thermal expansion with high precision in invar alloys (nickel–iron alloys). He was awarded 1920 Nobel Prize in Physics for the discovery of thermal expansion in invar alloys. In solids three types of thermal expansion occur depending on the dimensions that undergo change: linear expansion, areal expansion and volumetric expansion. The three types of expansion are related to one another: the coefficient of volumetric expansion is thrice while the coefficient of areal expansion is twice the coefficient of linear expansion in an isotropic solid: a solid which expands at the same rate in every direction is called isotropic solid. Theoretically speaking thermal expansion arises from the anharmonic terms in the potential energy function. The terms with third, fourth and fifth powers of displacement in potential energy describe phonon–phonon and phonon-strain couplings which contribute maximum to thermal expansion.

Helmholtz free energy provides all the information that is needed to obtain lattice parameter as a function of temperature and hence the thermal expansion coefficient. The Helmholtz energy is minimized to find the lattice parameter at a particular temperature T. In 1912, Eduard Gruneisen (1877–1949) proposed a theory of thermal expansion in terms of interatomic forces and atomic vibrations in solids. Although several developments had taken place in the understanding of crystalline solids but the Gruneisen theory still provides the basic framework and continues to dominate in the field of thermal expansion. There are some very rare materials which exhibited negative thermal expansion: they contract on heating. Such materials are of enormous industrial importance because they can be used to control thermal expansion in other materials.

Just like electrical conductivity, thermal conductivity in metals is also of equal importance and is related to anharmonicity in the potential function. In 1914, Peter Debye (1884–1966) formulated a theory of thermal conductivity for insulators. He assumed that on heating, the anharmonic lattice waves propagating through a solid are coupled due to anharmonicity. He introduced phonon mean free path τ_{ph} which measured coupling between lattice waves. Using the kinetic theory of gases thermal conductivity is given by the famous formula

$$\sigma_T = \frac{1}{3} C_v v \tau_{ph} \tag{8.7}$$

where v is the velocity of sound (lattice waves). From Eq. (8.7) it is evident that on the right side the only temperature dependent quantity is C_V, therefore, the temperature dependence of σ_T and C_V are nearly the same as observed experimentally. Debye applied the theory to study thermal conductivity in dielectrics. In 1929, Rudolf Ernst Peierls (1907–1995) extended the theory of Debye and proposed the existence of phonon–phonon interactions. In the presence of temperature gradient across a solid the phonon distribution cannot be in the equilibrium state, but the phonon–phonon interactions tend to restore the equilibrium. De Haas and Biermasz (1935) discovered the dependence of τ_{ph} on the scattering of phonons at the boundaries of a system. The thermal resistance is determined by the rate of the restoring processes, that is, the phonon–phonon interactions. In 1938, Hendrik Casimir (1909–2000) presented a detailed quantitative calculation of thermal conductivity and explained its temperature dependence. Peierls provided explanation for phonon scattering at boundaries and Casimir worked out the same. Berman (1953) had included explicitly the effect of umklapp processes in the calculation of thermal conductivity. The phonon mean-free path τ_{ph} depends on three factors and these are scattering of phonons at boundaries, the density of phonons, and the scattering by impurities and lattice interactions. In metals, major contribution to thermal conductivity comes from the electron scattering that has two contributions. The first contribution arises due to the scattering of electrons by lattice vibrations (electron–phonon interaction) and second due to the electron scattering from impurities. The second contribution is related to residual resistivity. The electronic contribution to thermal conductivity in metals had been calculated using the Boltzmann transport equation and it obeys the Wiedemann–Franz law. The Lorentz number obtained from this theory exhibits good agreement with the experimental data. In the late 1940s and early 1950s, Berman and coworkers and Rosenberg and coworkers measured experimentally the thermal conductivity of several metals, alloys and compounds as a function of temperature. It was found that electron and phonon contributions were of the same orders of magnitude.

In 1902, Josiah Willard Gibbs (1839–1903) introduced the concept of canonical ensemble which is composed of identical systems each having the same volume V, number of particles N, and temperature T. In an ensemble, the various systems are partitioned by isothermal walls to permit the flow of temperature but not particles. The ensemble itself is isolated from the surroundings by adiabatic walls. Later in 1902, Willard Gibbs proposed grand canonical ensemble which is the generalization of canonical ensemble where the restriction of definite number of particles is removed. With the help of ensemble theory, one can explain various properties of solids.

Satyandra Nath Bose (1894–1974), Public Domain, URL: https://en.wikipedia.org/wiki/Satyendra_Nath_Bose#/media/File:SatyenBose1925.jpg

The era of quantum statistics started with the proposals of distribution functions for quantum particles. In 1924, Satyandra Nath Bose (1894–1974) and Albert Einstein (1879–1955) proposed a distribution for photons called Bose–Einstein distribution. In 1924–1925, Einstein extended this distribution to atoms or particles with integral spin (called Bosons) and this led to the prediction of the existence of phenomenon of Bose–Einstein condensate, a dense collection of Bosons. In 1926, Enrico Fermi (1901–1954) and Paul Dirac (1902–1984) proposed a distribution function for half-integral spin particles (called Fermions) and it is called Fermi–Dirac distribution function. These distributions gave birth to the Bose–Einstein and Fermi–Dirac statistics. With the help of these statistics various properties of solids such as thermionic emission, phase transitions, electronic specific heat, Pauli paramagnetism, Landau diamagnetism, and photoelectron emission etc. have been studied using the ensemble approach. In 1995, Eric Allin Cornell (1961–present) and Carl Edwin Wieman (1951–present) were the first to create Bose–Einstein condensate in a gas of rubidium atoms at very low temperatures [see Anderson and coworkers (1995)]. Soon after Wolfgang Ketterle (1957–present) created Bose–Einstein condensate in a gas of sodium atoms [see Davis and co-workers (1995)]. Bose–Einstein condensation being one of the most important properties, Cornell, Wieman and Ketterle shared the 2001 Physics Nobel Prize for this discovery.

8.1 Specific Heat of Solids

The most extensively studied property of solids is specific heat which is expressed in calories per degree. Calorie is defined as the specific heat of water. In 1819, Pierre Louis Dulong (1785–1838) and Alexis Therese Petit (1791–1820) measured the specific heat of a number of solids at room temperature and proposed a law for molar specific heat known as *Dulong-Petit's law.* According to this law specific heat of solids is constant and is ≈ 6 cal per mole. At that time, the gas constant R had not yet been defined. Later the universal gas constant R was defined from the kinetic theory of gases and it was shown that specific heat of solids at high temperatures is $3R \approx 6$ cal per mole. In 1831, Franz Ernst Neumann (1798–1895) extended this law to compound materials and is known as Kopp-Neumann rule: the molecular heat of a compound is equal to the sum of the atomic heats of its constituents. It is worth mentioning that this rule was published 30 years later as Neumann was having a habit of scarcely publishing his findings. In 1872 and 1874, Heinrich Friedrich Weber (1842–1913) published two papers and found that the value of specific heat of diamond, graphite, boron and silicon were much lower than the Dulong-Petit's value. He further found that the specific heat of these elements approached the expected value as the temperature was increased. In 1911, Walther Nernst (1864–1941) discovered that specific heat decreased markedly at low temperatures and disappeared at absolute zero. At very low temperatures the specific heat varied as T^3, called T^3-law. In the medium temperature range, a gradual transition took place from T^3-law to linear variation of specific heat with T and finally at high temperatures it exhibited the Dulong-Petit's law.

In the beginning scientists tried to explain specific heat using kinetic theory of gases. As the simplest approximation, a solid was considered to be made of free atoms confined in a box which obeyed the Maxwell–Boltzmann distribution. Applying the kinetic theory of gases, C_V came out to be (3/2)R which was not in agreement with the Dulong and Petit's law. As an improvement over the free atom model, each atom was assumed to be an independent classical harmonic oscillator and it yielded $C_V = 3R$: Dulong and Petit's value. But the classical theories failed to explain the temperature dependence of specific heat. Albert Einstein (1879–1955) discarded the principle of equipartition of energy and adopted the Planck's revolutionary idea of quantization of energy. In 1907, Albert Einstein proposed a quantum theory of specific heat in which atoms were assumed to be independent quantum harmonic oscillators vibrating about their equilibrium positions with the same frequency called Einstein frequency (monochromatic approximation). These oscillators were assumed to obey the Maxwell–Boltzmann distribution (at that time Bose–Einstein and Fermi Dirac distributions were not known). Einstein theory explained the temperature dependence of specific heat of solids approaching zero exponentially at absolute zero. But this

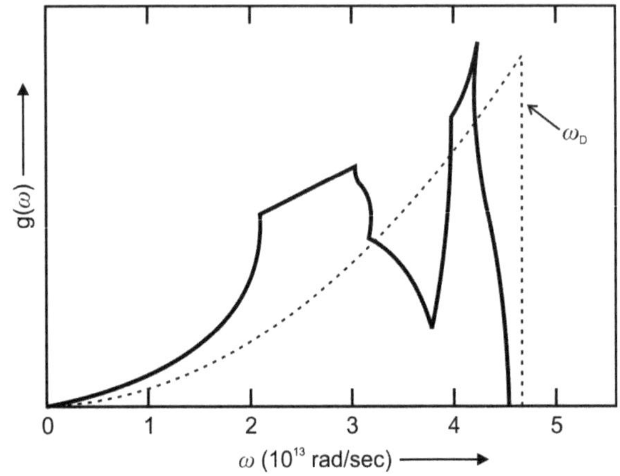

Fig. 8.1 Phonon density of states as a function of frequency ω in Cu metal. Solid curve is deduced from neutron scattering experiments by Svensson and co-workers (1967) and by Nilsson and Rolandson (1973). The dashed line represents the Debye approximation for the phonon density of states with $\omega_D = 4.5 \times 10^{13}$ rad s^{-1} (or with Debye temperature θ_D = 344 K)

work attracted little attraction of the scientists for several years because at that time the temperature dependence of specific heat was not fully known at low temperatures as mentioned above. Einstein theory explained the temperature dependence of specific heat at intermediate and high temperatures only but failed to explain T^3-behavior at very low temperatures. Einstein himself realized that atomic (molecular) oscillations were complex systems and the monochromatic approximation may not be good. The third law of thermodynamics (Nernst's theorem) states that entropy approaches zero at absolute zero, therefore the specific heat must also approach a limiting value independent of the nature of the material. Peter Debye (1884–1966) became familiar with the work of Einstein and he pointed out the shortcoming of monochromatic approximation. In 1912, Debye improved upon the Einstein theory by assuming atoms of a solid as quantum oscillators vibrating with frequencies ranging from zero to some maximum value called Debye frequency. Debye theory explained the T^3-behavior of specific heat at very low temperatures. Debye's formula for the specific heat has a very simple form and therefore is still used to calculate heat content of solids. Debye had used a very crude form of density of phonon states but later a number of attempts had been made to find better approximations for estimating the density of phonon states. Almost simultaneously Max Born (1882–1970) and Theodor von Karman (1881–1963) used lattice dynamical theory for the estimation of specific heat which is a more detailed and more accurate approach. These considerations yielded an expression for the specific heat in excellent agreement with the experimental results even at low temperatures. However, Max Born theory is much more complicated in comparison with the Debye formula. In 1937, Moses Blackman (1908–1983), for the first time, estimated the density of phonon states for simple crystal structures. Figure 8.1 shows the calculated density of phonon states for copper metal which is quite different from that of Debye approximation. Till date a number of scientists have evaluated the specific heat of a large number of materials using the actual form of density of phonon states.

Bibliography

Anderson MH, Ensher JR, Matthews MR, Wieman CE, Cornell EA (1995) Observation of Bose-Einstein condensation in a dilute atomic vapor. Science 269:198–201

Becher JJ (1667) Physica Subterranea, p 256

Berman R (1949) Thermal conductivity of glasses at low temperatures. Phys Rev 76:315

Berman R (1951) The thermal conductivity of some alloys at low temperatures. Phil Mag 42:642–650

Berman R (1953) The thermal conductivity of dielectric solids at low temperatures. Adv Phys 2:103–140

Berman R, Foster EL, Rosenberg HM (1955) The thermal conductivity of some technical materials at low temperatures. Brit Appl Phys 6:181

Berman R, Klemens PG, Simon FE, Fry TM (1950) Effect of neutron irradiation on the thermal conductivity of a quartz crystal at low temperature. Nature 166:864–866

Berman R, MacDonald DKC (1952) The thermal and electrical conductivity of copper at low temperatures. Proc Roy Soc (London) A211:122–128

Biot JB (1804) Mamoire sur la chaleur. Bibl Br Sci Arts 27:310–329; (1804) Memoire sur la propagation de la chaleur. J des Mines 13:203–224

Black J (1803) Lectures on elements of chemistry, vol 1. Mundall and Sons, Edinburgh

Blackman M (1937) On the vibrational spectrum of a three-dimensional lattice. Proc Roy Soc (London) A159:416–431

Born M, Von Karman T (1912) Uber Schwingungen im Raumgittern. Physikalishe Zeitschrift 13:297–309; (1913) Theory of specific heat. Physikalishe Zeitschrift 14:15–71

Boltzmann L (1872) Sitzungsberichte der Kaiserlichen Akademie der Wissenschaften in Wien. Mathematisch-Naturwissenschaftliche Classe 66:275–370; (1877) Sitzungsberichte derKaiserlichen Akademie der Wissenschaften in Wien. Mathematisch-Naturwissenschaftliche Classe 76:373–435

Boltzmann L (1884) Ableitung des Stefan'schen gesetzes, betreffend die abhängigkeit derwärmestrahlung von der temperatur aus der electromagnetischen lichttheorie [Derivation of Stefan'slaw, concerning the dependency of heat radiationon temperature, from the electromagnetic theory of light]. Ann Phys Chem (in German) 258:291–294

Bose SN (1924) Plancks Gesetz und Lichtquantenhypothese. Z Phys 26:178–181

Boyle R, Sharrock R (1662) A defence of the doctrine touching the spring and weight of the air. Thomas Robinson, London

Carnot S (1824) Réflexions sur la puissance motrice du feu (Reflections on the motive power of fire), reprinted by Dover in 1960

Casimir HBG (1938) Note on the conduction of heat in crystals. Physica 5:495–500

Celcius A (1942) Observationer om bestandiga grader pa en thermometer [Observation about two stable degrees on a thermometer]. Kungliga Svenska Vetenskaps Akademiens Handlingur, (Proceedings of the Royal Swedish Academy of Sciences (in Swedish) 3:171–180

Clapeyron E (1834) Memoire sur la puissance motrice de la chaleur, Journal de l'École Royale Polytechnique (in French). Paris: De l'Imprimerie Royale. Vingt-troisième cahier, Tome XIV:153–190

Clausius R (1850) Uber die bewegende kraft der warme und die gesetze, welche sich darausfur die warmelehre selbst ableiten lassen. Ann der Phys 79:368–397, 500–524; (1851) On the moving force of heat, and the laws regarding the nature of heat itselfwhich are deducible therefrom. Phil Mag 2:1–21, 102–119; (1854) Uber eineveranderte form des zweiten hauptsatzes der mechanischen warmetheoriein. Ann der Phys und Chem 93:481–506

Clausius R (1857) Uber die art der bewegung, die wir warme nennen. Ann Phys 100:353; (1862) Uber die warmeleitung gasformiger korper. Ann der Physik 115(1); (1864) Abhandlungen uber die mechanische warmetheorie, electronic manuscript from the "Bibliotheque nationale de France"

Clausius R (1865) Ueber verschiedene für die anwendung bequeme formen der hauptgleichungen der mechanischen wärmetheorie. Ann der Phys 125:353–400; (1870) On a mechanical theorem applicable to heat. Philos Mag 40:122–127
Dalton J (1802) Essay IV. On the expansion of elastic fluids by heat. Memoirs of the Literaryand Philosophical Society of Manchester (2nd.) 5:600
Davis KB, Mewes MO, Andrews MR, Van Druten MJ, Durfee DS, Kurn DM, Ketterle W (1995) Bose-Einstein condensation in a gas of sodium atoms. Phys Rev Lett 75:3969
De Haas WJ, Biermasz T (1935) Thermal conductivity of quartz at low temperatures. Communication 236C, from Kamerligh Onnes Laboratory at Leiden, The Netherlands; (1935) The thermal conductivity of quartz at low temperatures. Physica 2:673–682; (1937) The thermal conductivity of KBr, KCl and SiO_2 at low temperatures. Physica 4:752–756; (1938) The thermal conductivity of diamond and potassiumchloride. Physica 5:47–53
Debye P (1912) Zur theorie der spezifischen waerme. Ann der Phys 39:789–883
Debye P (1914) Vortrageuber die kinetische theorie der materie und der elektrizitat. Tueber, Berlin, p 40
Deluc JA (1772) Recherches sur les modifications de l'atmosphere [Modifications of the atmosphere], Vol. 2, Geneva; (1787) Neue ideen uber die meteorology [New ideas on meteorology], Erster Theil Aus dem Franzosischen ubersetzt, Fr. Nicolai, Berlin and Stettin, pp 143, 459
Dirac PAM (1926) On the theory of quantum mechanics. Proc Roy Soc (London) A112:661–677
Dulong PL, Petit AT (1819) Recherches sur quelques points importants de la théoriede la chaleur. Annales de Chimie et de Physique 10: 395. [English translation: (1819) Research on some important aspects of the theory of heat. Ann Philos 14:189]
Einstein A (1907) Theorie der strahlung und die theorie der spezifischen warme. Ann Phys 4:180–190
Fermi E (1926) Sulla quantizzazione del gas perfetto monoatomico. Rendiconti Lincei (in Italian) 3:145–149
Fahrenheit DG (1724) Experimenta circa gradum caloris liquorum nonnullorum ebullientium instituta [Experiments done on the digree of heat of a few boiling liquids]. Phil Trans Roy Soc 33:1–3; (1724) Experimenta et Observationes de Congelatione aquae in vacuo factae. Phil Trans Roy Soc 33:78–84
Fourier J (1807) Sur la propagation de la chaleur, manuscript; (1822) Théorie analytique de la chaleur. Firmin Didot Père et Fils, Paris
Gay-Lussac JL (1802) Recherches sur la dilatation des gaz et des vapeurs. Annales de Chimie XLIII:137
Gibbs JW (1876) On the equilibrium of heterogeneous substances. Trans Connecticut Acad Arts Sci 3:108–248; (1877) On the equilibrium of heterogeneous substances. Trans Connecticut Acad Arts Sci 3:343–524
Gibbs JW (1902) Elementary principles in statistical mechanics, developed with special reference to the rational foundation of thermodynamics. Charles Scribner's Sons, New York
Gruneisen E (1912) Theorie des festen zustandes einatomiger elemente. Ann Phys 344:257–306
Igenhouz J (1785) Nouvelle experiences et observations sur divers objects de physique. Barrois, Paris
Joule JP (1841) On the heat evolved by metallic conductors of electricity, and in the cells of a battery during electrolysis. Philos Mag 19:260; (1843) On the calorific effects of magneto-electricity and on the mechanical value of heat. Philos Mag 23:263–276, 347–355, 435–443
Kronig AK (1856) Grundzüge einer theorie der gase. Ann Phys 33:315
Lambert JH (1774/1776) Remarques sur le tempérament en musique. Nouveaux Mémoires de l'Académie royale des sciences de Berlin, année, pp 55–73; (1779) Pyrometrie. Haude und Spener, Berlin, pp 326–327
Lavoisier AL (1783) Reflexions sur le phlogistique, Academie des Sciences, Paris, [The paper was read before the Academy of Sciences, Paris]

Lavoisier AL, Laplace PS (1780) Memoire sur le chaleur. Mem Acad Sci Annee, 355–408; (1783) Memoir on heat. Read to the Royal Academy of Sciences by Messrs. Lavoisier & De La Place of the same Academy (New York: Neale Watson Academic Publications, 1982)
Leslie J (1804) An experimental enquiry into the nature and propagation of heat, London
Maxwell JC (1860) Illustrations of the dynamical theory of gases.—Part I. On the motions and collisions of perfectly elastic spheres. Lond Edinburgh Dublin Philos Mag J Sci 19:19–32; (1860) Illustrations of the dynamical theory of gases-II. Lond Edinburgh Dublin Philos Mag J Sci 20:21–37
Nernst W, Lindemann FA (1911) Spezifische warme und quantentheorie. Zeit f Elec Chem 17:817–827; (1911) Ber Berl Akad 1:494
Newton I (1701) VII. Scala graduum caloris. Phil Tans Roy Soc (London) 22:824–829
Nilsson G, Rolandson S (1973) Lattice dynamics of copper at 80K. Phys Rev B 7:2393
Peierls R (1929) Zur kinetischen theorie der wärmeleitung in kristallen. Ann Phys 395:1055–1101
Prévost P (1791) Memoire sur l'equilibre du feu. Observations Sur la Physique (in French) XXXVIII(1):314–323
Report from the Select Committee on Steam Carriages, Parliament of United Kingdom, pp 123–144 (1834)
Rosenberg HM (1955) The thermal conductivity of metals at low temperatures. Philos Trans Roy Soc (London) A247:441–497
Rumford (1792) Experiments upon heat. Phil Trans Roy Soc (London), 48–80; (1797) On the propagation of heat in fluids. Nicolson's J 1:298–341
Stefan J (1879) Uber die beziehung zwischen der warmestrahlung und der temperatur, sitzungsberichteder kaiserlichen akademie der wissenschaften: mathematisch-naturwissenschaftliche classe. Proc Imperial Philos Acad [Vienna]: Math Sci Class (in German) 79:391–428
Svensson EC, Brockhouse BN, Rowe JM (1967) Crystal dynamics of copper. Phys Rev 155:619
Thomson W (1848) On an absolute thermometric scale founded on Carnot's theory of motive power of heat and calculated from Regnault's observations. Mathematical and physical papers. Cambridge University Press, pp 100–106
von Helmholtz H (1882) Physical memoirs, selected and translated from foreign sources. Taylor & Francis
Weber HF (1874) Die spezifische wärme der elemente kohlenstoff, bor und silicium. Pogg Ann, 154
Wien W (1896) Über die energievertheilung im emissionsspectrum eines schwarzen körpers. Annalen der Physik und Chemie 294:662–669
Wilke JC (1772) Om snons kyla vid smaltningen [On the snow's cold at melting]. Proc R Swedish Acad Sci 33:97–120; (1981) Findings on the specific quantity of heat in solid bodies and on its degradation. Proc R Swedish Acad Sci 2:49–78

Chapter 9
Radioactivity and Nuclear Energy

Antoine Henri Becquerel (1852–1908), a French physicist and engineer, belonged to a rich family with four-generation legacy of producing physicists of eminence (his grandfather, father, and son). Becquerel was studying phosphorescence (emission of light of one color when a solid material is exposed to light of another color) and was investigating the connection between phosphorescence and X-rays. He was of the view that phosphorescent materials like uranium salt might emit X-ray radiations when illuminated with sun light. In 1896, Henry Becquerel observed that the photosensitive plates wrapped in black paper and kept in the drawer of his desk were found exposed to some unknown radiation while performing the experiment. The same experiment was repeated and again found that photosensitive plates were exposed. He concluded that some radiation came out from the uranium salt without providing energy from an external source and these radiations were named *Becquerel radiations*. Marie Curie (1867–1934) and her husband Pierre Curie (1859–1906) started investigating further the phenomenon discovered by Becquerel. Curies extracted uranium from its ore and to their surprise, found that the leftover ore showed more activity than pure uranium. In 1898, from the remaining ore, Curies along with Gustave Bémont (1857–1932) were able to extract two more elements named polonium and radium which emitted Becquerel radiations. The name polonium was given after the native country Poland of Marie Curie. In 1899, Becquerel further studied the behaviour of these radiations in the presence of electric and magnetic fields and found that one part of radiation was deflected in one direction, second part of it was deflected in the opposite direction while the third part remained undeflected (Fig. 9.1). It showed that three types of radiations are emitted: radiations with negative and positive charges and electrically neutral radiation. The term *radioactivity* was coined for the first time by Marie Curie for the phenomenon of emission of Becquerel radiations. The material which emits these radiations is called radioactive material. In the

J. S. Galsin, *History of Solid State Physics*,
https://doi.org/10.1007/978-981-95-0504-3_9

International System of Units, the unit of radioactivity is named Becquerel in honor of Henri Becquerel: single decay per second equals one Becquerel. But the more common unit is Curie named after Pierre Curie and Marie Curie. Henry Becquerel, Pierre Curie and Marie Curie shared the 1903 Physics Nobel Prize for the discovery of spontaneous radioactivity. In 1911, Marie Curie was again awarded Nobel Prize in Chemistry for the discovery of polonium and radium. At the time of Marie Curie, the damaging effects of radioactive materials were not known. She used to carry test tubes containing radioactive elements in her pockets and store them in the drawer of her desk. As a result, due to long exposure to radiation, Marie Curie developed aplastic anaemia and died in 1934. It is noteworthy that Marie Curie is the first and the only woman who won two Nobel Prizes in different subjects: 1903 Nobel Prize in Physics and 1911 Nobel Prize in Chemistry.

Antoine Henri Becquerel (1852–1908), Public Domain, URL: https://en.wikipedia.org/wiki/Henri_Becquerel#/media/File:Paul_Nadar_-_Henri_Becquerel.jpg

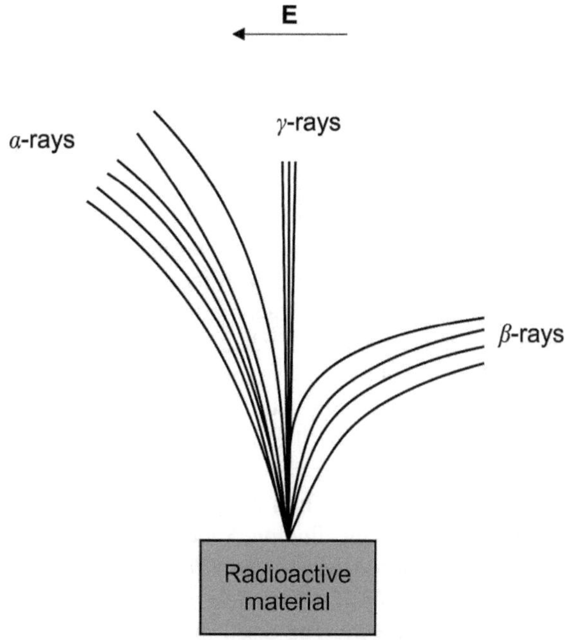

Fig. 9.1 Figure depicts deflection of Becquerel radiations in electric field **E**. The radiations split in three parts: radiation having positive and negative charges and neutral radiation

Ernest Rutherford (1871–1937), Public Domain, URL: https://en.wikipedia.org/wiki/Ernest_Rutherford#/media/File:Sir_Ernest_Rutherford_LCCN2014716719_-_restoration1.jpg

In 1899, Ernest Rutherford (1871–1937) studied the properties of radioactive decay and classified the emitted radiations as alpha, beta and gamma rays depending on their ability to penetrate matter. Alpha rays (with positive charge) have very small penetration, gamma rays (no charge) have large penetration while the penetration of beta rays (with negative charge) falls between alpha and gamma rays. In 1899, Julius Elster (1854–1920) and Hans Friedrich Geitel (1855–1923) found the law of exponential decay which states that the number of particles emitted per second decreases exponentially with time. From this law, they gave the concept of radioactive half life. In 1900, Becquerel measured the properties of beta particles, and found that they had the same measurements as high-speed electrons leaving the nucleus. In 1901, Henri Becquerel found that when a piece of radium was left in the pocket of his vest, he had been burnt by it. He immediately realized that radioactivity could be used for medicine. This discovery led to the development of *radiotherapy* which is now used to treat cancer. In 1902, Rutherford and his student Frederick Soddy (1877–1956) while studying radioactivity of thorium, found that many decay processes resulted in the transmutation of thorium into different elements which are isotopes of radon. At the moment of this finding, Soddy shouted: "*Rutherford, this is transmutation!*" Rutherford snapped back, "*For Christ's sake, Soddy, don't call it transmutation. They'll have our heads off as alchemists.*" Since then it has been customary to speak of *radioactive decay*. The discovery of radioactivity and its effects disproved the view of John Dalton (1766–1844) that an atom is indivisible. In 1909, Rutherford and Thomas Royds (1884–1955) proved, by collecting the rays in an evacuated tube, that alpha rays are nuclei of helium atoms as the neutralized alpha particles exhibit the yellow line of the helium spectrum. In general, alpha decay is mostly observed in heavy elements, with atomic number $Z = 52$ or more, but the other two types of decays are observed in almost all elements. In 1911, Johannes Wilhelm Hans Geiger (1882–1945) and John Mitchell Nuttall (1890–1958) proposed an empirical relation between the radioactive decay constant and the energy of the emitted alpha particles, known as *Geiger-Nattall law*. According to this law shorter is the half-life period of a radioactive element greater is the energy emitted in each disintegration series. In 1919–1920, Rutherford for the first time discovered induced radioactivity (artificial radioactivity) by bombarding alpha particles on light nuclei as a result of which new radioactive nuclei were produced. Actually, he bombarded alpha particles against nitrogen-14 nuclei producing oxygen-17 with the emission of a subatomic particle proton. Through this experiment Rutherford concluded that atoms of one specific element can be transformed into atoms of another element: first nuclear reaction in the laboratory. In 1928, George Gamow (1904–1968) gave the theory of alpha decay of a nucleus using the concept of quantum tunnelling. Gamow derived from first principles the relation between half-life of the alpha decay process and the energy of emission of alpha rays and found that their relation is similar to the empirical Geiger-Nuttall law. Actually, the problem of alpha decay was independently solved in the same year of 1928 by Ronald Wilfred Gurney (1898–1953) and Edward Uhler Condon (1902–1974) but they could not achieve quantitative results as Gamow.

On December 4, 1930, Wolfgang Pauli (1900–1958) wrote a letter to the Physical Institute of the Federal Institute of Technology, Zurich, in which he predicted the existence of a very small neutral particle, having little mass and ½ spin. He said to save the law of conservation of energy this particle must exist within the nuclei and be emitted along with the electron in beta decay. Later, in 1934, this small particle was given the name *neutrino* by Enrico Fermi. Neutrino provides a potential solution to solve the problem of the continuous beta decay spectrum. The letter in which Pauli made this suggestion is a document of considerable interest. In 1934, Enrico Fermi (1901–1954) worked on the phenomenon of radioactivity and he proposed the existence of a weak and very short ranged interaction in a nucleus usually known as Fermi interaction. In the same year, based on this weak interaction, Fermi gave an elegant theory of beta decay in which a neutron in a nucleus is transformed into three particles: a proton, an electron (beta particle) and a neutrino. Fermi developed a relationship which is now referred to as the Fermi *golden rule.* In 1934, Irene Joliot-Curie (1897–1956) [daughter of Marie Curie] and Frederic Joliot-Curie (1900–1958) [son-in-law of Marie Curie] bombarded natural stable isotope of aluminium with alpha particles which resulted in an unstable isotope of phosphorus, thus reporting alpha-induced radioactivity in them. This discovery is formally known as positron emission or beta decay and it allowed radioactive materials to be created quickly, cheaply, and plentifully. Irene Joliot-Curie and Frederic Joliot-Curie were awarded the 1935 Nobel Prize in Chemistry for the discovery of alpha-induced radioactivity. At the Nobel Prize Award Ceremony in Stockholm the chairman of Nobel Committee for Chemistry said in his speech, "*Thanks to your discoveries, it has become possible, for the first time, to transform artificially one element into another hitherto unknown*". Thus the Curie's family legacy includes five Nobel Prizes, a testament for their incredible contributions to science. Later in 1934, Fermi also reported the discovery of neutron-induced artificial radioactivity and was awarded the 1938 Nobel Prize in Physics for it. Rutherford's team, using protons from an accelerator, demonstrated artificially-induced nuclear reaction and transmutation. When a radioactive nucleus decays by the emission of either alpha or beta particles, the daughter nucleus obtained is usually left in the excited state which decays further to the lower energy state with the emission of a gamma ray photon. It is worth pointing out here that Rutherford's research work and work done under his supervision established the nuclear structure of an atom and radioactivity as a nuclear process. It is because of this reason Rutherford is called the *father of nuclear physics.* Radioactivity has immense applications in medical science such as in cancer treatment. In 1956, Clyde Cowan (1919–1974), Frederick Reines (1918–1998) and coworkers observed neutrinos experimentally. Frederick Reines alone got the 1995 Nobel Prize in Physics for the discovery of neutrino because Clyde Cowan died in 1974.

In 1789, Martin Klaproth (1743–1817) discovered a new element in the form of its oxide while analyzing pitchblende samples and named it *uranium* after the planet Uranus. He also discovered zirconium and cerium independently and was involved in the confirmation of titanium, tellurium, strontium and chromium elements. While studying the structure of an atom, Ernest Rutherford was possibly the first physicist who realized the power inherent in the atom. He was of the view that if this power is harnessed it could make bombs powerful enough to make this world vanish in smoke.

Neutrons being neutral particles could penetrate deep into the nucleus thus allowing one to study the effect of bombarding neutrons on atomic nuclei. In 1934, Enrico Fermi and his colleagues studied the bombardment of uranium with neutrons and concluded that new elements with 93 and 94 protons were created. However, most of the scientists were not convinced with the Fermi's analysis of results. In 1938, Otto Hahn (1879–1968) and Fritz Strassmann (1902–1980) discovered that thermal neutrons are easily absorbed by ${}_{92}U^{235}$ and after absorbing neutron, ${}_{92}U^{235}$ nucleus becomes unstable and splits up into two or more parts of comparable masses with a release of a large amount of energy. One of the possible splitting reactions is as follow:

$$ {}_{92}U^{235} + {}_{0}n^{1} \rightarrow {}_{56}Ba^{141} + {}_{36}Kr^{92} + 3{}_{0}n^{1} + 200\,\text{MeV} \tag{9.1} $$

Otto Hahn (1879–1968), CC BY-SA 3.0 nl, https://en.wikipedia.org/wiki/Otto_Hahn#/media/File:Otto_Hahn_1970.jpg

Figure 9.2 depicts the splitting of the ${}_{92}U^{235}$ nucleus into two parts. In 1929, George Gamow (1904–1968) first proposed the liquid drop model for the nucleus which was later developed by Niels Bohr (1885–1962) and John Archibald Wheeler (1911–2008). According to the liquid drop model, nucleus is a charged nonpolar liquid drop held together by nuclear force. In 1939, Lise Meitner (1878–1968)

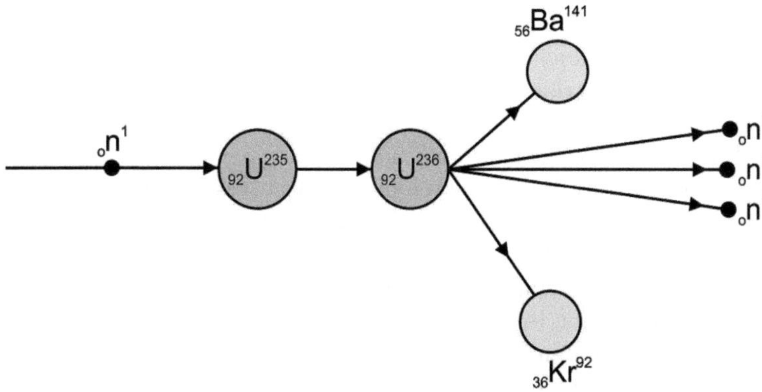

Fig. 9.2 Schematic representation of nuclear fission of $_{92}U^{235}$ atom consistent with Eq. (9.2)

along with her nephew Otto Robert Frisch (1904–1979) used liquid drop model to explain theoretically the phenomenon of splitting of a nucleus. Otto Frisch named the splitting of a nucleus as *nuclear fission* in analogy with biological fission of living cells. The fission process is based on Einstein's energy-mass relation $E = mc^2$ and during the fission process some of the mass is converted into energy. In 1940, Glenn Theodore Seaborg (1912–1999) and coworkers produced plutonium ($_{94}Pu^{244}$) by deuteron bombardment of uranium at the University of California, Berkeley. The name plutonium was given after Pluto, which was considered a planet at that time. In 1941, Joseph William Kennedy (1916–1957) and coworkers [Glenn Seaborg's group], discovered that an isotope of plutonium $_{94}Pu^{239}$ was a fissionable material just like $_{92}U^{235}$ and it was crucial to the decisions made in directing the Manhattan Project.

In general, in a nuclear fission process two or three neutrons per fission are produced (on average 2.5 neutrons per fission) and all of these neutrons are not available for continuing fission process further. The number of neutrons which can produce further fission depends on the mass of the fissionable material: more the mass of fissionable material, more the probability of neutrons to be captured by other nuclei to produce fission. A *chain reaction* refers to a process in which neutrons released in a nuclear fission produce additional fission in at least one nucleus and this nucleus in turn produces neutrons and the process repeats. In 1939, Hans von Halban (1908–1964), Frederic Joliot-Curie (1900–1958), Lew Kowarski (1907–1979) measured average number of neutrons emitted per nuclear fission and could perceive that under proper conditions, chain reaction is possible which continuously leads to more nuclear fissions so that the process goes on of its own accord at an ever-faster rate. The chain reaction may be a controlled one (nuclear power) or an uncontrolled one (nuclear weapon). If the conditions are such that more neutrons are lost than produced by a nuclear fission, the chain reaction will not be self-sustaining. The mass of fissionable material at which the chain reaction becomes self-sustaining is referred to as *critical mass*. In an atom bomb the fissionable material is arranged in different blocks such that each block has mass less than the critical mass. When an atomic bomb is to be detonated the different blocks are instantaneously brought together, with total mass greater than the critical mass, to permit the chain reaction to propagate and grow before the bomb explodes.

Otto Hahn was working on German nuclear weapon programme. Albert Einstein came to know that Germany was developing nuclear weapons. On August 2, 1939, Einstein wrote a letter to U. S. President Franklin Delano Roosevelt (1882–1945) in which he warned that Nazis were working on a new and powerful nuclear weapon based on nuclear fission. Immediately after receiving the letter, Franklin Roosevelt announced secret atomic bomb development programme *Manhattan Project* and set up its Advisory Committee. This project started modestly mainly for producing fissionable material. With the commencement of World War II, in December 1941, President Roosevelt elevated the Manhattan Project for the development of atomic bomb to the highest priority as National Security Project and the Advisory Committee was renamed as Office of Scientific Research and Development (OSRD). Facilities were setup in remote location of Los Alamos in northern New Mexico, Tennessee, and at some sites in Canada. Julius Robert Oppenheimer (1904–1967) studied in Europe at Cambridge University, London, and did Ph.D. under Max Born in University of Gottingen, Germany. Before the World War II, Robert Oppenheimer was working as professor of theoretical physics at the University of California, Berkeley and had expertise in the field of quantum mechanics. Nobel laureate Ernest O. Lawrence (1901–1958) knew about the ability of Oppenheimer and chose him for the development of atomic bomb. Oppenheimer was made the director of Los Alamos laboratory. On the other hand, Enrico Fermi (1901–1954) and Leo Szilard (1898–1964), at University of Chicago, were given the task of developing the self-sustaining nuclear chain reaction for enriching uranium to produce fissionable ${}_{92}U^{235}$, an important component for producing a functional atomic weapon. Enrico Fermi was one of the very few physicists who excelled in both theoretical and experimental physics. Sir James Chadwick (1891–1974) was appointed the Head of the British delegation to Manhattan Project in Los Alamos, New Mexico. Hans Albrecht Bethe (1906–2005) was appointed the chief of theoretical division for the Manhattan project. He played a key role in calculating the critical mass of the weapons and developing the theory behind the implosion method used. Toward the end of 1942, laboratory was set up at Chalk River, Ontario, Canada, with the help of United Kingdom and France. There scientists designed a prototype reactor called ZEEP (zero energy experimental pile) which had the potential to produced fissile plutonium ${}_{94}Pu^{239}$. In addition, there was a band of more scientists working for different aspects of the Manhattan project. On Dec. 2, 1942, Enrico Fermi designed and built the first nuclear reactor named Chicago Pile-1 which went critical. Fermi has been called, "*the architect of nuclear age*" and "*architect of atomic bomb*". Scientists working under Oppenheimer had developed three atomic bombs: uranium-based atomic bomb was code-named "the Little Boy", plutonium-based atomic bomb was code-named "the Fat Man" and the third one was code-named "The Gadget". A remote location near Alamogordo, New Mexico was chosen for the testing of atomic bomb—*the Trinity test*. The atomic bomb "The Gadget" was successfully detonated creating an enormous mushroom cloud about

40,000 feet high, thus entering the Atomic Age. The other two atomic bombs, Little Boy and Fat Man, were used in combat against the Japanese cities of Hiroshima and Nagasaki on Aug. 6 and 9, 1945, respectively. The bombing of Hiroshima and Nagasaki made the American scientists heroes. The photographs of Oppenheimer appeared on the front pages of newspapers and on the covers of magazines. Life magazine described Oppenheimer as "one of the most famous men in the world". Robert Oppenheimer is called the "*father of atomic bomb*". The Chicago Pile-1 was disassembled and transported to a more remote place in a forest preserve outside Chicago. After the War, equipment was reorganized and scientists were directed to begin investigating peaceful uses of atomic energy, becoming the nation's first national laboratory—Argonne National Laboratory. Henry Abraham Boorse (1904–2003) and his colleagues pointed out that the delay in the discovery of neutrons was a blessing in disguise because had the neutron been discovered in the early years of the twentieth century it was very likely that the atomic bomb had been discovered first by the Germans.

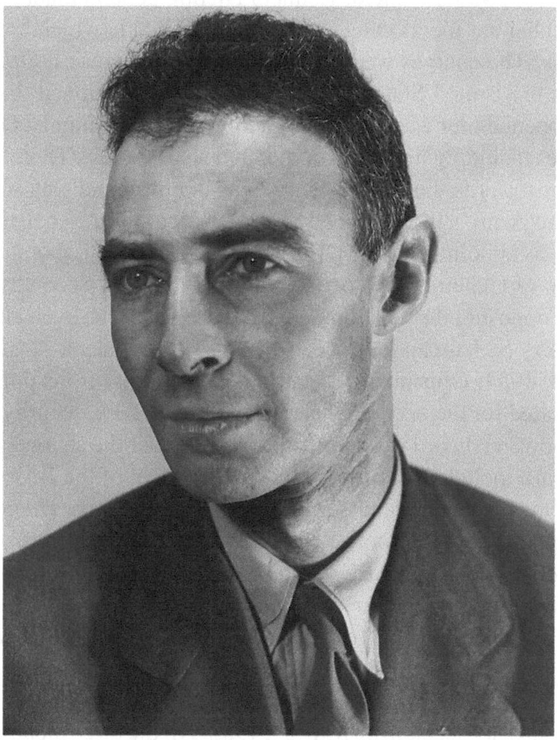

Julius Robert Oppenheimer (1904–1967), Public Domain, URL: https://en.wikipedia.org/wiki/J._Robert_Oppenheimer#/media/File:Oppenheimer_(cropped).jpg

After the World War II, in 1947, Oppenheimer was appointed the chairman of General Advisory Committee of Atomic Energy Commission (AEC) of United States. After the death of President Franklin D. Roosevelt in 1945, being Vice

President Harry S. Truman (1884–1972) took over as President of United States. In 1950, President Truman made the controversial decision to intensify research for the production of hydrogen bomb (thermonuclear weapon). At the time, David Lilienthal (1899–1981) was the chairman of AEC who was not in favour of pursuing the "*Super*" the hydrogen bomb. In 1953, Dwight David Eisenhower (1890–1969) became the president of United States. Lewis Lichtenstein Strauss (1896–1974) was a close friend and confidant of President Eisenhower who also played an important role in making atomic bomb. Lewis Strauss, Edward Teller (1908–2003) and some other scientists were in favour of developing hydrogen bomb but Oppenheimer was against it. Moreover, there was a technical discussion between Oppenheimer and scientists over the feasibility of hydrogen bomb. So Oppenheimer got involved in bureaucratic conflict between Army and Air Force and also in personal conflict with Lewis Strauss.

AEC starting exploring the background actions and associations of Robert Oppenheimer and ultimately Lewis Strauss levelled charges against Oppenheimer. On December 23, 1953, Major General Kenneth Nichol, general manager of AEC, sent a letter to Oppenheimer mentioning the charges levelled in which it was said that he is a security risk. The charges were divided into two categories. One was about his loyalty towards the United States. In the letter it was mentioned that in 1930s and early 1940s, Oppenheimer was a member of numerous Communist Organizations of the United States including his wife, brother and sister-in-law. Therefore, the loyalty of Oppenheimer towards the United States was doubtful and was accused of being a Soviet spy. But Army's intelligence wing knew about his associations at the time Oppenheimer was appointed as the Director of Los Alamos Laboratory in 1942 and as the chairman of General Advisory Committee in 1947. Secondly Oppenheimer was accused of opposing the development of hydrogen bomb even after the decision of President Harry S. Truman to proceed with its development.

On March 4, 1954, Oppenheimer sent a 43 page reply in response to the AEC letter and requested for the personal hearing before the AEC security board. A three member high-powered board was constituted to consider the charges against Oppenheimer. The board included Gordon Gray (as Chairman of the Board), president of University of North Carolina and former secretary of Army, Thomas A. Morgan, former president of Sperry Corporation, and Ward E. Evans, professor of chemistry at Loyola University, Chicago. AEC was represented by Roger Robb, an experienced prosecutor in Washington, and Arthur Rolander. Lloyad K. Garrison, a prominent New York attorney, and his team represented Oppenheimer. The hearing began on April 12, 1954 and several of Oppenheimer's colleagues testified in his favour. But Edward Teller, who clashed with Oppenheimer on hydrogen bomb, testified against him. The board gave its decision on May 27, 1954. All the three members of the board unanimously declared that Oppenheimer was a loyal citizen of the United States. But Gray and Morgan voted against the reinstatement of Oppenheimer as a consultant to AEC. Therefore with the majority vote of 2:1, Oppenheimer was stripped of his security clearance one day before his consultant contract was due to expire. The loss of security clearance marked the end of his formal relationship with the United States Government. Immediately after the decision there was a strong protest by the scientists of Los Alamos Laboratory. In general there was a considerable controversy over the decision of the board whether the trial was fair. Lewis Strauss often has been regarded as a villain in American history. In 2014, 60 years after the board's decision, the US Department of Energy declassified the proceedings of the hearing. The additional previously unknown material in the proceedings proved beyond doubt the loyalty of Oppenheimer. It proved that the nexus of bureaucracy has brought low the image of a brilliant scientist. On December 16, 2022, United States Secretary of energy, Jennifer Granholm nullified the 1954 decision of the board saying that it had been the result of a "flawed process".

The first Experimental Breeder reactor (EBR-I) was constructed at Argonne National Laboratory which went critical in August, 1951. A breeder reactor generates more fissile material than it consumes during its operation. In this type of reactor $_{92}U^{235}$ is mixed with $_{92}U^{238}$ and during fission process the bombardment of $_{92}U^{238}$ with neutrons converts it into fissionable $_{94}Pu^{239}$. On December 20, 1951, EBR-I lit a string of four light bulbs with the first usable electricity ever produced by the nuclear energy. After this a number of improved forms of EBR such as Chicago-Pile-2, Chicago-Pile-3, Chicago-Pile-4, and Chicago-Pile-5 were made. It is worth pointing here that fifteen years later on Dec. 2, 1957, America's first full-scale atomic electric plant went critical. The designs of nuclear reactors were immensely improved with the passage of time. Today large amount of electrical energy is produced with the help of nuclear reactors installed at various places in different countries of the world. But the disposal of the radioactive nuclear waste is one of the main problems in nuclear energy production.

Nuclear fusion is another very powerful process for producing nuclear energy. Nuclear fusion is a process in which two or more light atomic nuclei are fused together to form one or more different nuclei. In the fusion process, there is a definite mass deficit which is converted into a huge amount of energy. Before twentieth century the source of stellar energy was a mystery. In 1915, William Draper Hawkins (1873–1951), an American chemist, proposed the concept of nuclear fusion for the first time. In 1920, Arthur Stanley Eddington (1882–1944), a British astrophysicist, physicist and mathematician, was the first to predict that stars draw their huge and endless energy from the fusion of hydrogen into helium. In 1934, Marcus Laurence Elwin Oliphant (1901–2000) and Ernest Rutherford (1871–1937) performed early fusion experiments discovering tritium $_1H^3$, an isotope of hydrogen, and $_2He^3$, the rare helium isotope that holds the promise for nuclear fusion. Tritium is, however, radioactive and undergoes decay to $_2He^3$. For this reason, tritium does not occur naturally to a great extent and must be produced artificially. This experiment opened the way to present day fusion research. In 1937 and 1938, Carl Friedrich von Weizsäcker (1912–2007) made important theoretical discoveries regarding energy production in stars from nuclear fusion processes: proposed proton-proton cycle. He also did influential theoretical work on planetary formation in the early Solar System. In 1938, Hans Bethe (1906–2005) and Charles Louis Critchfield (1910–1994) analyzed the nuclear fusion of protons and deuterons and gave a series of nuclear reactions. In 1939, Hans Bethe (1906–2005) confirmed the proton-proton chain in which two protons fuse to form deuterium $_1H^2$ and then further fuse to produce $_2He^3$ and ultimately $_2He^4$. The basic proton-proton reactions are given below:

$$_1H^1 + {_1H^1} \rightarrow {_1H^2} + e^+ + \nu \quad (9.2)$$

$$_1H^1 + {_1H^2} \rightarrow {_2He^3} + \gamma \quad (9.3)$$

$$_1H^1 + {_2He^3} \rightarrow {_2He^4} + e^+ + \nu \quad (9.4)$$

$$_2He^3 + {_2He^3} \rightarrow {_2He^4} + 2{_1H^1} \quad (9.5)$$

where e^+ and ν represent positron and neutrino, respectively. To carry out reactions in the proton-proton cycle, very high temperatures are required, therefore these are called thermonuclear fusion reactions and take place in the interior of the sun where the temperature is approximately 1.5×10^7 K. At such high temperatures, the atoms are ionized and the collection of nuclei and electrons is generally called plasma. In this process Bethe led to the discovery of the carbon–nitrogen–oxygen (CNO) cycle. A number of fusion reactions are taking place in the sun but it is believed that proton–proton cycle is the one which generates most of the energy in the sun. The fusion reactions in case of deuterium are as under:

$$_1H^2 + {_1H^2} \rightarrow {_2He^3} + {_0n^1} \quad (9.6)$$

$$_1H^2 + {_1H^2} \rightarrow {_1H^3} + {_1H^1} \quad (9.7)$$

$$_1H^2 + {_1H^3} \rightarrow {_2He^4} + {_0n^1} \quad (9.8)$$

The advantage of nuclear fusion is that comparatively few radioactive byproducts are formed. Moreover the energy produced in nuclear fusion is much more than that produced in the nuclear fission. To have an idea, one gram of deuterium–tritium fuel mixture in the process of nuclear fusion produces 90,000 kWh energy or the equivalent of 11 tons of coal. Today it is well known that the huge energy of sun, in the form of light and heat, is produced by the fusion process taking place inside the sun. It is well known that 99.86% of total mass of sun constitutes hydrogen.

During the Manhattan Project, it was Enrico Fermi who suggested Edward Teller (1908–2003) that a weapon based on nuclear fission could be used to set off nuclear fusion reaction. He realized that even more explosive power would be released by the nuclear fusion of light elements such as hydrogen. While working on the nuclear fission project, Teller continued to push his ideas for fusion weapon even though it had been put on low priority during the World War II. On the suggestion of Hans Bethe (1906–2005), Robert Oppenheimer moved Teller out of Trinity division and made him in-charge of a special group for the fusion weapon as a part of Manhattan Project and asked Teller to report directly to him. This group included Stanislaw

Marcin Ulam (1909–1984), Geoffrey Chew (1924–2019), Maria Geoppert-Mayer (1906–1972), Jane Roberg, Harold Argo, and Mary Argo. By the end of World War II, the theoretical framework for fusion was well established. Edward Teller and other group scientists were given the responsibility for making the thermonuclear fusion device and the project was given the code name "Ivy Mike". Teller and Ulam were the main scientists in designing the fusion device for a full scale test and is usually called Teller-Ulam design. Ivy Mike was detonated successfully on November 1, 1952 on the island Eniwetok Atoll in the Pacific which is now an independent island nation called Marshall Islands. It was the first full scale test of a staged nuclear fusion device. It is pointed out here that Ivy Mike was not suitable as a nuclear weapon, rather it was a technical concept behind modern thermonuclear weapons known as Hydrogen bombs. Hydrogen bombs can be regarded as second-generation thermonuclear fusion weapons.

Enrico Fermi (1901–1954), Public Domain, URL: https://en.wikipedia.org/wiki/Enrico_Fermi#/media/File:Enrico_Fermi_1943-49.jpg

The efforts for the peaceful use of energy produced by nuclear fusion commenced simultaneously but it was a difficult task. Two advanced approaches were being developed to carry out controlled self-sustaining fusion reactions in the laboratory: inertial confinement (laser designs) and magnetic confinement (toroidal designs). In the

1950s, John Nickolls (1930–present) gave the concept of laser-driven inertial confinement fusion and the same he developed at Lawrence Livermore National Laboratory. The main idea of laser-driven fusion was to compress a small pallet containing fusion fuel: a mixture of deuterium and tritium. When the compression reaches very high values, fusion reactions begin to take place releasing alpha particles and neutrons which gives rise to self-sustaining fusion reactions. In 1950, Andrea Sakharov (1921–1989) and Tigor Tamm (1895–1971), Soviet physicists, proposed the design of a type of magnetic confinement device known as *Tokamak* for producing controlled self-sustaining fusion reactions. In 1946, Sir George Paget Thomson (1892–1975) and Moses Blackman (1908–1983) invented the first fusion reactor and filed the case for its patent but it could not be reproduced. The real research on fusion started in 1951 when Lyman Spitzer (1914–1997) proposed the "*stallarator*" concept for fusion. Stallarator is a device that confines plasma using external magnets and is one of the earliest fusion devices. In the 1950s, the stallarator concept dominated fusion research. The first Tokamak machine was made in 1954. Under the leadership of Lev Andreyevich Artsimovich (1909–1973), a Russian physicist, a number of improvements were made in the design of Tokamak and in 1968 the electronic plasma temperature of 1 keV was reached. It showed that Tokamak was a more efficient concept and ultimately dominated the stallarator concept. Artsimovich is called the father of Tokamak.

In the 1970s, it was realized that obtaining controlled self-sustaining fusion reactions and obtaining energy from them would be one of the greatest challenges in science. Scientists were of the view that collaboration might be the key to meet all the challenges. In 1975, European states proposed to setup Joint European Torus (JET) and after a long debate Culham in UK was chosen as the site for the design of the JET machine. In 1991, JET performed the first experiment using deuterium and tritium mix as the fuel. But JET failed to achieve the controlled self-sustaining fusion reaction. In 1982, another controlled fusion reaction project "Tokamak Fusion Test Reactor (TFTR)" came into existence and was operated at Princeton Plasma Physics Laboratory, USA. The TFTR project was shut down in 1997 after fifteen years of operation. In November 1985, a Super Power Summit was held in Geneva. Mikhail Gorbachev, General Secretary of the Soviet Union, suggested to Ronald Reagan, President of the USA, that the two countries should jointly undertake the construction of Tokamak. As a result, in 1988, the International Thermonuclear Experimental Reactor (ITER) project was initiated. All these multinational companies were set up with the purpose of designing and developing small fusion reactors to introduce fusion power into the grid by 2030. Tokamak constructed the first fusion reactor called ST40 while JET designed a fusion reactor producing 16 MW of fusion power. On July 3, 2024, ITER Director General Pietro Barabaschi (1965–present) announced that the first plasma production in the project will not take place until, at least, 2033. At present the production of electrical energy using nuclear fusion is still at its infancy.

From the discussions of the previous chapters one can conclude that three types of forces exist in the universe:

1. Gravitational force: It is a weak but long-range force and acts among large solid objects. It is responsible for large scale structure of the universe: the formation of galaxies, stars, planets and moons etc.
2. Electromagnetic force: Electromagnetic force exists inside the atoms and among the atoms. This force binds the atoms together to form solids and is a basis of all chemical reactions.
3. Nuclear force: Nuclear force exists inside the nucleus of an atom. It is further of two types: strong nuclear force and weak nuclear force. Strong nuclear force acts among neutrons and protons and binds them together in a nucleus. It is important in nuclear fission and nuclear fusion processes. The weak nuclear force determines radioactive decay that is the spontaneous emission of alpha and beta particles from within the nucleus.

In the end, it is pointed out that physics has ended up with two bodies of laws, one for the world of very small objects and other for the large bodies like galaxies and universe. But Einstein disliked this because he was of the view of unified fields.

Bibliography

Artsimovich LA et al (1969) Experimental studies on tokamak installations. In: Proceedings of the third international conference on plasma physics and controlled nuclear fusion research, held at Novosibirsk by IAEA, 1–7 Aug 1968, vol 1, pp 157–173

Becquerel H (1896) Sur les radiations emises par phosphorescence. Comptes Rendus 122:420–421; (1896) Sur les radiations invisibles emises par les corps phosphorescents. Comptes Rendus 122:501–503

Bethe HA (1939) Energy production in stars. Phys Rev 55:434–456

Bethe HA, Critchfield CL (1938) The formation of deuterons by proton combination. Phys Rev 54:248–254

Cowan CL, Reines F, Harrison FB, Kruse HW, McGuire AD (1956) Detection of the free neutrino: a confirmation. Science 124:103–104

Curie M (1998) Rays emitted by compounds of uranium and thorium. Comptes Rendus 126:1101–1103; (1902) Sur le poids atomique duradium. Comptes Rendus 135:161–163

Curie P, Curie M (1998) On a new radioactive substance contained in pitchblende. Comptes Rendus 127:175–178

Curie M, Curie P, Bémont MG (1898) Sur une nouvelle substance fortement radio-active contenue dans la pechblende. Comptes Rendus de l'Académie des Sciences, Paris 127:1215–1217

Eddington AS (1920) The internal constitution of the stars. A. S. Sci Monthly 11:297–303; (1916) On the radiative equilibrium of the stars. Monthly Notices Roy Astronom Soc 77:16–35

Elster J, Geitel H (1898) Experiments on Becquerel. Ann Phys Chem 302:735–740; (1899) Via the influence of a magnetic field on the Becquerel rays caused by the conductivity of the air. Negotiations German Phys Soc 1:136–138

Fermi E (1934) Radioattivita indotta da bombardamento di neutron. La Ricerca Scientifica (in Italian) 1:283; Fermi E, Amaldi E, d'Agostino O, Rasetti F, Segre E (1934) Artificial radioactivity produced by neutron bombardment. Proc R Soc A: Math Phys Eng Sci 146:483

Fermi E (1946) The development of the first chain reaction pile. Proc Am Philos Soc 90:20–24 [Although Pile-I was developed in 1942 but its publication waswithheld until the end of World War II]

Gamow G (1928) Zur Quantentheorie des Atomkernes. Z Angew Phys 51:204

Geiger H, Nuttall JM (1911) The ranges of the α particles from various radioactive substances and a relation between range and period of transformation. Philos Mag 22:613–621; (1912) The ranges of α particles from uranium. Philos Mag 23:439–445

Gurney RW, Condon EU (1928) Quantum mechanics and radioactive disintegration. Nature 122:439; (1929) Quantum mechanics and radioactive disintegration. Phys Rev 33:127

Hahn O, Strassmann F (1939) Über den Nachweis und das Verhalten der bei der Bestrahlung des Urans mittels Neutronen entstehenden Erdalkalimetalle [On the detection and characteristics of the alkaline earth metals formed by irradiation of uranium with neutrons]. Naturwissenschaften (in German) 27:11–15

Johansson CH, Linde JO (1925) Rontgenographische bestimmung der atoman ordnung in den mischkristallreihen Au-Cu und Pd-Cu [X-ray investigations of atomic order in binary alloys of Au-Cu and Pd-Cu]. Ann Phys 78:439–460; (1936) Roentgenographic and electrical investigations of the Cu-Au system. Ann Phys 25:1048

Joliot-Curie I, Joliot-Curie F (1934) Un nouveau type de radioactivite. Comptes Rendus hebdomadaries des séances de l'Academie des Sciences 198:254–256; (1934) Artificial production of a new kind of radio-element. Nature 133:201–202

Kennedy JW, Seaborg GT, Segre E, Wahl AC (1946) Phys Rev 70:555 [This letter was received in 1941 but its publication was withheld until the end of the World War]

Klaproth MH (1789) Chemische untersuchung des uranits, einer neuentdeckten metallische substanz. Chem Ann Freunde Naturl 2:387–403

Meitner L, Frisch OR (1939) Disintegration of uranium by neutrons: a new type of nuclear reaction. Nature 143:239–240

Nuckolls J, Wood L, Thiessen A, Zimmerman G (1972) Laser compression of matter to super-high densities: thermonuclear applications. Nature 239:139–142

Oliphant MLE, Rutherford E (1933) Experiments on the transmutation of elements by protons. Proc Roy Soc (London) A141:259–281; Oliphant MLE, Harteck P, Rutherford E (1934) Transmutation effects observed with heavy hydrogen. Proc Roy Soc (London) A144:692–703; Oliphant MLE, Kempton R, Rutherford E (1935) The accurate determination of the energy released in certain nuclear transmutations. Proc Roy Soc (London) A149:406–416

Rutherford E (1919) Collision of α particles with light atoms: an anomalous effect in nitrogen. Lond Edinburgh Dublin Philos Mag J Sci 37:581–587; (1920) Bakerian lecture. Nuclear constitution of atoms. Proc Roy Soc A Math Phys Eng Sci 97:374–400

Rutherford E, Royds T (1908) Spectrum of the radium emanation. Philos Mag 16:313; (1909) The nature of a particle from radioactive substances. Philos Mag 17:281

Rutherford E, Soddy F (1902) The cause and nature of radioactivity-I. Lond Edinburgh Dublin Philos Mag J Sci 4:370–396; (1902) The cause and nature of radioactivity-II. Lond Edinburgh Dublin Philos Mag J Sci 4:569–585

Seaborg GT, McMillan EM, Kennedy JW, Wahl AC (1946) Phys Rev 69:366–367 [This letter was received in 1941 but its publication was withheld until the end of World War]

Spitzer L (1958) Stellarator concept. Phys Fluids 1:253

Tammann G (1919) Die chemischen und galvanischen Eigenschaften von Mischkristallreihen und ihre Atomverteilung: Ein Beitrag zur Kenntnis der Legierungen [The Chemical and galvanic properties of mixed crystal series and their atomic distribution: a contribution to the knowledge of alloys] (in German). Z fur an Chem 107:1; (1924) Uber die atom verteilung in mischkristallreihen. Ann de Phys 75:S.212

Von Halban H, Joliot F, Kowarski L (1939) Nature 143:470, 680; Perrin F (1939) Calcul relatif aux conditions eventuelles de transmutation en chaine de l'uranium. Comptes Rendus 208:1394–1396

Von Weizsäcker CF (1937) Über elementumwandlungen im innern der sterne-I (On transformations of elements in the interiors of stars-I). Physikalische Zeitschrift 38:176–191; (1938) Über elementumwandlungen im innern der sterne-II (On transformations of elements in the interiors of stars. II). Physikalisch Zeitschrift 39:633–646

Chapter 10
Electronic Structure and Band Theory of Solids

The study of electronic properties of solids, particularly metals, commenced with the discovery of electrons in 1897 by Joseph John Thomson (1856–1940). Metal can be considered to be a sea of conduction electrons in which are embedded positively charged ions at the lattice positions. Therefore the theoretical study of electronic properties of metals is a many-body problem which is difficult to solve exactly. To simplify the calculations the ions are assumed to be smeared out to form a uniform positive background in which interacting electrons are distributed uniformly with density equal to that of the positive background so that the metal as a whole is electrically neutral. This model is generally called *jellium model* of metal. The term jellium was first coined by Conyers Herring in 1952 in analogy with positive jelly. If the interaction between the electrons is neglected one obtains a free electron gas moving in the uniform positive background. In 1900, Paul Karl Ludwig Drude (1863–1906), student of Woldemar Voigt (1850–1919), assumed that in a metal there is a sea of free electrons moving with the same thermal speed in which are embedded positively charged scattering centres making it electrically neutral: *Drude model*. He developed classical free-electron theory in which the free electrons were assumed to behave like gas molecules in a container obeying the laws of kinetic theory of gases: electrons were assumed to obey the Maxwell–Boltzmann distribution. Drude calculated electrical conductivity of metals and obtained the Wiedemann–Franz law. Drude's results were an outstanding indication of the validity of the electron gas concept in metals. It is noteworthy that the structure of atom was not known at the time Drude model was proposed and it is considered to be the first theory of solid state of matter. In 1905, Hendrik Antoon Lorentz (1853–1928), refined the Drude's free-electron model for metals in which he assumed that electrons are moving with different velocities obeying the Maxwell–Boltzmann distribution (*Lorentz model*). He used the Boltzmann transport equation for investigating the properties of metals and obtained results similar to those obtained by Paul Drude. The classical free-electron theory verified the Ohm's law and explained the electrical and thermal conductivities of metals. But it failed to explain a number of properties such as

J. S. Galsin, *History of Solid State Physics*,
https://doi.org/10.1007/978-981-95-0504-3_10

the electrical conductivity of semiconductors, specific heat of metals, temperature dependence of paramagnetic susceptibility and electrical conductivity at low temperatures etc. In 1913, Niels Bohr (1885–1962), gave the structure of an atom (*Bohr model*) in which electrons were assumed to move in quantized circular orbits about the nucleus. The electrons in the outermost orbit were called valence electrons and referred to as free electrons and the positively charged centres as the ions situated at the lattice positions. The various electronic properties such as electrical conductivity, electronic energy bands, superconductivity etc. are determined by the valence electrons. Therefore, Drude model provided good representation for metals. On the other hand, lattice properties such as lattice vibrations, lattice specific heat, thermal conductivity etc. are determined by the ions at the lattice positions.

In 1925, Wolfgang Pauli (1900–1958) gave exclusion principle according to which no two identical electrons in an atom can exist in the same state at the same time. In 1926, Enrico Fermi (1901–1954) and independently Paul Adrien Maurice Dirac (1902–1984) gave Fermi–Dirac distribution function which gives the probability of occupation of an electron state. These two discoveries changed the behavior of electrons in crystalline solids immensely. In an isolated atom, electrons have discrete energy levels. But in a solid, energy bands instead of discrete energy levels are formed which can be understood in terms of Pauli exclusion principle. When two same atoms are brought closer together, first the electrons in their outermost energy level interact with each other and due to the Pauli exclusion principle, it spits into two closely-spaced energy levels. Similarly, if N numbers of same atoms are brought closer to form a solid, the outermost energy level splits into very closely spaced N energy levels forming an energy band. Same is the case with electrons in other energy levels below the outermost level but their bandwidth decreases due to the decreasing electron interaction. Therefore, one it can be said that the energy band formation is a consequence of the Pauli exclusion principle. In 1927, Walter Heitler (1904–1981) and Fritz London (1900–1954) discovered bands for the first time and put forward a theory for energy bands called *Heitler-London theory*.

A solid is a many-body system and its quantum mechanical solution is quite difficult. Therefore, following assumptions are made to simplify the theoretical treatment of various properties of solids:

1. The electrons in an ion-core are least affected by the crystal field and therefore are assumed to move rigidly along with the nucleus: *rigid ion approximation*. In other words, the crystal field does not affect the core wave functions. Actually, rigid ion approximation was first used by Rudolf Peierls (1907–1995) in describing the interaction of electrons with vibrating lattice ions.
2. In 1927, Max Born (1882–1970) and Julius Robert Oppenheimer (1904–1967) proposed adiabatic approximation. In this approximation, the motion of conduction electrons is assumed to be independent of the motion of ions because electrons being very light and small particles as compared to the ions, respond instantaneously to the external field. Therefore, in the *Born–Oppenheimer approximation* the wave functions of electrons and of atomic ions can be treated separately. In other words, electronic and lattice properties can be evaluated independently.

3. In 1926, Werner Heisenberg (1901–1976) and Paul Dirac (1902–1984), independently, discovered the exchange interactions among the electrons in a solid. These interactions in fact arose due to the symmetry of the wave function of indistinguishable particles (here electrons). For Fermions, the exchange interaction is sometimes called Pauli repulsion as it is related to the Pauli exclusion principle. With the above assumptions, still the study of electronic properties of solids is a many-electron system. In 1928, Douglas Rayner Hartree (1897–1958) proposed *one-electron approximation* in which an electron is assumed to experience some sort of average potential due to all the ions and rest of the electrons. He gave a self-consistent method to evaluate such an averaged potential. Hartree took the total electronic wave function of a many-electron system as the product of the single-electron wave functions of individual electrons neglecting the exchange and correlation effects among the positions of the electrons. In 1930, Vladimir Fock (1898–1974) improved the Hartree theory by including the symmetry of the wave function: generally called Hartree–Fock theory. Both these methods make use of the variational technique to derive the equation of motion.
4. In 1934, Eugene Paul Wigner (1902–1995) found that the motion of an electron is influenced by the presence of other electrons in a solid. In 1955, Per-Olov Lowdin (1916–2000) coined the word '*correlation effects*' for the motion of electrons in a solid and these are of two types. First are the parallel spin correlation effects in which the motion of two electrons with parallel spins is coupled together. The exchange term in the Hartree–Fock theory includes these correlations. The change in energy due to these effects is called exchange–correlation energy. Here each electron is surrounded by a hole in which there is no electron with parallel spin and it is called exchange hole or Fermi hole. The second type of correlation effects arise from the correlation between the spatial positions of electrons due to the Coulomb repulsion: Coulomb correlations. The change in energy due to these effects is called Coulomb correlation energy. A number of researchers have evaluated the exchange and correlation effects in different approximations and some important findings are made by Ruderman and Kittel (1954), Kasuya (1956), Yoshida (1957), Hubbard (1964), Singwi and coworkers (1970), and Lindgren and Shwarz (1972) etc.

Within the above-mentioned approximations, one can solve one-electron Schrodinger equation to study the electronic properties of crystalline solids. In the energy band theory, one solves for the Schrodinger equation defined as

$$\left[-\frac{\hbar^2}{2m_e}\nabla^2 + V(\mathbf{r})\right]\Phi_{\mathbf{k}}(\mathbf{r}) = E_{\mathbf{k}}\Phi_{\mathbf{k}}(\mathbf{r}) \tag{10.1}$$

where m_e is the electron mass and $V(\mathbf{r})$ is the self-consistent crystal potential. $\Phi_{\mathbf{k}}(\mathbf{r})$ and $E_{\mathbf{k}}$ are the wave function and energy of an electron state with wave vector **k**. The calculation of the electron energy bands in solids is a central theoretical problem in solid state physics as the knowledge of electron energies and electron wave functions is required to evaluate a number of properties including vibrational spectra, magnetic order, electrical and thermal conductivities, optical dielectric function and so on. In the above equation, the only variable quantities are $V(\mathbf{r})$ and $\Phi_{\mathbf{k}}(\mathbf{r})$, therefore using different forms of $V(\mathbf{r})$ and $\Phi_{\mathbf{k}}(\mathbf{r})$, one can obtain different methods for the evaluation of energy band structure.

In 1926, Wolfgang Pauli (1900–1958) applied the Fermi–Dirac statistics and gave semi-classical theory of electrons in metals. In 1927, Llewellyn Hilleth Thomas (1903–1992) and Enrico Fermi (1901–1954) gave a semiclassical theory to derive the equation of state of a gas called Thomas–Fermi theory. Up to 1927 the classical and semi-classical theories were used to explain the properties of electrons in metals. Arnold Johannes Wilhelm Sommerfeld (1868–1951) did Ph.D. with Ferdinand von Lindemann (1852–1939) at the University of Konigsberg and acquired extraordinary mathematical skills, which he applied to a wide variety of physical problems. In 1906, Sommerfeld joined the University of Munich and soon after it became an international centre of research in the field of electrons in metals. A number of brilliant students joined his group and among these were Wolfgang Pauli (1900–1958), Peter Debye (1884–1966), Werner Heisenberg (1901–1976), Hans Bethe (1906–2005), Walter Heitler (1904–1981), and Gregor Wentzel (1898–1978). After completing their Ph.D., Peter Debye, Werner Heisenberg and Gregor Wentzel joined the Institute of Physics in Leipzing while Wolfgang Pauli joined the Institute of Theoretical Physics in Zurich. In 1927, Sommerfeld put forward quantum free-electron model of metals in which he used the Pauli exclusion principle and Fermi–Dirac distribution function for the distribution of free electrons. Sommerfeld's quantum free-electron model involved a very crisp and simple mathematics. The wave function $\Phi_{\mathbf{k}}(\mathbf{r})$ was expanded in terms of plane waves and Eq. (10.1) was used to obtain energy bands which were found to be parabolic in nature (Fig. 10.1). The free-electron approximation could explain successfully many aspects of metals such as electrical and thermal conductivities, Wiedemann–Franz law, electronic specific heat, magnetic susceptibility of metals, the range of values of binding energy, thermoelectric effects, thermionic emission of electrons, photoelectric effect, Compton effect, black-body radiations and shape of density of electron states etc. But the quantum free-electron theory had many limitations. The evaluation of electron energy bands in solids by expressing the wave function in terms of plane waves showed very poor convergence. It could not distinguish between conductors, semiconductors and insulators. It could not explain the positive value of the Hall coefficient and also could not explain why metallic solids crystallize into certain particular structures only. Initially Sommerfeld theory got mixed response from the scientific community. Albert Einstein said it is a correct revival of Drude's free-electron theory of metals. On the other hand, William Hume-Rothery (1899–1968) said quantum free-electron theory is wrong in spite of its success in explaining some properties.

Hans Bethe (1906–1995) studied theoretically the problem of electron scattering in crystals and found that electrons with some range of energy and in certain incident directions cannot propagate through the crystal and indicated the existence of forbidden energy gaps. In 1928, Maximilian Julius Otto Strutt (1903–1992) found the existence of energy band gap for weakly-bound electrons in metals but this result was not given attention by scientists. The simplest improvement over the free-electron

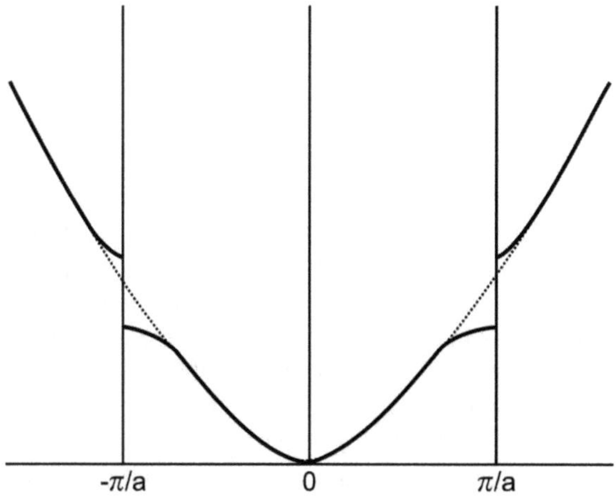

Fig. 10.1 Free-electron bands (dashed lines) and nearly free-electron bands (continuous lines) in a crystalline solid

theory was to assume the existence of a weak periodic crystal potential V(**r**). In 1930, Rudolf Ernst Peierls proposed the nearly free-electron model for a metallic system in which the crystal potential V(**r**) was assumed to be weak. He used plane wave functions for the nearly-free electrons which is a reasonably justified approximation. To his great surprise the presence of a weak potential gave rise to band gaps at the Brillouin zone (BZ) boundaries (see Fig. 10.1). In the same year of 1930 but a little later, Philip McCord Morse (1903–1985) also confirmed theoretically the existence of forbidden energy gaps between the bands in crystalline solids with periodic potential. In 1930, Leon Brillouin (1889–1969) generalized the Peierls idea of energy band gap to three-dimensional crystals and gave the concept of Brillouin zones (BZs). He also gave the method for the transfer of bands from the extended zone scheme to the first BZ (reduced zone scheme).

On the other hand, Werner Heisenberg (1901–1976) and his colleagues transformed the Institute in Leipzing to carry out research in the latest problems in solid state. Felix Bloch (1905–1983) was his first student and was given the problem, "How to incorporate the effect of ions in metals". Bloch approximated the lattice by a three-dimensional periodic potential and used the tight-binding approximation for representing the electrons. In 1928, Bloch derived a theorem, in one dimension, which obeyed the periodicity of the lattice: *Bloch theorem*. He further showed that the wave functions which satisfied the Bloch theorem were modified free-electron waves obeying the periodicity of the lattice: generally called Bloch wave functions. Figure 10.2 shows schematic representation of the Bloch wave function in a crystalline solid. Bloch put forward the full wave-mechanical theory for electrons in metals and the idea of electron energy bands emerged out of it. He developed the linear combination of atomic orbitals (LCAO) method, in the one-electron approximation, for determining energy bands in crystalline solids as a part of his Ph.D. dissertation: it is also called *tight binding approximation*. In the tight binding approximation, Bloch showed that for a simple cubic lattice the following relation gives the electron energy

$$E_{\mathbf{k}} = C + 2K\left[\cos k_x a + \cos k_y a + \cos k_z a\right] \tag{10.2}$$

where C and K are constants. Figure 10.3 shows the energy band, in one dimension, in the tight binding approximation [$E_{k_x} = C+2K\cos(k_x a)$]. He was able to demonstrate the difference between the behavior of a lattice electron and a free electron. The nature of energy bands in the nearly free-electron and tight binding approximations differ from each other in the sense that in the tight binding approximation the point of inflection in a band occurs in the centre of the BZ while in the nearly free electron approximation, it is near the BZ boundary. Bloch's theory laid down the foundations of many basic principles and techniques of the quantum theory of electrons in crystals. In 1937, Gregory Wannier (1911–1983) put the tight binding model on strong footing by showing how Bloch functions could always be summed together to obtain a complete set of wave functions centred at a single atomic site. It was thought that the solution of a number of properties of crystalline solids could be obtained using quantum mechanics and this fact made the development of quantum mechanics much faster.

Felix Bloch (1905–1983), , URL: https://en.wikipedia.org/wiki/Felix_Bloch#/media/File:Felix_Bloch,_Stanford_University.jpg

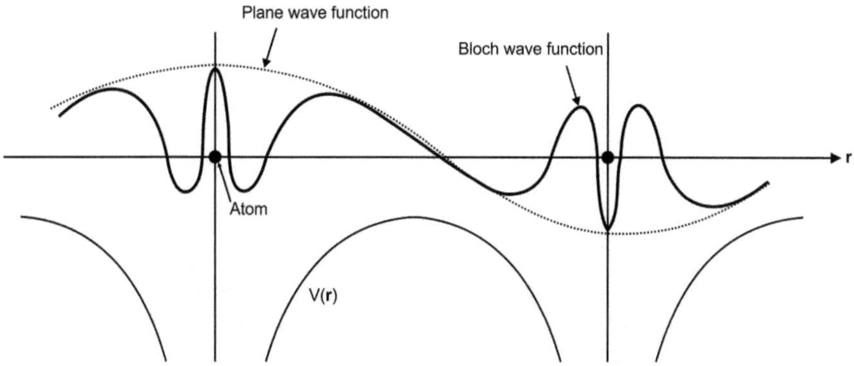

Fig. 10.2 Representation of the Bloch wave function and plane wave function in a crystalline solid. $V(r)$ represents the periodic crystal potential

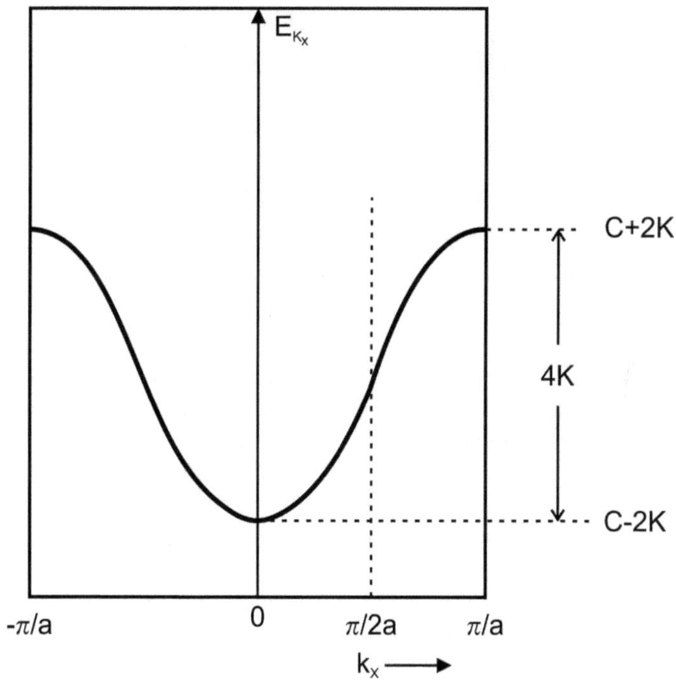

Fig. 10.3 Energy band, in one dimension, in the tight binding approximation

Rudolf Ernst Peierls (1907–1995), CC BY 3.0, URL: https://en.wikipedia.org/wiki/Rudolf_Peierls#/media/File:Peierls,Rudolf_1966_G%C3%B6ttingen.jpg

In 1928, Rudolf Ernst Peierls (1907–1995) also gave the theory of Hall effect and explained the positive values of the Hall coefficient. Using one-band model he proved that in a slightly filled band one obtains negative Hall coefficient. But in a nearly-filled band one obtains positive value of Hall coefficient as the band has a negative curvature near the top of the band. Therefore, Peierls almost made the idea of a hole although not yet clearly described. In 1930, Ralpf de L. Kronig (1904–1995) and William G. Penney (1909–1991) gave a model for energy bands in crystalline solids in one dimension. In the Kronig-Penney model they assumed periodic square-well potential for the linear lattice and Bloch functions for the wave function and solved the Schrodinger equation quantum mechanically. The Kronig-Penney model exhibited beautifully the existence of energy band gaps between different energy bands. In 1931, Werner Heisenberg for the first time introduced the concept of hole in a solid. Alan Herries Wilson (1906–1995) worked with Werner Heisenberg on the application of quantum mechanics to electrical conduction in metals and semiconductors. During the period 1931–1932, Alan Wilson formulated a theory explaining how the electronic energy bands can make a solid material conductor, semiconductor and insulator. In a semiconductor model, Wilson considered an impure insulator in which an electron belonging to the impurity atom has energy in the band gap but close to the conduction band. Such an electron can be easily thermally excited into the conduction band (n-type semiconductor) thus yielding finite electrical conductivity due to the motion of electron in the conduction band. In 1932, Peierls introduced the concept of acceptor level in a semiconductor where a valence electron can easily be excited to the acceptor level (p-type semiconductor). In the p-type semiconductor finite electrical conductivity is produced due to the motion of holes in the valance band and it is less than the conductivity in an n-type semiconductor. Sommerfeld, Pauli and Heisenberg considered the properties of solids as a testing ground for quantum mechanics and their study was confined mainly to Munich, Leipzig and Zurich in Europe. Heisenberg and Pauli were students of Sommerfeld and he acted as a fatherly figure. The trio formed a longstanding close relationship and functioned as a research family, sharing students, problems, seminars, celebrations and excursions. In this process, the subject of physics of solids developed rapidly during 1920s and 1930s. The attention of scientists shifted from qualitative and conceptual problems to more quantitative comparison of theory with experiment. In the due course of time, centres of research on properties of solids were also set up in United States and England.

The free-electron and nearly free-electron models are suited for the study of the electronic structure of simple metals as the valence electrons have s-character and behave as nearly free electrons. But no first principle method was available to calculate the actual electronic band structure of a real metal. In the early 1930s, three main centres, that are Physics Departments at MIT (USA), Princeton (USA) and Bristol (UK), came into existence for the development of electron band theory of metals. Regarding the band structure calculations there was a lot of exchange of ideas and students between MIT and Princeton. The real wave function in a crystalline solid is found to exhibit oscillatory behavior within the core of an atom and varies smoothly outside just like a plane wave (see Fig. 10.2). In 1933, Eugene Paul Wigner (1902–1995) and his first student Frederick Seitz (1911–2008) defined *Wigner–Seitz (WS) cell* and gave a simple method for evaluating the electron energy bands of monovalent sodium metal. Wigner–Seitz method, also called *cellular method*, is the first principle energy band structure method in metals. In case of simple structures, the WS cell resembles closely with a sphere and it is a good approximation to replace it with a sphere of equal volume generally called the *WS sphere*. They assumed atomic potential within the WS sphere which is cut off at the boundary of the WS sphere. Wigner and Seitz solved the Schrodinger equation within each WS sphere satisfying the boundary condition of continuity of wave function at its boundary and applied it to sodium metal. In 1934, in the second paper Wigner and Seitz improved the calculations of the electronic band structure by including the correlation effects among the electrons in the jellium model. Later this method was applied to study the band structure in lithium, copper and silver metals with refined approximations. With the formulation of Wigner–Seitz method began the modern era of electronic energy band calculations for real crystals. In 1934, Johan Clarke Slater (1900–1976) improved the cell model by considering the proper shape and symmetry of the WS cell and setting the boundary conditions at its surface. The main source of error in the cellular method comes from the various approximations involved in constructing a self-consistent crystal potential that would exhibit these symmetry properties.

Frederick Seitz (1911–2008) seated on the left side, Public Domain, URL: https://en.wikipedia.org/wiki/Frederick_Seitz#/media/File:Dr._Frederick_Seitz_&_Dr._H.A._Kramers_Oak_Ridge_(8365596582).jpg

In the d-band metals or transition metals, the conduction electrons possess both the s- and d-characters and the d-electrons are quasi-localized. The band-width of the five d-bands is much smaller than that of the s-band. Further, the s-electrons hybridize with the d-electrons causing finite s-d hybridization which may exist only in certain symmetry directions in a BZ. The s-electrons suffer resonant scattering from the quasi-localized d-electrons, yielding large crystal potential. Therefore, transition metals cannot be treated in the nearly free-electron approximation. Such a picture was already proposed by Nevill Francis Mott (1905–1996) and Harry Jones (1905–1986) and they did pioneering work in band theory of crystalline solids. In 1936, they wrote a classic book entitled, "*The Theory of Properties of Metals and Alloys*" in which they described the electronic structure of transition metals in reasonable detail. Frederick Seitz (1911–2008) said this book of Mott and Jones opened a new area for further work on solid state of materials. In 1937, Johan Clarke Slater (1900–1976) proposed augmented plane wave (APW) method for studying the electron energy band structure in crystalline solids. He, for the first time, introduced the concept of *muffin-tin potential* (Fig. 10.4) and represented the crystal potential as a linear combination of muffin-tin potentials: *muffin-tin approximation*. The trial wave

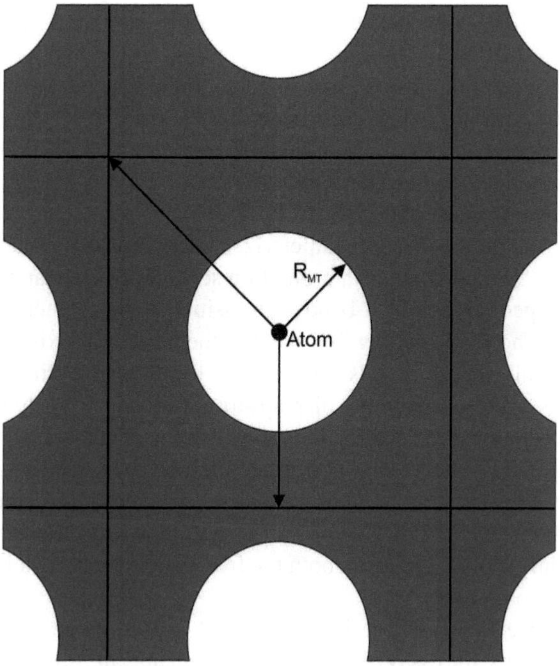

Fig. 10.4 Schematic representation of a muffin-tin sphere in the Brillouin zone. The white sphere shows the muffin-tin sphere with atom at its center and having radius R_{MT}. The spherical cutouts show the parts of muffin-tin spheres in the adjoining Brillouin zones

functions in each muffin-tin sphere are expanded in terms of augmented plane waves which contain as many angular momentum components as desired but outside the muffin-tin sphere it is a plane wave. Therefore, the APW method is applicable to all types of metals such as simple metals, transition metals and rare earth metals. The cellular and APW methods of energy band calculations are related to each other as the APW method is the advanced form of cellular method. In 1939, Marvin Chodorow (1913–2005) applied the APW method for the first time to study the energy bands of copper metal but the results were not very precise due to slow and small computing machines. After this APW method was almost completely ignored for about two decades. By the end of 1930s, the study of electronic band structure of crystalline solids had developed into a well-defined subfield of physics and a number of text books started appearing on the different topics of solids.

With the onset of World War II, the academic research suffered a huge set back as the leading scientists were assigned defence related projects. But during the World War II a number of centres on solid state research came into existence which invented a number of technologies, particularly computers, with the help of which one could obtain improved quantitative results. In 1947, John Bardeen (1908–1991) and Walter Brattain (1902–1987) were working together when they succeeded in creating a point-contact transistor that achieved amplification. In 1948, William

Shockley (1910–1989) introduced the improved bipolar junction transistor which entered mass production in the early 1950s and led to the first widespread use of transistors. This invention gave a great impetus to the band structure calculation in semiconductors. After the World War II, in 1952 and 1954, Frank Herman (1926–2021) calculated the band structure of diamond and germanium. After this many of the researchers studied the band structure of metals, semiconductors, and ferromagnetic and antiferromagnetic materials. In 1954, Johan Clarke Slater (1900–1976) and George Fred Koster (1927–2012) parameterized the tight-binding energy band theory and it resulted in a much simpler interpolation scheme for estimating the electronic band structure especially for the d-bands in transition metals: generally called Slater-Koster scheme. The tight binding approach is the basic approximation in strongly correlated materials, therefore, it is very suitable for highly localized 3d-electrons in transition metals as these sometimes display strongly correlated behavior. In this case the role of electron–electron interactions must be considered using the many-body theory. Asdente and Friedel (1961) performed the energy band structure calculations for Cr metal in the tight binding approximation considering only the d-bands. One should note that the conduction electrons in a crystalline solid are neither completely free nor are tightly bound, therefore both the free-electron and tight binding theories represent the extreme cases.

In 1828, George Green (1793–1841) invented Green's functions to solve partial differential equations; the essay of 1828 can be found in "*The Mathematical Papers of the late George Green*" edited by N. M. Ferrers. Green's function method provides us the single most powerful tool for solving the partial differential equations. In 1947, Jan Korringa (1915–2015) derived multiscattering theory using Green's functions. Later in 1954, Walter Kohn (1923–2016) and Norman Rostoker (1925–2014) extended Korringa's theory using variational technique in the muffin-tin approximation for studying electron band structure of crystalline solids, usually called Korringa-Kohn-Rostoker (KKR) method. The KKR method includes the multiple scattering processes to all orders. Korringa further showed that the KKR method could be used to find electron density as a function of wave vector (energy) in metallic crystals. Many physicists had applied KKR method to study the band structure of metallic solids. There are two main advantages of KKR method over the other methods. First, the KKR method includes the resonant scattering properly unlike other methods which makes it most suitable for the study of band structure of transition metals. Second, the KKR method includes the proper lattice symmetry of the metallic crystal. All the above-mentioned methods had also been applied to study the electronic band structure of dilute metallic alloys.

Walter Kohn (1923–2016), CC BY-SA 3.0, URL: https://en.wikipedia.org/wiki/Walter_Kohn#/media/File:Walter_Kohn.jpg

With the development of fast electronic computers, the interest in the energy band structure calculations received a fresh impetus. Burdick (1961) calculated the energy band structure of copper with improved accuracy and the results are shown in Fig. 10.5. In 1962, Wood (1962), student of John Slater, performed the APW calculations for the energy band structure of Fe metal with high accuracy. Mattheiss (1964) calculated the energy band structure of metals of iron series comprising of Ar, Co, Ni, Cu, V, Cr, Fe, Ti, Zn. Loucks (1965) performed the ab initio APW calculations of energy band structure of Cr, Mo. After this a number of research groups performed energy band structure calculations for transition metals and transition metal compounds using the APW method with increased accuracy [Evenson and coworkers (1969), Gupta and Sinha (1970, 1971), Liu and coworkers (1971), Cooke and coworkers (1974), Rath and Freeman (1975), Gupta and Freeman (1976), Freeman and coworkers (1975, 1976, 1977)]. It is worth pointing here that the APW method is found to be the most accurate method and is widely used in studying the electronic structure of metallic solids.

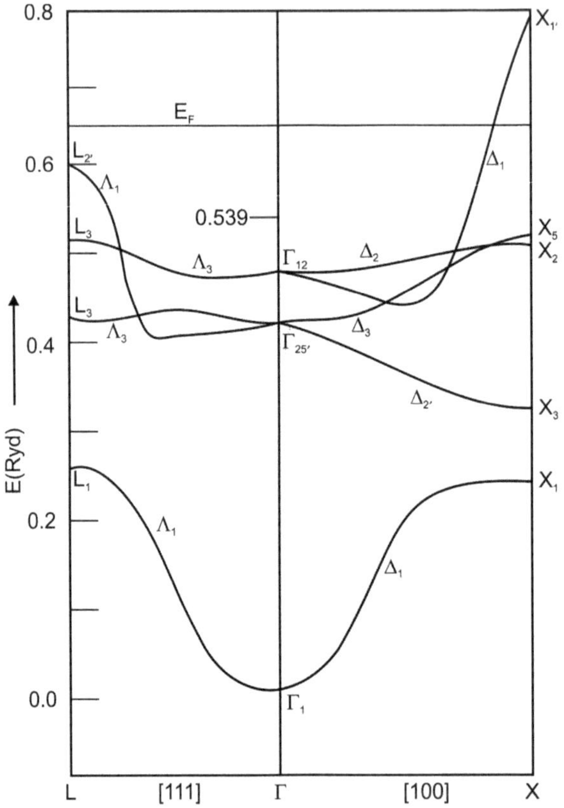

Fig. 10.5 APW band structure of copper in the first Brillouin zone along the [100] and [111] symmetry directions. The vertical axis shows band energy E in Rydbergs while the horizontal axis shows the wave vector. [modified from Burdick, G. A. (1961). Topology of the Fermi surface of copper, Phys. Rev. Lett. **7**, 156]

Dimmock and Freeman (1964) calculated the energy band structure of rare earth metals for the first time and demonstrated that their energy bands exhibit transition metal-like characteristics. But their results were not very precise because the approach used was nonselfconsistent, nonrelativistic and nonspin polarized. Keeton and Loucks (1966) developed relativistic APW method and calculated the relativistic energy band structure of a number of rare earth metals. Using the relativistic APW method, the energy band structure calculations were performed in a number of transition metals like Pb, W, Pd, Pt etc. and in rare earth metals like Sc, Y, Tm, Ho and Tb.

In 1934, Hans Gustav Adolf Hellmann (1903–1938) introduced for the first time the pseudopotential approximation in solids. William Conyers Herring (1914–2009) was working as a postdoctoral fellow with John Slater at MIT. In 1940, Conyers Herring gave the concept of orthogonalized plane wave (OPW) in which a plane wave is orthogonalized to the core states of the solid. Therefore, an OPW includes the advantages of the distinct features of plane wave function and of the tight binding wave functions. Further, Conyers Herring gave OPW method for evaluating the electronic energy band structure of crystalline solids in which the wave function of the solid is expressed as the linear combination of OPWs. Perturbation technique is employed in studying energy band structure using a limited number of OPWs in the wave function. Herring and Hill (1940) performed the energy band calculation for Beryllium, using the OPW method. The number of OPWs in the wave function is increased till one gets the lowest band energy. Paul Soven (1965) developed the relativistic OPW method by orthogonalizing the plane waves to the relativistic atomic wave functions for the core state. The OPW method is most useful when the wave function is expressed in terms of only a few OPWs. It is interesting to note that even a single OPW gave a remarkably good quantitative approximation to the correct crystal wave function. It is found that OPW method works well in simple metals but in transition metals due to resonant scattering the perturbation is strong that makes the use of perturbation theory unreliable. The OPW method was extensively used in studying the energy band structure and Fermi surface in a number of metals. Due to the finite s-d hybridization in transition metals, Hodges and co-workers (1966), Mueller (1967), Heine (1967) and Ehrenreich and Hodges (1968) proposed combined interpolation schemes (for example the combination of tight binding and OPW methods) for evaluating electron energy bands and these were widely used by the scientists.

In 1959, James Charles Phillips (1933–present) and Leonard Kleinman (1933–2022) showed that the OPW method is equivalent to solving Schrodinger equation with an effective potential called *pseudopotential* which is a weak potential. In the pseudopotential method, a wave function can be expanded in terms of plane waves. Due to these facts Phillips and Kleinman were able to demonstrate why the nearly free-electron theory worked in metals. One of the main advantages of the pseudopotential method is that one can use perturbation theory for studying the various electronic properties of crystalline solids. A number of first principle pseudopotentials have been proposed which are nicely reviewed by Walter A. Harrison (1921–2012) in his book, "*Pseudopotential in the Theory of metals*" published in 1966. In 1959, James Phillips formulated simplified pseudopotential theory of the electronic structure of semiconductors and evaluated, for the first time, the electronic structure of silicon and germanium that yielded good agreement with their known properties. Later in mid sixties, James Phillips and Marvin Cohen extended the pseudopotential theory to calculate the fundamental optical and photoemission spectra of many semiconductors with high precision and these are reviewed by James Phillips in 1973 and Marvin L. Cohen, and James R. Chelikowsky in 1988. In 1950, Thomas Samuel Kuhn (1922–1996) and John Hasbrouck Van Vleck (1899–1980) and in 1953, Harvey Brooks (1915–2004) noticed that the spectroscopic values of energy levels can be obtained from the Ritz interpolation formula $E_n = -1/n^2$ if one replaces the principal quantum number n by $n - \delta_n$, where δ_n is called the *quantum defect*. In the quantum defect method, one uses the pseudopotentials in place of real potential to evaluate the phase shifts and study the electronic structure of metals.

In the OPW method there is uncertainty in the peudowave function and psuedopotential and by exploiting the nonunique one can define a new class of parameterized pseudopotentials called model potentials. To simplify the problem of study of electronic properties of solids a number of ionic model potentials have been proposed for both the simple [Ashcroft (1966), Harrison (1966), Heine and Abarenkov (1964)] and transition metals [Harrison (1969), Borchi and Gennaro (1970), Animalu (1973)] but Heine and Abarenkov (1964) gave the most general and successful ionic model potential which is considered to be the best one. Cohen and Phillips (1961) have shown that the individual ionic pseudopotentials are screened by the conduction electrons, which is taken into account by introducing dielectric function. Model potentials have been used by a large number of physicists to study various electronic properties of solids. But the main problem with this method is that there is no single model potential which can explain all the properties of a crystalline solid.

The energy band theory has been extensively reviewed by a number of scientists [Callaway (1964), Loucks (1967), Heine and Weaire (1970), Cohen and Heine (1970), Dimmock (1971), Fletcher (1971), and Ziman (1971)] along with their computational procedures [Alder and co-workers (1968), Marcus and co-workers (1971)].

10.1 Fermi Surface

It is well-known that the valence electrons determine the electronic properties of metals and semiconductors. As the electrons obey the Pauli exclusion principle, two electrons possessing opposite spins can be accommodated in each **k**-state until all the electrons are exhausted. The highest filled state is called the Fermi state after the name of Italian physicist Enrico Fermi (1901–1954) and the energy of highest filled state is called Fermi energy. In a three-dimensional solid, all the electron states having energy less or equal to Fermi energy are filled. The surface on which the electron energy is equal to the Fermi energy is called Fermi surface. Therefore, Fermi surface is the surface in the reciprocal space, which separates the occupied states from the unoccupied states at absolute zero. The concept of Fermi surface though is abstract, but successfully depicts the nature of conduction electrons and their anisotropic distribution in the allowed energy bands in a crystalline solid. The existence of Fermi surface is a direct consequence of the Pauli exclusion principle. Although the concept of Fermi surface was introduced during the development days of quantum mechanics but its full significance was not appreciated for a long time due to the difficulty in determining the Fermi surface either theoretically or experimentally.

The theoretical determination of the Fermi surface involves the calculation of energy bands throughout the first BZ of the crystal. In late 1920s, Wolfgang Pauli (1900–1958) and Arnold Sommerfeld (1868–1951) defined Fermi surface within one-electron theory and constructed it for the free-electron gas (spherical Fermi surface). In 1933, Arnold Sommerfeld and Hans Bethe (1906–2005) published drawings of ideal Fermi surface of simple cubic metals and Bethe said it was clear to me that these surfaces would be important particularly for the study of magnetoresistance. Harrison (1960) discovered an ingenious method for constructing the shape of Fermi surface in the reduced zone scheme. It is well known that the Fermi surface of free-electron gas (empty-lattice metal) is a sphere with radius as Fermi wave vector (k_F) that may extend to higher order BZs. Later Harrison made corrections to the free-electron Fermi surface in terms of weak coupling between OPW waves. It yielded the Fermi surface, in the nearly free-electron approximation, the same as the free-electron Fermi surface except that the lattice potential rounds off the corners of the orbits predicted by the free-electron model. The earliest and most important contribution of free-electron and nearly free-electron models was in providing the outline of the Fermi surface in nontransitional metals. The actual shape of the Fermi surface can be obtained from the detailed calculations of the energy bands from ab initio and it may exhibit a complex shape.

Arnold Sommerfeld (1868–1951), Public Domain, URL: https://en.wikipedia.org/wiki/Arnold_Sommerfeld#/media/File:A._Sommerfeld_LCCN2014715010.jpg

In 1962, William Michael Lomer (1926–2013) proposed a model for the Fermi surface of Cr group metals. He did not perform any ab initio calculations but his model was based on the APW energy band calculations of John H. Wood performed in the same year. Burdick (1963) performed the APW band structure calculations for copper metal and obtained the Fermi surface for copper shown in Fig. 10.6. Loucks (1965) studied the Fermi surfaces of Cr, Mo and W using the ab initio APW calculations of energy band structure and these were almost identical and similar to the model postulated by Lomer for Cr group metals.

In 1930, W. J. de Haas (1878–1960) and P. M. Van Alphen found that magnetization and hence magnetic susceptibility of a single crystal of bismuth exhibit puzzling oscillations as a function of applied magnetic field at low temperatures. It was named *de Haas-van Alphen effect* and is observed in high magnetic fields. In the same year 1930, Lev Shubnikov (1901–1937) and W. J. de Haas found that quantum oscillations of the same periodicity also occur in the magnetoresistance (electrical resistivity in the presence of magnetic field) of single crystal of bismuth as a function of magnetic field: called *Shubnikov-de Haas effect*. In general, the oscillations in transport properties are periodic and occur because of the quantization of energy levels in a plane perpendicular to the magnetic field. In 1930 itself, Lev Landau (1908–1968) gave its theoretical interpretation in terms of *Landau levels* before the experiment was performed. Landau mentioned that such oscillations in

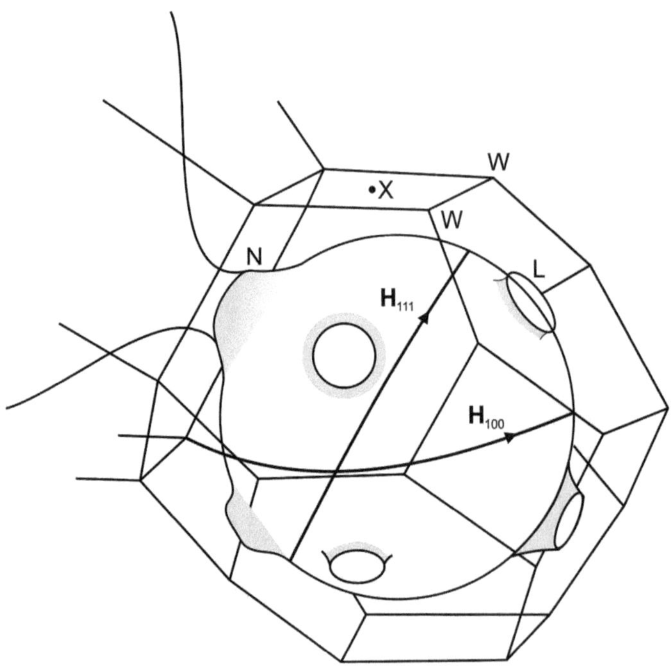

Fig. 10.6 Fermi surface of copper metal in the first BZ with necks at the centers of the zone faces. X, L, W and N are the symmetry points on the faces of the first BZ. [modified from Burdick, G. A. (1961). Topology of the Fermi surface of copper, Phys. Rev. Lett. **7**, 156; (1963). Energy Band Structure of Copper, Phys. Rev. **129**, 138]

magnetoresistivity may be observed in every other metal. In 1937, David Shoenberg (1911–2004) did pioneer work on the de Haas-van Alphen effect in bismuth, using the Landau's model. After about two decades, Lars Onsager (1903–1976), in 1952, and Ilya Mikhailovich Lifshitz (1917–1982) and Arnold Kosevich (1928–2006), in 1954, independently pointed out that de Haas-van Alphen effect could be used to find the shape of the Fermi surface of a metal. From the period of oscillations of magnetic susceptibility, one can directly determine the areas of extremal electron orbits and thus can construct Fermi surface. Similarly, the frequency of the magnetoresistance oscillations indicates the areas of extremal orbits around the Fermi surface. Applying magnetic field in different directions to a metal one can determine the oscillations in magnetoresistance and thus can construct the Fermi surface in three dimensions. Brandt and Rayne (1963) performed the de Haas van Alphen studies on chromium group (Cr, Mo, W) metals and their experimental measurements supported the Fermi surface model of Lomer. The major disadvantage of the de Haas-van Alphen effect is that high magnetic fields are needed to make observable the amplitude of oscillations.

Another method used to construct Fermi surface is anomalous skin effect. It is found that at low temperatures with the increase in frequency the skin depth decreased significantly: *anomalous skin effect*. In 1940, Heinz London (1907–1970) observed the anomalous skin effect for the first time but he did not analyze it in detail. In 1952, Robert G. Chambers (1924–2016) found that the surface impedance becomes large in the region in which the anomalous skin effect appeared. Historically, in 1957, Brian Pippard (1920–2008), a British physicist, made the most important early experimental investigations of the Fermi surface of Cu metal using the anomalous skin effect in the microwave region of the electromagnetic spectrum. He was the first to establish the finite shape of the Fermi surface of copper measuring the reflection and absorption of microwave radiations. Because the anomalous skin effect does not yield the direct features and shape of the Fermi surface therefore Pippard had to guess a shape of the Fermi surface of copper. The determination of the periods of oscillations in various field directions allowed the mapping of the Fermi surface. In general, the Fermi surfaces had been constructed through the experimental observations of different transport properties in magnet fields.

The shape of Fermi surface and the velocity of electrons on the surface can also be measured by *cyclotron resonance* method. The shape, width and anisotropy of cyclotron resonance depend on the shape of Fermi surface of a metal. In the cyclotron resonance method, an RF electric field is applied to a metallic solid. The electric field penetrates the surface of the metallic solid by a small distance (skin depth). Magnetic field can be applied either parallel or perpendicular to the surface. For magnetic field perpendicular to the surface the effect is usually very small and gives only indirect information about the electronic structure. Azbel and Kaner (1956) suggested that magnetic field should be applied parallel to the surface but perpendicular to the microwave field. They gave geometry for observing cyclotron resonance in metals in which the microwave impedance exhibited periodic oscillations as a function of magnetic field and its amplitude decayed as the field decreased. In all the magnetic phenomena in metals, the temperature must be very low since these methods require that the electron should not be scattered appreciably during its motion in an orbit in the presence of field. Gold (1958), a research student of Shoenberg, measured the Fermi surface of lead using de Haas-van Alphen effect and found that it was quite close to the spherical Fermi surface for free electrons. This fact is a triumph of Peierls' nearly free-electron model of a metal.

In 1887, Heinrich Hertz (1857–1894) observed the photoelectric effect which was explained by Albert Einstein (1879–1955) in 1905. In 1950, Kai Manne Borje Siegbahn (1918–2007) discovered *angle-resolved photoemission spectroscopy* (ARPES) which is the most powerful experimental technique to measure directly the electronic band structure of crystalline solids in the momentum-energy space and consequently the Fermi surface. He developed methods for achieving highly accurate measurements of energy levels in atoms by irradiating them with photons. He measured the kinetic energy and momentum distribution of the photoelectrons emitted which enabled him to investigate electronic band structure and map the Fermi surface. Kai Siegbahn shared the 1981 Nobel Prize in Physics for the discovery of ARPES with Nicolaas Bloembergen (1920–2017) and Arthur Leonard Schawlow (1921–1999) for their work on laser spectroscopy. Actually, this technique is best suited for two-dimensional solids. Later X-ray ARPES, time-resolved ARPES, spin-resolved ARPES and spatially-resolved ARPES were discovered.

Beringer and Montgomery (1942) performed the first *angular correlation* measurements on the two gama annihilation of positrons by the conduction electrons in Cu and Pb metals. Lock, Crisp and West (1973) invented *positron annihilation spectroscopy* for determining the Fermi surface of crystalline solids. In this method, it is possible to probe the electron momentum density in a solid which allows determining the Fermi surface. The precision with which the Fermi surface can be determined using either the positron annihilation or Compton scattering is, however, significantly lower than other methods. The advantage of positron annihilation method is that by using spin-polarized positrons, the momentum distribution of the two spin states in a magnetic material can be obtained and hence the Fermi surface for the two spins can be obtained separately.

10.2 Electronic Structure of Metals and Alloys

The study of electronic structure of solids, particularly the metals, is important from the point of view of studying various electronic properties of metals. There were two schools of thought regarding the study of electronic structure of crystalline solids. One school believed in wave mechanical theory given by Schrodinger equation where the potential and wave functions are assumed to have radial dependence and radial dependent electron density $n_e(\mathbf{r})$ is evaluated. In the first half of the twentieth century the wave mechanical approach was widely used by physicists to understand electronic structure of crystalline solids. According to the second school of thought, the energy of a crystalline solid could be assumed to be a functional of $n_e(\mathbf{r})$ and they believed that such a theory would be more appropriate in crystalline solids. In the beginning, the representation of the ground state of a system as a function of electron density faced the following questions:

1. The first and the basic question was the justification for the use of electron density as a variable.
2. The second question was whether one would be able to use the variational technique to evaluate ground state energy of the crystalline solids.

In 1964, Pierre Hohenberg (1934–2017) and Walter Kohn (1923–2016) established the unique representation of the ground state of a system in terms of electron density and developed the density functional theory (DFT) for crystalline solids. With time, DFT became the most powerful technique in studying the electronic structure of metallic systems and Walter Kohn was awarded 1998 Nobel Prize in Chemistry for the formulation of DFT. In DFT the effective one electron potential is given by

$$V^{eff}(\mathbf{r}, n_e(\mathbf{r})) = V^{HP}(\mathbf{r}, n_e(\mathbf{r})) + V_{xc}(\mathbf{r}, n_e(\mathbf{r})) \tag{10.3}$$

where $V^{HP}(\mathbf{r}, n_e(\mathbf{r}))$ and $V_{xc}(\mathbf{r}, n_e(\mathbf{r}))$ are the Hartree and exchange potentials and both are functions of electron density. Kohn and Sham (1965) extended the DFT by including exchange and correlation effects in the *local density functional approximation*. In this approximation, the exchange and correlation energy for a slowly varying electron density $n_e(\mathbf{r})$ is given by

$$E_{xc}[n_e(\mathbf{r})] = \int d^3r\, n_e(r)\, \varepsilon_{xc}(n_e(\mathbf{r})) \tag{10.4}$$

where $\varepsilon_{xc}(n_e(\mathbf{r}))$ is the exchange and correlation energy per electron in a homogeneous electron gas. Using the self-consistent procedure, one can determine the radial dependence of electron density $n_e(\mathbf{r})$. From the knowledge of $n_e(\mathbf{r})$ a number of electronic properties such as electrical conductivity, electronic specific heat, magnetic susceptibility and thermoelectric power etc., can be determined. In DFT, the general solution for a real metal is quite difficult therefore one makes the following two assumptions: First, the real metallic solid is approximated by a simple model and second a numerical procedure is used in the self-consistent solution of Schrodinger equation. A fairly large number of calculations had been performed for $n_e(\mathbf{r})$ approximating a metal either by the free-electron gas or by the jellium model.

In most of the initial calculations, the lattice effects were completely neglected which might be important in explaining the electronic properties of crystalline solids. Later researchers tried to include the lattice effects in the DFT calculations. As one of the simplifications, consider a person sitting over one of the ions of the lattice and the crystal as a whole is rotated about this ion. If the person looks at the crystal the ionic charge will appear to be spread out uniformly on spherical shells centred on the ion under consideration and the solid looks spherically symmetrical. Almbladh and Von Barth (1976) postulated a *spherical solid model* for a metal, which takes into account the effect of host lattice in an approximate manner as described above. In this model electron density can be evaluated either as a function of distance or as a function of wave vector (energy). Using the spherical solid model, Manninen and Nieminen (1979) had performed calculations for muon and hydrogen in metals. One should note that DFT gives exact results in two different cases: first when the electron density is slowly varying and second in the high electron density limit. The shortcoming of the DFT theory is that it includes only single scattering processes. Using DFT a large number of calculations have been performed for evaluating various electronic properties of a number of metals.

Recently, linearized energy band structure methods acquired an increasing importance for the treatment of complicated systems. Andersen (1973) and Koelling and Arbman (1975) developed linear augmented plane wave (LAPW) method for the determination of electronic structure of solids which solves the Schrodinger equation inside the muffin-tin spheres. Actually, it is the Slater's old APW method which has been linearized. It is well known that LAPW method enables most accurate calculation of electronic structure and magnetic properties of crystals and surfaces. It is based on the DFT for the treatment of exchange and correlations and uses local spin density approximation. A number of scientists have used LAPW method to evaluate electronic structure and electronic properties of metals. Andersen and coworkers (1987) also developed tight-binding version of linearized muffin-tin orbitals method which is convenient to apply to more complicated systems such as alloys, liquids and amorphous solids.

All the electronic properties of alloys are intimately related to their electronic structure. For the theoretical study, dilute alloys form the simplest systems because in them the separation between the impurities is quite large. Therefore, each impurity can be assumed to behave independent of other impurities (impurity-impurity interaction is negligible) and the problem reduces to the single impurity scattering processes. An impurity affects the host crystal in two ways:

1. The valency of the impurity atom may be different from that of the host atom. The excess impurity valency $\Delta Z = Z_I - Z_H$ scatters nearby conduction electrons of the host metal producing localized perturbation in the electron density. Here Z_I and Z_H are the valencies of the impurity and host metal atoms. It is usually called *valence effect*.
2. The different size and electronic structure of the impurity atom produces strain in the crystal lattice giving rise to additional perturbation in the conduction electron density. It is called *size effect*.

The quantum mechanical method of partial waves has been widely used in evaluating electron density in metallic solids. Kohn and Vosko (1960) and Blandin and Friedel (1960) independently applied the partial wave method to the impurity scattering problem and evaluated impurity-induced perturbation in conduction electron density $\Delta n_e(\mathbf{r})$ in the asymptotic case. They found that $\Delta n_e(\mathbf{r})$ varies as $1/r^3$ and is oscillatory in nature in the asymptotic limit: oscillations are named *Friedel oscillation* in honor of Jacques Friedel. Actually $\Delta n_e(\mathbf{r})$ consists of infinite number of oscillatory terms with radial dependence of the form $1/r^{2+p}$ with $p = 1, 2, 3, \ldots$. Alfred and Van Ostenberg (1967) improved upon the asymptotic theory by including the first preasymptotic contribution to $\Delta n_e(\mathbf{r})$ which varies as $1/r^4$. Suto and Vasvari

(1973) evaluated $\Delta n_e(\mathbf{r})$ for copper, in the muffin-tin approximation, arising from first three terms with radial variation $1/r^3$, $1/r^4$, and $1/r^5$ (see Fig. 10.7). The study of electronic structure of metallic alloys with transition metal impurities is more interesting from the theoretical point of view. The conduction electrons suffer resonant scattering from the d-state of a transition metal impurity atom which gives rise to a peak with finite width in the electron density around the impurity d-state. This peak is interpreted as virtual bound state (more appropriately as resonant state) because it lies in the conduction band (the real bound state lies above the conduction band). Due to the resonant scattering it is expected that the d-phase shift dominates over the other phase shifts. Brettell and Heeger (1967) evaluated the impurity-induced change in electron density in the asymptotic limit retaining only the d-phase shift for dilute Al alloys with 3d-impurities. The calculated amplitude of electron density shows a peak at Mn impurity. Tompa (1972) and Gruner (1972) have derived the amplitude of electron density from their experimental data and found that in nonmagnetic alloys it exhibit a single peak near Mn impurity while in magnetic alloys it exhibit two peaks which indicate the spin splitting in magnetic alloys. Rennert (1972) found that change in electron density near the impurity varies as $1/r^2$ due to the resonant scattering from the impurity. Lautenschlager and Mrosan (1979) evaluated the amplitude of electron density including the preasymptotic contribution in dilute Al alloys with 3d-impurities and found that it decreases as the impurity is approached. Gruner and Minier (1977) have reviewed the developments in the evaluation of change in electron density due to simple and transition metal impurities in metals.

On the other hand, a number of ab initio calculations had been performed using DFT of Hohenberg and Kohn in the free-electron gas approximation and the jellium model of metal but all these calculations neglected the effect of the crystal lattice [Popovic and co-workers (1976), Jena and co-workers (1978); Ponnambalam and Jena (1981, 1985)]. The crystal lattice effects were included in an approximate manner by performing ab initio calculations in the spherical solid model [Manninen and Nieminen (1979), Manninen and co-workers (1975a, b), Nieminen and Puska (1980)]. All the above mentioned calculations included single impurity scattering processes. The KKR method, which includes the effect of host lattice and the multi-scattering processes, has also been applied to study the conduction electron density in dilute alloys by many researchers and yields the most reliable results.

In metals the ions at the lattice positions are screened by the conduction electrons and the electrostatic screening has the effect of decreasing the strength and the range of ionic potentials. The electrostatic screening is one of the most common methods to study the electronic structure of metals and dilute alloys. In 1923, Peter Debye (1884–1966) and Erich Huckel (1896–1980) developed for the first time a theoretical treatment for the screening of an ion in an electrolyte comprising of positively and negatively charged ions. In 1927, Llewellyn Hilleth Thomas (1903–1992) and in 1928, Enrico Fermi (1901–1954) independently derived a theory of electrostatic screening of a point charge in a free-electron gas usually called *Thomas–Fermi theory*. In this theory the screened potential in the crystal space is nothing but the *Yukawa potential* postulated by Hideki Yukawa (1907–1981). Thomas–Fermi theory is considered as a predecessor to the modern DFT developed much later. In 1952

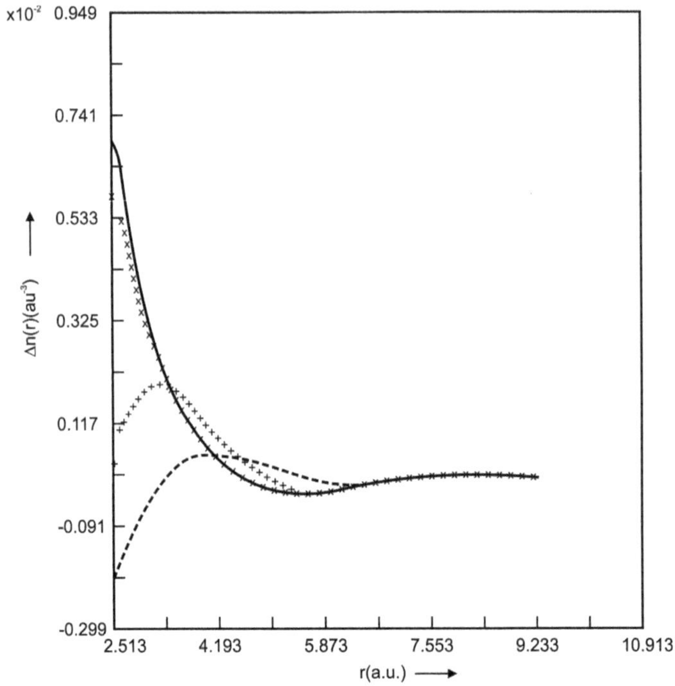

Fig. 10.7 $\Delta n(r)$ as a function of r for CuZn alloy obtained by Suto and Vasvari (1973) in the muffin-tin approximation, including the first three terms in the partial wave method. The dashed line, line with plus sign and line with cross sign show separately the results obtained from the first three terms having radial dependence as $1/r^3$, $1/r^4$, and $1/r^5$, respectively [modified from Suto, A., and Vasvari, B. (1973). A scheme for calculating charge density oscillation parameters. I, J. Phys. F: Metal Phys., 3, 1548]

and 1953, David Bohm (1917–1992) and David Pines (1924–2018) first introduced the random phase approximation in a series of seminar papers which accounts for the weak screened Coulomb interactions. It is commonly used for describing linear electron response in electron systems. In 1954, using random phase approximation, Jens Lindhard (1922–1997) derived quantum mechanical expression for electrostatic screening of a point charge in a free-electron gas generally called *Lindhard function*. The exchange and correlation effects were neglected in the derivation of the Lindhard function but these can be included through the electron–electron interactions. The Lindhard function exhibits a weak Kohn anomaly at a wave vector twice the Fermi wave vector ($2k_F$). The Thomas–Fermi screening function is obtained from the Lindhard function in the limit of very small wave vectors (much smaller than the Fermi wave vector) or in the long-distance limit. Fukai and Watanabe (1970) evaluated perturbation in conduction electron density using the Lindhard function, which showed the oscillations in the asymptotic limit (Friedel oscillations).

In transition metals the dielectric function becomes an infinite-dimensional matrix due to the nonuniform conduction electron density arising from the quasi-localized d-electrons. The nondiagonal elements of the dielectric matrix are the manifestations of local field effects. Sham and Ziman (1963) gave detailed derivation of the general expression for the dielectric matrix in the random phase approximation. Singh and coworkers (1975) gave a general formalism for the dielectric matrix in the *mixed band scheme* in which s-electrons were treated in the free-electron approximation and the d-electrons in the tight binding approximation. They estimated the dielectric matrix for Cr metal and found that the interband contribution is quite small in comparison with the intraband contribution. Singh and Prakash (1981) derived an expression for the dielectric matrix using the nonlocal model potential theory in which the Bloch wave effects were included through the depletion hole. Using the mixed band scheme, Singh and coworkers (1988) formulated a general theory for the conduction electron density in alloys with transition metal impurities in the mixed band scheme. Within this approximation, the formalism yielded exact results in both the reciprocal and crystal spaces for the excess impurity potential and the impurity-induced perturbation in conduction electron density. They estimated perturbation in electron density and screened impurity scattering potential in case of vanadium-based alloys with transition metal impurities (see Fig. 10.8).

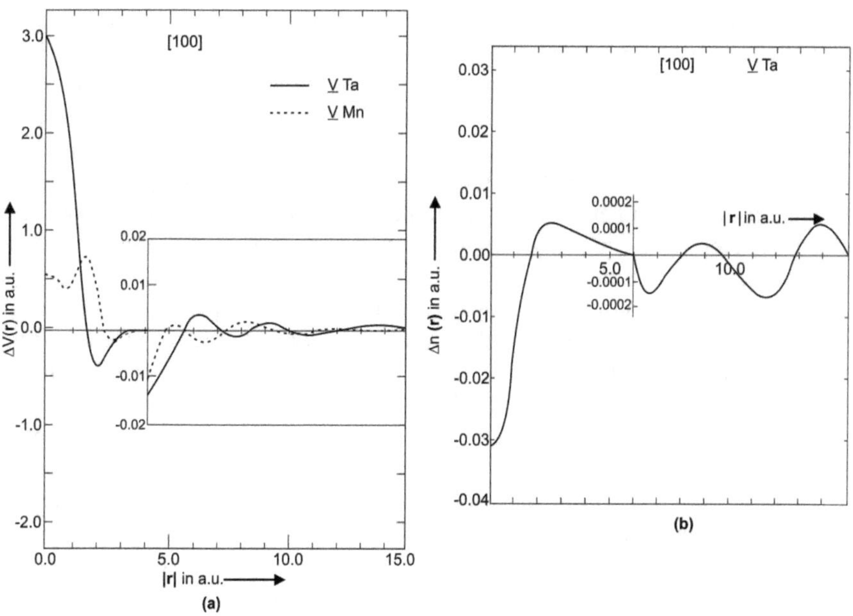

Fig. 10.8 **(a)** Screened excess impurity potential $\Delta V(\mathbf{r})$ due to Ta and Mn impurities in vanadium metal in the dielectric screening approach [modified from Singh, J., Rattan, S. K. and Prakash, S. (1988). Local field effects in the electric field gradient of dilute transition metal alloys, Phys. Rev. **B38**, 10440–10446]. **(b)** $\Delta n(\mathbf{r})$ due to Ta impurity in vanadium metal in the dielectric screening approach [modified from Singh, J., Rattan, S. K. and Prakash, S. (1988). Local field effects in the electric field gradient of dilute transition metal alloys, Phys. Rev. **B38**, 10440–10446]

In concentrated alloys the impurity-impurity interactions are finite, therefore the study of electronic structure becomes much more involved. March and Rousseau (1970) have reviewed the impurity-impurity interactions in concentrated alloys. The impurities are generally randomly distributed thus producing disordered alloys. The KKR method is most suited to many impurities in a crystalline solid as it includes multiscattering processes to all orders. In 1947 and 1958, Jan Korringa (1915–2015) and in 1967, Beeby gave *t-matrix* and *averaged t-matrix approximations* that had been used for evaluating the electronic structure of disordered substitutional alloys. Soven (1967) and Taylor (1967) independently, gave *coherent potential approximation* (CPA) and Velicky and coworkers (1968) applied it for the first time within the framework of tight binding approximation to study electronic structure of disordered binary alloys. Schoen (1969) gave virtual crystal approximation (VCA) and developed a new perfect crystal technique based on APW method. In this approximation, the actual excess crystal potential is replaced by some sort of averaged excess potential yielding the simplest approximation for concentrated alloys. Herman and Dyke (1968) have used VCA to evaluate electron states in amorphous germanium. The electronic structure of concentrated alloys had been studied by many scientists in the virtual crystal lattice approximation. Stocks and coworkers (1978) combined CPA with the KKR theory to obtain KKR-CPA method. Faulkner and Beaulac (1982) develop a quadratic version of KKR method in addition to the linear KKR method. Till date KKR method forms one of the most powerful techniques in studying the electronic energy bands in metals and their alloys. These techniques were extensively applied by different scientists to study the electronic structure of concentrated alloys.

On the other hand, the lattice strain produced by an impurity had been studied in two different approaches: elastic continuum approach, and lattice static approach. In a dilute metallic alloy an impurity ion interacts with its neighboring host ions and the impurity-host ion interactions are naturally different from the host-host ion interactions. Therefore, an impurity ion produces local distortion of the lattice around itself in the form of contraction or expansion of the lattice depending on the size and nature of the impurity. The impurity-induced lattice displacement field $\mathbf{u}_I(\mathbf{r})$ changes the size of the impurity cell as a result of which effective valency of the impurity changes. Blatt (1957) calculated the change in impurity valency in the jellium model of a metal and is named *Blatt correction* to the impurity valency. In 1956 and 1957, John Douglas Eshelby (1916–1981) used continuum elastic theory, the simplest approach, to evaluate $\mathbf{u}_I(\mathbf{r})$. In this approximation, a solid is considered to be a continuous elastic material. Substitutional and interstitial impurities are represented by an elastic material having spherical or ellipsoidal shape and vacancy is represented by a spherical cavity in an elastic material. In 1957, Eshelby used the *misfit model* in which the size and shape of the impurity is assumed to be different from that of the cavity thus producing stress and strain in the surrounding elastic material. Using linear continuum elastic theory, Eshelby evaluated $\mathbf{u}_I(\mathbf{r})$ and found that it varies as $1/r^2$ where r is the distance from the impurity. Ponnambalam and Jena (1981) modified $\mathbf{u}_I(\mathbf{r})$ obtained from elasticity theory by incorporating oscillatory nature in it. In 1984, they evaluated the conduction electron density in a relaxed and unrelaxed lattice yielding the change in conduction electron density due to the impurity-induced strain field. This size effect contribution to conduction electron density is significant and should be added to that obtained from the valence effect to study any electronic property of a crystalline solid.

The second and more reliable method for evaluating the impurity-induced displacement field is the *lattice static method*. Kanzaki (1957) proposed a lattice static method, named Kanzaki method, which employed Born-von Karman theory to evaluate $\mathbf{u}_I(\mathbf{r})$ for a vacancy. In the lattice static method, the impurity-induced strain field was found to exhibit the symmetry of the lattice in contrast to that evaluated in the continuum elastic theory. Singhal (1973) calculated the strain field due to a vacancy using improved lattice statics method based on Born-von-Karman theory. Rattan and coworkers (1993) generalized the Kanzaki method for the substitutional impurities and applied it to evaluate $\mathbf{u}_I(\mathbf{r})$ in simple and transition metals. Later this method has been applied by a number of researchers to evaluate $\mathbf{u}_I(\mathbf{r})$ due to substitutional transition metal impurities in Cu, Ni, Pd, Fe metals. Flinn and Maradudin (1962) used Green functions in the lattice static method and evaluated $\mathbf{u}_I(\mathbf{r})$ for a substitutional impurity in an fcc crystal structure. The impurity-induced strain field had also been estimated with the help of computer simulations by Girifalco and Weizer (1960), Gibson and co-workers (1960), Hafner and Punz (1983) and Kalia and Vashishta (1990).

The impurity-induced change in conduction electron density gives rise to change in crystal potential around the impurity and both exhibit oscillatory behavior in the asymptotic limit. The impurity-induced electric field gradient (negative of double derivative of impurity-induced potential $V_{\alpha\beta}^{\text{imp}}$) is useful in studying the electronic structure of dilute alloys. Schuler and Schmidt (1935) reported the first experimental evidence of nonspherical nuclei and the nonsphericity led to finite nuclear quadrupole moment. Later, in 1935, Hendrik Brugt Gerhard Casimir (1909–2000) gave theory of nuclear quadrupole moment. The nuclear quadrupole moment is useful in solid state physics as it acts as a probe for measuring impurity-induced electric field gradient at the nuclear site. $V_{\alpha\beta}^{\text{imp}}$ has two contributions, first arising from the valence effect and second arising from the size effect. A large number of workers have evaluated the electric field gradient arising from the valance effect which has been summarized by Galsin (2002). Sagalyn and coworkers (1961) estimated the size effect contribution to electric field gradient arising from the introduction of an impurity in a metal considering it an elastic material. They proposed that the impurity-induced electric field gradient in an elastic material is linearly proportional to the impurity-induced strain field and gave the expression

$$V_{\alpha\beta}^{\text{imp}} = \sum_{\mu,\nu} F_{\alpha\beta\mu\nu}\varepsilon_{\mu\nu} \tag{10.6}$$

where $F_{\alpha\beta\mu\nu}$ are the components of fourth-order tensor. Sagalyn and coworkers (1961) evaluated axially symmetric $V_{\alpha\beta}^{\text{imp}}$. Sagalyn and Alexander (1977) evaluated axially asymmetric $V_{\alpha\beta}^{\text{imp}}$ in a number of metals with cubic symmetry. Beal-Monod and Kohn (1968) put forward the theory of size effect electric field gradient in the free electron theory but it has not been applied to any metal.

Thermionic emission of electrons and photoelectron emission are important surface effects which depend on the work function of the material and can be calculated from the electronic structure of the surface. Lang and Kohn (1970), using the self-consistent method, evaluated the charge density normal to the surface $\rho_s(z)$ and found that it decreases rapidly with the increase in distance from the surface. In the asymptotic limit the charge density varied as $1/z^2$ and exhibited the Friedel oscillations. From the charge density, work function can be evaluated. Later a number of self-consistent calculations have been performed for the electronic charge density perpendicular to the surface of the semiconductors, simple and transition metals (see Inglesfield 1982). The calculations show that the surfaces reconstruct themselves with atoms moving away from their positions occupied in the bulk.

Bibliography

Alfred LCR, Van Ostenberg DO (1967) Preasymptotic form of impurity screening in a metal. Phys Lett 26A:27–28

Alder B, Fernbach S, Rotenberg M (eds) (1968) Methods in computational physics, vol 8. In: Energy bands of solids. Academic Press, New York

Almbladh CO, Von Barth U (1976) The spherical-solid model: an application to x-ray edges in Li, Na, and Al. Phys Rev B13:3307

Andersen OK (1973) Simple approach to the band-structure problem. Solid State Commun 13:133–136; (1975) Linear methods in band theory. Phys Rev B12:3060–3083

Andersen OK, Jepsen O, Sob M (1987) In Yusouff M (ed) Lecture Notes in Physics, vol 283. Springer, Berlin, p 1

Animalu AOE (1973) Electronic structure of transition metals. I. Quantum defects and model potential. Phys Rev B8:3542; (1973) Electronic structure of transition metals. II. Phonon spectra. Phys Rev B8:3555

Asdente M, Friedel J (1961) 3d band structure of Cr. Phys Rev 124:384

Ashcroft NW (1966) Electron-ion pseudopotentials in metals. Phys Lett 23:48

Azbel MI, Kaner EA (1956) J Exptl Theoret Phys (U.S.S.R.) 30:811; (1956) Sov Phys JEPT 3:772; (1957) J Exptl Theoret Phys (U.S.S.R.) 32:896; (1957) Theory of cyclotron resonance. Soviet Phys JETP 5:730–744; (1958) Cyclotron resonance in metals. J Phys Chem Solids 6:113–135

Bardeen J, Brattain WH (1948a) Three-electrode circuit element utilizing semiconductive materials. US Patent No. 2,524, 035 (Washington, DC); (1948b) The transistor, a semi-conductor triode. Phys Rev 74:230–231; (1949) Physical principles involved in transistor action. Phys Rev 75:1208–1225

Beal-Monod MT, Kohn W (1968) Strain effects in dilute alloys. J Phys Chem Solids 29:1877–1887

Beeby JL (1967) The density of electrons in a perfect or imperfect lattice. Proc Roy Soc (London) A302:113–136

Beringer R, Montgomery CG (1942) The angular distribution of positron annihilation radiation. Phys Rev 61:222

Blandin A, Friedel J (1959) Propriétés magnétiques des alliages dilués. Interactions magnétiques et antiferromagnétisme dans les alliages du type métal noble-métal de transition. J Phys Radium 20:160–168; (1960) J Phys Radium 21:869

Blatt FJ (1957) Residual resistivity of copper and silver alloys: dependence on periodic table. Phys Rev 108:285

Bloch F (1928) Quantum mechanics of electrons in crystal lattices. Z Phys 52:555–600

Bohm D, Pines D (1951) A collective description of electron interactions. I. Magnetic interactions. Phys Rev 82:625–634; (1952) A collective description of electron interactions: II. Collective vs individual particle aspects of the interactions. Phys Rev 85:338–353; (1953) A collective description of electron interactions: III. Coulomb interactions in a degenerate electron gas. Phys Rev 92:609–625

Borchi E, De Gennaro S (1970) Electron-ion pseudopotentials in the noble metals. Phys Lett 32A:301–302

Brandt GB, Rayne JA (1963) Low-field de Haas-van Alphen studies of chromium group transition element. Phys Rev 132:1945

Brettell JM, Heeger AJ (1967) Nuclear magnetic resonance in dilute alloys of Mn, Fe, and Cu in aluminum. Phys Rev 153:319

Brooks H (1953) Cohesive energy of alkali metals. Phys Rev 91:1027

Burdick GA (1961) Topology of the Fermi surface of copper. Phys Rev Lett 7:156; (1963) Energy band structure of copper. Phys Rev 129:138

Callaway J (1964) Energy band theory. New York, Academic Press

Chamber RG (1952) The anomalous skin effect. Proc Roy Soc (London) A215:481–497

Cohen ML, Chelikowsky JR (1988) Electronic structure and optical properties of semiconductors. Springer, Berlin

Cohen MH, Phillips JC (1961) Dielectric screening and self-consistent crystal fields. Phys Rev 124:1818

Cohen ML, Heine V (1970) Fitting of pseudopotentials to experimental data and their subsequent applications. Solid State Physics 24:38–248

Cooke JF, Davis HL, Mostoller M (1974) Electronic susceptibility of niobium. Phys Rev B9:2485

De Haas W, Van Alphen PM (1930) The dependence of the susceptibility of diamagnetic metals upon the fields, Koninklitje Academie van Wetenschappen Amsterdam. Proc Section Sci 33:1106–1118 (Communication 212a, Physical Laboratory, Leiden)

Debye P, Huckel E (1923) Zur theorie der elektrolyte. I. Gefrier punktserniedrigung und verwandte erscheinungen. Z Phys 24:185–206; (1923) Zur theorie der elektrolyte.II. Das grenzgesetz fur die elektrische leitfahigkeit. Physikalische Zeitschrift 24:305–328

Dimmock JO (1971) The calculation of electronic energy bands by the augmented plane wave method. Solid State Physics 26:103–274

Dimmock JO, Freeman AJ (1964) Band structure and magnetism of gadolinium metal. Phys Rev Lett 13:750

Ehrenreich H, Hodges L (1968) Interpolation scheme for the band structure of transition metals with ferromagnetic and spin-orbit interactions. Methods Comput Phys 8:149

Einstein A (1905) Über einen die erzeugung und verwandlung des lichtes betreffenden heuristischen gesichtspunkt [On a heuristic viewpoint concerning the production and transformation of light]. Ann. der Physik (in German), Leipzig, Germany 17(series 322):132–148

Eshelby JD (1956) In Solid state physics, vol 3 (eds: Seitz F, Turnbull D). Academic Press, New York; (1957). The determination of the elastic field of an ellipsoidal inclusion, and related problems. Proc Roy Soc (London) A241:376–396

Evenson WE, Fleming GS, Liu SH (1969) Generalized susceptibilities and magnetic properties of some bcc transition metals. Phys Rev 178:930

Evenson WE, Liu SH (1968) Generalized susceptibilities and magnetic ordering of heavy rare earths. Phys Rev Lett 21:432; (1969) Theory of magnetic ordering in the heavy rare earths. Phys Rev 178:783

Faulkner JS, Beaulac JP (1982) Multiple-scattering approach to band theory. II. Fast band theory. Phys Rev B26:1597

Fermi E (1928) Eine statistische methode zur bestimmung einiger eigenschaften des atoms und ihre anwendung auf die theorie des periodischen systems der elemente. Z Phys 48:73–79

Ferrers NM (ed) (1871) The mathematical papers of the late George Green. Macmillan Publ. (The 1828 essay can be found in it)

Fletcher GC (1971) The electron babd theory of solids. North Holland, Amsterdam

Flinn PA, Maradudin AA (1962) Distortion of crystals by point defects. Ann Phys 18:81–109

Fradin FY, Koelling DD, Freeman AJ, Watson-Yang TJ (1975) Phys Rev B12:5570–5574
Freeman AJ (1977) Conduction electron polarization, magnetization densities and neutron magnetic form factors in metals. Physica 91B:103–116
Freeman AJ, Watson-Yang TJ, Rath J (1979) Generalized susceptibilities and phonon anomalies in Pd and Pt metals. J Magn Magn Mater 12:140–148
Fukai Y, Watanabe K (1970) Nuclear magnetic resonance in aluminum alloys. Phys Rev B2:2353; (1974) Nuclear magnetic resonance in aluminum alloys. II. Phys Rev B10:3015
Galsin JS (2002) Impurity scattering in metallic alloys. Kluwer Academic/Plenum Publishers, New York
Gibson JB, Goland AN, Milgram M, Vineyard GH (1960) Dynamics of radiation damage. Phys Rev 120:1229
Girifalco LA, Weizer VG (1960) Vacancy relaxation in cubic crystals. J Phys Chem Solids 12:260–264
Gold AV (1958) An experimental determination of the Fermi surface in lead. Philos Trans Roy Soc (London) A251:85–112
Gupta M, Freeman AJ (1976) Direct correlation of phonon anomalies in NbC with Fermi-surface-induced maxima in generalized susceptibilities. Phys Rev Lett 37:364
Gupta RP, Sinha SK (1970) Exchange enhanced generalized susceptibility function for paramagnetic chromium including band structure effects. J Appl Phys 41:915–916; (1971)Wave-number-dependent susceptibility function for paramagnetic chromium. Phys Rev B3:2401
Gruner G (1972) Temperature dependence of charge perturbation around 3d-transition metal impurities in aluminum. Solid State Commun 10:1039–1044
Gruner G, Minier M (1977) Charge perturbation around impurities in metals. Adv Phys 26:231–284
Hafner J, Punz G (1983) Static lattice distortions in substitutional alloys: a computer simulation. J Phys F13:1393
Harrison WA (1960) Electronic structure of polyvalent metals. Phys Rev 118:1190
Harrison WA (1966) Pseudopotentials in the theory of metals. Benjamin, New York
Harrison WA (1969) Transition metal pseudopotentials. Phys Rev 181:1036
Heine V (1967) S−d interaction in transition metals. Phys Rev 153:673
Heine V, Abarenkov I (1964) A new method for the electronic structure of metals. Phil Mag 9:451–465
Heine V, Weaire (1970) Pseudopotential theory of cohesion and structure. Solid State Phys 24:250–463
Hellmann H (1934) A combined approximation procedure for calculation of energies in the problem of many electrons (in German). Acta Physicochim U.R.S.S 1:913–940; Hellmann H, Gelman G (1934) On a combined perturbation procedure in the problem of many electrons. C R Acad Sci URSS NS (Dokl Akad Nauk SSSR) 4:442–444 (in Russian), 444–446 (in German)
Herman F (1952) Electronic structure of the diamond crystal. Phys Rev 88:1210–1211; (1954) Calculation of the energy band structures of the diamond and germanium crystals by the method of orthogonalized plane waves. Phys Rev 93:1214–1225
Herman F, Van Dyke JP (1968) New interpretation of the electronic structure and optical spectrum of amorphous germanium. Phys Rev Lett 21:1575
Herring WC (1940) A new method for calculating wave functions in crystals. Phys Rev 57:1169
Herring WC, Hill AG (1940) The theoretical constitution of metallic beryllium. Phys Rev 58:132
Hertz HR (1887a) Ueber sehr schnelle electrische schwingungen. Ann Phys 267:421–448; (1887b) Ueber einen einfluss des ultravioletten lichtes auf die electrische entladung. Ann Phys 267:983–1000
Hodges L, Ehrenreich H, Lang ND (1966) Interpolation scheme for band structure of noble and transition metals: ferromagnetism and neutron diffraction in Ni. Phys Rev 152:505
Hohenberg P, Kohn W (1964) Homogeneous electron gas. Phys Rev B136:864
Hubbard J (1964) Electron correlations in narrow energy bands. II. The degenerate band case. Proc Roy Soc (London) A277:237–259; (1964) Electron correlations in narrow energy bands III. An improved solution. Proc Roy Soc (London) A281:401–419
Inglesfield JE (1982) Surface electronic structure. Rep Prog Phys 45:223

Jena P, Gupta AK, Singwi KS (1978) Nonlinear electron-density distribution around point defects in simple metals. II. Applications. Phys Rev B18:2723

Kalia RK, Vashishta P (1990) Molecular dynamics simulations of classical and quantum systems. In: Prakash S, Pathak KN (eds) Advances in statistical physics of liquids and solids. Wiley Eastern Ltd, New Delhi, pp 142–151

Kanzaki H (1957) Point defects in face-centered cubic lattice—I distortion around defects. J Phys Chem Solids 2:24–36

Kasuya T (1956) A theory of metallic ferro- and antiferromagnetism on Zener's model. Prog Theort Phys 16:45–57

Keeton SC, Loucks TL (1966) Relativistic energy bands for thorium, actinium, and lutetium. Phys Rev 146:429

Koelling DD, Arbman GO (1975) Use of energy derivative of the radial solution in an augmented plane wave method: application to copper. J Phys F: Met Phys 5:2041

Kohn W, Vosko (1960) Theory of nuclear resonance intensity in dilute alloys. Phys Rev 119:912

Kohn W, Sham LJ (1965) Self-consistent equations including exchange and correlation effects. Phys Rev 140A:1133

Korringa J (1947) On the calculation of the energy of Bloch wave in a metal. Physica 13:392–400

Korringa J (1958) Dispersion theory for electrons in a random lattice with applications to the electronic structure of alloys. J Phys Chem Solids 7:252–258

Kuhn TS, Van Vleck JH (1950) A simplified method of computing the cohesive energies of monovalent metals. Phys Rev 79:382

Landau L (1930) Diamagnetismus der Metalle [Diamagnetism of metals]. Z für Phys 64:629–637

Lang ND, Kohn W (1970) Theory of metal surfaces: charge density and surface energy. Phys Rev B1:4555

Lautenschlager G, Mrosan E (1979) Localization of additional screening charge around 3d-transition metal impurities in Al. Phys Status Solidi B96:183

Lifschitz IM, Kosevich AM (1954) On the theory of the de Haas–Van Alphen effect for particles with an arbitrary dispersion law. Dokl Akad Nauk SSSR 96:1954; (1956) Theory of magnetic susceptibility in metals at low temperatures. Sov Phys JETP 2:636–645

Lindgren I, Schwarz K (1972) Analysis of the electronic exchange in atoms. Phys Rev A5:542–550

Lindhard J (1954) On the properties of a gas of charged particles. Danske Matematish-fysiske Meddeleiser 28(8): Kobenhavn

Liu SH, Gupta RP, Sinha SK (1971) Generalized susceptibility function for rare earths and thorium and their alloys. Phys Rev B4:1100

Lomer WM (1962) Electronic structure of chromium group metals. Proc Phys Soc (London) 80:489

Lock DG, Crisp VHC, West RN (1973) Positron annihilation and Fermi surface studies: a new approach. J Phys F: Metal Phys 3:561

London H (1940) The high-frequency resistance of superconducting tin. Proc Roy Soc (London) A176:522–533

Loucks TL (1965) Fermi surfaces of Cr, Mo, and W by the augmented-plane-wave method. Phys Rev 139:A1181; (1967) Augmented plane wave method. Benjamin, New York

Malhotra B (1986) M. Phil. (Physics) thesis, Guru Nanak Dev University, Amritsar, Punjab (India)

Manning MF, Chodorow MI (1939) Electronic energy bands in metallic tungsten. Phys Rev 56:787–798; Chodorow MI (1939) Phys Rev 55:675

Manninen M, Nieminen RM (1979) Spherical solid model for muon and hydrogen in metals. J Phys F: Met Phys 9:1333

Manninen M, Nieminen RM, Hautojärvi P, Arponen J (1975) Electrons and positrons in metal vacancies. Phys Rev B12:4012

March NF, Rousseau JS (1970) Crystal lattice defects, vol 2, 1. Gordon and Breach, Belfast

Marcus PM, Janak JF, Williams AR (eds) (1971) Computational methods in band theory. Plenum, New York

Mattheiss LF (1964) Energy bands for solid argon. Phys Rev 133:A1399; (1964) Energy bands for the iron transition series. Phys Rev 134:A970; (1965) Energy bands for V_3X compounds. Phys Rev 138:A112

Mott NF, Jones H (1936) The theory of the properties of metals and alloys. Dover, New York
Mueller FM (1967) Combined interpolation scheme for transition and noble metals. Phys Rev 153:659
Nieminen RM, Manninen M, Hautojärvi P, Arponen J (1975) Self-consistent electron densities in metal vacancies. Solid State Commun 16:831–834
Nieminen M, Puska M (1980) 3d impurities in Al: density functional results. J Phys F: Met Phys 10:L123
Onsager L (1952) Interpretation of the de Haas-Van Alphen effect. Phios. Mag. 43:1006–1008
Pal B, Mahajan S, Raj SD, Singh J, Prakash S (1984) Electric field gradients in nonmagnetic dilute alloys of aluminum and copper. Phys Rev B30:3191
Phillips JC (1973) Bonds and bands in semiconductors. Academic Press, New York
Phillips JC, Kleinman L (1959) New method for calculating wave functions in crystals and molecules. Phys Rev 116:287; (1959) Crystal potential and energy bands of semiconductors. I. Self-consistent calculations for diamond. Phys Rev 116:880
Pippard AB (1957) An experimental determination of the Fermi surface in copper. Philos Tran Roy Soc (London) A250:325–357
Ponnambalam MJ, Jena P (1981) Theory of electric field gradient due to a point defect in a cubic metal. Phys Rev Lett 46:610; (1984) Distribution of electric field gradients around impurities in aluminum. Hyperfine Interact 20:65–117
Popovic ZD, Stott MJ, Carbotte JP, Piercy GR (1976) Theory of the heat of solution of hydrogen in Al and Mg using nonlinear screening. Phys Rev B13:590
Rath J, Freeman AJ (1975) Generalized magnetic susceptibilities in metals: application of the analytic tetrahedron linear energy method to Sc. Phys Rev B11:2109
Rattan SK, Singh P, Prakash S, Singh J (1993) Strain field due to point defects in metals. Phys Rev B47:599
Rennert P (1972) Influence of a resonance state on the density oscillations around an impurity. Phys Status Solidi 50:K37
Ruderman MA, Kittel C (1954) Indirect exchange coupling of nuclear magnetic moments by conduction electrons. Phys Rev 96:99–102
Sagalyn PL, Alexander MN (1977) Electric-field-gradient tensor in nonmagnetic dilute alloys of copper. Phys Rev B15:5581
Sagalyn PL, Paskin A, Harrison RJ (1961) Contribution of strain to nuclear quadrupole interactions in dilute alloys of copper. Phys Rev 124:428
Schoen JM (1969) Augmented-plane-wave virtual-crystal approximation. Phys Rev 184:858–863; Schoen JM, Denker SP (1969) Band structure, physical properties and stability of TiO by the augmented plane wave virtual crystal approximation. Phys Rev 184:864
Schubnikov L, De Haas WJ (1930) Magnetic resistance increase in single crystals of bismuth at low temperatures. Proc Roy Netherlands Acad Arts Sci (in German) 33:130–133; (1930) New phenomena in the change in resistance of bismuth crystals in a magnetic field at the temperature of liquid hydrogen. I. Proc Roy Netherlands Acad Arts Sci 33:363–378; (1930) Neue erscheinungen bei der widerstandsanderung von wismuthkristallen im magnetfeld bei der temperature von flussigem wasserstoff. II. Proc Roy Netherlands Acad Arts Sci 33:418–432; (1930) The change in resistance of bismuth crystals in a magnetic field at the temperature of liquid nitrogen. Proc Roy Netherlands Acad Arts Sci 33:433–439
Schüler H, Schmidt T (1935) Über Abweichungen des Atomkerns von der Kugelsymmetrie. Z Phys 94:457–468; (1935) Über die abweichung des cassiopeiumatomkerns von der kugelsymmetrie. Z Phys 95:265–273
Sham LJ, Ziman JM (1963) The electron-phonon interaction. Solid State Phys 15:221–298
Shockley W (1949) The theory of *p-n* junctions in semiconductors and *p-n* junction transistors. Bell Syst Tech J 28:435–489
Shoenberg D (1939) The magnetic properties of bismuth. III. Further measurements on the de Haas-Van Alphen effect. Proc Roy Soc (London) A170:341; (1962) The Fermi surface of copper, silver and gold. I. The de Haas-Van Alphen effect. Phil Trans Roy Soc (London) A255:85; Shoenberg

D, Stiles PJ (1964) The de Haas-Van Alphen effect in alkali metals. Proc Roy Soc (London) A281:62
Siegbahn K, Hedgran A, Svartholm N (1950) Proc Phys Soc A63:960; Siegbahn K (1954) Ark Fys 7:86 and in the Proceedings of conference of the Swedish National Committee for Physics, Stockholm, 1952; Siegbahn K (1954) Ark Fys 8:19 and in the Proceedings of conference of the Swedish National Committee for Physics, Stockholm, 1953; Siegbahn K (1952) Physica 18:1043, Introductory Talk at the International Conference on beta and gamma radioactivity, Amsterdam, 1952; Siegbahn K (ed) (1965) Alpha-, beta and gamma ray spectroscopy. North-Holland, Amsterdam
Singh J, Singh N, Prakash S (1975a) Noninteracting band model for dielectric screening and local field corrections in bcc transition metals: Application to paramagnetic chromium. Phys Rev B12:3159; (1975b) Phonon frequencies of paramagnetic chromium. Phys Rev B12:3166
Singh J, Prakash S (1981) Dielectric matrix using the nonlocal model potential approach. J Phys F11:2409–2416
Singh J, Rattan SK, Prakash S (1988) Local field effects in the electric field gradient of dilute transition metal alloys. Phys Rev B 38:10440–10446
Singhal SP (1973) Lattice relaxations near a vacancy or interstitial in aluminum. Phys Rev B8:3641
Singwi KS, Sjolander A, Tosi MP, Land RH (1970) Electron correlations at metallic densities. Phys Rev B1:1044–1053
Slater JC (1934) Electronic energy bands in metals. Phys Rev 45:794; (1934) The electronic structure of metals. Rev Mod Phys 6:209
Slater JC (1937) Wave functions in a periodic potential. Phys Rev 51:846; (1953) An augmented plane wave method for the periodic potential problem. Phys Rev 92:603–608
Slater JC, Koster GF (1954) Simplified LCAO method for the periodic potential problem. Phys Rev 94:1498–1524
Sommerfeld A, Bethe HA (1933) The electron theory of metals. Handbuch der Physik 24(Part 2)
Soven P (1965a) Relativistic band structure and Fermi surface of thallium. I. Phys Rev 137:A1706; (1965b) Relativistic band structure and Fermi surface of thallium. II. Comparison with experiment. Phys Rev 137, A1717.
Soven P (1967) Coherent-potential model of substitutional disordered alloys. Phys Rev 156:809
Stocks GM, Temmerman WM, Gyorffy BL (1978) Complete solution of the Korringa-Kohn-Rostoker coherent-potential-approximation equations: Cu-Ni alloys. Phys Rev Lett 41:339
Suto A, Vasvari B (1973) A scheme for calculating charge density oscillation parameters. I. J Phys F: Metal Phys 3:1548
Taylor DW (1967) Vibrational properties of imperfect crystals with large defect concentrations. Phys Rev 156:1017
Thomas LH (1927) The calculation of atomic fields, Proc Chamb. Phil Soc 23:542–548
Tompa K (1972) The asymptotic region of charge density oscillations around impurities in dilute copper-based alloys. J Phys Chem Solids 33:163–175; (1971) Conference on electrical and magnetic properties of dilute alloys (Tihany)
Velicky B, Kirkpatrick S, Ehrenreich H (1968) Single-site approximations in the electronic theory of simple binary alloys. Phys Rev 175:747
Watson-Yang TJ, Freeman AJ, Koelling DD (1977) Magnetic field induced conduction electron polarization, magnetization density, and neutron magnetic form factor of Pt metal. J Mag Magn Mater 5:277–286
Watson-Yang TJ, Harmon BN, Freeman AJ (1976) Electronic structure, field-induced magnetization density and neutron magnetic form factor of Pd. J Magn Magn Mater 2:334
Wood JH (1962) Energy bands in iron via the augmented plane wave method. Phys Rev 126:517–527
Yosida K (1957) Magnetic properties of Cu-Mn alloys. Phys Rev 106:893–898
Yukawa H (1935) On the interaction of elementary particles. Proc Phys-Math Soc Jpn 17:48
Ziman JM (1971) The calculation of Bloch functions. Solid State Physics 26:1–101

Chapter 11
Electricity, Magnetism, Electromagnetism and Magnetic Solids

11.1 Electricity

The natural phenomenon of static electricity had been known since ancient times. The credit of discovery of static electricity goes to Thales of Miletus around 600BC. Thales noticed that amber rubbed with wool attracted lighter objects and it is because amber gained some substance by rubbing. Ancient Greeks discovered that rubbing amber with silk also attracted lighter objects. For many years Greeks thought that this phenomenon was due to the unique property of amber. William Gilbert (1544–1603), an English physician and physicist, studied static electricity using amber. Gilbert found that when amber is rubbed with silken cloth, it removes some substance called "effluvium" which causes attraction effect on other objects. But he could not realize that this substance (electric charge) was universal to all materials. Amber in Greek is called "elektron", so he called the attractive force as electric force. Electroscope is the first scientific instrument for detecting the charge on an object. Around 1600, William Gilbert made first electroscope which was simply a pivoted needle. In 1600, he wrote a book, "*Electricus*", meaning like amber, in which he described all his findings about static electricity. In 1646, Sir Thomas Browne (1605–1682), first used the word electricity which was derived from Gilbert's book "Electricus". In general, static electricity is caused when certain materials are rubbed against each other, like wood on plastic or the soles of shoes on carpet etc. Electrostatics involves the build up of some electric charge on the surface of one object due to contact with the surface of another object. Electrostatic phenomenon arises from the forces that electric charges exert on each other. Gilbert was of the view that electricity and magnetism are two independent phenomena and is credited as the "*Father of electricity and magnetism*".

Till the beginning of the eighteenth century, scientists studied the generation of static charges and the static phenomena. Stephen Gray (1666–1736), an English

J. S. Galsin, *History of Solid State Physics*,
https://doi.org/10.1007/978-981-95-0504-3_11

dyer and astronomer, was the first person who started experimental study of the phenomenon of electrical conduction in 1729. He discovered that charge can flow through the metallic objects which were generally regarded in those days as nonelectrics because they could not generate or hold static charge. He assumed charge a fluid that could flow over a distance. He was the first person to make distinction between conduction and insulation. Jean-Antoine Nollet (Abbe Nollet) (1700–1770), French physicist, and Charles Francois du Fay (1698–1739), French chemist, were knowing Stephen Gray personally and his experimental findings and formulated *two-fluid theory* for the electricity. They assumed that there are negative and positive fluids moving in an object. When these two liquids came into contact with one another, they would produce a neutral fluid. Many scientists in Europe had advocated the two-fluid theory. Charles du Fay also discovered that objects with like-fluids repel and those with unlike-fluids attract each other. Du Fay's observations on electricity were reported in a paper written in December, 1733 and printed in Volume 38 of the Philosophical Transactions of the Royal Society in 1734. Benjamin Franklin (1706–1790), American scientist and statesman, was having interest in the study of static electricity and performed many experiments. John Bevis (1695–1771) and William Watson (1715–1787), both English researchers, were in correspondence with Benjamin Franklin's group and in 1746 they jointly proposed a *single-fluid theory.* According to them there is abundance of fluid on one side of the material and deficit on the other side. In 1747, Benjamin Franklin independently developed the single-fluid theory. He proposed that an object with excess fluid acquired one type of charge (say positive charge) but an object with deficit of fluid acquired opposite charge (that is negative charge). A neutral object was therefore thought to contain a "normal" amount of this fluid. But this theory could not explain fully the attraction and repulsion between two bodies. For example how two bodies with too little fluid can repel each other? Further, how two objects one with too large fluid and other with too small fluid can attract each other? After a long debate extending over a century, one fluid theory of Franklin was accepted. Many people believe that Franklin discovered electricity in 1752 with his famous kite-flying experiment. He thought that electricity from the clouds was transferred to kite and then it flowed down through the string and gave him a shock.

The precise determination of the amount of electric charge on a body was one of the main problems. In 1745, Pieter van Musschenbroek (1692–1761), a Dutch physicist, invented a device, for the first time, which could store large amounts of charge. Charles du Fay gave it the name *Leyden jar* after the place Leiden where it was invented. In the same year Ewald Georg von Kleist (1700–1748), a German cleric and physicist, independently developed the idea for such a device but did not investigate it as thoroughly as Musschenbroek did. Both Pieter van Musschenbroek and Ewald Georg von Kleist are considered to be the discoverer of static electricity. When the Leyden jar first arrived in United Kingdom, John Bevis and William Watson worked together in refining it. They also tried to measure the speed of electricity but in vain because of its very large value. The invention of the Leyden jar confirmed the single-fluid theory. In 1748, Jean Nollet (1700–1770), a clergyman and physicist, invented one of the first electroscopes which detected the presence of charge using the concept of electrostatic repulsion and attraction. Later in 1754, John Canton (1718–1772), a British physicist, invented pith-ball electroscope. He was the first to verify Benjamin Franklin's hypothesis of lightening and electricity. In 1786, Abraham Bennet (1749–1799), an English clergyman and physicist, invented *gold-leaf electroscope* which is a very sensitive instrument. It consists of an upright brass rod whose lower end holds two very thin gold-leaf strips and the upper end is fitted with a ball. It is used for detecting and measuring the quantity of charge on an object with precision.

In 1753, John Canton (1718–1772), discovered the phenomenon of electrostatic induction. In 1762, Johan Carl Wilcke (1732–1796), a Swedish physicist, found that charge is redistributed in an object due to the presence of another nearby charge. The induced charges disappear when the external charge is removed. The electrostatic induction is responsible for the attraction of light objects such as paper scraps, balloons etc. to the static charge. From the study of force among electrical charges, Wilcke proposed three laws: like charges repel each other, opposite charges attract each other and charged objects attract neutral objects. In 1762, Wilke invented an electrostatic generator that was the first version of the electrophorus, a device named and popularized in 1775 by Alessandro Volta (1745–1827).

In 1767, Joseph Priestley (1733–1804) experimentally studied the force between two charges and was the first to propose that electrical force followed the inverse square law just like the gravitational force. In 1767, Priestley published a book on the History of Electricity which became a standard text for over a century. In 1768, he also wrote a popular version of history of electricity titled, '*A familiar Introduction to the Study of Electricity*'. Charles-Augustin de Coulomb (1736–1806), French engineer and physicist, performed a number of experiments for measuring the force between electrical charges using torsion balance. Based on his experimental observations, he proposed in 1785 the inverse square law for electrical forces between charged particles known as *Coulomb inverse square law* in his honor. The Coulomb law states that the magnitude of force of attraction or repulsion between two stationary charges (q_1 and q_2) is directly proportional to the product of their charges, inversely proportional to the square of the distance r between them and directed along the straight line joining the two charges. Mathematically the Coulomb inverse square law can be written as

$$\mathbf{F} = K\frac{q_1q_2}{r^2}\hat{\mathbf{r}} \tag{11.1}$$

Charles-Augustin de Coulomb (1736–1806), Public Domain, URL: https://en.wikipedia.org/wiki/Chales-Augustin_de_Coulomb#/media/File:Charles_de_Coulomb.png

Here K is the Coulomb constant and $\hat{\mathbf{r}}$ is a unit vector along r. In CGS system of units, K = 1 but in MKS system $K = 1/4\pi\varepsilon_0$ where ε_0 is permittivity: K is a system related constant. The unit of charge is named Coulomb in his honor. With the discovery of Coulomb law, the static electricity attained the rank of a separate subject called *electrostatics*. The Coulomb law is the fundamental law in electrostatics and later became one of the four Maxwell equations.

In the eighteenth century, the phenomenon of electrification by contact, also known as contact electrification, was discovered. When two objects were touched and then pulled apart, sometimes the objects became spontaneously charged. One object developed a negative charge while the other developed an equal positive charge. The charge is transferred to the solid with higher resistivity and is locked out on the surface in such solids. As a result a weak potential difference was generated between two dissimilar metals called contact potential difference (the *Volta effect*). This fact led to the development of batteries, fuel cells, electroplating and thermocouples etc. Contact electrification is responsible for lightening in thunderstorms, sandstorms and volcanic plumes. In 1780, Luigi Galvani (1737–1798), an Italian physician, physicist and biologist, discovered animal electricity. He and his wife Lucia discovered that the muscles of the leg of a dead frog got contracted when struck by an electric arc. It was the first evidence of electric oscillations and Galvani supposed that these were manifestations of animal electricity. Alessandro Volta (1745–1827), an Italian scientist, carefully investigated the Galvani's experiment and realised that frog's leg served as a conductor (electrolyte) but is irrelevant to the electric current. The flow of current was caused by the two different metals contrary to the animal electricity proposed by Galvani.

By the second half of eighteenth century a number of metals and insulators were known. But in some elements the strength of electrical conduction was found to lie between the conductors and insulators. In 1782, Alessandro Volta (1745–1827) used the term semiconducting for these elements and hence the name semiconductors. Volta studied the relation between electrical potential difference and charge and found that for a given object, they are proportional and the proportionality constant is called capacitance. This is called *Volta's Law of Capacitance*, and for this work the unit of electrical potential has been named volt. Up to the eighteenth century the scientists had studied the sources of charges and their different properties (static electrostatics) but there was no dependable source of electricity for conducting experiments. In 1800, Alessandro Volta made a great invention of *electric cell*. In the original form, it was a paper soaked in salt which was placed between zinc and copper sheets serving as two electrodes: it was named Volta cell. Later the electrolyte was replaced by sulphuric acid. Volta cell is a direct current (dc) cell which provides almost constant current and its strength is measured in terms of volts. The invention of electric cell gave birth to the subject of *current electricity*. By connecting these cells together, Volta created an electric battery with increased voltage.

On the other hand, simultaneously efforts were being made to establish quantitative relations between different electrostatic quantities. In 1777, Joseph-Louis Lagrange (1749–1827) introduced the concept of *potential* and gave Laplace equation that could be used to evaluate gravitational force of an extended body. In 1812, Simeon-Denis Poisson (1781–1840) adopted Laplace's idea and obtained, using the Coulomb law, an appropriate expression which related potential function $V(\mathbf{r})$ to the electric charge density $\rho(\mathbf{r})$: Poisson equation, which was first published in the *Bulletin de la sociétéphilomatique*. Poisson pointed out that the Laplace equation is valid only outside the solid system.

In 1806, Humphry Davy (1778–1829), a British chemist, created an early form of arc lamp (arc light) by passing an electric arc through two charcoal rods. He invented what is called *Davy lamp* which was Miner's safety lamp. Humphry Davy's major work was the discovery of a number of elements: potassium and sodium in 1807; calcium, barium, and magnesium in 1808; boron in 1809 and chlorine and iodine in 1810. Davy invented laughing gas which is nothing but nitrous oxide. In 1789, Antoine-Laurent de Lavosier (1743–1794), French chemist, tried to group the available elements as metals and nonmetals. In 1829, Johann Wolfgang Dobereiner (1780–1849), German chemist, observed similarities in physical and chemical properties of certain elements which led to the anticipation of periodic table. While writing a book on Chemistry, Dmitri Ivanovitch Mandeleev (1834–1907), Russian chemist, made an important discovery on grouping the elements. In 1869, Mandeleev constructed periodic table in which elements were arranged in the order of increasing atomic weights and valence. Lothar Meyer (1830–1895), German chemist, constructed the same Periodic Table in 1868 but unfortunately it was published in 1870 after the publication of Mandeleev. Dmitri Mendeleev is often referred to as the *Father of the Periodic Table*. Later, some difficulties were found in arranging the elements in it. In 1913, Henry Moseley (1887–1915), a British physicist, observed a systematic mathematical relation between the wave length of X-rays emitted and the atomic number of metals that were used as targets: this relation is known as *Moseley law*. He thought that atomic number is a more suitable fundamental property as compared with the atomic weight. Therefore, Moseley arranged elements in the order of increasing atomic numbers (number of protons or electrons) in the Periodic Table and is considered to be the final Periodic Table. In doing so he predicted a number of elements which were later found experimentally. Moseley was able to demonstrate correctly that lanthanide elements must have exactly 15 members—no more no less. It has been found that in Moseley's Periodic Table, elements which lie on the left side are metallic in nature, elements on the right side are insulators, but those lying in the middle of the periodic table are either semiconductors or semimetals.

Michael Faraday (1791–1867), Public Domain, URL: https://en.wikipedia.org/wiki/Michael_Faraday#/media/File:Michael_Faraday_sitting_crop.jpg

Michael Faraday (1791–1867), an English physicist and chemist, had no formal education but attended only Sunday church classes. At an early age he distributed newspapers, worked as an apprentice of a bookbinder and the joined as assistant in the laboratory of Humphry Davy (1778–1829). During this time he learnt science by reading and self-study and then became a most brilliant experimenter of the nineteenth century physics. He started his career as a chemist and in 1833 gave laws of electrolysis. Faraday coined the words like ion, electrode, anode and cathode. Faraday's main contribution is, however, in electricity and magnetism. Before Faraday only basic properties of electricity and magnetism were known. He worked on the force between two bodies: force between two charges or between two magnetic poles etc. He was of the view that force cannot act directly between two distant bodies and thought that there must be some physical entity diffused throughout the space around the electric and magnetic bodies. In 1845, Faraday discovered this physical entity and named it '*field*'. He said that an electrical charge is surrounded by an electric field represented by lines, which carry force with them and these are called *lines of force*. Therefore, the space around an electric charge is filled with large number of lines of electric force. Similarly the space around a magnetic body or magnetic pole is filled with lines of magnetic force. These lines transmit electric or magnetic force from one body to another. Figure 11.1 depicts electric field around positive and negative charges in terms of lines of force (lines of force are directed away from a positive charge but these are directed towards the negative charge). Two charges separated by a finite distance interact with each other through their electric fields thereby producing attractive or repulsive force between them. The magnetic lines of force of north and south poles are just similar to those of positive and negative charges: lines of force are directed away from a north pole and these are directed towards the south pole. Two magnetic poles interact through their magnetic fields causing attraction or repulsion. Faraday's concept of field changed the scientific world because a system of particles (say a system of charged particles) is now made of particles and their fields. So, Newton's picture of space, time and particles has changed to **space, time, particles** and **fields** associated with them. Although it seems to be a minor change but it changed whole of the Newton's world. The theory of relativity as described earlier has further changed the scientific picture to **spacetime**, **particles** and **fields**. Faraday never learnt mathematics, yet he made discoveries that scientists later explained using mathematics. We want to point out that Faraday was offered knighthood but declined saying he preferred "plain Mr. Faraday". Further, he chose a simple grave instead of burial in the Abbey, despite being offered this high honor.

Carl Friedrich Gauss (1777–1855) was a genius German mathematician and physicist and is considered to be the greatest mathematician of all times. He was pioneer in the fields of electricity and magnetism. In 1813, Gauss derived a relation between the electric force that follow inverse square law and the charges which produce it. In 1839, Gauss gave the rigorous proof of Poisson equation with variable density.

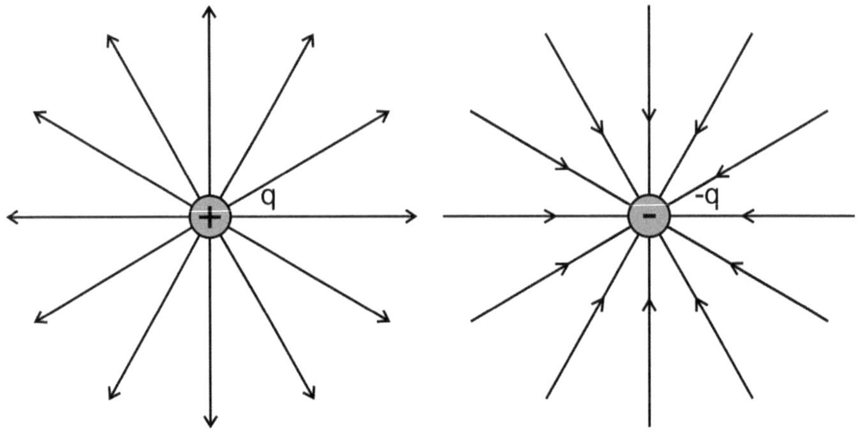

Fig. 11.1 Lines of electric force around a positive and negative charge

It is noteworthy that Poisson equation is valid in electricity, magnetism and gravitation. Gauss defined electric flux as the total number of electric lines of force passing through a closed surface. Electric field intensity **E** is defined as the number of electric lines of force per unit area crossing perpendicular to the surface. In 1867, Friedrich Gauss gave a relation between the electric flux passing through a closed surface and the charge enclosed by it: *Gauss law in electrostatics.* Mathematically the Gauss law in the C. G. S. System, is written as

$$\oint \mathbf{E} \cdot d\mathbf{s} = 4\pi q \tag{11.2}$$

where d**s** is an infinitely small surface. Therefore, the Gauss law describes the static electric field generated by the distribution of static charges. It can be easily shown that the Gauss law for electric fields is equivalent to the Coulomb law. In 1828, George Green (1793–1841), a British mathematician, introduced the idea of potential functions currently used in physics. Green's work on potential theory ran parallel to that of Carl Friedrich Gauss. In the same year of 1828, Green formulated a theorem called Green theorem in an Essay on the "*Application of Mathematical Analysis on the Theories of Electricity and Magnetism*", and it is widely applied to study properties of electric and magnetic potentials. Green was the first person to formulate mathematical the theory of electricity and magnetism, which formed the basis for the work of other scientists such as James Clerk Maxwell. The remarkable part of Green's life story is that he was almost entirely self-taught. He attended formal schooling only for about one year as a child between the age of 8 and 9 years.

In 1820, Johann Schweigger (1779–1857) designed the earliest form of electromagnetic galvanometer naming it after Luigi Galvani. It was used to detect and measure small currents. This instrument was initially called multiplier. In 1825, Leopoldo Nobili (1784–1835) developed improved form of galvanometer and later a number of improved forms of galvanometer came into existence. Ammeter and voltmeter are galvanometers with very small and very large resistance, respectively. Using the Volta cell, Georg Simon Ohm (1789–1854), a school teacher, discovered electrical resistance while performing a series of experiments. He designed his own equipment for studying the flow of current through a conducting wire. He found that there is a direct proportionality between the potential difference applied across the conductor and the current flowing through it. This relation is called *Ohm's law* in his honor and forms the backbone of electricity. The Ohm's law first appeared in his book: "*Die Galvanische Kette, Mathematisch Bearbeitet (The Galvanic Circuit Investigated Mathematically)*" published in 1827. Through a series of experiments, Ohm demonstrated that there was no perfect electrical conductor as every conductor offered some resistance. This resistance is constant for a particular element at a constant temperature. The unit of resistance was named '*Ohm*'. Georg Ohm came from a very humble family and struggled financially throughout his life but is well-known today for Ohm law.

Georg Simon Ohm (1789–1854), Public Domain, URL: https://en.wikipedia.org/wiki/Georg_Ohm#/media/File:Georg_Simon_Ohm_(1789-1854).jpg

Electrical resistance is one of the most important properties of solids. The resistance R of a solid wire is directly proportional to its length ℓ and inversely proportional to its area of cross section A. Therefore, one can write

$$R = \rho \frac{\ell}{A} \tag{11.3}$$

The constant of proportionality ρ is the specific resistance or resistivity of a material. The reciprocal of resistivity ρ is conductivity σ. After the discovery of Ohm's law a number of instruments were designed to measure the electrical resistance of metals and some of these are mentioned here. In 1833, Samuel Hunter Christie (1784–1865), a British physicist and mathematician, invented a bridge for measuring an unknown resistance in a circuit. In 1841, Johann Christian Poggendorff (1796–1877), a German physicist, invented *slide wire potentiometer* which is used for measuring voltage and it became a standard laboratory measuring technique. In 1843, Sir Charles Wheatstone (1802–1875) improved and popularized this bridge and is called Wheatstone bridge in his honor. The Wheatstone bridge is widely used even today to measure unknown resistance in laboratories. In 1872, George Carey Foster (1835–1919) invented another bridge known as Carey Foster Bridge for measuring small resistances; it is a modified form of Wheatstone bridge.

The work of Ohm marked an early beginning of the field of circuit theory although the importance of this field was not realized until the end of nineteenth century. Gustav Robert Kirchhoff (1824–1887), a German physicist, contributed to the fundamental understanding of electrical circuits. Kirchhoff was a student when he, in 1845, gave two laws of electrical circuits usually called *Kirchhoff's laws*. The first law states that the algebraic sum of currents in a network of current-carrying conductors meeting at a point is zero. According to the second law, the sum of directed voltages in a closed circuit is zero. These laws generalized the work of Georg Ohm and the field of *circuit theory* emerged in electricity and electronics. Gustav Kirchhoff also worked on the emission of thermal radiations. In 1859, he proposed the law of thermal radiations according to which for an arbitrary body emitting and absorbing thermal radiations, emissivity is equal to absorptivity in thermodynamical equilibrium. In 1861, he gave the proof of the law of thermal radiations. In 1862, Kirchhoff coined the term *black-body radiations* for thermal radiations emitted by a black body. Kirchhoff also contributed greatly to the field of spectroscopy and gave three laws in spectroscopy mentioned below:

1. A solid, liquid, or dense gas when excited produces a continuous spectrum.
2. A low-density gas when excited produces an emission spectrum with discrete lines.
3. If light with a continuous spectrum passes through a cool, low-density gas, an absorption spectrum is produced.

At the time of Kirchhoff, the existence of energy levels in atoms was not known but Fraunhofer had already discovered discrete spectral lines in 1814. Kirchhoff and Robert Bunsen (1811–1899) invented spectroscope that was used in analyzing and identifying elements present in the sun.

Thomas Alva Edison (1847–1931) was America's one of the greatest and genius inventors. He attended a public school in Port Huron for a few months (about 12 weeks) but he did not like the traditional education system. His teachers described him a slow learner and difficult to deal with. Edison's mother was a school teacher who recognized his potential. She quickly pulled him out from the school and taught him reading, writing and arithmetic at home. In his childhood Edison sold candies, newspapers and vegetables on the train running from Huron to Detroit. Whatever he used to earn was spent on buying equipment for electrical and chemical experiments. Even as a child, he was very much interested in engineering and technology and used to perform experiments at home for hours together. Edison developed the problem of hearing at the age of twelve: he was completely deaf from one ear and hard of hearing from the other ear. As far as electricity is concerned he invented electric light bulb in 1879: an amazing invention of all times. In the first ever electric bulb he used ordinary carbonized cotton thread as the filament. The bulb glowed with pale yellow light and became the first incandescent electric bulb. Edison developed the power generation and distribution system and established the Pearl Street Power Station in New York City. It was one of the world's first central direct current (dc) electric power plants which could power 5000 lights. In addition to electric bulb Edison had developed many other devices in different fields such as phonogram (sound recording) in 1877, the motion picture camera (called kinetoscope) in 1891 etc. He had 1093 US patents in his name as well as patents in other countries. Edison established the first industrial research laboratory where many researchers and employees were working together. He was one of the first to apply the principles of organized science and teamwork to the process of invention. He was one of the first inventors to apply the principle of mass production. All his inventions had a wide spread impact on modern industrial world.

Nikola Tesla (1856–1943), a Croatian-born scientist, made dozens of breakthroughs in the production, transmission and application of electric power. In 1882, Nikola Tesla worked with Edison in Paris with the Continental Edison Company and later in 1884 migrated to USA to work at Edison's Machine Works at New York. Tesla left Edison's company after some time as he was not paid promised bonus of $50,000, 'but it turned out to be a practical joke'. Tesla started working independently and in 1882, discovered rotating magnetic field which produced alternating current (ac) and alternating voltage: a physical principle which figured prominently in many of his future inventions. In 1885, Galileo Ferraris (1847–1897) independently invented rotary magnetic field and made the working model of single phase induction motor. In 1887, Tesla built his working two-phase induction motor which ran on ac and obtained its patent. In 1888, Mikhail Dolivo-Dobrovolsky (1862–1919) invented the first three-phase motor and three-phase generator to work on three-phase ac system. The turning point in electric age came a few years later when Tesla developed ac systems. In 1890, he constructed alternators capable of producing undamped waves at 39 kHz. Tesla collaborated with engineer and businessman George Westinghouse to patent the ac systems and provided the nation with ac power that could travel longer distances: a direct competition with Thomas Edison's dc system. The power plants using ac current and voltage could transport electricity much faster to longer distances and with much less power loss. On the other hand, Edison's dc power plant could transport electrical energy only within one square mile of his Pearl Street Power Station. In 1891, he invented Tesla coil and got it patented. It is a new kind of transformer and is used in science laboratories even today. In the same year of 1891, Mikhail Dolivo-Dobrovolsky invented first three-phase transformer and short-circuited induction motor. In 1891 itself, Dolivo-Dobrovolsky designed the world's first three-phase hydroelectric power plant and made the electric transport system which was able to transport 75% energy generated. Tesla later went on to form the Tesla Electric Company. Unfortunately Tesla's laboratory caught fire in which his early notes, models and other important material got burnt and it was a great set back to his ongoing projects. Tesla tried to develop a wireless lighting system based on near-field inductive and capacitive coupling but could not succeed. In 1893, he pronounced that there is a possibility of wireless communication and undertook *Wardenclyffe Tower Project*, an intercontinental wireless communication and power transmitter. In 1897, Lord Kelvin visited Nikola's laboratory for wireless communication but he was initially skeptical of Tesla's idea of wireless energy transmission, dismissing it as an illusion. But after a more detailed explanation from Tesla, the next day, Kelvin's perspective dramatically shifted in its favour. He worked on this project for a number of years but had to abandon it due to lack of funding. After Wardenclyffe Project, Tesla made a number of inventions with varying degrees of success and died in New York City in January 1943.

The pioneer work of Georg Ohm and Gustav Kirchhoff created interest in the study of electrical conductivity or resistivity of metals. In 1853, Rudolph Franz (1826–1902) and Gustav Heinrich Wiedemann (1826–1899), both German physicists, gave an empirical law, known as Wiedemann–Franz law, which states that the ratio of electrical conductivity to thermal conductivity has approximately a constant value for different metals at a particular temperature. In 1872, Ludvig Lorenz (1829–1891) found experimentally the value of this constant of proportionality as a function of temperature. In 1898, Carl Victor Eduard Riecke (1845–1915) conducted experiments on electrical conduction and realized that in a metal current arises due to the flow of electrons. After this theoretical models were developed for the conductivity of metals. In 1900, Paul Karl Ludwig Drude (1863–1906) developed a classical model for electrons in metals, called *Drude model*, in which interactions between the electrons were neglected (see Chap. 10 also). Thermal and electrical conductivities were evaluated using the Drude model but it did not yield the experimental value of the constant given by the Wiedemann–Franz law. In 1905, Hendrick Antoon Lorentz (1853–1928) improved upon the Drude model by including interactions between the electrons and evaluated thermal and electrical conductivities. He obtained the correct value of the constant of Wiedemann–Franz law and it is usually called Lorentz number. It is worth noting that Robert Andrews Millikan (1868–1953), an American physicist, started measured the charge on an electron in 1906 using oil drop experiment. Later he performed a series of oil drop experiments to improve the determination of electron charge and in 1909, for the first time, made a direct and convincing measurement of an electron's charge, as a natural unit of electric charge using the Millikan oil-drop experiment. Millikan also worked on the photoelectric effect and in 1914, experimentally confirmed the Einstein's interpretation of photoelectric effect. Millikan was awarded 1923 Physics Nobel Prize for his study of the elementary electronic charge and the photoelectric effect. In 1913, Wilhelm Wien (1864–1928) put forward one of the most important classical theory of metal conduction in which he assumed that the mean free path is inversely proportional to mean square of the amplitude of the atomic oscillations. He was able to explain the change in resistance with temperature.

Later resistivity of a number of metals was measured experimentally and Eduard Gruneisen (1877–1949), in 1928, has compiled the data for resistivity of metals. It has been found that the resistivity of metals depends on temperature. At low temperatures, $\rho(T)$ of metals varies as T^5. But at high temperatures, $\rho(T)$ increases with temperature T for most of the metals but not in a linearly fashion. For example in Cu, Ag, Au and W, the slope R/T of the resistivity curve increases with the increase in T while in Pd, Pt and Ta metals, R/T decreases. The calculation of resistivity using Einstein model predicts the linear dependence of $\rho(T)$ on T at high temperatures. At intermediate temperatures, Eduard Gruneisen, in 1933, has suggested a formula for the variation of resistivity introducing Gruneisen function. Percy Williams Bridgman (1882–1961) measured the resistivity under high pressures for a number of metals and found that the resistance of most of the metals decreased with the increase in pressure. On physical grounds one can expect that under high pressure the atoms in a metal come closer thereby increasing the attractive forces among them. Therefore, atoms will vibrate with smaller amplitude under high pressure. The Wien's theory also predicted the decrease in resistance with the increase in pressure on the solids. Eduard Gruneisen was also able to explain the change in resistance with pressure.

Arnold Sommerfeld (1868–1951) was the first to apply the ideas of quantum mechanics to the problem of metallic conduction (see Chap. 10). In 1927, Sommerfeld gave the free-electron theory in which the Fermi–Dirac distribution function was used for the distribution of electrons in metals. He was able to explain the specific heat of conduction electrons. In 1928, Felix Bloch (1905–1983) gave quantum theory of electrons in crystals. He showed that in a perfect periodic crystal there is no hindrance to the flow of electrons. Finite hindrance (resistance) arises due to the breakage of periodicity of the lattice which may be brought either by the atomic vibrations of the crystal or by the presence of the impurities. In 1930, Felix Bloch gave the quantum theory of electrical resistance by considering the scattering of electrons from the lattice (electron–phonon interactions). He showed that at low temperatures resistivity varies as T^5 and at high temperatures it is proportional to T (linear temperature variation). After this a number of physicists including Hans Bethe (1906–2005), Rudolf Peierls (1907–1995), Werner Heisenberg (1901–1976), Lev Landau (1908–1968), John Clarke Slater (1900–1975), Alan Herries Wilson (1906–1995) and others developed further the theories of metallic conduction. In transition metals, the resistivity has high value as compared to the simple metals which is due to the overlap of s- and d-bands and the anisotropic Fermi surface. In 1933, Sommerfeld and Bethe gave a comprehensive account of various theories of metallic conduction.

The field of electrical conduction in alloys is enormous as one can produce large number of alloys by mixing different elements. Therefore, electrical conduction in all the alloys cannot be reviewed but we will mention the important discoveries here. The electrical resistivity of a perfectly ordered alloy is zero but in real alloys perfect order does not exist. Therefore the electrical resistivity in alloys is always finite and depends on their electronic structure. Electrical resistivity comprises of two contributions: electrical resistivity of pure metal $\rho_L(T)$ and the impurity resistivity $\Delta\rho(C)$. Therefore, the electrical resistivity of an alloy becomes

$$\rho(T, C) = \rho_L(T) + \Delta\rho(C) \tag{11.4}$$

It is well known that $\rho_L(T)$ approaches zero as T approaches absolute zero. On the other hand, $\Delta\rho(C)$ is independent of temperature but depends on the impurity concentration C. Therefore, in an alloy the electrical resistivity $\rho(T, C)$ approaches the value $\Delta\rho(C)$ at absolute zero that is why $\Delta\rho(C)$ is called residual resistivity. It is found that $\Delta\rho(C)$ obeys some rules listed below:

1. Matthiesen rule (1861): $\Delta\rho(C)$ is independent of temperature for low impurity concentrations.
2. Linde's rule (1931, 1932): $\Delta\rho(C)$ is proportional to the square of excess impurity valency ΔZ.
3. Nordheim rule (1931): $\Delta\rho(C)$ is a function of impurity concentration and $\Delta\rho \propto C(1 - C)$ which is an inverted parabola.
4. Mott's rule (1934): $\Delta\rho(AB) = \Delta\rho(BA)$ which says that residual resistivity due to 1% of B metal dissolved in A metal is equal to the residual resistivity when 1% of A metal is dissolved in B metal.

In 1921, A. L. Norbury had summarized the available experimental data on resistivity. In 1931 and 1932, Linde measured experimentally the resistance of dilute solid solutions in Cu, Ag, and Au. In 1927, Hilding Faxen (1892–1970) and John Peter Holtsmark (1894–1975) gave the exact solution for the evaluation of residual resistivity of metals. In 1936, Mott derived the expression for $\Delta\rho(C)$ in the Born approximation and showed its explicit dependence on the square of the excess impurity valency $\Delta Z = Z_I - Z_H$ where Z_I and Z_H are impurity and host atom valencies. The method of partial waves has been applied extensively to evaluate $\Delta\rho(C)$ of dilute alloys. In 1952, Jacques Friedel (1921–2014) evaluated the change in electron density in metallic alloys and showed that it exhibits oscillatory behavior at large distances (asymptotic limit) named Friedel oscillations. He also formulated the sum rule in terms of phase shifts named *Friedel sum rule*. The Friedel sum rule established the relation between the excess impurity valency and the phase shifts and is an important self-consistency condition for determining the change in electron density and the impurity potential. The expression for $\Delta\rho(C)$ was derived using the partial wave analysis which does not exhibit explicit dependence on the excess impurity valency. But the results obtained in the partial wave method are in reasonable agreement with the experimental results. It shows that the dependence on the excess impurity valency has been taken care of indirectly through the phase shifts. In 1966, Walter A. Harrison (1930-present) derived the expression for $\Delta\rho(C)$ in dilute simple alloys in the pseudopotential approximation and calculated it in a number of simple metal alloys. It was found that the agreement with the experimental results was crude. In 2002, Galsin (1947–present) has reviewed the partial wave method as applied to dilute alloys.

The behavior of residual resistivity $\Delta\rho(C)$ in dilute metallic alloys with transition metal impurities is more interesting. $\Delta\rho(C)$ due to the transition metal impurities does not vary as the square of excess impurity valency. In dilute Al metal alloys with 3d-impurities, $\Delta\rho(C)$ exhibits a broad peak near Cr impurity which shows the occurrence of resonant scattering of conduction electrons. Friedel (1956) explained that the resonant scattering of conduction electrons is caused by the atomic d-states near the Fermi surface and is interpreted as the virtual bound state (vbs) [Friedel effect]. In dilute Cu metal alloys with 3d-impurities, $\Delta\rho(C)$ exhibits two peaks which show the existence of spin splitting of the 3d-state (Friedel 1956; Friedel 1958; Gruner 1974). The presence of spin-splitting is supported by the data on thermoelectric power of these alloys (Friedel 1956; Mott and Jones 1936). Spin splitting gives rise to finite magnetic moment on the d-band impurity. Mn impurity corresponds to the minimum in $\Delta\rho(C)$ between the two peaks and has magnetic moment of five effective Bohr magnetrons ($5\mu_B$) in agreement with the experiment. In Chap. 10 it has been explained that due to resonant scattering the d-phase shift dominates other phase shifts. Bela Malhotra (1986) calculated residual resistivity $\Delta\rho$ in Al-, Cu- and V-based dilute alloys with transition metal impurities retaining only the d-phase shift. She found that the calculated residual resistivity is less than the experimental values but it exhibits correctly the observed trend as found in all the alloys. The lower values of calculated residual resistivity may be because of two reasons. First the other phase shifts also contribute to $\Delta\rho$ and second the size effect contribution to $\Delta\rho$ should also be taken into care of. In 1988, Singh and coworkers evaluated impurity-induced change in conduction electron density due to transition metal impurities including the local field effects. John Michael Ziman (1925–2005) gave a detailed account of electron conduction in metals and metallic alloys.

11.2 Magnetism

A story goes that roughly 2000 BC a Greek shepherd boy named Magnes of Magnesia discovered magnetic material in the form of a rock usually called *loadstone*. The writings of Thales (625BC–545BC) of Miletus mentioned that loadstone was known to attract iron pieces and because of this property the loadstone was called *magnet*. Later it was found that the naturally occurring magnet was nothing but a mineral called magnetite. The Roman poet Lucretius Carus (99BC–55BC) mentioned about the phenomenon of magnetic induction in his philosophical poem "*De rerum natura*". He was perhaps among the first persons to notice, by accident, that magnetic materials could also repel. On the other hand, in one of the Chinese books written in about 400BC, reference is made of magnetism and in about 200BC it was known that loadstone attracted iron pieces. In eleventh century, Shen Kuo (1031–1095), a Chinese scientist, talked about magnetic needle in his book "*Dream Pool Essays*" published in 1088. He performed experiments with magnetic needle and discovered the direction of true North which made magnetic compass more useful for navigation. In the twelfth century Chinese used loadstone compass for navigation. This concept was unknown in Europe at that time but in 1187, Alexander Neckam (1157–1217), an English scientist, was the first to describe about the magnetic compass and its use for navigation in his book "*De utensilibus* (*On instruments*)". In 1269, Peter Peregrinus de Maricourt (1240–1299), French mathematician and physicist, performed experiments with a spherical loadstone which he called "*terrella*". He placed an oblong piece of iron on terrella and studied its motion on its surface. He found that the iron piece moved along the lines parallel to the surface crossing it at two points. He called them two poles of the loadstone in analogy with the poles of the earth. He described the properties of magnets and discussed about freely pivoted compass needles in his book "*Epistola Petri Peregrini de Maricourt ad Sygerum de Foucoucourt Militem de Magnete*", an earliest known treatise of experimental physics. Maricourt found that like poles repel and unlike poles attract each other in a magnet. He also described the attraction of iron by loadstone and magnetization of iron by loadstone.

William Gilbert (1544–1603), was an English physician but physics was his hobby. He spent 17 years in writing a master-piece book on magnetism named, "*De Magnete*" which was published in 1600. This book contained all the trustworthy knowledge of his time on magnetism together with his own major contributions. Gilbert was the first to argue that the centre of earth is made of iron, a magnetic material. Therefore, earth behaves like a big magnet as a result of which the magnetic compass points towards north and south (geomagnetism). He also found that if a magnet is cut into two or more pieces, each piece forms a new magnet with north and south poles. This argument can be extended further that is if one goes on cutting a magnet into smaller and smaller pieces ultimately the smallest piece will be an atom. Therefore, each atom will behave as a complete magnet with finite magnetic dipole moment. Gilbert was the first person to explain and generalize the laws of magnetism in his book on magnetism. Gilbert believed that earth has a soul, therefore, loadstone, being a part of earth, also has a soul. In 1750, John Canton (1718–1772) read a paper before the

Royal Society of London on the method of making artificial magnets without the use of natural magnets.

Rene Descartes (1596–1650) said there is no soul in a loadstone thus laid the foundation of rational theory of magnetism. He gave the first theory of magnetism which is stated in his Principia, Part IV, Sections 133–183. Descartes gave theory of geomagnetism and assumed that earth comprises of two types of one-way threaded parts forming channels. The first threaded part enters the North Pole and leave by the South Pole while in the second part it is the other way round. The return trip was by air in either case. Since loadstone is a part of the earth so the same explanation also applies to the loadstone. Descartes claimed that all the magnetic phenomena known at that time could be explained with his theory. Descartes reputation was such that his theory was accepted as fact.

Up to the early eighteenth century efforts were made to understand the phenomenon of magnetism on metaphysical grounds. It was only in the second half of the eighteenth century that rational hypotheses and theories of magnetism were proposed. In 1759, Franz Maria Aepninus (1724–1802) put forward one-fluid theory for magnetism in analogy with Franklin's one-fluid theory in electricity. His theories of electricity and magnetism appeared in "*Tentamen Theoriae Electricitatis et Magnetism* (An Attempt at a Theory of Electricity and Magnetism)" published in1759 by St. Petersburg. After some years, Anton Brugmans (1732–1789) proposed the two-fluid theory in magnetism in his book, "*Tentamina philosophica de materia magnetica eiusque actione in ferrum et magnetem*, (in Latin), Franeker: Willem Coulon (1765), in analogy with the two-fluid theory in electricity originally proposed by Charles Francois de Cisternay du Fay (1698–1739) in 1733. The two fluids were denoted as *austral* and *boreal* in correspondence with the two magnetic poles. In this theory, it was assumed that in the nonmagnetic state of a material the two fluids saturate iron equally. But in the magnetic state, magnetization separates the two fluids and pushes them slightly towards their respective poles where they accumulate. In 1777, Johan Carl Wilke (1732–1796), a former student and collaborator of Aepninus and Anton Brugmans presented the two-fluid theory independently.

In 1785, Charles-Augustin de Coulomb (1736–1806) was a great promoter of the two-fluid theory but he proposed an important modification in it. He said that magnetic fluids are not free to flow like the electric fluids but are limited to the individual molecules. He supposed that molecules become polarized somewhat in the presence of magnetization. In this way it can be explained why two new poles always appear when a magnet is cut. He extended the inverse square law to magnetic force between two magnetic poles: it is named Coulomb law of magnetic force in his honor. According to it the force of attraction or repulsion $\mathbf{F}_m$ between two magnetic poles is directly proportional to the product of their pole strengths m_1 and m_2 and inversely proportional to the square of distance r between them, that is

$$\mathbf{F}_m = K_m \frac{m_1 m_2}{r^2} \hat{\mathbf{r}} \tag{11.5}$$

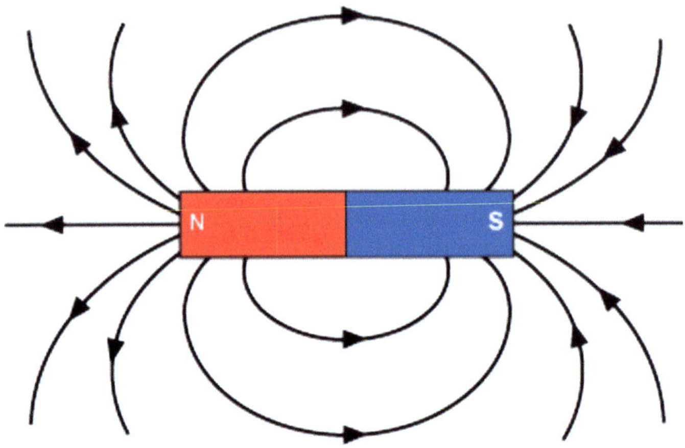

Fig. 11.2 Magnetic field due to a bar magnet

K_m is a constant which depends on the system of units used. The magnetic lines of force due to north and south poles are just similar to those of positive and negative charges. Figure 11.2 represents the magnetic field due to a magnet. From Eqs. (2.2), (11.1) and (11.4) it is evident that gravitational force, electrostatic force and magnetic force follow exactly similar inverse square laws. Carl Friedrich Gauss (1777–1855) was also a pioneer in the field of magnetism and made significant contributions. In 1833, he made the first magnetometer with the help of which one can determine the direction and strength of magnetic field. He studied the earth's magnetic field and determined the intensity of horizontal component of earth's magnetic field with the help of magnetometer (terrestrial magnetism). Gauss defined magnetic field and magnetic flux on the similar lines as in the case of electrostatics and in 1835 extended the Gauss law to magnetic fields. Because the magnetic charge (magnetic monopole) does not exist independently, therefore the Gauss law in magnetostatics becomes

$$\oint \mathbf{B} \cdot \mathrm{ds} = 0 \tag{11.6}$$

Later Gauss' law for magnetism became one of the four Maxwell equations. Gauss developed a consistent system of units in magnetism and the unit of magnetic flux density is named Gauss in his honor.

11.3 Electromagnetism

The similarity between the theories of electricity and magnetism was pointing towards the connection or relation between the electric and magnetic fields. Hans Christian Oersted (1777–1851), a Danish physicist, had been seeking to find such a relation since 1807 but always remained unsuccessful. Ultimately in 1820, Oersted found that when a magnetic compass is brought near a wire carrying current it gets deflected. On July 21, 1820, Oersted published a memoir in Latin, which was sent to scientists and scientific societies around the world. Translations of his paper were published in many languages and journals of every country. Oersted could not give any explanation of his experiment but it created a revolution throughout the scientific community. This discovery marked the beginning of *electromagnetism.* Dominique Francois Jean Arago (1786–1853) was the first to report to the Academy the news of Oersted's discovery, on September, 11, 1820. Shortly after his initial report to the Academy, Arago performed experiments of his own and established that a current through a wire acts like a magnet, both in attracting iron filing and its ability to induce permanent magnetism in iron needles. Seven days after Arago's report, Andre Marie Ampere (1775–1836) read a paper before the Academy, in which he suggested that internal electrical currents were responsible for the existence of ferromagnetism. On June, 1821, Augustin-Jean Fresnel (1788–1827) wrote a letter to Ampere suggesting that Amperian currents causing ferromagnetism should be molecular currents but Ampere had hesitated on this point. One can explain Oersted's experiment in terms of the concept of field. One can say that electric field is produced by static electrical charges while magnetic field is produced by moving electrical charges that is electric current. The electric current flowing through a conducting wire naturally produces electric field parallel to the wire but in addition it produces magnetic field with concentric circular magnetic lines of force around the wire perpendicular to the direction of current. Therefore, the flow of current through a wire produces both the electric and magnetic fields perpendicular to each other. Figure 11.3 shows the magnetic field produced by a current carrying straight conductor. Jean-Baptiste Biot (1774–1862) and Felix Savart (1791–1841) experimentally investigated the force exerted by an electric current on a nearby magnetic needle. In 1820, they established a quantitative relation between the magnetic field strength and the current flowing through the wire generally called *Biot-Savart law*. Biot also worked on geomagnetism and was the first scientist who boarded a hot balloon to study how the earth's magnetic field varied with height. Biot was also the first scientist to report that meteorites are rocks which have extra-terrestrial origin. In 1822, Louis Gay-Lussac (1778–1850) observed for the first time that a piece of iron is magnetized by a current flowing in a wire coiled around it: *first electromagnet*. In 1823, Andre-Marie Ampere (1775–1836) derived a mathematical relation between magnetic field intensity **B** and current density **J** for steady state phenomena, usually named *Ampere circuital law*, and is written as

$$\oint_C \mathbf{B} \cdot \mathrm{d}\mathbf{l} = \frac{4\pi}{c}\mathbf{J} \tag{11.7}$$

Hans Christian Oersted (1777–1851), Public Domain, URL: https://en.wikipedia.org/wiki/Hans_Christian_%C3%98rsted#/media/File:Hans_Christian_%C3%98rsted_daguerreotype.jpg

The symbol C represents that the integral over a closed contour. Just as Gauss law can be used for the evaluation of electric field, the Ampere law can be used to evaluate magnetic field. Later Ampere law became one of the Maxwell equations. In the same year of 1823, Ampere studied the force between two parallel current-carrying conductors and found that it obeys the inverse square law. Ampere gave his familiar *swimmer rule* for indicating the direction of magnetic field of an electrical circuit. He discovered that two parallel wires attract each other if the currents passing through them are in the same direction but repel if the directions of currents are opposite. In 1824, William Sturgeon (1783–1850) invented electromagnet and the first electromagnet was a horseshoe-shaped magnet. In 1826, Francois Arago established the relation between electricity and magnetism.

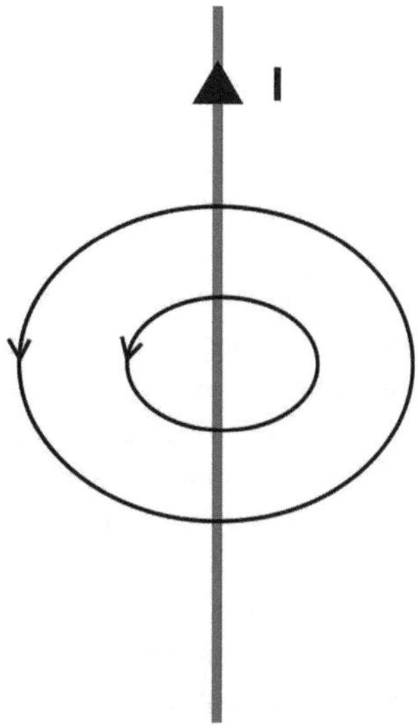

Fig. 11.3 Magnetic field due to a current-carrying straight conductor

In 1831, Michael Faraday (1791–1867) discovered *laws of electromagnetic induction* after conducting a long series of experiments. Faraday explained the electromagnetic induction using the concept of lines of force. However, scientists at that time widely rejected his ideas mainly because they were not formulated mathematically. He found that when a conductor is placed in a varying magnetic field, an electromotive force (*emf*) is induced. If the conductor's circuit is closed, a current is induced in the circuit called induced current. Faraday's law states that emf induced in an electrical circuit is proportional to the rate of change of magnetic flux and acts in a direction, which opposes the change in flux. Mathematically one can write

$$\oint_C \mathbf{E} \cdot d\mathbf{l} = -k \frac{d}{dt} \int_S \mathbf{B} \cdot d\mathbf{s} \qquad (11.8)$$

The value of constant k depends on the system of units and for C. G. S. system k is unity. The left side of Eq. (11.8) gives induced emf and symbol C indicates the integral over a closed loop. Faraday's law of electromagnetic induction is the basic law and forms the pillar of electromagnetism. In 1834, Heinrich Friedrich Emil Lenz (1804–1865) formulated a law named after him (*Lenz's law*) which gave the direction of induced emf and induced current resulting from the electromagnetic induction. In December 1840, James Prescott Joule (1818–1889) first published an abstract in the Proceedings of the Royal Society, suggesting that heat could be generated by an electrical current; Joule's law. In 1842, Lenz also independently discovered Joule's heating law and to honor his efforts on the problem, it is also given the name the Joule–Lenz law. The generation and transportation of electric power is based on the principle of electromagnetic induction. Electromagnetic induction is the greatest discovery of all times as it forms the basis of modern world. In 1830s, Faraday invented electric motor and dynamo, which are based on the principle of electromagnetic induction. The invention of electric motor revolutionized the modern industry.

Magnetic materials were categorized with regard to their properties as diamagnetic, paramagnetic, ferromagnetic, antiferromagnetic and ferrimagnetic materials. In 1778, Anton Brugmans (1732–1789) was the first to observe that bismuth (Bi) was repelled by magnetic field. In 1845, Michael Faraday showed that it is a property of the material and first referred to this phenomenon as diamagnetic for the first time and later changed it to diamagnetism. Faraday concluded that all the materials are either attracted or repelled by the magnetic field. Therefore, he classified materials into two categories: paramagnetic materials which are attracted by magnetic field and diamagnetic materials which are repelled by magnetic field. Most of the elements in the periodic table are paramagnetic in nature. It was found that in diamagnetic materials, atoms do not possess any magnetic moment and so these are nonmagnetic in nature. But when magnetic field is applied externally, finite magnetic moments are induced on their atoms, which produce magnetic field in a direction opposite to the applied field, thus causing repulsion. On the other hand, in paramagnetic materials each atom possesses intrinsic magnetic moment. With the application of external magnetic field the atomic magnetic moments tend to align parallel to each other producing a weak magnetic field in the direction of the applied field thus causing attraction. The magnetic materials found in early days, about 600BC, contained iron (for example lodestone) and therefore these were called ferromagnetic materials. Ferro had been taken from a Latin word *'ferrum'* which means iron. Most of the early studies in magnetism had been performed on lodestone. Later iron, nickel, cobalt and some of their alloys and compounds were found to exhibit ferromagnetism. In ferromagnetic materials, the atomic magnetic moments exhibit parallel alignment even in the absence of applied field yielding finite spontaneous magnetization. Although Faraday was having no formal education but he was a genius and a great experimentalist: he had lot of ideas which he materialized through experiments. Seeing the influence of Faraday's work on the progress of science and industry one concludes that Faraday was one of the greatest scientific discoverers of all times. Wilhelm Eduard Weber (1804–1891) did some theoretical work on ferromagnetism which appeared in the *'Elektrodynamische Massenbestimmungen'*, seven long works published from 1846 to 1878. In 1852, he worked out for the first time the idea that a magnet is made of very small magnets (magnetic needles): each atom or molecule acts as a small magnet with finite magnetic moment. In a magnetic material magnetization is defined as the magnetic moment per unit volume. In the unmagnetized state the magnetic moments of atoms/molecules form closed chains neutralizing each other's field thus yielding zero magnetization. But with the application of magnetic field these atomic magnets tend to align in the direction of magnetic field thereby producing finite magnetization and ultimately saturation magnetization is achieved with the increase in magnetic field. Weber assumed that atomic/molecular magnets produce local magnetic field. Later, James Alfred Ewing (1855–1935) confirmed that the molecular magnets can arrange in geometric arrays so that they might be free to turn and assume various magnetic configurations.

James Clerk Maxwell (1831–1879), a Scottish physicist, was specialized in the field of mathematical physics. In 1861, Maxwell presented the first durable color photograph at the Royal Institution. He is most famous for formulating classical theory of electromagnetism. In 1865, Maxwell realized the importance of Gauss law, Ampere law, and Faraday law [given by Eqs. (11.2), (11.6)–(11.8)] and summarized them in the form of a set of equations known as Maxwell equations and these are defined in the differential form (in Gaussian units) below:

$$\nabla \cdot \mathbf{D} = 4\pi\rho \quad \text{Gauss law in electricity} \tag{11.9a}$$

$$\nabla \cdot \mathbf{B} = 0 \quad \text{Gauss law in magnetism} \tag{11.9b}$$

$$\nabla \times \mathbf{E} = -\frac{1}{c}\frac{\partial \mathbf{B}}{\partial t} \quad \text{Faraday law} \tag{11.9c}$$

$$\nabla \times \mathbf{H} = \frac{4\pi}{c}\mathbf{J} + \frac{1}{c}\frac{\partial \mathbf{D}}{\partial t} \quad \text{Ampere's law} \tag{11.9d}$$

where $\mathbf{D} = \varepsilon\mathbf{E}$ is displacement field, $\mathbf{B} = \mu\mathbf{H}$ is magnetic induction and ρ and $\mathbf{J}$ are charge density and current density. The displacement current (last term) in the Ampere's law was introduced by Maxwell. Here $\mathbf{H}$ and $\mathbf{E}$ are applied magnetic and electric fields, ε and μ are permittivity and permeability of the medium. The refractive index n of the material is related to ε and μ as

$$\mathrm{n} = \sqrt{\mu\varepsilon} \tag{11.10}$$

The velocity of electromagnetic radiation in vacuum is the velocity of light c given by $\mathrm{c} = \sqrt{1/\varepsilon_0\mu_0}$, where ε_0 and μ_0 are permittivity and permeability of vacuum. In the above equations, Gauss law in electrostatics and Ampere law were modified to include polarization of the medium. In these equations Maxwell brought together, for the first time, electricity and magnetism as different manifestations of the same phenomenon. Maxwell equations predicted that the Faraday's lines of force can vibrate (tremble) with certain frequencies thereby producing wavy motion—that is the electric and magnetic fields travel through space as waves. The vibrations of the Faraday's lines of force carry energy along with them. Maxwell further proved that light is a transverse electromagnetic wave in which the electric and magnetic field waves are perpendicular to each other and propagate in space in a direction perpendicular to both the fields (Fig. 11.4). It shows that light is nothing but rapid vibration of the Faraday's lines and the frequency of vibration of the Faraday' lines is the frequency of light. The color of light is determined by the frequency of light. Actually, Newton and Maxwell theories contradicted each other. Maxwell equations determine velocity while the Newton's equations of motion involve acceleration. In Newtonian mechanics velocity of a body is measured with respect to another body: relative velocity as absolute velocity does not exist. Maxwell calculated the speed of Faraday's wavy motion and it turned out to be the velocity of light. In the phenomenon of electromagnetic induction Faraday found that lines of force vibrate with much lower frequency. Thus, James Clerk Maxwell provided the theoretical foundation to the physics of electromagnetism by showing that electricity and magnetism represent different aspects of the same fundamental force field. Maxwell also contributed toward the field of statistics and developed a statistical distribution called Maxwell–Boltzmann distribution that is used in describing kinetic theory of gases.

In 1873, James Clerk Maxwell extended Weber's theory to explain residual magnetization. He suggested that if a molecule is shifted too far from its equilibrium direction, it would fail to return back to the original direction after magnetic field is removed and so some magnetization would remain. Sir James Alfred Ewing (1855–1935), a Scottish physicist and engineer, is known for his work on the magnetic properties of metals. Ewing began his research on magnetic hysteresis through a project to study the effect of stress on the thermoelectric properties of metals. In 1881, he discovered that the thermoelectric effect lags behind the applied stress. He next suggested that other pairs of variables might also be related in such a cyclic manner, and he studied the transient currents produced by twisting a magnetized

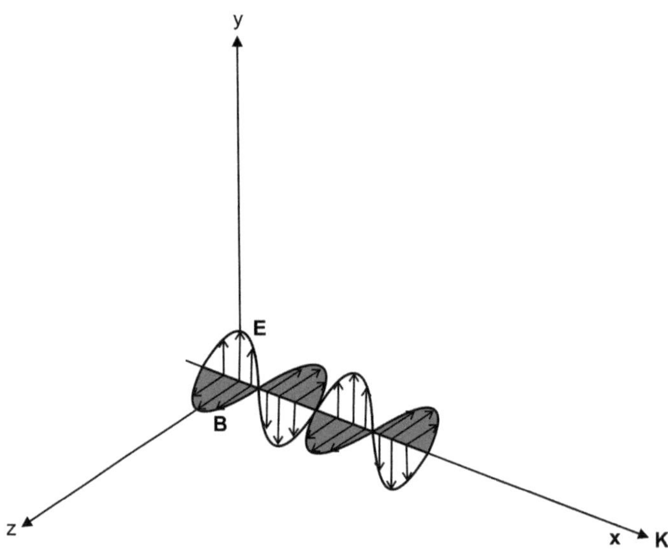

Fig. 11.4 Electromagnetic wave in vacuum moving with velocity c. **E** and **B** represent electric and magnetic field vectors and **K** represents the propagation vector

wire. He found a lag here also and introduced the term *hysteresis* to describe it: hysteresis is a Greek word meaning "to be late". Ewing then turned to the study of hysteresis in magnetization. He observed in 1882 that the area enclosed by the hysteresis loop is proportional to the work done during a complete cycle of magnetization and demagnetization. In 1885, he presented an important paper on this topic to the Royal Society, London. From the experimental results, he conjectured that all molecules are like tiny magnets and further showed that ferromagnetism is a collective phenomenon produced by the mutual action of the entire system of molecules. In 1890, Alfred Ewing found that there is finite resistance of molecules to rearrange themselves along the new direction of magnetic force as a result magnetization lagged behind the applied alternating current. The lag, in some processes, between a force and its effect was known in Germany even before Ewing's experiments. In his 1885 paper, Ewing noted that Emil Warburg (1846–1931) had independently discovered magnetic hysteresis and had emphasized the physical importance of the area of the hysteresis loop. The systems exhibiting hysteresis are nonlinear in nature and their model construction is relatively difficult. A number of models for hysteresis such as Preisach model in 1935 and Bouc-Wen model in 1971 were proposed. In 1970s, Mark Aleksandrovich Krasnoselskii (1920–1997) and his group developed a formal mathematical theory of systems with hysteresis.

Hendrik A. Lorentz (1853–1928) was a respected Dutch theoretical physicist of his time and contributed in many fields. In the year 1895, Lorentz gave an expression for electromagnetic force, which included contributions to the force from both the electric and magnetic fields, called *Lorentz force*, and is defined as

$$\mathbf{F} = q\left(\mathbf{E} + \frac{1}{c}\mathbf{v} \times \mathbf{B}\right) \tag{11.11}$$

The first term gives contribution to force acting on charge q, moving with velocity **v**, due to electric field **E** while the second term is the force due to magnetic field **B**. Electromagnetism is fully described by the four Maxwell equations along with the fifth equation for the Lorentz force. By the end of nineteenth century two main forces were known to exist in the universe that is gravitational and electromagnetic forces. Electromagnetic force governs almost all the electric and magnetic phenomena. Electromagnetic force holds electrons in an atom, atoms in a molecule and the atoms in a solid. It transmits signals in our body to the brain. It creates friction, it runs electric motors, combustion engines, allows us to turn on lights and listen to radio etc.

Heinrich Hertz (1857–1894), a German physicist, was working on the electromagnetic radiations. He designed a dipole antenna for the transmission and reception of electromagnetic radiations. Between 1885 and 1888, Hertz performed a number of experiments and in 1888, he created history becoming the first person to successfully produce, transmit and detect electromagnetic waves in the laboratory. He measured the electric field intensity, wavelength, velocity, polarization, and reflection of the waves. These experiments established that both light and these waves were a form of electromagnetic radiation obeying the Maxwell equations. Later the electromagnetic waves were called radio waves and the unit of frequency was named Hertz in his honor. When asked about the potential uses of his groundbreaking discovery, Hertz humbly replied, "*It is of no use whatsoever ... just an experiment that proves Maxwell was right*". Hertz did not know that his "*useless*" discovery would revolutionize human communication. That the waves he detected would eventually enable radio, television, cell phones and virtually all wireless technology we rely on today. Giglielmo Giovanni Maria Marconi (1874–1937), an Italian inventor and electrical engineer, was aware of the work of Heinrich Hertz (production and detection electromagnetic radiations) and got interested in extending his work. At that time, a number of scientists were working on wireless telegraphy; Marconi also started working on this subject. In 1895, Marconi developed a *wireless telegraph system* for long-range transmission and reception of electromagnetic radiations and got it patented. It was capable of radiating signal up to 2 miles (3.7 km) over the hills. Later he improved the system which could transmit signal up to 3.7 miles (6 km). Marconi was an entrepreneur and businessman and set up the Wireless telegraph and Signal Company in the United Kingdom in 1897, which later became Marconi Company. In 1898, Karl Ferdinand Braun (1850–1918) started working on wireless telegraphy by attempting to transmit Morse signals in water by means of high frequency currents. Actually, Marconi used some of the Patents of Braun in developing Wireless Telegraphy and in 1901, developed the first successful long-distance wireless telegraph and broadcast the first transatlantic radio signal. Marconi shared the 1909 Nobel Prize in Physics with Karl Ferdinand Braun in recognition of their contributions to the development of wireless telegraphy.

11.4 Magnetic Materials

Mainly French physicists developed the early theories of magnetic materials. Pierre Curie (1859–1906), a French physicist, studied ferromagnetism, paramagnetism and diamagnetism for his doctoral thesis, '*Propriétés magnétiques des corps à diverses temperatures*' submitted in 1895. Pierre Curie found that in ferromagnetic materials there exists a critical temperature T_c above which ferromagnetism disappears and the material becomes paramagnetic. This transition temperature was named *Curie temperature* in his honor. Magnetic susceptibility χ_M is a measure of strength of magnetization in a magnetic material. In 1895, Pierre Curie observed that the magnetic susceptibility of oxygen is inversely proportional to temperature, a behavior called *Curie law*. The Curie law says that above the Curie temperature, χ_M is inversely proportional to temperature, in absolute degrees and is given by

$$\chi_M = \frac{C_M}{T} \tag{11.12}$$

Here C_M is the Curie constant. The Curie law was formulated only from the experimental observations on paramagnetic materials. Pierre Curie's work can be considered as the starting point in the history of modern magnetism. In 1903, Joseph John Thomson (1856–1940) and at the same time Woldemar Voigt (1850–1919) tried to explain magnetism in terms of damped motion of electrons but satisfactory explanation did not come. Paul Langevin (1872–1946) did Ph.D. under the supervision of Pierre Curie and in 1905, he developed the famous classical theories of diamagnetism and paramagnetism. Langevin associated magnetic dipole field (magnetic dipole moment) with each atomic electron using the Lorentz equation for the magnetic field and the net magnetic dipole field associated with an atom/molecule is the combined effect of these fields. In a diamagnetic material, the atomic/molecular magnetic moment is zero and finite magnetic moment is induced only in the presence of applied magnetic field. He concluded that all matter exhibits the property of diamagnetism whether or not the initial net magnetic moment of its atom/molecule is zero. He showed that diamagnetism is nearly independent of temperature in agreement with the Curie's results. Langevin theory of diamagnetism is applicable to atomic electrons. Quantum theory of diamagnetism of atoms can be derived using Lorentz force but one obtains the same expression for diamagnetic susceptibility as was obtained by Langevin. In 1920, Wolfgang Pauli (1900–1958), while he was a student, gave the fundamental formula for diamagnetic susceptibility in terms of the sum over the mean square radii of the electron orbits which gave results comparable with the measured atomic and ionic susceptibilities. In 1930, Lev Landau (1908–1968), a Russian physicist, by solving Schrodinger equation showed that a free-electron gas displayed a weak diamagnetism which contradicted the classical theory. In 1933, Rudolf Peierls (1907–1995) gave theory of diamagnetism of conduction electrons in metals. Langevin gave theory of paramagnetism in which each atom/molecule was associated with finite magnetic moment whose exact origin was not known at that time. It exhibited the Curie law at high temperatures (above the Curie temperature). Langevin theory of paramagnetism is the first theory in magnetism which opened up the way towards explaining ferromagnetism. Langevin came to the conclusion that "only one or at the most a few electrons contributed to producing paramagnetism, whereas all the electrons cooperated in producing diamagnetism. In 1920, Wolfgang Pauli had shown that paramagnetic susceptibility can be explained if only limited number of orientations of magnetic moment is assumed. In 1927, Leon Nicolas Brillouin (1889–1969), a French physicist, gave quantum theory of paramagnetism and introduced Brillouin function. In the same year of 1927, Wolfgang Pauli derived an expression for spin susceptibility of a paramagnetic free-electron gas using the Fermi–Dirac distribution function (called Pauli spin susceptibility) and it was a test case in statistics of gases.

Paul Langevin (1872–1946), CC BY 4.0, URL: https://en.wikipedia.org/wiki/Paul_Langevin#/media/File:Paul_Langevin_Wellcome2.jpg

Pierre-Ernest Weiss (1865–1940), a French physicist, was specialized in magnetism. In 1907, Weiss suggested that, in ferromagnetic materials, spontaneous magnetization occurs in small domains (regions) known as Weiss domains: domain theory of ferromagnetism. He further assumed that some sort of internal magnetic field arises from the interactions between magnetic molecules and called it molecular field $\mathbf{H}_{\mathrm{m}}$. Molecular field, usually called Weiss molecular field, is present in the ferromagnetic materials which is proportional to spontaneous magnetization $\mathbf{M}$ and is given by $\mathbf{H}_{\mathrm{m}} = \lambda_{\mathrm{M}}\mathbf{M}$ where λ_{M} is the temperature-independent molecular constant. Weiss confessed to Edmond Bauer (1880–1963) that the idea of molecular field came to his mind as an extraordinary intuition and he immediately understood that it could provide key to the properties of ferromagnetic Fe, Ni, and Co. Weiss simply introduced the molecular field into the Langevin theory keeping the theory unchanged. According to the Weiss molecular field theory of ferromagnetism magnetic susceptibility χ_{M} is inversely proportional to temperature and is given by

$$\chi_M = \frac{C_M}{T - T_c} \tag{11.13}$$

with $T > T_c$. Here C_M is the Curie constant and T_c is the transition temperature above which ferromagnetism disappears and the material becomes paramagnetic in nature. Equation (11.13) is called *Curie–Weiss law*. In 1919, Heinrich Georg Barkhausen (1881–1956) observed experimentally that magnetization in a magnetic material increased in a series of jumps producing cracking sound when placed in a continuously increasing magnetic field: called *Barkhausen effect*. In 1926, Weiss interpreted the Barkhausen effect in terms of the existence of domains in a magnetic material and each domain possesses finite magnetization called spontaneous magnetization: so Barkhausen effect provided experimental evidence to the concept of magnetic domains. In the absence of applied magnetic field the domains are oriented randomly yielding zero magnetization. He further concluded that in the presence of applied magnetic field the domains align themselves along the direction of the field thereby producing finite magnetization. With the increasing magnetic field, ultimately one obtains saturation magnetization when all the domains get aligned in the direction of the field. It is worthwhile to remark here that Weiss domain theory is analogous to Weber's theory of magnetism except that atomic magnetic moments are replaced by magnetic domains. Weiss theory could explain the anomalous behavior of specific heat near the Curie temperature in the ferromagnetic crystals. Further, Weiss developed strong electromagnets in the beginning of the twentieth century. One of the limitations of the Weiss theory of ferromagnetism is that it does not provide any explanation of the origin of molecular field. With the formulation of theory of ferromagnetism Weiss became the first leading theorist of *collective phenomena* in the solid state of materials. After this, magnetism of solids acquired some independent authority. In 1919, Weiss established Physics Institute at the University of Strassbourg, France that became the leading centre for research in magnetism. It is noteworthy that Curie in 1895, Langevin in 1905, and Weiss in 1907, all French physicists, clarified the ideas about magnetism to a great extent as previously there were simply confusions and inconsistencies in understanding magnetism. In 1927, William Francis Giauque (1895–1982) explained the possibility of adiabatic demagnetization to cool a substance to a temperature considerably below 1 degree Kelvin. In 1933, William Giauque with the help of his student D. P. MacDougall, demonstrated experimentally the adiabatic demagnetization method of cooling. Shortly after this, the same method was independently demonstrated by W. J. de Haas and E. C. Wiersma. In 1931, Francis Bitter (1902–1967) introduced the powder pattern method for visualizing magnetic domain structure. The improvements in the powder pattern technique were required in conforming the domain theory of ferromagnetism and in 1930s, Elmore had considerably improved Bitter's crude initial demonstrations. In 1951, Richard M. Bozorth (1896–1981) published his pioneer work on Barkhausen effect in relation with the magnetic domains.

Pierre-Ernest Weiss (1865–1940), Public Domain, URL: https://en.wikipedia.org/wiki/Pierre_Weiss#/media/File:ETH-BIB-Weiss,_Pierre_(1865-1940)-Portrait-Portr_01257.tif_(cropped).jpg

Hendrika Johanna van Leeuwen (1887–1974), a Dutch physicist and a student of Hendrik. A. Lorentz, obtained her Ph.D. degree in 1919 and is known for her contribution to the theory of magnetism. In 1919, Van Leeuwen gave a theorem which states that the thermal average of magnetization in a free electron gas, evaluated by applying statistical mechanics and classical mechanics consistently, is always zero. Actually, Van Leeuwen theorem was discovered by Niels Bohr (1885–1962), in 1911, in his doctoral dissertation and was later rediscovered in 1919 by Van Leeuwen in her doctoral thesis. The significance of this theorem is that diamagnetism, paramagnetism and ferromagnetism are quantum mechanical effects.

Ernest Ising (1900–1998) did his Ph.D. in 1924 under Wilhelm Lenz who suggested him to work on a one-dimensional model of ferromagnetism. In 1925, a part of his thesis was published in the journal Zeitschrift fur Physik. Ising considered a linear chain of atomic magnetic moments interacting only with its first nearest neighbor magnetic moments and developed a model for ferromagnetism. He showed that the one-dimensional model would not lead to a phase transition between ferromagnetic and paramagnetic states. Therefore, this model did not receive much attention in relation to phase transitions. Without extending, Ernest Ising further declared that his model would not lead to phase transitions even in two or three dimensions. In 1936, Rudolf Ernst Peierls (1907–1995) gave this model the name *Ising model* in his publication and extended the one-dimensional Ising model to two-dimensions considering a square lattice and found that it does lead to phase transition. Therefore, as regards the phase transitions, Peierls pointed towards the dimensionality of a system. In 1941, Elliot Waters Montroll (1916–1983) developed a method of summation of diagrams to all orders in solid state theory even before the invention of Feynman diagrams. In 1942, Montroll gave a general mathematic procedure for evaluating the partition function for any system with nearest neighbor interactions by means of solving linear homogeneous operator equations which he illustrated by applying it to the two-dimensional Ising lattice. In 1944, Lars Onsagar (1903–1976) gave the exact and complete solution of the Ising model and demonstrated the existence of a phase transition and associated a critical behavior in the thermodynamic limit. The Onsagar model yielded singularity in specific heat of the form $-\log|T - T_c|$ at the critical temperature. This model can be extended for hexagonal and triangular lattices. Onsagar model established the suitability of equilibrium statistical mechanics for describing the phase transitions and critical phenomena. In 1941, Edwin N. Lassettre (1912–1990) and John P. Howe (1910–1999) had already published similar work on the partition function in binary systems and Onsagar was aware of their work beforehand.

As described above most of the experiments on magnetic materials were performed between 1908 and 1927 and these indicated that the phenomenon of magnetism has quantum mechanical origin. It was established that the atomic magnetic moment is basically due to the spin of its electrons. Arthur Holly Compton (1892–1962) used X-rays to study ferromagnetism and concluded that it was the result of alignment of electron spins. In 1926, Paul Adrien Maurice Dirac (1902–1984) and Werner Heisenberg (1901–1976), independently discovered exchange interactions. Shortly after the development of quantum mechanics by Heisenberg and Schrodinger, in 1927, Walter Heitler (1904–1981) and Fritz London (1900–1954) developed quantum theory for hydrogen molecule and found that in addition to the Coulomb interaction contribution there is electron-exchange interaction contribution to the energy. The electron-exchange energy is fundamentally electrostatic in nature and represents a many-body quantum effect, which has no classical analogue. Warner Heisenberg wanted Felix Bloch to work on the problem of ferromagnetism but he had already chosen the field of electrons in metals. Therefore, Heisenberg himself started working on the theory of ferromagnetism following the approach of Heitler and London. He worked on ferromagnetism for two years and had many discussions with Wolfgang Pauli. In 1928, Werner Heisenberg and Jacov Frenkel (1894–1952) independently, for the first time, gave quantum theory of ferromagnetism and explained the physical nature of the Weiss molecular field. Heisenberg showed that the spin–spin exchange interactions are responsible for the Weiss molecular mean field. In a many-body system the electron energy is given by

$$E = E_{ij}^{c} \pm J_{ij} \tag{11.14}$$

E_{ij}^{c} and J_{ij} represent the Coulomb and exchange energies, respectively, between the i and j spins. Here the negative sign gives the energy of a magnetic state while the positive energy represents the energy of a nonmagnetic state. It is found that the Heisenberg theory is invalid at low temperatures. Bloch calculated the exchange energy of a free-electron gas and found that at low electron densities the exchange interaction comes out to be positive to give ferromagnetic state. During these days Bloch was at Zurich and wanted to investigate the role of conduction electrons in ferromagnetism using the tight-binding method. In 1930, John Clarke Slater (1900–1976) found that ferromagnetism depends on the existence of incomplete shells within the atoms and metals like Fe, Ni, and Co exhibit ferromagnetism.

In the early 1930s, most of the scientists shifted to other fields and only a handsome of them continued research in quantum theory of magnetism: worth mentioning are Edmund Clifton Stoner (1899–1968) and John Hasbrouck Van Vleck (1899–1980). Van Vleck worked on the quantum theory of paramagnetism and made significant contributions towards developing quantum theory of magnetism. In 1932, Van Vleck wrote a book entitled, "*Theory of Electric and Magnetic Susceptibilities*" and it was widely acclaimed among scientists on the subject of magnetism. In 1930, it was realized that there is close relationship between the electronic band structure and quantum theory of ferromagnetism which guided Stoner to work in this direction. In 1934, Stoner published a book, "*Magnetism and Matter*" and was one of the first books fully devoted to magnetism in solids. In 1936, Felix Bloch (1905–1983) and in 1939 Otto Halpern (1899–1982) and M. H. Johnson measured the magnetic part of neutron scattering with good accuracy and it allowed to determine distribution, direction, order and magnitude of magnetic moments. In 1954, Stoner studied ferromagnetism in terms of band model and derived a criterion for ferromagnetism to occur in *itinerant electron model.* According to Stoner, ferromagnetic ordering is obtained by the divergence of magnetic susceptibility and this condition is obtained when

$$\mathrm{J}n_e(E_F) \geq 1 \tag{11.15}$$

where J is the electron-exchange integral and $n_e(E_F)$ is the density of electron states at Fermi energy E_F. Equation (11.15) is called the *Stoner criterion* for ferromagnetism to occur. According to Eq. (11.15) ferromagnetism occurs only in those metals for which both J and $n_e(E_F)$ are large in magnitude. Therefore, transition metals and rare-earth metals are the most favorable candidates for ferromagnetism to occur. A large number of theoretical calculations had been performed to investigate the occurrence of ferromagnetism in metals. In 1977, J. F. Janak estimated both J and $n_e(E_F)$ for 32 elements using fully self-consistent ground state method. He found the occurrence of magnetism near the end of 3d-transition metal series and no indication of ferromagnetism in 4d-transition metal series.

In a ferromagnetic material, all the magnetic moments are aligned parallel to each other at absolute zero. At a finite temperature, some disturbance in the alignment of magnetic moment is produced by the thermal energy which causes decrease in spontaneous magnetization. Both the Weiss molecular field theory and Heisenberg theory of ferromagnetism were found to be invalid at low temperatures. With a view to improve the Heisenberg model at low temperatures, in 1930, Felix Bloch (1905–1983) used Slater determinants to study ferromagnetism and discovered *spin waves*—a state corresponding to a single spin or a few spin flips in the fully aligned ground state. The disturbance in the ordering of atomic magnetic moments or their spins propagates as a spin wave in a material. In the Bloch theory of spin waves only the first nearest neighbor interactions between the atomic spins on a primitive cubic lattice were considered. In 1940, Theodore Holstein (1915–1985) and Henry Primakoff (1914–1983) and in 1956, Freeman Dyson (1923–2020) developed a quantitative theory of quantized spin waves and gave $T^{3/2}$-law for magnetization. The spin waves are determined experimentally through the following techniques:

1. Inelastic neutron scattering
2. Inelastic light scattering (Brillouin scattering, Raman scattering, inelastic X-ray scattering).
3. Inelastic electron scattering (spin-resolved electron energy loss spectroscopy)
4. Spin wave resonance (ferromagnetic resonance).

The quanta of spin waves are *magnons* and the light scattering experiments had confirmed that magnons are bosons that obey the Bose–Einstein statistics. The $T^{3/2}$-law for magnetization was observed experimentally in ferromagnetic materials. At higher temperatures, a ferromagnetic material goes to the paramagnetic state but still small oscillations of magnetic moments around their natural direction propagate as waves. These waves are known as *paramagnons* and propagate through short distances. The above studies confirmed that ferromagnetism is a collective phenomenon and single-electron theory is insufficient to explain it. Therefore, a reliable theory is required to take into account the collective interactions between electrons (many-electron theory). For detailed discussion on magnetism in metals and alloys reader is referred to Walter Marshall (1967).

Until the nineteenth century all the magnetic materials were called ferromagnetic materials. Louis Neel (1904–2000), a French physicist, was specialized in the field of magnetism and he advanced the idea of molecular field of Weiss further. In 1932, Neel considered a binary alloy AB with two types of atoms A and B. He introduced local molecular field and considered the possible interactions between the two types of atoms: A–A, B–B, and A–B interactions. This is usually called *two-sublattice model* and with its help he predicted the existence of *antiferromagnetic state* even before such materials had been discovered. For the first time, Neel assumed that antiferromagnetic order can be described in terms of "*two spin-sublattices*" with adjacent spins in opposite directions but in one sublattice all spins are aligned in one direction. In other words, one can say that in antiferromagnetic materials the adjacent magnetic moments are directed antiparallel to each other, an arrangement which is opposite to that found in ferromagnetic materials and so is their name. In antiferromagnetic materials, the net magnetization is expected to be zero at absolute zero. But with the rise in temperature the antiferromagnetic order is disturbed and therefore the magnetization becomes finite. Using two-sublattice model Louis Neel showed that these materials exhibit antiferromagnetic order below some characteristic temperature, called *Neel temperature* and at and above Neel temperature the material makes a transition to the paramagnetic state. Therefore, with the help of two-sublattice model Neel explained both the magnetic susceptibility and Neel temperature in an antiferromagnetic material. In 1938, Henri Bizette and coworkers observed experimentally the phenomenon of antiferromagnetism in MnO for the first time and Francis Bitter (1902–1967) proposed the name antiferromagnetism to this new phenomenon. In 1949, Clifford Glenwood Shull (1915–2001) and J. Samuel Smart performed the neutron diffraction experiments on MnO and confirmed the hypothesis of two sublattice model in antiferromagnetism. In 1952, Cornelius Jacobus Gorter (1907–1980) with Johannes Haantjes (1909–1956) developed theoretical double lattice model of antiferromagnetism.

Louis Neel (1904–2000), Public Domain, URL: https://en.wikipedia.org/wiki/Louis_N%C3%A9el#/media/File:Louis_N%C3%A9el_1970b.jpg

In 1947, through experiments Evert Johannes Willem Verwey (1905–1981) showed that ferrites are inverse spinels. In 1948, Louis Neel discovered materials in which the adjacent magnetic moments are unequal but directed opposite to each other as in antiferromagnetic materials. These materials yielded finite spontaneous magnetization just like ferromagnetic materials and Neel proposed the general term *ferrimagnetic materials*. Before the discovery of ferrimagnetic materials the naturally occurring magnetite, that is loadstone, was considered a ferromagnetic material. Actually, magnetite is the oldest known ferrimagnetic material. Further, it was found that ferrimagnetic order exists only below a transition temperature T_c. It is found that ferrimagnetism occurs mostly in magnetic oxides known as *ferrites*. The above facts show that the strength of ferrimagnetic materials is generally less than that of the pure ferromagnetic materials. Neel used two-sublattice model to find magnetic susceptibility and transition temperature in ferrimagnetic materials. He used the local molecular field in the same way as in binary alloys. Although ferrimagnetic materials are ferromagnetic in nature but these differ in some aspects. For example, at high temperatures the magnetic susceptibility does not obey the Curie–Weiss law. One can conclude that all forms of magnetism involve magnetic moments (or spins) of all the atoms and thus constitute cooperative phenomena. It is worth mentioning that the phenomenological theories of Langevin, Weiss and Neel were closely related to the bulk properties of the magnetic materials.

In 1957, Bertrand Neville Brockhouse (1918–2003) detected experimentally, for the first time, the magnons (magnon dispersion relations) in ferrites by inelastic neutron scattering technique. Since then magnon dispersion relations have been measured in many ferromagnetic, ferrimagnetic, and antiferromagnetic materials. The Brillouin and Raman scattering methods probe low energy magnons and have superior energy resolution for detecting magnons with energy in the range of meV. Broadly speaking ferromagnetic resonance is similar to electron paramagnetic resonance and also somewhat similar to nuclear magnetic resonance (see Chap. 12). It is a standard tool for the investigation of spin waves.

All the time physicists were engaged in finding materials with higher and higher magnetic strength. The strength of ferromagnetism can be increased by making alloys of suitable elements. Up to the middle of twentieth century it was well established that stronger are the exchange interactions larger will be the spin splitting of energy bands in magnetic materials to yield larger atomic magnetic moments. In 1955, Clifford Glenwood Shull (1915–2001) and M. K. Wilkinson (1921–2013) measured for the first time the magnetic moments on the different components of iron-chromium alloy as a function of impurity concentration using neutron scattering technique. In a metallic alloy the Stoner criterion for ferromagnetic ordering of impurity magnetic moments, in analogy with that in pure elements, is given by

$$\mathrm{J}\mathrm{n}_{\mathrm{alloy}}(\mathrm{E}_{\mathrm{F}}) \geq 1 \tag{11.16}$$

where $n_{alloy}(E_F)$ is the density of electron states at the Fermi energy E_F in an alloy. Magnetic alloys are classified with respect to the nature of host metal and impurity. One has magnetic impurities in simple metals and magnetic impurities in transition metals and rare-earth metals. Moriya (1965, 1967) studied the occurrence of ferromagnetism and antiferromagnetism in transition metal-based alloys using the tight binding theory and gave some general rules given below:

1. If the impurity has d-band less than or nearly half filled, it is coupled antiferromagnetically to the transition metal host. But if the impurity has d-band more than half filled it is coupled ferromagnetically to the transition metal host.
2. If the d-band of the host metal is less than or nearly half filled the transition metal impurities are coupled antiferromagnetically to the host metal. But if the d-band of the host metal is more than half filled there is increasing tendency for the transition metal impurities to couple ferromagnetically to the host metal.

Dederichs and coworkers (1981, 1985, 1991) performed a number of ab initio calculations using self-consistent approach for 3d-, 4d- and 5d-transition metal impurities in transition metal hosts and calculated the local and total magnetic moments on the impurity atoms. Their calculations supported the rules given by Toru Moriya.

It is interesting to note that although Mn and Cr metals are antiferromagnetic metals but these form numerous ferromagnetic alloys. The most suitable example is that of alloys or compounds having cubic structure with general formula Cu_2MnX where X can be Al, In, Sn, ... These are called heusler alloys or compounds as these were discovered in 1903 by Carl Ludwig David Friedrich Heusler (1866–1947). In 1964, Sugikuchiand and Endo performed NMR studies on heusler alloys and indicated the appearance of magnetic moments on Mn ions without any local magnetic moment on Cu ion. Another special class is the Pd-based alloys with 3d-series transition metal impurities. Pure Pd metal is nearly ferromagnetic in nature but its alloys with 3d-series transition metal impurities are magnetic alloys. In 1975, Nieuwenhuys and in 1984, Peter and coworkers observed *giant magnetic moments* on Fe atoms in PdFe alloy and on Co atoms in PdCo alloy: the magnetic moments at the impurity site are of the order of $10\mu_B$ or more. In 1991, Van Acker and coworkers and in 1993, Rudolf Zeller estimated theoretically the magnetic moments in Pd-based alloys using KKR Green's function method in the local spin density approximation and obtained giant magnetic moments on Fe and Co impurities in Pd metal.

In 1964, Jun Kondo (1930–present) discovered that resistivity in dilute magnetic alloys exhibit a minimum at a particular low temperature. This phenomenon was named *Kondo effect* and the alloys, which exhibit a resistance minimum, are called Kondo alloys. Kondo effect describes the scattering of conduction electrons from magnetic impurities in a metal. In 1961, Philip Warren Anderson (1923–2020) proposed a model for magnetic impurity while Kenneth Geddes Wilson (1936–2013) gave renormalization theory to explain the Kondo effect. Another important property of magnetic materials is the *giant magnetoresistance* observed in multilayered systems composed of alternate ferromagnetic and nonmagnetic conductive layers of nanometer thickness. Therefore, giant magnetoresistance is a quantum mechanical effect. In 1988, two physicists, Albert Fert (1938–present) [see Baibich and coworkers (1988)] and Peter Grunberg (1939–2018) [see Binasch and coworkers (1989)] independently discovered giant magnetoresistance in multilayered systems and both were awarded 2007 Nobel Prize in Physics because of large number of applications of giant magnetoresistance in sensors and memory storage systems.

Magnetic alloys of rare-earth metals are used in making permanent magnets of superior quality: magnets having large coercivity, retentivity and large energy storage as compared to the usual AlNiCo magnets. In 1984, General Motors and Sumitomo Special Metals independently discovered NdFeB compound simultaneously. It is a material used for making permanent magnets which are much stronger than already existing rare earth magnets with many times larger coercivity. Some of the rare earth metal compounds with formula RX_2 exhibit large magnetic moments: here R is a rare earth metal and X is an element which may or may not have magnetic moment. For example, in $ErAl_2$ the magnetic moment is $9.2 \pm 0.2\mu_B$ and in $GdNi_2$ it is $7.1\mu_B$. Legvold and coworkers (1976, 1977 and 1978) studied both experimentally and theoretically the properties of REM alloys and their compounds.

In 1879, Edwin Herbert Hall (1855–1938) discovered that when a conductor or semiconductor with current flowing in one direction (say x-direction) was placed in a magnetic field directed perpendicular to the current (say z-direction), finite voltage is produced in a direction perpendicular to both the current and magnetic field (in y-direction). He discovered this effect while working on his doctoral thesis and in his honor named *Hall effect*. In the original experiment Edwin Hall took a thin gold leaf and passed current along its length by tapping and applied magnetic field perpendicular to the sheet. He found that finite voltage was generated on the opposite sides of the leaf. Later he performed experiment with other materials. It is noteworthy that the Hall effect was discovered about 18 years before the discovery of electron. But at that time, it was well known that current arises due to the flow of small charge carriers which experience force due to the presence of both electric and magnetic fields (later called Lorentz force). One of the main advantages of the Hall effect is that it gives information about the sign of the charge carriers in a solid. Klaus Von Klitzing and coworkers (1980) were studying the Hall effect in a two-dimensional free-electron gas formed by an inversion layer at a Si/SiO_2 interface. They observed that for certain combinations of magnetic fields and surface electron density, the Hall conductance exhibited plateaus at values that are integral multiples of e^2/h: e is the electron charge and h is the Planck constant. These plateaus extended over a finite range of electron density. This effect is named *quantum Hall effect*. It is noteworthy that quantum Hall effect is observed only in two-dimensional systems kept at a very low temperature (liquid helium temperature) and subjected to very high magnetic fields ($\approx$ 10 Tesla). Actually, Ando and co-workers (1975) were the first to observe this effect but their results were crude to make sound conclusions. It is also called the *integral quantum Hall effect* because the plateaus of the Hall conductance correspond to the complete filling of integral number of Landau levels. The importance of the integral quantum Hall effect is that it yielded basic quantum of resistance in terms of electronic charge and the Planck constant (fundamental quantities) that is e^2/h. Later, Tsui and co-workers (1982) found that in a two-dimensional electron gas in GaAs-AlGaAs heterojunction (having very high values of electron mobility) the Hall conductance plateaus also occur at 1/3rd and 2/3rd filling of the Landau levels. It is named *fractional quantum Hall effect* and is observed in samples with very high mobility of charge carriers and in the extreme quantum limit (at very high magnetic fields). Stormer and co-workers (1983) further observed that plateaus in general can be found at filling factors corresponding to the simple fraction p/q where q is an odd number. In the same year, Laughlin (1983) gave the most appealing explanation of fractional quantum Hall effect. According to him the Coulomb interactions between the electrons cause the electron system to condense into a highly correlated incompressible quantum liquid. The quasi-particles of the quantum liquid behave as particles having fractional charge of e/q, where q is an odd number, and these fractional charges give rise to plateaus at fractional quantum numbers. Saminadayar and co-workers (1997) have experimentally tested this idea and showed that the charge carriers in systems with filling fraction 1/3 have charge e/3.

Klaus von Klitzing, CC BY 4.0, URL: https://en.wikipedia.org/wiki/Klaus_von_Klitzing#/media/File:Klaus_von_Klitzing_2015.jpg

Magnetic susceptibility is a measure of strength of magnetic materials. The evaluation of static magnetic susceptibility in the free-electron approximation is simple and is just similar to the evaluation of electric susceptibility. Evenson and Liu (1968, 1969) calculated static spin susceptibility of a number of rare earth metals, for the first time, using relativistic APW band structure calculations. They related the peaks in the spin susceptibility with the magnetic ordering wave vector. Evenson and co-workers (1969) further evaluated spin susceptibility for Cr-group (Cr, Mo, W) and V-group (V, Nb, Ta) metals. They found that prominent and narrow peak near the H symmetry point in the Brillouin zone is related to antiferromagnetism in Cr metal.

Hebborn and March (1970) presented, for the free-electron gas, the general expressions for wave vector dependent dynamical magnetic susceptibility in his review paper: both spin and orbital contributions. Gupta and Sinha (1970, 1971) evaluated the spin susceptibility of Cr metal using APW calculations of energy band structure with larger mesh size in the Brillouin zone. They also tried to include local field effects in the spin susceptibility of metals. Gupta and Freeman (1976) performed the APW band structure calculations and then evaluated spin susceptibility for transition metal compounds. A number of other calculations were performed for estimating static spin susceptibility of different elements using APW band structure calculations. In general it has been found that the peaks in spin magnetic susceptibility are related to the magnetic ordering in magnetic materials. On the other hand, the static orbital magnetic susceptibility is much less studied. Singh and coworkers (1980) calculated the spin and orbital magnetic susceptibilities for transition metals. Further, Singh and coworkers (1981) studied the temperature variation of dynamical spin magnetic susceptibility. Singh (1991) formulated theory for the temperature variation of static orbital magnetic susceptibility in transition metals and estimated it for gold metal.

Bibliography

Anderson P (1961) Localized magnetic states in metals. Phys Rev 124:41–53

Ando T, Matsumoto Y, Uemura Y (1975) Theory of Hall effect in a two-dimensional electron system. J Phys Soc Japan 39:279–288

Arago (1826) Note concernant les phenomenes magnetiques ausquels le movement donne naissance (Note concerning magnetic phenomena that motion creates). Annales de chimie et de physique 32:213–223

Baibich MN, Broto JM, Fert A, Van Dau FN, Petroff F, Etienne P, Creuzet G, Friederich A, Chazelas J (1988) Giant magnetoresistance of (001)Fe/(001)Cr magnetic superlattices. Phys Rev Lett 61:2472–2475

Binasch G, Grünberg P, Saurenbach F, Zinn W (1989) Enhanced magnetoresistance in layered magnetic structures with antiferromagnetic interlayerexchange. Phys Rev B 39:4828–4830

Biot JB, Savart F (1820) Note sur le magnétisme de la pile de Volta" was published in the Annales de chemie et de physique 15:222–223

Bitter F (1931) On inhomogeneities in the magnetization of ferromagnetic materials. Phys Rev 38:1903–1905; Bitter F (1932) Experiments on the nature of ferromagnetism. Phys Rev 41:507–515

Bizette H, Squire CF, Tsai B (1938) The λ transition point of the magnetic susceptibility in the manganosite MnO. C R Acad Sci.(Paris) 207:449

Bloch F (1928) Uber die quantenmechanik der electronen in kristallgittern. Z Phys 52:555–600

Bloch F (1929) Bemerkung zur elektronentheoriedes ferromagnetismus und de elektrischen leitfähigkeit. Z Physik 57:545–555

Bloch F (1930) Zur theorie des ferromagnetismus. Z Phys 61:206–219

Bloch F (1930) Zum elektrischen widerstandsgesetz bei tiefen temperaturen. Z Phys 59:208–214

Bloch F (1936) On the magnetic scattering of neutrons. Phys Rev 50:259

Bohr N (1911) Dissertation, Copenhagen

Bozorth RM (1951) Ferromagnetism. Van Nostrand, New York

Bridgman PW (1931) The physics of high pressure. G. Bells and Sons Ltd., London; (1933) The physics of high pressure. Nature 131:259

Brockhouse BN (1957) Scattering of neutrons by spin waves in magnetite. Phys Rev 106:859–864

Christie SH (1833) The Bakerian lecture: experimental determination of the laws of magneto-electric induction in different masses of the same metal, and its intensity in different metals. Philos Trans Roy Soc (London) 123:95–142

De Coulomb CA (1785a) Premier mémoire sur l'électricité et le magnétisme. Histoire de l'Académie Royale des Sciences, 569–577; (1785b) Second memoire sur l'electricite et lemagnetisme. Histoire de l'Académie Royale des Sciences, 578–611; (1785c) Troisieme memoire sur l'electricite et le magnetisme. Histoire de l'Académie Royale des Sciences, 612–638

Dederichs PH, Zeller R (1981) Festkorperprobleme XXI:243; Dederichs PH, Zeller R, Akai H, Blugel S, Oswald A (1985) Ab initio calculations for impurities in Cu and Ni. Phil Mag B 51:137–150; Dederichs PH, Zeller R, Akai H, Ebert H (1991) Ab-initio calculations of the electronic structure of impurities and alloys of ferromagnetic transition metals. J Mag Magn Mater 100:241–260

Dirac PAM (1926) On the theory of quantum mechanics. Proc Roy Soc (London) A112:661–677; (1928) The quantum theory of electron. Proc Roy Soc (London) A117:610–624

Döbereiner JW (1829) Versuch zu gruppirung der elementaren stoffe nach ihrer analogie [An attempt to group elementary substances according to their analogies]. Annalen der Physik und Chemie (in German) 15:301–307

Drude P (1900) Zur electronen theorie der metalle. Ann de Phys 1:566

Dyson FJ (1956) General theory of spin-waves interactions. Phys Rev 102:1217, 1230

Evenson WE, Fleming GS, Liu SH (1969) Generalized susceptibilities and magnetic properties of some bcc transition metals. Phys Rev 178:930

Evenson WE, Liu SH (1969) Theory of magnetic ordering in the heavy rare earths. Phys Rev 178:783; (1968) Generalized susceptibilities and magnetic ordering of heavy rare earths. Phys Rev Lett 21:432

Ewing JA (1882) On effects of retentiveness in the magnetization if iron and steel. Proc Roy Soc (London) 34:39; (1885) Experimental researches in magnetism. Philos Trans Roy Soc (London) 176:523–640; (1890) Contributions to the molecular theory of induced magnetism. Proc Roy Soc 48:342–358; (1892) Magnetic induction in iron and other metals. The Electrician Printing and Publishing Company, London

Faxén H, Holtsmark J (1927) Beitrag zur theorie des durchganges langsamer elektronen durch gase. Z Phys 45:307–324

Franz R, Wiedemann G (1853) Ueber die wärme-leitungsfähigkeit der metalle. Ann der Phys 165:497–531

Frenkel J (1928) Elementare theorie magnetischer und elektrischer eigenschaften der metalle beim nullpunkt der temperature. Z Phys 49:31–45

Friedel J (1952) The distribution of electrons around impurities in monovalent metals. Phil Mag 43:153–189; (1956) On some electrical and magnetic properties of metallic solid solutions. Can J Phys 34:1190–1211; (1958) Metallic alloys. Nuovo Cimento (Suppl. 2) 7:287–311

Galsin JS (2002) Impurity scattering in metallic alloys. Kluwer Acad./Plenum Publishers, New York, pp 206–238

Giauque WF (1927) A thermodynamic treatment of certain magnetic effects. A proposed method of producing temperatures considerably below 1° absolute. J Am Chem Soc 49:1864–1870

Giauque WF, MacDougall DP (1933) Attainment of temperatures below 1° absolute by demagnetization of $Gd_2(SO_4)_3{\cdot}8H_2O$. Phys Rev 43:768

Gorter CJ, Haantjes J (1952) Anti-ferromagnetism at the absolute zero of temperature in the case of rhombic symmetry. Physica 18:285–294

Gruneisen E (1913) Berichte deutsche physikal ges 15:186; (1928) Handbuch d. Phys 13:22; (1933) Die abhängigkeit des elektrischen widerstandes reiner metalle von der temperatur. Ann de Phys 16:530

Gruner G (1974) Experimental evidence for many-body effects in dilute alloys. Adv Phys 23:941–1024

Gruner G, Minier M (1977) Charge perturbation around impurities in metals. Adv Phys 26:231–284

Gupta M, Freeman AJ (1976) Direct correlation of phonon anomalies in NbC with Fermi-surface-induced maxima in generalized susceptibilities. Phys Rev Lett 37:364

Gupta RP, Sinha SK (1970) Exchange enhanced generalized susceptibility function for paramagnetic chromium including band structure effects. J Appl Phys 41:915–916; (1971) Wave-number-dependent susceptibility function for paramagnetic chromium. Phys Rev B3:2401

Hall EH (1879) On a new action of the magnet on electric currents. Am J Math 2:287–292

Halpern O, Johnson MH (1939) On the magnetic scattering of neutron. Phys Rev 55:898

Harrison WA (1966) Pseudopotentials in the theory of metals. Benjamin Inc., Massachusetts, p 149

Hebborn JE, March NH (1970) Orbital and spin magnetism and dielectric response of electrons in metals. Adv Phys 19:175–215

Heisenberg W (1928) Zur Theorie des Ferromagnetismus. Z Phys 49:619–636

Heitler WH, London FW (1927) Wechselwirkung neutraler atome und homoopolare bindung nach der quantenmechanik [Interaction between neutral atoms and homopolar binding according to quantum mechanics]. Z fur Phys 44:455–472

Hertz HR (1887) Ueber sehr schnelle electrische schwingungen. Annalen der Physik 267:421–448; (1887) Ueber einen einfluss des ultravioletten lichtes auf die electrische entladung. Annalender Physik 267:983–1000; (1888) Ueber die einwirkung einer geradlinigen electrischen schwingung auf eine benachbarte strombahn. Annalen der Physik 270:155–170; (1888) Ueber die ausbreitungsgeschwindigkeit der electrodynamischen wirkungen. Annalen der Physik 270:551–569

Heusler F (1903) Verh. Deutsche Physikalische Gesellschaft 5:219–223

Holstein T, Primakoff H (1940) Field dependence of the intrinsic domain magnetizationof a ferromagnet. Phys Rev 58:1098–1113

Ising E (1925) Beitrag zur theorie des ferromagnetismus. Z Phys 31:253–258

Janak JF (1977) Uniform susceptibilities of metallic elements. Phys Rev B16:255

Kondo J (1964) Resistance minimum in dilute magnetic alloys. Progress Theoret Phys 32:37–49

Krasnosel'skii MA, Pokrovskii AV (1989) Systems with hysteresis. Springer, Berlin-Heidelberg, New York, Paris, Tokyo, p 410

Landau LD (1930) Diamagnetismus der metalle. Zeitschrift für Physik A: Hadrons and Nuclei 64:629–637

Langevin P (1905) Sur la theorie du magnetisme. Journal de Physique Théorique et Appliquée (in French) 4:678–693; (1905) Magnetisme et theorie des electrons. Annals de Chimie et Physique 5(8th series):70–127.

Laughlin RB (1983) Anomalous quantum Hall effect: an incompressible quantum fluid with fractionally charged excitations. Phys Rev Lett 50:1395–1398

Legvold S, Beaudry BJ, Ostenson JE, Harmon BN (1976) Superconducting electron pair breaking in DHCP LA-EU and LA-GD. Solid State Commun 21:1061–1062; Legvold S, Harmon BN, Beaudry BJ, Burgardt P, Younkin DR, White HW (1977) Magnetic properties of gadolinium-rich alloys. Phys Rev B16:4986; Legvold S, Green RW, Ostenson JE, Beaudry BJ (1978) Superconductivity of fcc-La-heavy-rare-earth alloys. Phys Rev B18:1162

Linde JO (1931) Elektrische Eigenschaften verdünnter Mischkristallegierungen I. Goldlegierungen. Ann der Phys 10:52; (1932) Elektrische Eigenschaften verdünnter Mischkristallegierungen. II. Widerstand von Silberlegierungen. Ann Der Phys 14:353; (1932) Elektrische Eigenschaften verdünnter Mischkristallegierungen III. Widerstand von Kupfer- und Goldlegierungen. Gesetzmäßigkeiten der Widerstandserhöhungen. Ann der Phys 15:219, 239

Liu SH, Gupta RP, Sinha SK (1971) Generalized susceptibility function for rare earths and thorium and their alloys. Phys Rev B4:1100

Lorentz HA (1892) The relative motion of the earth and the aether. Zittingsverlag Akad V Wet 1:74–79

Lorentz HA (1895) Versuch einer theorie der electrischen und optischenerscheinungen in bewegten korpern. E.J. Brill, Leiden [Attempt of a theory ofelectrical and optical phenomena in moving bodies]

Lorentz HA (1899) Simplified theory of electrical and optical phenomena in moving systems. Proc Roy Netherlands Acad Arts Sci 1:427–442
Lorentz HA (1904) Electromagnetic phenomena in a system moving with any velocity smaller than that of light. Proc Roy Netherlands Acad Arts Sci 6:809–831
Lorentz HA (1905) The motion of electrons in metallic bodies-I. Proc Roy Netherlands Acad Arts Sci 7:438–453; (1905) The motion of electrons in metallic bodies-II. Proc Roy Netherlands Acad Arts Sci 7:585–593; (1905) The motion of electrons in metallic bodies-III. Proc Roy Netherlands Acad Arts Sci 7:684–691; (1909) The theory of electrons. Teubner, Leipzig, pp 63–67, 266–273
Lorenz L (1872) Bestimmung der wärmegrade in absolutem maasse. Ann der Phys Und Chem 223:429–452
Makeham J (2008) China: The world's oldest living civilization revealed. Thames and Hudson, p 239
Marconi G (1897) Telegraphy without wires. Sci Am 76:55–56 (1897); (1901) Wireless telegraphy. Smithsonian Annual Report, 294
Marshall W (ed) (1967) Proceedings of the international school of physics "Enrico Fermi" Course XXXVII: Theory of magnetism in transition metals. Acad. Press, New York
Matthiessen A (1862) On standards of electrical resistance. Rept Brit Ass 32:144
Maxwell JC (1856) On Faraday's lines of force. Trans Camb Philos Soc 10:155
Maxwell JC (1861) The theory of the primary colours. Br J Photogr 8&9 (Available on Internet Archive)
Maxwell JC (1865) A dynamical theory of the electromagnetic field. Philos Trans Roy Soc (London) 155:459–512
Maxwell JC (1873) A treatise on electricity and magnetism, vols I and II. Clarendon Press, Oxford, England
Millikan RA (1910) A new modification of the cloud method of determining the elementary electrical charge and the most probable value of that charge. Phil Mag 19:209; (1913) On the elementary electric charge and the Avogadro constant. Phys Rev 2:109–143
Millikan RA (1914) A direct determination of h. Phys Rev 4:73–75; (1916) A direct photoelectric determination of Planck's h. Phys Rev **7**:355–388
Moriya T (1965) Ferro- and antiferromagnetism of transition metals and alloys. Prog Theor Phys 33:157–183; (1965) Spin polarization in dilute magnetic alloys: with particular reference to palladium alloys. Prog Theor Phys 34:329–356; (1967) Localized magnetic moments in transition metals and alloys. In: Proceedings of international school of physics, "Enrico Fermi" Course **XXXVII**. Acad. Press, London, pp 206–254
Mott NF (1934) The conductivity of metals. Proc Phys Soc 46:680; (1936) The electrical resistance of dilute solid solutions. Proc Camb Phil Soc 32:281–290
Neel L (1932) Influence des fluctuations du champ moléculaire sur les proprieties magnétiques des corps. Ann Physique 17:5–105; (1971) Magnetism and local molecular field. Science 174:985–992
Neel L (1948) Propriétés magnétiques des ferrites; ferrimagnétisme et antiferromagnétisme. Ann Physique 3:137–198
Nieuwenhuys GJ (1975) Magnetic behaviour of cobalt, iron and manganese dissolved in palladium. Adv Phys 24:515–591
Nobili L (1825) Sur un nouvean galvanometer presente a Academie des Sciences [On a new galvanometer presented at the Academy of Sciences]. Bibliothèque Universelle (in French) 29:119–125
Norbury AL (1921) The electrical resistivity of dilute metallic solid solutions. Trans Faraday Soc 16:570
Nordheim L (1931) Zur elektronentheorie der metalle. I. Ann der Phys 9:607–640
Oersted JC (1820) Experiments on the effect of a current of electricity on the magnetic needle. In: Thomson T (ed) Annals of philosophy; or, magazine of chemistry, mineralogy, mechanics,

natural history, agriculture, and the arts, vol XVI. Baldwin, Cradock, and Joy, London, pp 273–276

Onsager L (1944) Crystal statistics: a two-dimensional model with an order-disorder transition. Phys Rev 65:117–149

Pauli W (1921a) Über das Modell des Wasserstoff-Molekülions (Ph.D. thesis). Ludwig-Maximilians-Universität München; (1921b) Encyclopädie der mathematis-chen wissenschaften. B.G. Teubner, Leipzig, Germany; (1927) Über gasentartung und paramagnetismus. Z fur Physik, hadrons and nuclie 41:81–102

Peierls R (1933) Zur theorie des diamagnetismus von leitungselecktronen (On the theory of diamagnetism of conduction electrons). Z Phys 80:763–791; (1933) Zur theorie des diamagnetismus von leitungselektronen. II Starke magnetfelder [On the theory of diamagnetism of conduction electrons. II. Strong magnetic fields]. Z Phys (in German) 81:186–194; (1936) On Ising's model of ferromagnetism. Proc Camb Phil Soc 32:477–481

Peter RP, Buchal C, Kubota K, Mueller RM, Pobell F (1984) Phys Rev Lett 53:1108

Priestley J (1767) The History and present state of electricity, with original experiments. London, England, p 732; (1768) A familiar introduction to the study of electricity. London: Printed for Dodsley J, Cadell T, Johnson J

Rath J, Freeman AJ (1975) Generalized magnetic susceptibilities in metals: Application of the analytic tetrahedron linear energy method to Sc. Phys Rev B11:2109

Riecke E (1898) Theorie des galvanismus und der warme. Ann de Phys und Chem 66:353, 545; (1900) Uber das verhaltms der leitfahigkeiten der metalle fur warme und fur elektrizitat. Ann Phys 2:835

Saminadayar L, Glattli DC, Jin Y, Etienne B (1997) Observation of the e/3 fractionally charged Laughlin quasiparticle. Phys Rev Lett 79:2526–2529

Shull CG, Smart J (1949) Detection of antiferromagnetism by neutron diffraction. Phys Rev 76:1256

Shull CG, Wilkinson MK (1955) Neutron diffraction studies of the magnetic structure of alloys of transition elements. Phys Rev 97:304–310

Singh J (1991) Temperature variation of the dynamical orbital susceptibility of metals. Physica B 172:383–391

Singh R, Prakash S, Singh J (1980) Model potential theory of the dynamical susceptibility of metals: application to vanadium, copper and aluminium. J Phys F10:1231; (1981) Temperature dependence of dynamical spin susceptibility of transition metals. Phys Rev B23:2357–2366

Singh J, Rattan SK, Prakash S (1988) Local field effects in electric field gradient of dilute transition metal alloys. Phys Rev B38:10440–19446

Slater JC (1930) Atomic shielding constants. Phys Rev 36:57; (1936) The ferromagnetism of nickel. Phys Rev 49:537; (1936) The ferromagnetism of nickel. II. Temperature effects. Phys Rev 49:931; (1937) The theory of ferromagnetism: lowest energy levels. Phys Rev 52:198

Sommerfeld A (1927) Zur elekronentheorie der metalle. Naturwissenschaften 15:825; (1928) Zur elekronentheorie der metalle, Naturwissenschaften 16:374

Sommerfeld A, Bethe H (1933) Electronentheorie der metalle. In: Handbuch der Physik 24/2:333

Stoner EC (1934) Magnetism and matter. Methuen, London

Stoner EC (1938) Collective electron ferromagnetism. Proc Roy Soc (London) A165:372–414; (1939) Collective electron ferromagnetism II. Energy and specific heat. Proc Roy Soc (London) A169:339–371; (1954) The magnetic susceptibility and electronic specific heat of transition metals in relation to their electronic structure. Acta Metall 2:259–273

Stormer HL, Chang AM, Tsui DS, Huang JCM, Gossard AC, Weigmann W (1983) Fractional quantization of the Hall effect. Phys Rev Lett 50:1953–1956

Sturgeon W (1825) Improved electro magnetic apparatus. Trans Roy Soc Arts Manuf Comm 43:37–52

Sugikuchi K, Endo K (1964) J Phys Chem Solids 25:217

Tesla N (1891) Experiments with alternate currents of very high frequency and their application to methods of artificial illumination, lecture delivered before the American Institute of Electrical Engineers, Columbia College, New York

Teala N (1896) Tesla on electricity without wires. Electrical Engineer, New York, p 52 (Refers to letter by Tesla in the New York Herald, 31 December 1895.) The letter of Petrus Peregrinus on the magnet. (1269). Translated by brother Arnold (New York, McGraw Publishing, 1904) (online)

Tesla N (1898). High frequency oscillators for electro-therapeutic and other purposes. In: Proceedings of the American Electro-theraeutic Association. American Electro-Therapeutic Association, p 25

Tsui DC, Stormer HL, Gossard AC (1982) Two-dimensional magnetotransport in the extreme quantum limit. Phys Rev Lett 48:1562

Van Acker JF, Speier W, Zeller R (1991) Local perturbation and induced magnetization originating from 3*d* impurities in Pd. Phys Rev B43:9558

Van Leeuwen HJ (1919) Vraagstukken uit de electronentheorie van het magnetisme (in Dutch); (1921) Problemes de la theorie electronique du magnetisme. Journal de Physique et le Radium 2:361–377

Van Vleck JH (1932) The theory of electric and magnetic susceptibilities. Clarendon Press, Oxford; Oxford University Press, London

Volta A (1800) On the electricity, excited by the mere contact of conducting substances of different kinds. Philos Trans Roy Soc (in French) 90:403–431

Von Klitzing K, Dorda G, Pepper M (1980) New method for high-accuracy determination of the fine structure constant based on quantized Hall resistance. Phys Rev Lett 45:494–497

Warburg E (1881) Magnetische untersuchungen. Ann der Phys und Chem 13:141–164

Weiss P (1907) L'hypothèse du champ moléculaire et la propriété ferromagnétique. J de Physique Theor Appl 6:661–690

Wheatstone C (1843) The Bakerian lecture: an account of several new instruments and processes for determining the constants of a voltaic circuit. Philos Trans Roy Soc (London) 133:303–327

Wien W (1913) Zur theorie der elekrischen leitung in metallen. Sitz Preuss. Akad. Wiss (Berlin), pp 184–200

Wilcke JC (1762) "Ytterligare rön och försök om *contraira electriciteterne* vid laddningen och därtil hörande delar" (Additional findings and experiments on the opposing electric charges [that are created] during charging, and parts related thereto). Proc Roy Swedish Sci Acad 23:206–229, 245–266; (1777) Rön om Magnet-Nålens årliga och dageliga ändringar i Stockholm. Kongl. Vetenskps Academiens Handlingar 38:274–300

Wilson K (1975) The renormalization group: Critical phenomena and the Kondo problem. Rev Mod Phys 47:773–840

Zeller R (1993) Modelling Simul. Mater Sci Eng 1:553

Ziman JM (1972) Electrons and phonons. Clarendon Press, Oxford

Chapter 12
Magnetic Resonance in Solids

It is well known that static magnetic field produces magnetization in a crystalline solid and one can study a number of magnetic properties. Much more information can be obtained by applying alternating magnetic field to magnetic materials. The basic principle of magnetic resonance is to apply both the static and alternating magnetic fields perpendicular to one another. The static magnetic field splits the nuclear states into substates and the radio frequency (rf) magnetic field stimulates transition between these substates with the photon absorption. The magnetic resonance occurs when the energy hν (or $\hbar\omega$) of the radio frequency photons coincides with the energy difference between the magnetically-induced energy separation of the two substates of a nucleus. There are different magnetic resonance techniques with which one can study a number of properties, such as fine structure of nuclear levels, the electronic structure of an impurity, interaction between magnetic dipoles and the lattice and the line-width of nuclear transitions etc.

Pieter Zeeman (1865–1943), a Dutch physicist, did Ph.D. in 1893 with Heike Kamerlingh Onnes (1853–1926). In 1890, even before doing Ph.D., he became an assistant of Hendrik Antoon Lorentz (1853–1928). Zeeman was lucky enough to study physics from Kammerlingh Onnes and Hendrik Lorentz, two great scientists, at the University of Leiden. In 1896, Zeeman observed that in strong magnetic fields spectral lines got split into equally spaced substates and measured the splitting of spectral lines. This phenomenon was named *Zeeman effect*. Lorentz came to know about the observations made by Zeeman while attending the meeting of the Royal Netherlands Academy of Arts and Sciences in Amsterdam and called him to his office. Lorentz discussed about the splitting of spectral lines in the presence of magnetic field and gave its explanation based on the Lorentz theory of electromagnetic radiations. The energy of Zeeman levels E_m in the presence of magnetic field H is given by

J. S. Galsin, *History of Solid State Physics*,
https://doi.org/10.1007/978-981-95-0504-3_12

Pieter Zeeman (1865–1943), Public Domain, URL: https://en.wikipedia.org/wiki/Pieter_Zeeman#/media/File:Pieter_Zeeman.jpg

$$E_m = -\gamma_I \hbar H m_I \qquad (12.1)$$

where m_I is the magnetic quantum number of a nucleus with spin I and γ_I is the gyromagnetic ratio. Figure 12.1 shows the Zeeman levels for a nucleus with spin I = 3/2. It is noteworthy that Zeeman effect is the first experiment on nuclear magnetic resonance and served as the basis of other magnetic resonance experiments performed in future. Zeeman effect made it clear that the oscillating particles, which according to Lorentz emit light, are negatively charged and a thousand-fold lighter than the hydrogen atom (electron at that time was not discovered). In 1902, Pieter Zeeman shared the Nobel Prize in Physics with Hendrik A. Lorentz for the discovery of the Zeeman effect.

The idea of nuclear magnetic resonance was proposed by Cornelius Jacobus Gorter (1907–1980), a Dutch physicist, in 1936 but he was unable to materialize the phenomenon due to the limitations of his experimental set up. In 1938, Isidor Isaac Rabi (1898–1988), for the first time, performed nuclear magnetic resonance experiment for molecular beams, where all the molecules were independent of one

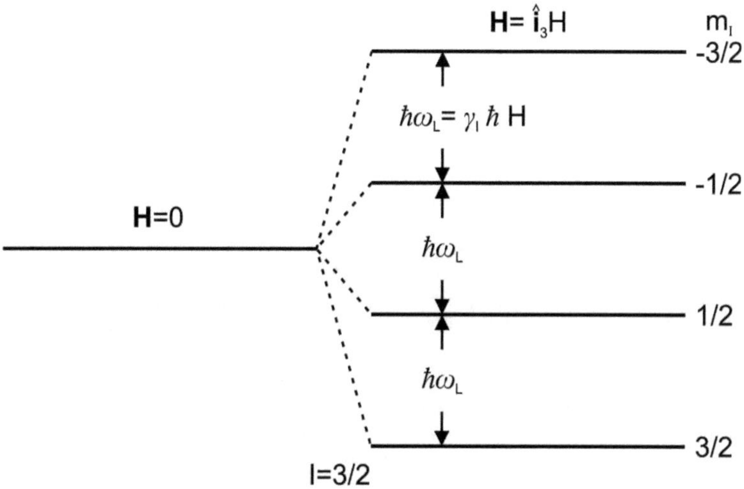

Fig. 12.1 Zeeman splitting of a cubic crystal with nuclear spin I = 3/2

another, by extending the Stern-Gerlach experiment and made accurate measurements of nuclear magnetic moments. Later they measured the magnetic moments of a number of elements including proton and deuteron. In 1941, Yevgeny Zavoisky (1907–1976), a Russian physicist, observed nuclear magnetic resonance in liquids and solids but he discarded the results by saying that they are not reproducible. Before discussing the development of nuclear magnetic resonance techniques, the geometry of the experimental set up should be understood. A constant and uniform (dc) magnetic field **H** is applied to a magnetic material, say in the z-direction. In addition, rf magnetic field $\mathbf{H}_0$ is applied perpendicular to **H** that is in the xy plane, here in the x-direction (see Fig. 12.2). In a magnetic material with noninteracting nuclear spins (or magnetic moments) the rate of change of magnetization **M** is given by the standard relation

$$\frac{d\mathbf{M}}{dt} = \gamma_I \mathbf{M} \times \mathbf{H} \tag{12.2}$$

In the presence of dc magnetic field the magnetic moment precesses about **H** with Larmor frequency. If both the dc and rf magnetic fields are applied perpendicular to one another, it is found that the transverse component of magnetization and susceptibility become infinite at Larmor frequency. As the infinite values of susceptibility and magnetization are not physically possible it was suggested that some kind of relaxation phenomena must exist which make their values finite. In a magnetic material two types of relaxation phenomena exist:

1. A nuclear spin interacts with other nuclear spins yielding finite spin–spin interaction. Such an interaction relaxes exponentially the transverse component of

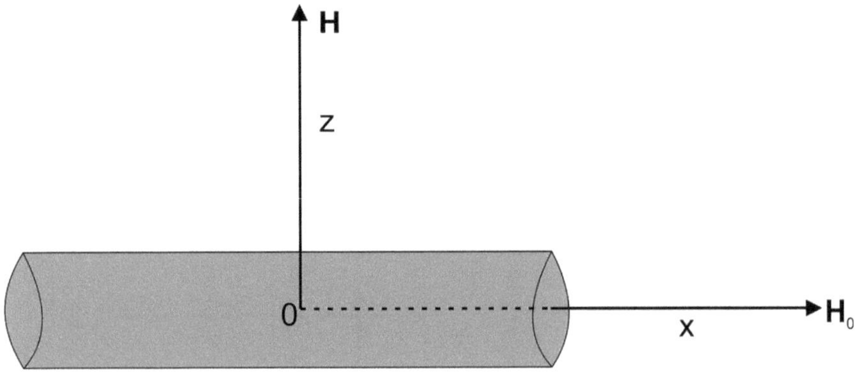

Fig. 12.2 Schematic set up for observing magnetic resonance in a solid where dc magnetic field H and rf magnetic field H_0 are acting perpendicular to each other

magnetization $\mathbf{M}_{xy}$ to its equilibrium value. The spin–spin relaxation is characterized by a time constant called spin–spin relaxation time t_{ss} and involves energy transfer between neighboring spins via dipole and exchange interactions.

2. The nuclear spin also interacts with the lattice yielding spin–lattice interaction. This interaction relaxes the magnetization parallel to the dc magnetic field **H** to its thermodynamical equilibrium value with the surroundings. Spin–lattice interaction is characterized by spin–lattice relaxation time t_{sl}.

In 1946, Felix Bloch (1905–1983) derived phenomenological equations called Bloch equations which included both spin–spin and spin–lattice relaxations and obtained finite values of magnetization and susceptibility. In 1946, Felix Bloch and co-workers measured experimentally the transverse magnetization $\mathbf{M}_y$ the behavior of which is described well by the Bloch equations. The corresponding magnetic susceptibility exhibited Lorenzian-shaped resonance absorption peak at Larmor frequency. In 1946 itself, Edward Mills Purcell (1912–1997) and co-workers independently demonstrated nuclear magnetic resonance in condensed matter using different instruments and techniques. It also forms the basis of magnetic resonance imaging (MRI), one of the most important medical advances of the twentieth century. For this discovery, Felix Bloch and Edward Purcell were awarded 1952 Nobel Prize in Physics. Later Roald K. Wangsness and Felix Bloch (1953) derived the equations rigorously from the first principles for certain experimental circumstances. Experimentally it has been observed that resonance absorption of rf magnetic field is not very sharp but exhibits a peak with finite width called the line-width. A number of factors contribute towards the width of the resonance peak. These are nonhomogeneous magnetic field arising from the crystal structure of solid, the quadrupole moment of nuclei, the spin–lattice interactions and the presence of point defects. Among these, inhomogeneous magnetic field is the most important factor. In 1948, John Hasbrouck Van Vleck (1899–1980) has shown that it is primarily produced by the interaction between the magnetic dipoles. He further pointed out that it is difficult to determine the shape of the resonance line but the root mean square value of its width can be calculated. Van Vleck has shown that the inclusion of exchange interactions in paramagnetic substances causes narrowing of the absorption peaks. This *exchange narrowing* explains why microwave absorption lines in paramagnetic substances are much narrower as compared to that from dipolar coupling.

John Hasbrouck Van Vleck (1899–1980), CC BY-SA 3.0 nl, URL: https://en.wikipedia.org/wiki/John_Husbrouck_Van_Vleck#/media/File:JH_van_Vleck_1974.jpg

It has been observed experimentally that the width of the resonance peak is much less in liquids than that in solids. Further, it is found that the resonance peaks are narrower in a solid in which the atoms are in rapid motion: larger is the motion of atoms in a solid, narrower is the resonance peak. The motion of atoms in solids increases with the increase in temperature yielding narrower resonance peak. This phenomenon is called *motional narrowing* and was discovered by Nicolaas Bloembergen (1920–2017) in 1948 during his thesis, *Nuclear Magnetic Relaxation* [see Bloembergen and coworkers (1948)]. The motional narrowing in a solid can be explained from the fact that the time-averaged local magnetic field becomes nearly constant thereby decreasing the width of resonance peak. In 1949, Gregory J. Hatton and Bernard V. Rollin extended the nuclear magnetic resonance studies down to below 1 °K and observed that the width of resonance absorption changes with temperature. He found that these effects are related to changes in the nature of motions and the state of molecular order.

Nicolaas Bloembergen (1920–2017), CC BY-SA 3.0, URL: https://nl.wikipedia.org/wiki/Nico_Bloembergen#/media/Bestand:Nico_2006_Tucson_Wiki.png

In light elements, the atomic nuclei are nearly spherical in shape. With the increase in atomic number, the shape of atomic nuclei gets distorted and acquires ellipsoidal shape. Such nuclei develop finite *electric quadrupole moment*, the value of which is a measure of the deviation of charge distribution from spherical symmetry. In 1935, Hermann Schuler (1894–1964) and Theodor Schmidt (1908–1986) discovered and measured the electric quadrupole moment of a nucleus. Schuler was one of the pioneers in determining the magnetic moments of nuclei from hyperfine structure observations. In 1935 itself, Hendrik Brugt Gerhard Casimir (1909–2000) formulated quantum mechanical theory for nuclear quadrupole moments. The quadrupole interaction energy of a nucleus with finite quadrupole moment is given by

$$E_{QI} = \frac{1}{6} \sum_{\alpha,\,\beta=1}^{3} Q_{\alpha\beta} V_{\alpha\beta} \tag{12.3}$$

where $Q_{\alpha\beta}$ are the components of quadrupole tensor and $V_{\alpha\beta}$ are the components of double derivative of crystal potential: $V_{\alpha\beta}$ are negative of the components of electric field gradient in solids. The finite quadrupole moment of a nucleus changes significantly the energies of nuclear substates in the presence of magnetic field and hence the transition frequencies. In 1950, Robert Vivian Pound (1919–2010) got the post of a tenured professor of physics at Harvard without even having received a graduate degree. He was the first who reported *nuclear quadrupole resonance* (NQR) in crystalline solids: another very powerful magnetic resonance technique. In magnetic solids with finite quadrupole moment both the Zeeman and quadrupole effects are present. When the magnetic field is high, the Zeeman splitting is large as compared with the quadrupole splitting. Therefore, the quadrupole coupling can be treated as a perturbation in the nuclear magnetic resonance technique and it causes shifting of the Zeeman levels. Such a situation exists in many metals and metallic alloys where the Zeeman splitting is at least 10 times larger than the quadrupole splitting. On the other hand, if the magnetic field is very small and the Zeeman splitting is quite small relative to the quadrupole splitting, the Hamiltonian corresponding to Zeeman effect can be treated as a perturbation in the quadrupole resonance theory. In the limit of zero magnetic field, one obtains pure NQR. The beginning of NQR in solids dates back to the beginning of NMR in the late 1940s or early 1950s. Dehmelt and Kruger (1950) independently discovered NQR and made the first measurement on trans-dichloroethylene in the solid form. NQR spectroscopy provides information about the ground state charge distribution about a quadrupolar nucleus because the NQR spectrum is determined by the components of electric field gradient tensor. One should note that NQR is applicable only to solids because in liquids the quadrupole moment averages out to zero. In 1953, George Daniels Watkins (1924–present) and Robert Vivian Pound (1919–2000) found that in cubic crystals the electric field gradient vanishes due to symmetry reasons. But the presence of point defects destroys the local cubic symmetry and therefore produces finite electric field gradient at the impurity site. Both the NMR and NQR methods are used in estimating impurity-induced electric field gradients in cubic metals and hence the impurity-induced charge distribution. George Watkin basically had done work on defects in semiconductors using different techniques.

In 1936, Cornelis Jacobus Gorter (1907–1980) discovered paramagnetic relaxation in solid paramagnetic salts. Realising the effect of relaxation phenomenon on a nuclear spin system, Gorter was of the view that transitions between the Zeeman components might be observed as a resonance effect. In 1936 and 1942, he made attempts for observing such resonances in a solid but unfortunately could not succeed. In 1967, Gorter said that he had bad luck in making scientific discoveries. Yevgeny Konstantinovich Zavoisky (1907–1976), a Russian physicist, got acquainted with Gorter's experiment on paramagnetic resonance and became interested in the resonance phenomenon. He first reproduced Gorter's experiment and then started doing his own experiments on magnetic resonance. In 1945, Zavoisky discovered *electron paramagnetic resonance* (EPR) while studying a manganese salt $MnSO_4$. Actually, EPR was discovered by him on July 12, 1944, the date when paper was submitted for publication to JEPT. He further found that there was a distinct shift in the peak of sinusoidal signal with the increase in static magnetic field and the peak shifted in the opposite direction with the reversal of magnetic field. In 1946, Brebis Bleaney (1915–2006) independently developed and studied the phenomenon of EPR at low temperatures. He showed that resonance could not be detected at room temperature in many paramagnetic substances due to strong interaction between magnetic ions

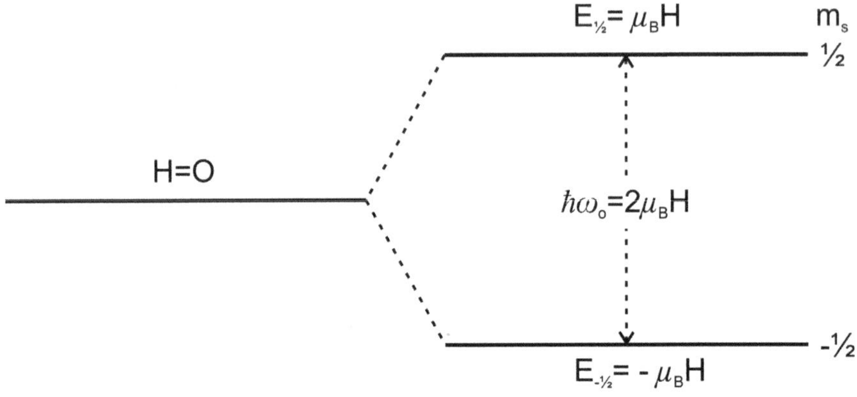

Fig. 12.3 Splitting of an electron energy state with z-component of spin $s_z = ½$, into two substates in the presence of dc magnetic field applied in the z-direction

and lattice vibrations. Bleaney's first paper on EPR was published in 1948 jointly with Roger Penrose (1931–present). In EPR, the energy levels of unpaired electrons are split and the splitting is directly proportional to the strength of magnetic field (Fig. 12.3). In EPR spectroscopy an unpaired electron can change its spin either by emitting or absorbing a photon having energy in the microwave region. Therefore, EPR is also called *electron spin resonance* (ESR). The resonance condition can be satisfied either by varying the magnetic field while keeping the frequency of microwaves constant or doing the reverse: it is usually the frequency that is kept constant. In 1948, Bernard V. Rollin and Gregory J. Hatton studied nuclear paramagnetism at low temperatures in solids. It is noteworthy that the basic concepts of EPR are analogous to NMR but in EPR it is the electron spins that are exited instead of spins of atomic nuclei. Griswold and co-workers (1952) discovered spin resonance of conduction electrons in metals where the skin depth effects are important in studying nuclear absorption resonance. EPR is a very powerful technique and the detailed work of Bleaney on EPR stimulated the study of point defects, color centers and other systems with unpaired electrons in organic, chemical, geological and physical systems.

Ferromagnetic resonance is a coupling between electromagnetic waves and magnetization of a ferromagnetic material through which it passes. The external magnetic field **H** produces magnetization in a ferromagnetic material which starts precessing around the direction of magnetic field with Larmor frequency. The precession frequency of macroscopic magnetization **M** depends on strength of magnetic field, the value of magnetization and on the orientation and shape of the material because saturation magnetization and demagnetization fields are large in the ferromagnetic materials. The resonance absorption takes place when the precession frequency of magnetization and the microwave frequency of ac magnetic field become the same and it gives rise to resonance absorption peak. The resonance absorption of microwave energy is indicated by the decrease in intensity of the signal and it produces local heating of the ferromagnetic material. In 1911, *ferromagnetic resonance* was experimentally discovered by V. K. Arkadyev when he observed the absorption of ultrahigh frequency (UHF) radiation by a ferromagnetic material. The ferromagnetic resonance spectroscopy is a standard technique to probe spin waves and magnetization of ferromagnetic materials. In 1923, Ykov G. Dorfman gave a qualitative explanation of ferromagnetic resonance along with the explanation of results of Arkadyev and suggested that it could provide a method to study ferromagnetic structure. In 1935, Lev Landau (1908–1968) and Evgeny Lifshitz (1915–1985) predicted the existence of ferromagnetic resonance of the Larmor precession, which was experimentally verified by Griffiths (1946) and Zavoisky (1946) in thin films of iron, nickel and cobalt. Griffiths found that the ferromagnetic resonance frequency in thin films is proportional to the square root of the product of magnetic induction **B** and the external field strength **H** as is the case in nuclear and paramagnetic resonance. Charles Kittel (1916–2019) was mainly interested in magnetism and magnetic resonance. In 1947, Kittel gave the theory of ferromagnetic resonance which, in 1948, he extended to include the effect of the shape and orientation of the crystal. Kittel's work strengthened our theoretical understanding of ferromagnetism and ferrimagnetism and helped in giving the shape of a separate field. In 1950, James Ross MacDonald (1923–present) carried out subsequent experiments in thin films of nickel in his thesis work. It is noteworthy that ferromagnetic resonance is broadly similar to EPR and also somewhat similar to NMR. Ferromagnetic resonance occurs at high values of magnetic fields while EPR occurs at low values of magnetic field. Further the precession frequency in a ferromagnetic material is much lower than the frequency observed for electrons in EPR.

As mentioned above Charles Kittel showed that the resonance absorption in a ferromagnetic material depends on the shape of the material. The interesting situation occurs in thin films of ferromagnetic materials in which the magnetic characteristics such as magnetization, magnetic anisotropy and exchange length etc. are in general nonuniform along the growth direction perpendicular to the surface. In studying resonance absorption, the dc magnetic field is applied perpendicular to the surface and ac magnetic field is applied parallel to the surface of the film. In the film, long wavelength spin waves are produced which is nothing but disturbance in the magnetic order propagating in the material. In fact, the surface spins are pinned and the uniform ac magnetic field excites spin waves with odd number of half-wavelengths. The waves with even number of half-wavelengths have no interaction energy with the field. In thin ferromagnetic films, commonly several and often more than 10 resonance-absorption peaks are observed in contrast to the usual ferromagnetic material. Each resonance peak corresponds to the excitation of a distinct spin wave and this phenomenon is called *spin wave resonance*. In 1958, Charles Kittel first predicted the possibility of observing spin wave resonance in thin ferromagnetic films. Immediately after in the same year, Seavey and Tannenwald (1958) confirmed experimentally the propagation of spin waves in permalloy films with thickness down to approximately 15 A. Soon after, the spin wave resonance was made apparent in thin films of pure ferromagnetic elements Fe, Ni, Co and ferrites like $NiFe_2O_4$. Till date spin wave resonance has been studied in numerous ferromagnetic materials. The spatial magnetic profiles in thin films can be determined by means of ferromagnetic resonance which reveals its fine structure in a multipeak resonance spectrum.

Fine structure is defined as the splitting of spectral lines into closely-spaced lines. In 1887, Albert Abraham Michelson (1852–1931) and Edward Williams Morley (1838–1923) identified *fine structure* for hydrogen atom but they could not provide any explanation for it. Later it was found that fine structure arises from the interaction between the electron spin and electron orbital angular momentum (spin–orbit interaction). Therefore, fine structure describes the splitting of spectral lines due to electron spin and relativistic effects. Nonrelativistic quantum mechanics predicted the line spectra but the energy levels depended on the principal quantum number only and could not explain fine structure. In 1916, Arnold Sommerfeld (1868–1951) gave theoretical treatment of fine structure incorporating the relativistic effects, and introduced the fine structure constant α having value 1/137. It was found that even fine structure does not explain the finer details of line spectra of elements. Later it was realized that there exists additional interaction between the nuclear multipole moments and the internal electric and magnetic fields arising from the motion of electrons in an atom: it includes the interaction of nuclear magnetic moment with magnetic field and the interaction of nuclear electric quadrupole moment with electric field gradient. This interaction is called *hyperfine interaction* as it causes further shift in the energy of spectral lines which is smaller, by orders of magnitude, than those of the fine structure. Each of these energy levels may be assigned a quantum number thus yielding quantized levels. In addition, the spinning motion of electrons gives rise to spin current which produces contact hyperfine interaction. In 1930, Enrico Fermi (1901–1954) gave the first theory of hyperfine interactions in an atom containing a single valence electron with an arbitrary angular momentum.

In 1949, Walter D. Knight (1919–2000) found that magnetic resonance absorption in paramagnetic metals occurred at a slightly different value of dc magnetic field as compared to the Zeeman value. This shift in magnetic field is called *Knight shift* which is also a measure of shift in energy and hence in the resonance frequency. In 1950, Charles Hard Townes (1915–2015) and coworkers gave a theory for the Knight shift and it is found to have much in common with the Fermi contact interaction. In a paramagnetic metal, in the presence of external magnetic field, the conduction electron spin is polarized at the atomic nucleus giving rise to extra magnetization which produces extra local magnetic field at the atomic nucleus site. The net magnetic field at the nucleus obtained by adding local magnetic field and the applied magnetic field, is responsible for the shift in magnetic resonance frequency in a paramagnetic metal. The shift in resonance comes from two sources: one from Pauli spin susceptibility and the other from finite s-wave function at the nucleus. Therefore, the Knight shift provides knowledge about the distribution of conduction electrons in a paramagnetic metal.

In an antiferromagnetic material, in addition to the applied magnetic field, there is dipolar field, field due to hyperfine interactions and exchange field due to exchange interactions. In antiferromagnetic materials in which ions with d-electrons occupy low symmetry positions, the anisotropy energy is comparable to the exchange energy. Therefore, anisotropic magnetic field B_A is comparable with the exchange field B_{ex}

in antiferromagnetic materials. In 1941, John Hasbrouck Van Vleck (1899–1980) put forward a theory of antiferromagnetism in which he assumed that the magnetic domains parallel and perpendicular to the applied magnetic field are distributed in the ratio of 1:2. So he gave the formula for magnetic susceptibility as

$$\chi = (\chi_{\|} + 2\chi_{\perp})/3 \tag{12.4}$$

Nagamiya (1951) formulated theory for antiferromagnetism including anisotropy energy and justified the formula for magnetic susceptibility given by Van Vleck. Kittel (1951) independently gave a theory including both anisotropy and exchange energy and have shown that the *antiferromagnetic resonance* frequency ω_0 is given by

$$\omega_0^2 = \gamma^2 \mathrm{B_A}(\mathrm{B_A} + 2\mathrm{B_{ex}})^{1/2} \tag{12.5}$$

where $\gamma = \mathrm{ge}/2\mathrm{mc}$. For an antiferromagnet in which the exchange magnetic field is dominating over the anisotropy field, one can write

$$\omega_0 = \gamma(2\mathrm{B_A}\mathrm{B_{ex}})^{1/2} \tag{12.6}$$

The above equation yields resonance frequency of the same order of magnitude as observed in case of MnF_2. From the above equation, it is evident that if one field (B_A or B_{ex}) is known independently, the other can be estimated from the observed resonance frequency. Keffer and O'Sullivan (1957) have evaluated resonance frequency in terms of magnetic susceptibility $\chi_{\perp}$ (perpendicular to the applied magnetic field) and is given by

$$\omega_0 = \gamma\left(\frac{3\,\mathrm{K_A}}{\chi_{\perp}}\right)^{1/2} \tag{12.7}$$

where K_A is an anisotropic constant. The presence of $\chi_{\perp}$ shows the dependence of resonance frequency on the exchange interactions. Suhl (1958) and Nakamura (1958) independently have shown that there exists indirect nuclear spin–spin interaction in addition to the direct coupling of a nuclear magnetic moment with its own atomic electrons. These indirect interactions are relatively strong and therefore contribute significantly toward line-width in nuclear magnetic resonance. In the presence of indirect coupling, a nucleus may excite a spin wave via its atomic hyperfine interaction which is absorbed by another nucleus through its hyperfine interaction. The comparison of calculated line-widths with the experimental results clearly shows that the indirect coupling is the main source of line-width in antiferromagnetic materials. Experimental measurements of resonance absorption frequencies have been performed on some antiferromagnets. Kondoh (1960) studied the temperature dependence of absorption peak in antiferromagnetic NiO. He found that with the increase in temperature, the height of the peak is diminished and its position is shifted to the longer wavelength side. Keffer and coworkers (1961) and Sievers and Tinkham (1963) have observed resonances in NiO and MnO in the infrared region at low temperatures and obtained similar results. Ohlman and Tinkham (1961) had made antiferromagnetic resonance measurements in FeF_2 in the infrared region. They found that line-width follow the T^4-law above 15 °K.

The perturbed gamma-gamma angular correlation, usually named perturbed angular correlation (PAC) is a method of nuclear solid-state physics in which magnetic and electric fields can be measured at the atomic site. Originally PAC had its origin in the field of nuclear physics for measuring spins and parities. But with the development of experimental techniques, it was found that PAC has also a great potential in solid state physics. In 1940, Donald Ross Hamilton (1914–1972) did theoretical work on PAC and derived the correlation function $W(\theta)$ which gives the probability of the second photon to be emitted at an angle θ with respect to the first photon. In 1947, Edward L. Brady (1919–1987) and Martin Deutsch (1917–2002) carried out the first successful experiment on PAC. In 1953, Anatole Abragam (1914–2011) and Robert Vivian Pound (1919–2010) gave theory of PAC including extra nuclear fields. In PAC, radioactive probe atoms ($\approx 10^{10}-10^{12}$) are introduced in a material under study. The study of materials involves time differential perturbed angular correlation (TDPAC) between a pair of gamma rays emitted by the radioactive atoms which serve as probes of local charge distribution and magnetic fields in the material around the probe nuclei on the atomic scale. A number of experiments were performed mainly for evaluating the electric and magnetic fields in crystals in which probe nuclei were introduced. In 2008, Christian H. Herden and coworkers developed the first fully digitized PAC instrument with which one can analyze extensive data and can use multiple probes. After this, in 2010, Nagl and coworkers and in again in 2010, Jager and co-workers made improvements in the PAC instrument. PAC has been extensively employed to investigate the structure of point defects, diffusion phenomenon, magnetic and electronic properties at the atomic scale in condensed matter. Electric field gradients and its asymmetry on the probe atoms can also be estimated.

Another interesting phenomenon is the nuclear resonance absorption of gamma rays in solids. The nuclear resonance absorption of gamma rays could not be observed in gases where nuclei are free. During the emission of gamma ray a free nucleus gets recoiled due to the conservation of momentum. Therefore, the energy of emitted gamma ray is less than the nuclear transition energy. But for the resonance absorption to occur the energy of gamma ray must be equal to the nuclear transition energy. This fact prevented the resonance absorption in free nuclei or gases. On the other hand, in a solid the atoms (nuclei) are bound together to form a lattice and the effective mass of a nucleus is very much larger. During the emission of gamma ray the recoiling mass is effectively the mass of whole of the solid system making recoil energy very small. If the energy of the emitted gamma ray is small enough the recoil of the nucleus is too weak to be transmitted as a phonon in the lattice. Therefore, recoil of whole of the solid system makes the recoil energy practically zero which amounts to recoilless emission of gamma ray. In this case the energies of emitted and absorbed gamma rays by a nucleus of the solid become the same which makes the resonance absorption a reality and such a situation occurs in isotopes of the nucleus. In 1958, Rudolf Ludwig Mossbauer (1929–2011) during his Ph.D. work discovered a recoilless nuclear resonance phenomenon in the nuclei of solid iridium and it is named Mossbauer effect in his honor. Mossbauer was awarded the 1961 Nobel Prize in Physics for the discovery of Mossbauer effect. To observe the Mossbauer effect two factors are important: first, it is observed in isotopes with very low energy excited states and second the resolution depends on the life-time of the excited state. These two factors limit the number of isotopes that can be used successfully for Mossbauer spectroscopy. The most used isotope is $_{57}Fe^{26}$ which has both the low energy gamma ray and long-lived excited state. Mossbauer spectroscopy is used in studying composition of iron-containing specimens and in geology, phase transitions and bioinorganic chemistry etc. It is noteworthy to mention here that Mossbauer spectroscopy is completely different from nuclear magnetic resonance spectroscopy.

One of the most important applications of magnetic resonance is magnetic resonance imaging (MRI). In 1969, Raymond Damadian (1936–2022) hypothesized and demonstrated that magnetic resonance could be applied to differentiate cancer cells from noncancerous cells. In 1972, Damadian was the first to file the patent of MRI technology which was approved in 1974. Damadian designed and built a full-body MRI machine after the approval of the patent and named it '*Indomitable*'. In 1977, he performed the full-body scan of a human being (of his own postgraduate assistant named Larry Minkoff) and this method became to be known as MRI. In 1978, Damadian founded FONAR Corporation which produced first commercially available MRI machine in 1980. During 1970s, Paul Christian Lauterbur (1929–2007), an American chemist, and Peter Mansfield (1933–2017), a British physicist, developed a method to convert the signal obtained from the scanner into an image. In 1977, Mansfield and co-workers created the first MRI image of the finger of a Ph.D. student Andrew Maudsley and later this technique was improved. The first MRI scanner for clinical use was built in 1980 which was marketed in 1984. The MRI technique was non-evasive and gave excellent soft tissue contrast which led it to dramatic developments in diagnostic radiology. Paul Lauterbur and Peter Mansfield were awarded 2003 Nobel Prize for medicine.

Bibliography

Abragam A, Pound RV (1953) Influence of electric and magnetic fields on angular correlations. Phys Rev 92:943–962

Bleaney B, Penrose RP (1948) Paramagnetic resonance at low temperatures in chromic alum. Proc Phys Soc 60:395

Bloch F (1946) Nuclear induction. Phys Rev 70:460–474

Bloch F, Hansen WW, Packard M (1946) The nuclear induction experiment. Phys Rev 70:474

Bloembergen N, Purcell EM, Pound RV (1948) Relaxation effects in nuclear magnetic resonance absorption. Phys Rev 73:679

Brady EL, Deutsch M (1947) Angular correlation of successive gamma-ray quanta. Phys Rev 72:870–871; (1950) Angular correlation of successive gamma-rays. Phys Rev 78:558–566

Casimir HBG (1935) Uber die hyperfeinstruktur des europium. Physica A 2:719–723

Damadian R (1969) Letter to George. S. Mirick, scientific director. The Health Research Council of the City of New York, Department of Health; (1971) Tumor detection by nuclear magnetic resonance. Science 171:1151–1153

Damadian RV, Goldsmith M, Minkoff L (1977) NMR in cancer: XVI. FONAR image of the live human body. Physiol Chem Phys 9:97–100

Das TP, Hahn EL (1958) Nuclear quadrupole resonance spectroscopy. Solid State Physics, suppl. 1

Dehmelt HG, Kruger H (1950) Kern quadrupol frequenzen im festem dichlorathylen. Naturwissenschaften 37:111–112

Fermi E (1930) Uber die magnetischen momente der atomkerne. Z Phys 60:320–333

Gorter CJ (1936) Negative result of an attempt to detect nuclear magnetic spins. Physica 3:995–998

Gorter CJ (1967) Bad luck in attempts to make scientific discoveries. Phys Today 20:76

Griffiths JHE (1946) Anomalous high-frequency resistance of ferromagnetic metals. Nature 158:670–671

Griswold TW, Kip A, Kittel C (1952) Microwave spin resonance absorption by conduction electrons in metallic sodium. Phys Rev 88:951

Hamilton DR (1940) On directional correlation of successive quanta. Phys Rev 58:122–131

Hatton J, Rollin BV (1949) Nuclear magnetic resonance at low temperatures. Proc Roy Soc A199:222–237

Herden C, Roder J, Gardner JA, Becker KD (2008) Fully digital time differential perturbed angular correlation (TDPAC) spectrometer. Nucl Instrum Methods Phys Res Section A Accelerators, Spectrometers Detectors Assoc Equip 594:155–161

Jäger M, Iwig K, Butz T (2010) A user-friendly fully digital TDPAC spectrometer. Hyperfine Interactions 198:167–172

Keffer F, O'Sullivan W (1957) Problems of spin arrangements in MnO and similar antiferromagnets. Phys Rev 108:637; (1958) Problem of spin arrangements in MnO and similar antiferromagnets. Phys Rev 110:1484

Keffer F, Siever AJ, Tinkham M (1961) Infrared antiferromagnetic resonance in MnO. J Appl Phys 32:65S

Kondoh H (1960) Antiferromagnetic resonance in NiO in far-infrared region. J Phys Soc Japan 15:1970–1975

Kittel C (1947) Interpretation of anomalous Larmor frequencies in ferromagnetic resonance experiment. Phys Rev 71:270; (1948) On the theory of ferromagnetic resonance absorption. Phys Rev 73:155; (1949) Physical theory of ferromagnetic domains. Rev Mod Phys 21:541–583; (1951) Theory of antiferromagnetic resonance. Phys Rev 82:565; (1958) Excitation of spin waves in a ferromagnet by a uniform rf field. Phys Rev 110:1295

Knight WD (1949) Nuclear magnetic resonance shift in metals. Phys Rev 76;1259; Landau L, Lifshitz EM (1935) On the theory of the dispersion of magnetic permeability in ferromagnetic bodies. Phys Z Sowjetunion 8:153

Lauterbur PC (1973) Image formation by induced local interaction; examples employing nuclear magnetic resonance. Nature 242:190–191; (1986) Cancer detection by nuclear magnetic resonance zeugmatographic imaging. Cancer 57:1899–1904

MacDonald JR (1950) Thesis, Oxford

Mansfield P, Maudsley AA (1977) Medical imaging by NMR. Brit J Radiol 50:188–194

Michelson AA, Morley EW (1887) On a method of making the wave-length of sodium light the actual and practical standard of length. Am J Sci 34:427–430; (1887) On the relative motion of the earth and the luminiferous aether. Phil Mag 24:449–463

Mossbauer RL (1958) Kernresonanzfluoreszenz von Gammastrahlung in Ir^{191}. Z Phys A151:124–143

Nagamiya T (1951) Theory of antiferromagnetism and antiferromagnetic resonance absorption. Progr Theoret Phys 6:342

Nagl M, Vetter U, Uhrmacher M, Hofsäss H (2010) A new all-digital time Differential γ-γ angular correlation spectrometer. Rev Sci Instrum 81:073501–073509

Nakamura T (1958) Indirect coupling of nuclear spins in antiferromagnet with particular reference to MnF_2 at very low temperatures. Progr Theoret Phys (Kyoto) 20:542–552

Ohlman RC, Tinkham M (1961) Antiferromagnetic resonance in FeF_2 at far-infrared frequencies. Phys Rev 123:425

Pound RV (1950) Nuclear electric quadrupole interactions in crystals. Phys Rev 79:685

Purcell E, Torrey H, Pound R (1946) Resonance absorption by nuclear magnetic moments in a solid. Phys Rev 69:37–38

Rabi II, Zacharias JR, Millman S, Kusch P (1938) A new method of measuring nuclear magnetic moment. Phys Rev 53:318; (1939) The molecular beam resonance method for measuring nuclear magnetic moments: the magnetic moments of ${}_3Li^6$, ${}_3Li^7$ and ${}_9F^{19}$. Phys Rev 55:526–535

Rollin BV, Hatton J (1948) Nuclear paramagnetism at low temperatures. Phys Rev 74:346

Schuler H, Schmidt T (1935) Uber abweichungen des atomkerns von der kugelsymmetrie. Z Phys 94:457–468; (1935) Uber die abweichung des cassiopeiumatomkerns von der kugelsymmetrie. Z Phys 95:265

Seavey MH, Tannenwald PE (1958) Direct observation of spin-wave resonance. Phys Rev Lett 1:168; (1958) Ferromagnetic resonance in ultra-thin films. J Appl Phys 29:292–293; (1959) Direct observation of spin wave resonance. J Appl Phys 30:S227–S228

Sievers AJ, Tinkham M (1963) Far infrared antiferromagnetic resonance in MnO and NiO. Phys Rev 129:1566–1571

Sommerfeld A (1916) Zur quantentheorie der spektrallinien. Ann Phys 4:51–52

Suhl H (1958) Effective nuclear spin interactions in ferromagnets. Phys Rev 109:606; (1959) Nuclear spin interactions in ferromagnetics and antiferromagnets. J Phys Radium 20:333–335

Townes CH, Herring C, Knight WD (1950) The effect of electronic paramagnetism on nuclear magnetic resonance frequencies in metals. Phys Rev 77:852

Wangsness RK, Bloch F (1953) The dynamical theory of nuclear induction. Phys Rev 89:728–739

Watkins GD, Pound RV (1953) Intensities of nuclear magnetic resonances in cubic crystals. Phys Rev 89:658

Van Vleck JH (1941) On the theory of antiferromagnetism. J Chem Phys 9:85–90

Van Vleck JH (1948) The dipolar broadening of magnetic resonance lines in crystals. Phys Rev 74:1168

Zavoisky EK (1944) Paramagnetic absorption in perpendicular and parallel fields for salts, solutions and metals. Ph.D. Thesis, Kazan State University

Zavoisky EK (1945) Paramagnetic relaxation of liquid solutions for perpendicular fields. J Phys USSR 9:211–216; (1945) Paramagnetic relaxation of liquid solutions for perpendicular fields. Zhur Eksperiment i Theoret Fiz 15:344–350; (1945) Spin-magnetic resonance in paramagnetics. Fizicheskiĭ Zhurnal 9:211–245

Zavoisky EK (1946) Paramagnetic absorption in some salts in perpendicular magnetic fields. Zhur Eksperiment i Theoret Fiz 16:603–606; (1946) Paramagnetic absorption in some salts in perpendicular magnetic fields. J Phys 10:170–173; (1946) Spin magnetic resonance in the decimetre-wave region. J Phys 10:197–198

Zeeman P (1896) Over de invloed eener magnetisatie op den aard van het door een stof uitgezonden licht [On the influence of magnetism on the nature of the light emitted by a substance]. Rep Ordinary Sessions Math Phys Section (Royal Academy of Sciences in Amsterdam) 5:181–184 and 242–248; (1896) The influence of a magnetic field on radiation frequency. Proc Roy Soc (London) 6:513–514, 514–515; (1897) On the influence of magnetism on the nature of the light emitted by a substance. Philos Mag 43:226–239

Chapter 13
Dielectric Properties of Solids

Dielectric materials are insulators which exhibit reasonably high electric polarization in the presence of externally applied electric field: polarizability and insulation are essential properties of a dielectric. The term dielectrics is used to indicate the energy storing capacity of the material by means of polarization. Physically insulators form a subgroup of dielectrics because these exhibit very weak polarization. Dielectrics in general may include insulators, semiconductors and other materials with finite polarizability. In 1834, Michael Faraday published a paper 'On electrical decomposition' which was published in Philosophical Transactions of Royal Society, London. In this paper, he used the terms ion, electrode, anode, cathode and dielectric as suggested by William Whewell (1791–1867). The term dielectric originated from the combination of two words that is dia + electric. Faraday put forward the idea of polarization of dielectrics for the first time. All dielectrics can be classified in to two categories: Linear and nonlinear dielectrics.

1. In linear dielectrics, electric polarization is linearly proportional to the applied electric field. The linear dielectrics are further of two types: polar dielectrics in which each atom or molecule possesses intrinsic finite electric dipole moment and nonpolar dielectrics comprising of neutral atoms or molecules. The symmetrical molecules are nonpolar because in their case the centres of gravity of both the positive and negative charges of the molecule coincide with the centre of symmetry of the molecule. On the other hand, asymmetric molecules are always polar in nature.
2. Nonlinear dielectrics are materials in which electric polarization is not linearly proportional to applied electric field rather depends on the second or higher powers of the electric field. The familiar example of nonlinear dielectrics is ferroelectric materials.

The fact that the presence of a dielectric material decreases the force between two charges, can be understood from the basic laws of electrostatics. If the two charges q_1 and q_2 are situated in a dielectric medium at a distance r from each other, the force between them is given by the famous Coulomb law as

J. S. Galsin, *History of Solid State Physics*,
https://doi.org/10.1007/978-981-95-0504-3_13

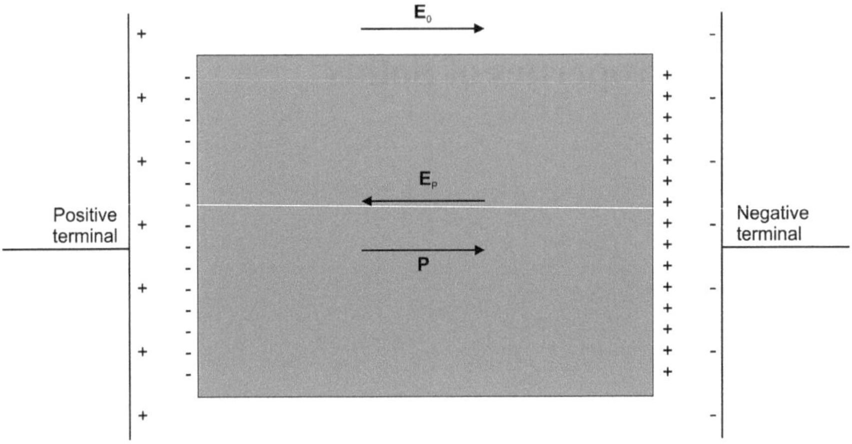

Fig. 13.1 Parallel plate capacitor with a dielectric between the plates

$$\mathbf{F} = \frac{q_1 q_2}{\varepsilon r^2}\hat{\mathbf{r}} \tag{13.1}$$

Here ε is a dimensionless parameter and is a property of the material itself. ε is called relative permittivity (or permittivity) or the dielectric constant of the material and is always greater than one. If one substitutes $\varepsilon = 1$, one obtains the usual Coulomb law in vacuum. From Eq. (13.1) the electric field due to the charge q in a dielectric material is given by

$$\mathbf{E} = \frac{q}{\varepsilon r^2}\hat{\mathbf{r}} \tag{13.2}$$

Therefore, the electric field due to a charge in a dielectric material is less than that in vacuum. The behavior of electric field inside a dielectric can be understood better with the help of a parallel plate capacitor with a dielectric in between the plates (Fig. 13.1). In the presence of an externally applied electric field $\mathbf{E}_0$, each atom or molecule of the dielectric material gets distorted due to small drift of negative and positive charges inside each atom or molecule. As a consequence, finite electric dipole moment $\mathbf{p}$ is induced on each atom or molecule. Polarization $\mathbf{P}$ in a dielectric material is defined as the electric dipole moments per unit volume. The induced polarization $\mathbf{P}$ in a dielectric gives rise to additional electric field $\mathbf{E}_p$ called *depolarization field* as it opposes the applied electric field. The average electric field $\mathbf{E}$ inside the dielectric material is called macroscopic electric field and is given by

$$\mathbf{E} = \mathbf{E}_0 - \mathbf{E}_p = \mathbf{E}_0 - N\mathbf{P} \tag{13.3}$$

The factor N in the depolarization field, depends on the shape of the dielectric material: 4π for a rectangular plate (parallel plate capacitor) and $4\pi/3$ for a spherical shape of the dielectric material. James Clark Maxwell (1831–1879) derived the formula of the depolarization factors N_x, N_y and N_z for an ellipsoid. Osborn (1945) calculated depolarizing factors of a general ellipsoid covering a number of special cases. Stoner (1945) calculated the values of the depolarization factor N for different shapes of the dielectric material. Aharoni (1998), Chen and coworkers (2005) and Beleggia and coworkers (2005) put forward the detailed discussion about depolarization factors for a few other geometries such as rectangle and cylinder. It is pointed out here that the depolarizing factor and the demagnetizing factor have the same value for a particular shape of the material. In a linear dielectric material, polarization produced is proportional to the macroscopic electric field, that is

$$\mathbf{P} = \chi_e \mathbf{E} \tag{13.4}$$

The constant of proportionality χ_e is called electric susceptibility and is a measure of the degree of polarization of a dielectric material. In a dielectric material, one usually defines displacement vector $\mathbf{D}$ as follows:

$$\mathbf{D} = \varepsilon \mathbf{E} = \mathbf{E} + 4\pi \mathbf{P} \tag{13.5}$$

From Eqs. (13.4) and (13.5) one can write the relation between ε and χ_e as

$$\varepsilon = 1 + 4\pi\, \chi_e \tag{13.6}$$

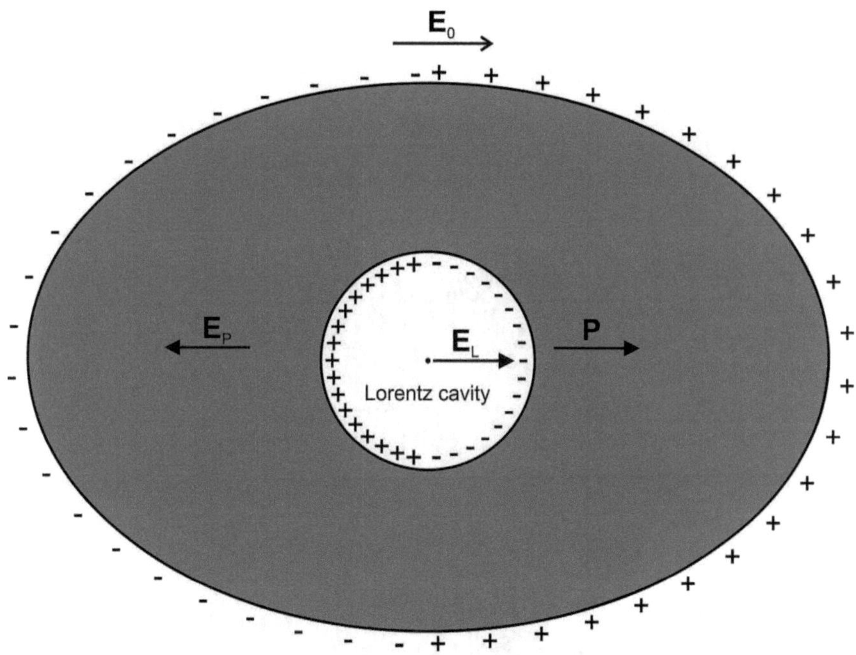

Fig. 13.2 Lorentz cavity in a dielectric material. The distribution of charge on the external surface of the material and inner surface of the cavity are shown

The dielectric constant can also be defined in terms of capacitance of a capacitor as

$$\varepsilon = \frac{C_{diel}}{C_{vac}} \tag{13.7}$$

where C_{diel} and C_{vac} is the capacitance of a capacitor with a dielectric material and with vacuum, respectively. One usually determines the dielectric constant experimentally in terms of capacitance of a capacitor.

The field acting on an atom or molecule inside the dielectric is called *local electric field* $\mathbf{E}_{loc}$ which may not necessarily be the same as the macroscopic field $\mathbf{E}$. The knowledge of local electric field is necessary to study the properties of dielectric materials from the microscopic point of view. In 1887, Hendrik Antoon Lorentz (1853–1928) gave an elegant method to evaluate local electric field at the position of an atom. He created a fictitious spherical cavity around the atom under consideration (called *Lorentz cavity*), see Fig. 13.2, and evaluated the electric field at its centre. The local electric field, in the Lorentz approximation, is given by

$$\mathbf{E}_{loc} = \mathbf{E} + \frac{4\pi}{3}\mathbf{P} \tag{13.8}$$

Hendrik Antoon Lorentz (1853–1928), CC BY 3.0, URL: https://www.researchgate.net/publication/309449683/figure/fig1/AS:570806994051077@1513102400889/Hendrik-Antoon-Lorentz-in-1911-Fotoboek-Solv

The above equation is called Lorentz equation and the second term in it is called *Lorentz field*. Equation (13.8) is a fairly good approximation in nonpolar dielectrics but it may not work in polar dielectrics.

In a linear dielectric material, the induced electric dipole moment **p** on an atom or molecule is proportional to the local electric field and is given by

$$\mathbf{p} = \alpha^{\mathrm{a}} \mathbf{E}_{\mathrm{loc}} \tag{13.9}$$

The proportionality constant α^{a} is called atomic polarizability and is the most important microscopic quantity in a dielectric. There are three contributions to atomic polarizability. First is the electronic contribution arising from the distortion of electron distribution in atoms or molecules and it is finite in all types of solids. Second is the ionic contribution which arises due to the polarization in the distribution of ions and is finite only in ionic solids. Third is orientational polarizability which is finite in dipolar dielectric solids. In 1912, Peter Debye (1884–1966) proposed that atoms/molecules in a certain class of materials possess permanent electric dipole moments in analogy with the magnetic moments of atoms/molecules in paramagnetic materials. The unit of dipole moment was named Debye in his honor. Debye gave a theory for orientational polarizability in a way similar to that of the Langevin paramagnetism and is called *Langevin-Debye theory*. Electric susceptibility shows Curie type behavior at high temperatures and dipolar polarizability is inversely proportional to temperature. In 1850, Ottaviano Mossotti (1791–1863) and in 1879, Rudolf Clausius (1822–1888) independently derived a relation between atomic polarizability and dielectric constant, usually called *Clausius–Mossotti relation*, and is given by

$$\frac{\varepsilon - 1}{\varepsilon + 2} = \frac{4\pi}{3} \sum_{\mathrm{i}} \rho_{\mathrm{i}} \alpha_{\mathrm{i}}^{\mathrm{a}} \tag{13.10}$$

where $\alpha_{\mathrm{i}}^{\mathrm{a}}$ and ρ_{i} are atomic polarizability and density of ith type of atoms. Equation (13.10) was the first and most important relation relating the microscopic and macroscopic parameters in a dielectric material. It is noteworthy that Eq. (13.10) works well in gases but only approximately true in liquids and solids particularly if the dielectric constant is large.

Peter Debye (1884–1966), Public Domain, URL: https://en.wikipedia.org/wiki/Peter_Debye#/media/File:Debye100.jpg

Optical properties of solids are determined by the interaction of solids with electromagnetic radiation whose wavelength lies in the infrared to ultraviolet regions. Some of the important optical properties of solids are refractive index, dispersion, absorption and reflectance etc. and these are directly related to the dielectric function of a solid. The refractive index n of a solid is related to its permittivity and permeability as $n = \sqrt{\mu \varepsilon}$ and for a solid with unit permeability ($\mu = 1$) the refractive index becomes $n = \sqrt{\varepsilon}$. Ludvig Lorenz (1829–1891) is known for his work on refraction of light and its relation to the molecular state of the transparent medium. In 1869, Lorenz derived a phenomenological relation between the refractive index and the density of the medium. Because the paper was in Danish so the scientists did not come to know about it. In 1878, Hendrik Antoon Lorentz (1853–1928) derived independently the same relation using the Helmholtz's electromagnetic theory of light. The Lorenz paper came to light in 1880 when it was translated from Danish to German language and published in Annalen der Physik. In1909, Lorentz referred this relation as Lorenz–Lorentz relation. This relation is similar to the Clausius–Mossotti relation where dielectric function ε is replaced by the square of refractive index n.

Rudolf Clausius (1822–1888), Public Domain, URL: https://en.wikipedia.org/wiki/Rudolf_Clausius#/media/File:Rudolf_Clausius_01.jpg

A number of models have been used to study optical properties of solids. In 1878, Hendrik Lorentz gave a model in which the binding between a nucleus and electrons was assumed to be similar to that of the mass-spring system and is usually called *Lorentz oscillator model*. In this model electrons vibrate like damped harmonic oscillators on interacting with the electromagnetic radiation. Lorentz evaluated the frequency-dependent complex dielectric function and polarizability function. In 1900, Paul Karl Ludwig Drude (1863–1906) developed a classical model, called *Drude model*, for metals for the evaluation of complex dielectric function, complex refractive index, reflectivity of metals and their variation with frequency. In the Drude model the interaction between the electrons was neglected. In 1903, Ernst Bessel Hagen (1851–1923) and Heinrich Rubens (1865–1922) obtained an empirical relation between reflectivity and frequency from the experimental measurements of reflectivity in the infrared region and is named *Hagen-Rubens relation* in their honor.

In 1905, Hendrik Lorentz improved upon the Drude model, called *Drude-Lorentz model*, by including the interactions between the electrons. He treated electrons as damped harmonically bound particles subject to the external electric field. He studied complex dielectric function, complex refractive index and reflectivity of electromagnetic radiation and found significant improvement in the results due to the interaction among electrons. It is a classical model used to calculate thermal and electrical conductivities of metals and the relation between them. Later this was supplemented with the results of quantum theory in 1933 by Arnold Sommerfeld and Hans Bethe leading to the Drude–Sommerfeld model.

In 1926, Ralph Kronig (1904–1995) derived a general relation connecting the real and imaginary parts of a complex function. In 1927, Hendrik Anthony Kramers (1894–1952) independently derived the same relation and named it *Kramers–Kronig relation*. This relation enabled us to compute real part of a general response function (particularly electric and magnetic susceptibilities) from its imaginary part and vice versa. Kramers–Kronig relation is also used in electron energy-loss spectroscopy and angle-resolved spectroscopy to calculate real and imaginary parts of absorption coefficient, reflectivity and permittivity. In 1941, R. H. Lyddane, R. G. Sachs (1916–1999) and Edward Teller (1908–2003) derived a relation in an ionic solid which expresses the ratio of natural frequency of longitudinal optical lattice vibrations ω_{LO} to the natural frequency of the transverse optical lattice vibrations ω_{TO} in terms of the dielectric function $\varepsilon(\omega)$. It is named *Lyddane-Sachs-Teller relation* in their honor and is written as

$$\frac{\omega_{\mathrm{LO}}^2}{\omega_{\mathrm{TO}}^2} = \frac{\varepsilon(0)}{\varepsilon(\infty)} \tag{13.11}$$

Dielectric material in the presence of an external electric field dissipates some quantity of electrical energy that is converted into heat. This phenomenon is commonly known as *dielectric loss*. If dc voltage V is applied to a dielectric material, in general to any material, the power loss P is given by the standard relation $\mathrm{P} = \mathrm{V}^2/\mathrm{R}$ where R is the resistance of the dielectric material. If ac voltage is applied to a dielectric material the power loss depends on a number of physical quantities. In 1886, Ivan Borgman (1849–1914) studied the dielectric loss, for the first time, and established the factors contributing towards it. He found that the dielectric loss increases with the increase in voltage, frequency of voltage and capacitance. It also depends on the nature of the dielectric material. The dielectric loss is especially high around the resonance frequency of polarization. As the polarization lags behind the applied electric field in a dielectric material it causes interaction between the field and polarization which results in heating. The dielectric loss can be parameterized in terms of either the loss angle or the tangent of loss angle. Another physical quantity of interest is *dielectric breakdown*. With the increase in voltage the insulation property of a dielectric material is affected causing some leakage current, capacitative current and dielectric loss. There is some maximum finite value of voltage at which insulation of a dielectric material can withstand. If the voltage is progressively increased, at some stage the insulation will be broken causing dielectric breakdown. At this point the flow of conduction current through the dielectric will sharply rise.

Pyroelectric, piezoelectric and ferroelectric materials are also dielectric materials. Theophrastus (371BC–287BC) was a student of Aristotle and became his successor at his school. He mentioned about the existence of pyroelectric effect. He found that tourmaline crystal when heated could attract sawdust or small pieces of straw. In 1707, Johann Georg Schmidt found that tourmaline crystal attracted only hot ashes not the cold ones. In 1717, Louis Lemery (1677–1743) found that small pieces of a nonconducting material were attracted by the tourmaline crystal but after contact these were repelled by it. In 1747, Carl Linnaeus (1707–1778) said that when the temperature of some materials including tourmaline is changed, an electric potential is generated but he could not establish this fact. In 1756, Franz Ulrich Theodor Aepinus (1724–1802) established that electric potential is produced in these materials with the change in temperature. As early as 1785, Rene Just Hauy (1743–1822) detected the same effect in calamine, an oxide of zinc. In 1820, Antoine Cesar Becquerel (1788–1878) found that pressure can induce electricity in every material and attributed this effect to surface interactions. He put forward a relation between mechanical stress and electric charge but his experiments remained inconclusive. In 1824, Sir David Brewster (1781–1868) observed the phenomenon of generation of electric potential with the change in temperature in some salts and gave it the name *pyroelectricity* from the Greek word '*pyr*' meaning fire. The possible explanation for the appearance of electric voltage in the pyroelectric materials is as follows: The change in temperature modifies the positions of the atoms within the crystal which causes change in polarization of the material. The polarization change gives rise to a voltage across the crystal. If the temperature stays constant at a particular higher value, the pyroelectric voltage disappears due to the leakage current. In 1878, William Thomson (1824–1907) and in 1897, Woldemar Voigt (1850–1919) developed the theory of pyroelectricity. It is found that all the polar crystals are pyroelectric, therefore the ten polar crystal classes are sometimes referred to as pyroelectric classes. One should not confuse pyroelectricity with thermoelectricity as the origins of the two are entirely different.

Sir David Brewster (1781–1868), CC BY-SA 4.0, URL: https://en.wikipedia.org/wiki/David_Brewster#/media/File:David_Brewster_1618.jpg

Elie Seignette (1632–1698) was living in La Rochelle, a city of France and was a pharmacist dealing with natural medicines made from plants. He found a salt called "*selpolychresta*", having chemical formula $NaKC_4H_4O_6 \cdot 4H_2O$, which was later, in 1675, named "*Rochelle salt*" after the name of the city. In 1824, Sir David Brewster (1781–1868) found that in Rochelle salt electric polarization is produced with the application of electric field which remains finite even in the absence of electric field. In 1912, Ervin Schrodinger (1887–1961) coined the word *ferroelectric* for nonlinear dielectric like Rochelle salt. In 1920, Joseph Valasek (1897–1993) found that permanent polarization is a natural state of Rochelle salt. He further stated that **D**, **E** and **P** in ferroelectric materials are analogous to **B**, **H**, and **M** in magnetism. In 1920 itself, Valasek discovered hysteresis in ferroelectric materials and published the first hysteresis curve of a ferroelectric material. Before publication hysteresis curve in ferroelectrics was presented in April, 1920 to American Physical Society. Valasek found that hysteresis in ferroelectric materials is just parallel with the hysteresis in ferromagnetic materials, a discovery made three decades earlier by James Alfred Ewing (1855–1935). It is experimentally found that spontaneous polarization persists up to certain temperature called Curie temperature above which it vanishes and the material becomes paraelectric. In 1922, Valasek found, for the first time, the existence of two Curie temperatures in Rochelle salt. In 1935, Georg Busch (1908–2000) and Paul Scherrer (1890–1969) found that $K_2H_2PO_4$ was also ferroelectric material with critical temperature $T_c = 123$ °K. Busch suggested that the hydrogen bonds between adjacent oxygen atoms are responsible for ferroelectricity. In 1946, Vitaly Ginzburg (1916–2009) wrote a classic paper on the subject and called it the "Seignetto electric effect" as it had been first observed by Seignette. By the end of 1950s, several hundred oxide-based ferroelectric materials had been discovered.

In 1880, Pierre Curie (1859–1906) and his brother Paul-Jacques Curie (1855–1941) performed systematic investigations on the Rochelle salt and established that an electric charge is generated in response to the applied mechanical stress. Such materials are called piezoelectric materials and the phenomenon is called *piezoelectric effect* and the electricity generated is called *piezoelectricity.* This word is derived from the Greek word "*piezein*" which means squeeze or press. In 1881, Jonas Ferdinand Gabriel Lippman (1845–1921) predicted mathematically, from the principles of thermodynamics, the converse piezoelectric effect: that is solids become stressed or strained when subjected to external electric field. In the same year the Curie brothers confirmed the existence of converse piezoelectric effect. This fact shows that the piezoelectric effect is a reversible process. The converse piezoelectric effect is used in producing ultrasonic waves. It is interesting to note that all the ferroelectric materials are also piezoelectric but the reverse is not true that is all the piezoelectric materials may not be ferroelectric. For example, quartz is piezoelectric but not ferroelectric. On the other hand, barium titanate is both ferroelectric and piezoelectric. All the pyroelectric crystals are also piezoelectric in nature. In 1899, Thomas Alva Edison (1847–1931) used Rochelle salt in constructing phonograph: the first commercial use of the piezoelectric effect. At the beginning of the World War I, Nicolson from USA and Paul Langevin (1872–1946) from France, worked independently and were able to make transducers to produce ultrasonic waves. In 1917, Langevin and coworkers developed ultrasonic detectors for the detection of submarines.

Ferroelectric materials are the most typical and most studied nonlinear dielectrics and a brief account of these materials is presented here. In 1839, Gustav Rose (1798–1873) discovered naturally occurring mineral calcium titanate with chemical formula $CaTiO_3$. This mineral was named perovskite in honor of Russian mineralogist Lev Perovski (1792–1856). In 1942, Wainer and Solomon from USA and in 1944, Ogawa from Japan and Wul from Russia independently prepared the first man-made ferroelectric material barium titanate with chemical formula $BaTiO_3$ possessing perovskite structure. This material can exist in five phases including cubic, hexagonal, orthorhombic, tetragonal and rhombohedral structures. The dielectric constant of $BaTiO_3$ was very large: about 1000 times than any other ceramic known at that time. The discovery of $BaTiO_3$ was extremely important as it demonstrated for the first time that ferroelectricity could exist in simple oxide materials. It had number of advantages over the previously known ferroelectric materials. It is easy to manufacture, exhibits high mechanical strength, resistant to heat and moisture, and exhibits ferroelectric properties over a broad range of temperature. By the end of 1960s a number of similar ferroelectric materials were discovered.

Early theories of ferroelectricity were aimed at explaining the properties of Rochelle salt in which it was assumed that H_2O molecules possess permanent electric dipole moments. Kobeko and Kurchatov (1930) proposed dipolar theory analogous to the Weiss theory of ferromagnetism. They assumed that there exists some internal field $\mathbf{E}_{int}$ in a ferroelectric material, in addition to the applied electric field, which is proportional to electric polarization, that is, $\mathbf{E}_{int} = \gamma_i \mathbf{P}$. It was shown that above the Curie temperature T_c, the electric susceptibility obey Curie–Weiss law defined as

$$\chi_E = \frac{C_E}{T - T_c} \tag{13.12}$$

where Curie constant $C_E = T_c/\gamma_i$. But there are some objections to the dipolar theory as it did not explain two transition temperatures in Rochelle salt and secondly it gave very high values of polarizability in ferroelectrics. In 1936, Lars Onsagar (1903–1976) made an improvement in the dipolar theory by assuming γ_i to be dependent on the dielectric function ε.. In 1941, John Clarke Slater (1900–1976) put forward first molecular theory of ferroelectricity and established that ferroelectricity in Rochelle salt is primarily due to the ordering of H-bonds. One can talk about electric polarization catastrophe that is very large value of polarization in ferroelectric solids. It was proved that near the polarization catastrophe condition, the dielectric function was very large in value, ideally speaking infinity, and obeyed the Curie–Weiss law. From the Clausius–Mossotti relation [Eq. (13.10)] one can find the condition at which the dielectric function of a ferroelectric solid goes to infinity.

In 1951, Sawaguchi and co-workers and Shirane and co-workers first reported the existence of antiferroelectricity in lead zirconate possessing perovskite structure. Double hysteresis loop has been observed in antiferroelectric materials but it alone does not provide conclusive proof of a phase being antiferroelectric. Simultaneously, in 1951, Charles Kittel put forward a theory for the existence of antiferroelectricity in the materials. Later antiferroelectricity was discovered in over 100 materials over the last 60 years and among these, materials like $PbHfO_3$ and $NaNbO_3$ etc. possess perovskite structure. Antiferroelectric materials in dielectrics are analogous in behavior to the antiferromagnetic materials. These materials exhibit antiferroelectric phase below a transition temperature T_c and makes a transition to the paraelectric phase above T_c. On the application of an electric field of sufficient strength, the material can switch from the low temperature antiferroelectric phase to a ferroelectric phase. Antiferroelectric materials exhibit domain structure just like antiferromagnetic materials. In antiferroelectric materials, there are two sublattices with the same magnitude of spontaneous polarization but in opposite directions, leaving a net-zero macroscopic polarization within the crystal at zero field. One can say that antiferroelectric materials are similar in behaviour to the ferromagnetic materials, the only difference is that the atomic magnetic moment is replaced by atomic electric dipole moment.

There are number of properties of crystalline solids in which they exhibit phase transition at a particular temperature such as superconductivity, ferromagnetism, ferroelectricity etc. Thermodynamics of solids is of vital importance in studying phase transitions in solids. According to Paul Ehrenfest (1880–1933), phase transitions are of two types: first order and second order phase transitions. *First order phase transition* exhibits a discontinuity in the first derivative of free energy with respect to the thermodynamical variable. It is characterized by discontinuous jump in entropy and latent heat is liberated. In the second order phase transition, the second order derivative of free energy exhibits discontinuity but the first derivative is continuous. In the second order transition the entropy changes continuously but there is discontinuous change in specific heat without any release of latent heat. In ferroelectric crystalline solids, with the decrease in temperature there occurs a phase transition at the Curie temperature from symmetric to asymmetric phase. Therefore, it is of immense importance to study the behavior of ferroelectric materials near the Curie temperature. In 1950, Vitaly Ginzburg (1916–2009) and Lev Landau (1908–1968) proposed a phenomenological theory for studying phase transition in superconductors. In the *Ginzburg–Landau theory*, free energy of a solid was expressed in terms of a series with terms having increasing powers of order parameter and variational technique was used to obtain differential equation. Based on the Ginzburg–Landau theory, Devonshire (1951) developed a phenomenological theory of phase transitions in ferroelectric solids by choosing polarization **P** as the order parameter. Free energy was expressed in terms of a series in powers of **P** and the equilibrium state was obtained by minimizing free energy with respect to **P**. Electric susceptibility was estimated in the first- and second order phase transitions in ferroelectric solids which obeyed Curie–Weiss law. In 1959, William Cochran (1922–2003) gave a general lattice dynamical theory of ferroelectricity and studied soft modes in the phonon dispersion relations. His theory had provided a major understanding of the ferroelectric phenomena. In 1962, Roger Arthur Cowley (1939–2015) had given a microscopic theory of ferroelectricity in which the temperature dependence of soft modes arose from the anharmonic interactions between the normal modes although the anharmonic interactions in these materials are quite small. In late 1960s, Ivan Paul Kaminow (1930–2013) and coworkers confirmed experimentally the existence of soft modes in ferroelectric solids. In 1971 and 1972, Gen Shirane (1924–2005) and coworkers reconfirmed the existence of soft modes in ferroelectric crystals.

A ferroelectric crystal splits into domains because such splitting lowers the free energy by reducing the electrostatic energy of the spontaneous polarization charges. As far as the experimental studies are concerned, the X-ray and optical studies confirmed the existence of domains in ferroelectric materials. The ferroelectric domains are electrical analogue of ferromagnetic domains but there are a number of differences between them. In each ferroelectric domain, all the electric diploes are in the same direction yielding finite and uniform polarization in it: polarization in a domain is called spontaneous polarization. The direction of spontaneous polarization is not the same throughout rather the domains are polarized in opposite directions. Merz (1954) had shown that in $BaTiO_3$ when electric field is applied in a direction opposite to the spontaneous polarization, a large number of new needle-shaped domains appear. The domains grow essentially in the forward direction rather than sideways, a behavior quite different from that observed in ferromagnetic materials. Ferroelectrics have large number of applications. They are used in making high permittivity capacitors, sensing devices, storage devices, optoelectronic devices, automobile industry and defence related sectors etc.

In the recent years ferrielectric materials in the polar phase have been discovered. They may have two or more spontaneous polarization directions which can be flipped with the application of electric field. In 2012, Masatomo Yashima and Shota Matsuyama showed that $AgNbO_3$ exhibit ferrielectric structure and found the off-center displacements of Nb and Ag cations along the c-direction induce ferrielectricity in it. In 2020, Fu and co-workers showed the ferrielectric nature of $PbZrO_3$-based antiferroelectric materials. In 2014, Dong and co-workers predicted theoretically that $BaFe_2Se_3$ in the polar phase possesses ferrielectric structure which has been confirmed by Aoyama and co-workers in 2019 by the neutron diffraction experiments. It should be noted that $BaFe_2Se_3$ is the first reported iron-based material to become nonlinear ferrielectric and superconductor. In 2019, Lin and co-workers proposed that compounds with formula MO_2X_2 (M = Mo/W, X = Br/Cl) exhibit nonlinear ferrielectric behavior. In 2020, Zhang and co-workers predicted the ferrielectric order in $BaFe_2Te_3$ (iron telluride). They further pointed out that the structural similarity with $BaFe_2Se_3$ suggest that $BaFe_2Te_3$ could become a superconductor under high pressure.

In the end, from Chaps. 11 and 13, we conclude that electric and magnetic materials are of similar types. In magnetic materials we have diamagnetic, paramagnetic, ferromagnetic, antiferromagnetic and ferrrimagnetic materials. In electric materials also we have dielectric, paraelectric, ferroelectric, antiferroelectric and ferrielectric materials. In magnetic materials, magnetic dipole moment arises from the combination of north and south poles of a magnet while in electric solids the electric dipole moment arises from the combination of equal and opposite charges (positive and negative charges). We have seen that the theories of the corresponding electric and magnetic materials have also been founded on the similar grounds. The nuclear resonance phenomenon in solids is also of two types: nuclear magnetic resonance arising from the finite magnetic moment of the nucleus (magnetic effect) and nuclear quadrupole resonance arising from finite electric quadrupole moment of the nucleus (electrostatic effect). It shows that the magnetic and dielectric materials are closely related with each other.

Bibliography

Aepinus FUT (1756) Memoire concernant quelques nouvelles experiences électriques remarquables [Memoir concerning some remarkable new electrical experiments]. Histoire de l'Académie royale des sciences et des belles lettres (Berlin) 12:105–121

Aharoni A (1998) Demagnetizing factors for rectangular ferromagnetic prisms. J Appl Phys 83:3432–3434

Aoyama T, Imaizumi S, Togashi T, Sato Y, Hashizume K, Nambu Y, Hirata Y, Matsubara MM, Ohgushi K (2019) Polar state induced by block-type lattice distortions in $BaFe_2Se_3$ with quasi-one-dimensional ladder structure. Phys Rev B 99:241109(R)

Beleggia M, De Graef M, Millev YT, Goode DA, Rowlands G (2005) Demagnetization factors for elliptic cylinders. J Phys D: Appl Phys 38:3333; Beleggia M, De Graef M, Millev YT (2006) The equivalent ellipsoid of amagnetized body. J Phys D: Appl Phys 39:891

Brewster D (1824) Observation on the pyro-electricity in minerals. Edinbg J Sci 1:208–218

Busch G, Scherrer PP (1935) Eine neue seignetteelektrische substanz. Naturwiss 23:737

Chen DX, Pardo E, Sanchez A (2005) Demagnetizing factors for rectangular prisms. IEEE Trans Magn 41:2077–2088

Cochran W (1959) Crystal stability and the theory of ferroelectricity. Phys Rev Lett 3:412; (1960) Crystal stability and the theory of ferroelectricity. Adv Phys 9:387–423

Cowley RA (1962) Temperature dependence of a transverse optic mode in strontium titanate. Phys Rev Lett 9:159

Curie J, Curie P (1880) Développement par pression de l'électricité polaire dans des crystaux hémiédriques à faces inclines. Compt Rend 91:294–295, 383–387

Curie J, Curie P (1880) Développement par compression de l'électricité polaire dans les cristaux hémièdres à faces inclinées. Bulletin de la Société Minéralogique de France 3:90–93

Debye P (1912) Einige resultate einer kinetischen theorie der isolatoren (Vorläufige mitteilung). Physik Zeitschr XIII:97–100

Devonshire AF (1949) XCVI. Theory of barium titanate-Part I. Phil Mag 40:1040–1063; (1951) CIX, Theory of barium titanate-Part II. Phil Mag 42:1065–1079; (1954) Theory of ferroelectrics. Adv Phys 3:85–130

Dong S, Liu JM, Dagotto E (2014) $BaFe_2Se_3$: a high T_c magnetic multiferroic with large ferroelectric polarization. Phys Rev Lett 113:187204

Drude P (1900a) Zur elektronentheorie der metalle. Annalen der Physik 306:566–613; (1900b) Zur elektronentheorie der metalle, II. Teil, Galvanomagnetische undthermomagnetische effecte, Annalen der Physik 308:369–402

Ewing JA (1882) On effects of retentiveness in the magnetization if iron and steel. Proc Roy Soc (London) 34:39; (1885) Experimental researches in magnetism. Philos Trans Roy Soc (London) 176:523–640; (1890) Contributions to the molecular theory of induced magnetism. Proc Roy Soc 48:342–358; (1892) Magnetic induction in iron and othermetals. The Electrician Printing and Publishing Company, London

Fu Z, Chen X, Li Z, Hu T, Zhang L, Lu P, Zhang S, Wang G, Dong X, Xu F (2020) Unveiling the ferroelectric nature of $PbZrO_3$-based antiferroelectric materials. Nat Commun 11:3809

Ginzburg VL, Landau L (1950) On the theory of superconductivity. JEPT (USSR) 20:1064–1082; (1950) Zh Eksp Teor Fiz 20:1064–1082

Gregg J (1998) The Ehrenfest classification of phase transitions: introduction and evolution. Arch Hist Exact Sci 53:51–81

Kittel C (1951) Theory of antiferroelectric crystals. Phys Rev 82:729–732

Kobeko P, Kurchatov I (1930) Z Phys 66:192

Kramers HA (1927) La diffusion de la lumière par les atomes. Atti Cong Inter Fisici (Trans Volta Centenary Congr) Como 2:545–557

Kronig RL (1926) On the theory of the dispersion of X-rays. J Opt Soc Am 12:547–557

Langevin P (1920) Procédé et appareils d'émission et de réception des ondes élastiquessous-marines à l'aide des propriétés piézoélectriques duquartz. Brevet d'Invention No. 505, 703; (1918) Demandé le 17; (1920) Délivré le

Lin LF, Zhang Y, Moreo A, Dagotto E, Dong S (2019) Frustrated dipole order induces noncollinear proper ferrielectricity in two dimensions. Phys Rev Lett 123:067601

Linnaeus C (1747) Flora Zeylanica: Sistens Plantas Indicas Zeylonae Insulae [The Flora of Ceylon: consisting of Indian plants of the Island of Ceylon]. Laurentii Salvii, Stockholm, Sweden

Lippmann G (1881) Principe de la conservation de l'electricite. Annales de chimie et de physique (in French) 24:145

Lorentz HA (1878) Ueber die theorie der reflexion und refraction des lichtes. Dritte mittheilung. Z Math Phys 23:197–210

Lorentz HA (1905) The motion of electrons in metallic bodies, KNAW. Proceedings 7:438–453

Lorentz HA (1909) The theory of electrons. Dover Publications, New York, p 145; (1909) The theory of electrons and its applications to the phenomena of light and radiant heat, vol Bd. XXIX, New York; B.G. Teubner, Leipzig

Lorenz L (1880) Ueber die Reflexion des Lichts an der Grenzfläche zweier isotropen durchsichtigen Mittel. Annalen der Physik 111:460–473; (1880) Über die refractionsconstante. Ann Phys 11:70–103

Lyddane RH, Sachs RG, Teller E (1941) On the polar vibrations of alkal halides. Phys Rev 59:673

Maniwa H, Sawaguchi E, Hoshino S (1951) Antiferroelectric structure of lead zirconate. Phys Rev 83:1078–1078

Matthias BT, von Hippel A (1948) Domain structure and dielectric response of barium titanate single crystals. Phys Rev 73:1378; Forsbergh PW (1949) Domain structures and phase transitions in barium titanate. Phys Rev 76:1187; Blattner H, Kanzig WJ, Merz WJ, Sutter H (1948) Die domanen struktur von $BaTiO_3$—kristallen. Helv Phys Acta 21:207

Maxwell JC (1873) A treatise on electricity and magnetism, vol 2. Clarendon Press, Oxford

Merz WJ (1954) Domain formation and domain wall motions in ferroelectric $BaTiO_3$ single crystals. Phys Rev 95:690

Nicholson AM, Piezophony US. Patent 1, 495, 429. Filed on April 10, 1918 and patented on May 27, 1924; Nicholson AM. Generating and transmitting electric currents, U.S. Patent 2, 212, 845. Filed on April 10, 1918 and patented on August 27, 1940

Ogawa T, Waku S (discovered 1944); Ogawa T (1947) Busseiron. Kenkyu 6:1

Onsagar L (1936) Electric moments of molecules in liquids. J Am Chem Soc 58:1486

Osborn JA (1945) Demagnetizing factors of the general ellipsoid. Phys Rev 67:351–357

Schmidt JG (1707) Curiöse speculationes bey schalflosen nächten [Curious speculations during sleepless nights], Chemnitz and Leipzig, Conrad Stössen pp 269–270; An English translation of the relevant passage appears in Lang, S. B. (1974) Sourcebook of pyroelectricity vol 2. Gordon and Breach, New York, p 96

Schrödinger E (1912) Studien üder kinetik der dielektrika, dam schmelzpunkt, pyro-und piezoelekrizität, Akademie der Wissenschaften in Wien: Mathem.-Naturw. Klasse; Bd. CXXI, Abt. Iia

Shirane G, Sawaguchi E, Takagi Y (1951) Dielectric properties of lead zirconate. Phys Rev 84:476–481

Slater JC (1941) Theory of the transition of KH_2PO_4. J Chem Phys 9:16–33

Sommerfeld A (1928) Zur elektronentheorie der metalle auf grund der fermischen statistic. Zeitschrift für Physik 47:1–32; Sommerfeld A, Bethe H (1933) Elektronentheorie der metalle. Springer, Berlin Heidelberg

Stoner EC (1945) The demagnetizing factors for ellipsoids. Phil Mag 36:803–821

Thomson W (1878) On the thermoelastic, thermomagnetic and pyroelectric properties of matter. Philos Mag 5:4–26

Valasek J (1920) Piezoelectric and allied phenomena in Rochelle salt. Phys Rev 15:537–538; (1921) Piezoelectric and alliedphenomena in Rochelle salt. Phys Rev 17:475–480; (1922) Piezoelectric activity of Rochelle salt under various conditions. Phys Rev 19:478

Voigt W (1897) Versuch zur bestimmung des wahren specifischen electrischen moments eines turmalins (Experiment to determine the true specific electric moment of a tourmaline). Ann Phys 60:368–375

Wainer E, Solomon AN (1943) Titanium alloy manufacturing co. Report No. 8 (September 1942) and No. 9

Wul B, Goldman JM (1945) CR Acad Sci URSS 46:139

Yashima M, Matsuyama S (2020) Origin of ferrielectricity and visible-light photocatalytic activity of silver niobate $AgNbO_3$. J Phys Chem C 116:24902–24906

Zhang Y, Lin LF, Moreo A, Dong S, Dagotto E (2020) Iron telluride ladder compounds: predicting the structural and magnetic properties of $BaFe_2Te_3$. Phys Rev B 101:144417

Chapter 14
Lattice Vibrations in Crystalline Solids

It has been described in Chap. 8 that atoms in a crystalline solid start vibrating about their mean positions at finite temperatures. As the atoms are bound with each other, their vibrations produce a wave (sound wave) propagating through the lattice of a solid: a phenomenon called lattice vibrations. To begin with, the propagation of sound waves had been studied using mechanical models. The first reference of mechanical models was found in Newton's "*Philosophiae Naturalis Principia Mathematica*" commonly known as "*Principia*" published in 1687. In the mechanical model, the particles of a medium are interconnected with harmonic mechanical springs. Sir Isaac Newton (1643–1727) gave the formula for the velocity of sound waves in a medium under the isothermal conditions. But actually, sound waves (crests and troughs) propagate in the medium under adiabatic conditions. In 1816, Pierre-Simon Laplace (1749–1827) modified the Newton's formula for the velocity of sound under adiabatic conditions. Leon Brillouin (1889–1969) in his book "*Wave Propagation in Periodic Structures*", published in 1953, explained that such a model had been used in studying the propagation and properties of sound waves.

Sir William Rowan Hamilton (1805–1865) did a lot of work on the propagation of vibrations in a dispersive medium particularly the way light propagates in a crystal. He was aware of the concept of coupled harmonic oscillators and the distinction between phase and group velocities. Hamilton worked on the problem of one-dimensional lattice including first and second nearest neighbor interactions and on the square lattice including only first nearest neighbor interactions. He published only a small part of his work during his life time and the major part of the work remained unpublished which was in the form of notebooks and therefore it did not get much attention. William Thomson (1824–1907) or Lord Kelvin, in his lectures 'Popular Lectures and Addresses' delivered in 1891 and 'Baltimore Lectures' delivered in 1904 discussed the propagation of waves in lattices. He studied the lattices of particles with different masses particularly the diatomic lattices and obtained the forbidden gap in the dispersion relations at the Brillouin zone boundary. Mathematical frame work in terms of normal coordinates, diagonalization of quadratic form

J. S. Galsin, *History of Solid State Physics*,
https://doi.org/10.1007/978-981-95-0504-3_14

and relation of normal coordinate transformation to the positive definite quadratic forms, was developed for the study of lattice vibrations.

In 1912 and 1913, Max Born (1882–1970) and Theodore von Karman (1881–1963) used matrices in their work on lattice theory. They developed a rigorous mathematical formulation for lattice dynamics in crystalline solids which still forms the basis of majority of activities in this field. They applied their theory to one-dimensional lattices considering only the first nearest neighbor interactions and to a simple cubic lattice considering first and second nearest neighbor interactions and derived the dispersion relations. The phonon dispersion curve for a one-dimensional monatomic lattice is shown in Fig. 14.1, which is nothing but a sine curve. They showed how the force constants could be related to the experimentally observed elastic constants. Figure 14.2 shows the phonon dispersion curves for a diatomic one-dimensional lattice. It is evident that in the diatomic one-dimensional lattice, in addition to acoustic phonons, one obtains optical phonons (phonons having frequency in the optical region). The dispersion curves exhibit forbidden gap at the Brillouin zone boundary as obtained by William Thomson. In the same year of 1912, Peter Debye (1884–1966) gave theory of specific heat in which he used a simplified frequency distribution in the vibrations of lattice atoms behaving as quantum oscillators (see Fig. 8.1). However, due to the simplicity of the Debye model, the Born-von Karman theory remained eclipsed for about 20 years approximately. In 1923, Ivar Waller (1898–1991) introduced a factor called *Debye–Waller factor* in his Ph.D. work and showed that the atomic vibrations had significant effect on the intensity of Bragg reflection of X-rays. Actually in 1913, Debye pointed out the existence of Debye–Waller factor and calculated it. In 1932, Igor Tamm (1895–1971) introduced the concept of *phonon* as the quantum of lattice waves in analogy with photons in case of light: phonon is a Greek letter. In 1954, Max Born and Kun Huang published a book on "*Dynamical Theory of Crystal Lattices*" in which they described the lattice dynamics from the fundamentals. Historically the interest in studying lattice vibrations arose with a view to understand thermodynamical properties of solids but later shifted in studying the nature of interatomic forces because atomic vibrations depend on the nature and strength of interactions between atoms: a central quantity in studying the electronic properties of solids.

14.1 Experimental Studies

Mainly two methods are used for studying the phonon dispersion relations in crystalline solids.

1. Thermal Diffused Scattering of X-rays.
2. Inelastic Neutron Scattering Method.

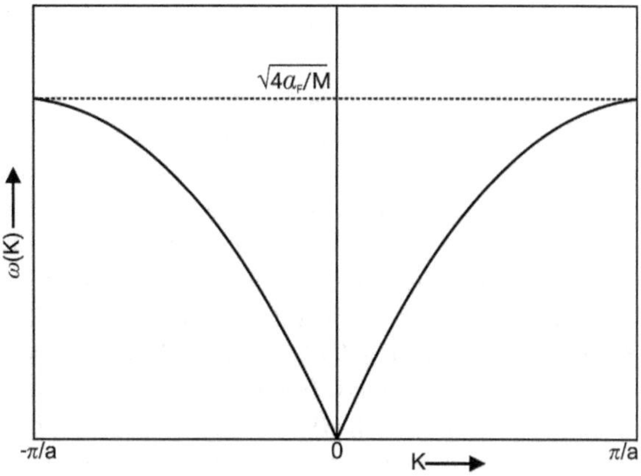

Fig. 14.1 Phonon dispersion curve for a one-dimensional monatomic lattice, in the first Brillouin zone, considering only the first nearest neighbour interactions [the phonon frequency is given by the expression $\omega(\mathrm{K}) = \sqrt{4\alpha_F/\mathrm{M}}|\sin(\mathrm{Ka}/2)|$]. M is the mass of an atom and α_F is the force constant between the nearest neighbour atoms and K is the wave vector

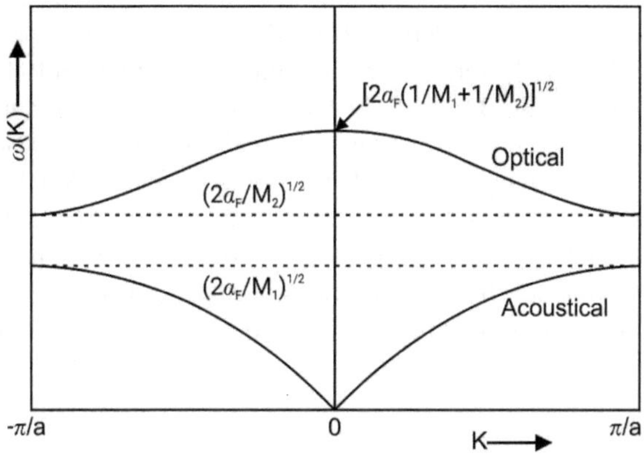

Fig. 14.2 Phonon dispersion curves for a one-dimensional diatomic lattice, in the first Brillouin zone, considering only the first nearest neighbour interactions [the phonon frequency is given by the expression $\omega^2 = \alpha_F\left(\frac{1}{M_1} + \frac{1}{M_2}\right) \pm \alpha_F\left[\left(\frac{1}{M_1} + \frac{1}{M_2}\right)^2 - \frac{4}{M_1M_2}\sin^2(\mathrm{Ka})\right]^{1/2}$]. Here α_F is the force constant between the nearest neighbour atoms, K is the wave vector and M_1 and M_2 are the masses of two types of atoms with $M_1 \rangle M_2$

Laval in (1938) and William Houlder Zachariasen (1906–1979) in (1940) established the relation between thermal diffused scattering (TDS) of X-rays and lattice vibrations. In 1942, Max Born developed theoretically the relation between X-ray scattering and crystal dynamics. In 1945, Zachariasen published his most important monograph, "*Theory of X-ray Diffraction in Crystals*". In 1948, Par Ph. Olmer measured experimentally the phonon dispersion relations of Al metal using TDS of X-rays, for the first time, but the cross section for TDS was found to be small. Early TDS measurements of X-rays were performed using X-ray tubes, and the measurements were slow with limited accuracy, therefore measurements were performed only in high symmetry directions. Thereafter lot of improvements were made and the X-ray spectroscopy was established. Using TDS of X-rays only a few studies were carried out on V, InAs and VSi etc. Phonon dispersion relations can also be studied by inelastic X-ray scattering technique.

Thermal neutron scattering had proven to be the most important technique for studying the structure and dynamics of solids and fluids. Thermal neutrons undergo both elastic and inelastic scattering from crystalline solids. The coherent elastic scattering gives information about the structure of crystalline solids. On the other hand, inelastic scattering of thermal neutrons (ISTN) yields information about the lattice vibrations in crystalline solids. In 1936, Hans von Halban (1908–1964) and Peter Preiswerk (1907–1972) and D. P. Mitchell and P. N. Powers demonstrated the phenomenon of neutron diffraction soon after the discovery of neutrons. In the same year of 1936, Felix Bloch (1905–1983) gave theory of magnetic scattering of neutrons and found that it could lead to a beam of polarized neutrons. In 1939, Halpern and Johnson gave detailed account of magnetic scattering of neutrons from solids which allowed us the determination of distribution, direction, order and magnitude of atomic magnetic moments. After completing his Ph.D. on X-ray diffraction studies, Ernest Omar Wollan (1902–1984) started work on neutron diffraction. In 1944, Ernest Omar Wollan and Lyle Benjamin Borst (1912–2002); and Walter Henry Zinn (1906–2000) were able to observe neutron diffraction from single crystals. Zinn also played a very important role in developing Chicago Pile-3 during the World War-II and bringing it to the critical state for the chain reaction. Later in 1946, Clifford G. Shull (1915–2001) joined Wollan's group at Oak Ridge and laid the foundation of neutron diffraction as an important research tool. In inelastic neutron scattering technique, a neutron either creates or annihilates one quantum of vibration (phonon) in a crystal. In the coherent scattering process the following energy and momentum conservation laws must be satisfied

$$\mathrm{E} - \mathrm{E}' = \pm \mathrm{h}\nu(\mathbf{Q}\mathrm{j}) = \pm \hbar\omega(\mathbf{Q}\mathrm{j}) \tag{14.1}$$

$$\mathbf{k} - \mathbf{k}' = \mathbf{Q} \tag{14.2}$$

where $\left(\mathrm{E}, \mathrm{E}'\right)$ and $\left(\mathbf{k}, \mathbf{k}'\right)$ are the (incoming, outgoing) neutron energies and wave vectors, respectively. ν(or ω) is the frequency of a phonon and j gives its polarization. Here one can write

$$\mathbf{Q} = \mathbf{q} + \mathbf{G} \tag{14.3}$$

where $\mathbf{q}$ is a wave vector restricted to the first Brillouin zone, called reduced wave vector, and $\mathbf{G}$ is the reciprocal wave vector.

Ernest Omar Wollan (1902–1984) sitting on the left side and Clifford G. Shull (1915-2001) standing on the right side, CC BY 3.0, URL: https://en.wikipedia.org/wiki/Ernest_O._Wollan#/media/File:Wollan_and_Shull_1949.jpg

Clifford G. Shull (1915–2001), Public Domain, URL: https://en.wikipedia.org/wiki/Clifford_Shull#/media/File:Clifford_Shull_(cropped).jpg

The first method to measure the phonon dispersion curves is time-of flight (TOF) method. In 1935, John Ray Dunning (1907–1975) and coworkers employed the TOF method for the first time for energy-dependent neutron measurements. Initially natural neutron sources were used to obtain a beam of neutrons. After the World War II, reactors were able to produce sufficiently intense beam of neutrons and these were more suitable for the study of elastic and inelastic scattering of neutrons. After the invention of nuclear reactors, the TOF method used pulsed neutron beam sources, but the principle remained the same. In TOF technique, a pulse of monochromatic neutrons from the source is allowed to fall on the sample of a crystalline solid. After interacting with the solid, neutrons either gain or lose energy resulting in their velocity change. The time of flight of neutrons from source to the counter is measured. The TOF spectra are measured by a number of detectors arranged at various angles to the incoming beam. So, in TOF method velocity of neutrons is measured by timing over a known distance and then the lattice vibrations are evaluated from the data so obtained. In the TOF method phonons are measured at random values of wave vector $\mathbf{Q}$ in a symmetry plane. Bertram Neville Brockhouse (1918–2003) decided to add a third axis to the two-axis spectrometer used for studying neutron scattering from solids. In 1955, Brockhouse built and installed the first *triple axis spectrometer* at Chalk River Laboratory in Canada. In the same year, Brockhouse and Stewart reported the first successful phonon spectrum of aluminium metal measured through inelastic neutron scattering. Later, in 1958, Brockhouse gave the constant momentum ($\mathbf{Q}$) method for determining the phonon dispersion curves in crystalline solids in which one can chose the desired value of $\mathbf{Q}$. In the constant $\mathbf{Q}$ method, the desired values of $\mathbf{k}$ and $\mathbf{k}'$ can be set by the geometry of the apparatus and successive measurements can be made in which $\mathbf{Q}$ is held constant while the energy is varied. Therefore, the advantage of this method is that one can study in detail the dispersion relations at that values of $\mathbf{Q}$ at which phonon anomalies are observed in high symmetry directions. Brockhouse himself performed the measurements of phonon dispersion curves on a number of crystalline materials. The triple-axis spectrometry became the most powerful and standard technique and using it a large number of scientists measured phonon dispersion curves in various crystalline materials. In 1965, Yarnell and co-workers measured the phonon frequencies in Al metal and Fig. 14.3 shows these along the [110] direction. Both Shull and Brockhouse shared the 1994 Nobel Prize in Physics for their contribution to neutron scattering in crystalline solids. Till date phonon dispersion curves of a large number of magnetic and nonmagnetic crystalline materials had been measured using the ISTN technique.

The two techniques used for determining the phonon dispersion relations have merits and demerits. The X-ray methods are much suited for small crystals and materials that are available only in polycrystalline forms. On the other hand, the neutron scattering requires single crystals with large sizes (preferably in the range of 1 cm). The results obtained from ISTN technique are most reliable in those crystalline materials in which the coherent scattering dominates the incoherent scattering. In case of crystals with large cross sections for incoherent neutron scattering (for example V metal) the ISTN technique yields puzzling results. The phonon spectra of such materials can be investigated with reasonable accuracy using TDS of X-rays. In the

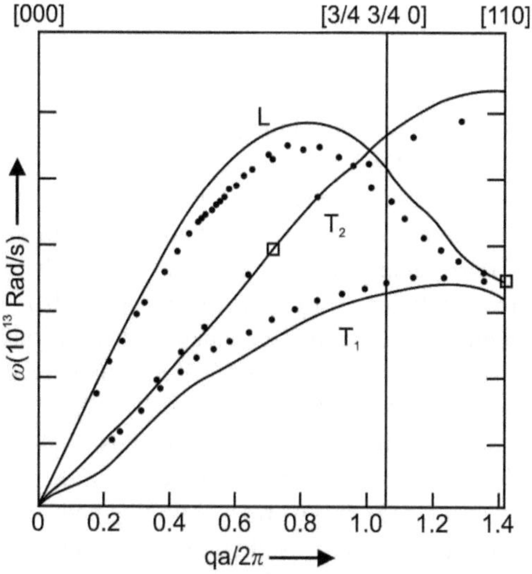

Fig. 14.3 Phonon frequencies of Al metal in the [110] direction (Harrison 1966). L represent the longitudinal branches while T the transverse branches. In the [110] direction there are two transverse branches represented by T_1 and T_2. The continuous lines represent the theoretical results obtained by using the Harrison model potential and dots represent the experimental results due to Yarnell and co-workers (1965). The point at which the pseudopotential was fit to the experiment is indicated by a square

neutron scattering experiments the neutron beam is obtained from a nuclear reactor which may not be exactly monochromatic in nature because there is finite spread in energy of neutrons. The spread in energy in the neutron beam is much larger than that found in an X-ray beam. The TDS of X-rays gives information about the chemical structure while the neutron scattering yields information about both the chemical and magnetic structures of a crystalline solid.

14.2 Theory of Lattice Dynamics

The Born von Karman (BvK) theory is a rigorous mathematical formulation of lattice dynamics of crystalline solids. The concept of electronic energy band structure did not exist when BvK theory was formulated. Therefore, in the initial stages, the experimental dispersion relations were usually analyzed using force constant models based on the BvK theory. In general the force constants are of two types: bond stretching and bond bending force constants. In these models, the force constants are taken as parameters and varied to obtain the best fit of the experimental results. Begbie and Born (1947) proposed general tensor force (GTF) model. They assumed atomic interactions as two-body interactions without any restrictions on the nature of such interactions. GTF has been mainly used to analyze the experimental dispersion curves of crystalline solids. To simplify the calculations Lehman and coworkers (1962) introduced axially symmetric (AS) model in which it is assumed that the restoring forces between the ions have the character of bond stretching and axially symmetric bond bending. They derived the phonon dispersion curves of aluminium and copper using this model. DeWames and coworkers (1965) introduced modified axially symmetric (MAS) model, an improvement over the AS model, in which some restrictions on the force constants were removed. Out of all the force constant models, GTF is the most general and complete model. In the absence of detailed energy band structure calculations for metals the physicist also developed a number of phenomenological models taking into account the effect of conduction electrons in the lattice dynamical study [de Launey model (1953), Bhatia model (1955), Sharma-Joshi model (1963), Krebs model (1964) Lax model (1965), etc.]. These models were used to obtain the phonon frequencies for simple metals (metals with nearly free s-conduction electrons) and the dispersion curves were found to be smoothly varying with momentum (wave vector **Q**). It is worth mentioning here that the force constants so obtained may not give any physical insight into the problem of lattice vibrations on the microscopic scale. Joshi and Rajagopal (1968) have extensively reviewed all these phenomenological models. It has been found that phonon dispersion curves exhibit phonon anomalies especially in the d- and f-band metals and their compounds. The central force models are not able to explain the occurrence of these anomalies. The phonon anomalies can be interpreted in terms of special points on the Fermi surface obtained from the energy band structure calculations of elements. The anomalous behaviour in phonon dispersion curves can occur at the extremal portions of Fermi surface (Kohn anomalies) or at the separation of the nesting portions of the Fermi surface. It is still not very clear whether the phonon anomalies depend mainly on valence, electronic structure, and geometry of the lattice or on all these factors.

The second types of lattice dynamical studies are performed from the first principles. Toya (1958) was a pioneer to calculate the normal mode frequencies of mono-valent metals taking into account the influence of the conduction electrons from the first principles. He formulated the lattice dynamical theory including the electron-phonon interactions given by Bardeen (1937). It was found that phonon frequencies have contributions from the following three interactions present in a solid:

1. The Coulomb ion-ion interactions.
2. The ion-ion exchange-overlap interactions.
3. The ion–electron-ion interactions that is the interaction between ions via electrons.

The first contribution arising from the direct Coulomb interaction between ions is evaluated from the standard method of *Ewald's θ-function transformation* [Ewald (1921); Kellermann (1940)] if the atomic mass and ionicity are known. It is difficult to evaluate the overlap interaction between ion-cores from the first principles but for simplicity it is usually evaluated using Born–Mayer pair-ion potential. The accurate determination of the third contribution is encountered with many difficulties as it involves the treatment of electron–ion and electron–electron interactions. All the lattice dynamical calculations revolve around the evaluation of the third contribution.

The solution of secular equation of dynamical matrix for cubic structures is quite simple. Figure 14.3 shows the theoretical results of phonon frequencies in Al metal obtained by Walter Harrison (1966) using Harrison model potential. It is evident from the figure that phonon dispersion curves exhibit parabolic behaviour near the center of the BZ as found in a linear chain of atoms (see Figs. 14.1 and 14.2) but it departs significantly as the boundary of the BZ is reached. Czachor (1965) worked on the lattice dynamics of metals with hcp structure and factorized the secular equation of dynamical matrix along the symmetry directions to yield the expressions for phonon frequencies for optical and acoustic modes. There exist a variety of methods for calculating electronic contribution to the phonon dispersion relations but their success is limited to only a particular class of materials.

In case of transition metals (TMs) and their compounds and alloys, the conduction electrons possess both the s- and d-characters and d-electrons are quasilocalized. Similar situation exists in rare-earth metals (REMs) where the conduction electrons exhibit s- and f-characters. In these metals, the dielectric function becomes an infinite-dimensional tensor therefore lattice dynamical theory from the first principles becomes quite complex. In 1960s and 1970s, a number of phenomenological models were purposed which involve basically parameter fitting of phonon dispersion curves. Among the most well-known and successful models are *shell models* (simple shell model, generalized shell model, breathing shell model etc.), *bond charge models* (simple bond charge model, dielectric bond charge model, generalized bond charge model etc.), *deformation model* and *valence force model.* Shell model was proposed by Cochran (1959), breathing shell model by Schroder (1966), deformation dipole model by Karo and Hardy (1967) and bond charge model by Johnson (1974). But in these models many parameters were required to fit the dispersion curves and in this process, it was difficult to interpret precisely the atomic forces.

Microscopic theory for lattice dynamics of metals provides a neat picture of the various microscopic processes involved in the dynamical properties. Therefore, microscopic approach is much suitable for the investigation of origin of phonon anomalies both qualitatively and quantitatively. Golibersuch (1964) and Sinha (1968) developed the augmented plane wave theory to investigate the electron–phonon interactions in transition metals. Das (1973) also calculated electron–phonon interactions in noble metal copper using augmented plane wave phase shift analysis scheme introduced by Ziman (1965) and Mueller interpolation scheme (1967) to calculate the mass enhancement and relaxation time for the quasiparticles. But none of these theories has yet been applied to study the lattice dynamical properties of the partially filled d-band metals. Microscopic theory of lattice dynamics is generally formulated in the linear density response function approximation using different representations for the wave functions and the energy bands. The density response function is in turn determined by the dielectric screening function. There exist many theories in literature for the dielectric screening. In 1958, Philippe Nozieres (1932–2022) and David Pines (1924–2018) and in 1959 Henry Ehrenreich (1928–2008) and Morrel H. Cohen (1927–present) gave quantum mechanical treatment of the frequency and wave vector dependent longitudinal component of dielectric function and evaluated it in the free electron approximation. Adler (1962) and Nathen Wiser (1963) extended the formalism to the full dielectric function and applied it to an insulating cubic solid. The formal microscopic theory of dielectric screening was also given in 1959 by Lu Jeu Sham (1938–present) for transition metal compounds and in 1969 by Hayashi and Shimizu (1969) for a transition metal in a single band model for d-electrons and in a two-band model for s- and d-electrons and applied it to calculate the impurity screening and induced spin density in ferromagnetic nickel. In 1970, Robert M. Pick and coworkers gave most general and rigorous microscopic theory for evaluating electronic contribution in terms of density response function. They factorized the dielectric matrix and inverted it completely to study the lattice dynamics of semiconductors. In simple metals the density response function and hence the dielectric function becomes a scalar. A large number of research workers have calculated the phonon dispersion relations of simple metals using this approach and obtained fairly good agreement with the experimental results. In 1971, Maradudin and coworkers presented a comprehensive review of lattice dynamical theory of crystalline solids. Sinha and coworkers (1971) put forward a generalized screening model based on factorization ansatz, which allows the exact inversion of infinite-dimensional dielectric function. In this model, the contributions to phonon frequencies arising from the interactions between monopoles and dipoles at the atomic sites are included. These authors applied the formalism to study he lattice dynamics of germanium and silicon. Hanke and Bilz (1972) and Hanke (1973) used Wannier representation for the wave function to invert the dielectric matrix and studied the lattice dynamics of Pd and Ni metals. In the limiting case, one obtains screened dipolar model and shell model thus providing quantum mechanical justification for the shell model. Singh and coworkers (1975) used the mixed band scheme in inverting the dielectric matrix and developed the lattice dynamical theory for the TMs. But in the estimation of phonon frequencies both Hanke and Singh used noninteracting band scheme, which is not a very good approximation for the energy bands in TMs.

Moriarty (1970, 1972) used first principle pseudopotential method to estimate phonon frequencies in noble metals. The difficulties faced in this method were overcome by replacing pseudopotential by an intuitive fictitious model potential. A large number of calculations of phonon frequencies were performed by researchers using screened model potentials for the ions but the problem is that no unique model potential was available for a particular metal which could explain all the properties of solids. All the above-mentioned methods, although explained the overall behavior of the phonon frequency curves but failed to explain phonon anomalies in metals.

It is well known that the phonon spectra is intimately related to the screening function and particularly to the structure in electronic susceptibility function $\chi^0(\mathbf{q})$. Keeton and Loucks (1968) first pointed out that the peaks in $\chi^0(\mathbf{q})$ are intimately related to the phonon spectrum and these are caused by the nesting of Fermi surface that is by those states having nearly the same energy but different occupancies. The evaluation of $\chi^0(\mathbf{q})$ requires the accurate determination of electronic energy bands and wave functions at a large number of points in the Brillouin zone. Evenson and Liu (1968) calculated $\chi^0(\mathbf{q})$, in the rigid atom model, for the first time using the relativistic APW energy band structure. They further pointed out that, in superconducting TMs, the positions of the peaks in $\chi^0(\mathbf{q})$ arise due to the nesting of Fermi surface and these are reflected as dips in the phonon dispersion relations. Later Evenson and coworkers (1969) calculated $\chi^0(\mathbf{q})$ for a number of metals and found that those bands contribute most to $\chi^0(\mathbf{q})$ which cross the Fermi energy (surface). In an antiferromagnetic metal like Cr, the peak is related to the wave vector of spin density wave (SDW). Liu and coworkers (1971) calculated $\chi^0(\mathbf{q})$ in REMs like Sc, Y, ... and found that the positions of the peaks may be correlated with magnetic ordering wave vector in REMs and their dilute alloys. Gupta and Freeman (1976) calculated from ab initio $\chi^0(\mathbf{q})$ in transition metal compounds (TMCs) such as NbC, TaC, ZrC and HfC and related the positions of the peaks to the phonon dispersion curves. Afterwards, a number of physicists performed the calculation of $\chi^0(\mathbf{q})$ for a number of metals and their compounds with a view to correlate them to their phonon spectra.

As the ions move about their equilibrium positions their charge density gets distorted. The charge density distortion (CDD) gives rise to various types of charge distributions such as monopoles, dipoles and quadrupoles etc. In TMs and REMs, there is charge on the ion below the d- or f-band that is moving rigidly with the nucleus but the quasilocalized d- or f-electron charge suffers distortion. Therefore, there are two contributions to the force constants: first is the direct force constant contribution between rigidly moving parts of the ions and second is the indirect force constant contribution from CDD and latter is responsible for long range forces in the d- and f- band materials. Allen (1972) proposed the *CDD model* for studying lattice dynamics of TMs. This model includes quadrupole interactions, CDD-displacement interactions and CDD-CDD interactions. The CDD theory is applied to study lattice dynamics of Nb metal and it reproduces anomalies in the transverse branches which are associated with the quadrupole distortion. Werner Weber (1945–2014) and coworkers (1972) interpreted the observed anomalies in the phonon spectra of some TMCs as resonances in the electronic polarization and their results strongly support the covalent model. They discussed the possible connections of these anomalies to lattice instabilities and superconductivity. In 1973, Werner Weber (1945–2014) formulated double shell model and applied it to study phonon dispersion curves of TMCs and found that phonon anomalies arise due to the resonance-like increase in the polarizability due to the d-band in TMs. Phonon anomalies are manifested as spin density wave (SDW) and *charge density wave* (CDW) in structural phase transitions. The microscopic theories of Sinha and Harmon (1975, 1976) and of Hanke and coworkers (1976) were also applied to evaluate phonon dispersion curves of transition metal carbides and the anomalous structure was found to arise from the peaks in electron–phonon coupling.

Sinha and coworkers (1977) proposed *charge fluctuation model* in which electron density was divided into two parts: One part of charge density was expressed as a sum of rigid ion charges inside muffin-tin volume and second was deformable charge outside muffin-tin sphere. In the general charge fluctuation theory, the expressions are quite complicated and not physically transparent. They calculated phonon dispersion curves for Nb metal and NbZr alloy considering only the charge fluctuations of monopoles and were able to reproduce phonon anomalies in Nb metal and NbZr alloys. Later the charge fluctuation theory was applied to superconducting TMCs and the position and dips of the anomalies were accounted for. It was found that the size and the depth (strength) of these anomalies are very sensitive to the density of electron states at the Fermi energy. Wakabayashi (1977) gave a phenomenological charge fluctuation model based on the microscopic model of Sinha and Harmon which included monopolar charge fluctuations. It yielded a much better fit to the observed anomalous structure in the longitudinal branches for Nb but could not explain observed anomalies in the transverse branches. Allen (1977) developed a similar model for Nb incorporating general symmetry. He found that monopole-type charge fluctuations were important in explaining anomalies in the longitudinal branches, in agreement with Wakabayashi, but the quadrupole charge fluctuations determined anomalous upward curvature behavior in the transverse branches.

Varma and Weber (1977, 1979) calculated from the first principles phonon frequencies of Nb and NbMo alloys using the *nonorthogonal tight binding scheme.* Varma et al. (1979) calculated the electron–phonon matrix elements in nonorthogonal tight binding scheme and applied it for the calculation of phonon spectra of TMs and TMCs. All the observed anomalies were correctly reproduced both in Nb and Mo and their alloys. Almost all the anomalies were found to originate from the scattering of d-like states (d-d scattering) close to the Fermi surface. The nonorthogonal tight binding scheme had been applied further to other superconducting materials and phonon anomalies were explained. Varma and Weber have shown that the conventional and nonorthogonal tight binding methods are formally equivalent in the random phase approximation (RPA).

It is worthwhile to make a comparison of different microscopic approaches regarding the origin of phonon anomalies. The first principle pseudopotential calculations showed that phonon anomalies arise due to s-d hybridization while the model potential calculations showed that phonon softening was produced by the nonlocal character of model potentials in the TMs. Weber's double-shell model calculation showed that phonon anomalies arise due to the resonance-like increase in polarizability function in TMs. Sinha and Harmon explained that phonon anomalies arise due to polarization and the cancellation of Coulomb potential due to charge fluctuations. Hanke and Bilz pointed out that some anomalies might be related to local field effects in d-band metals. At the same time, Hanke attributed the anomalies to the correlation effects beyond RPA in the d-d interaction in the Wannier representation. From the above discussion it is evident that different microscopic theories attribute phonon anomalies to different physical processes going on in d-band metals which create some doubts. It is believed that superconducting transition temperature is related to the phonon softening and this aspect of lattice vibrations will be discussed in the next Chap. 15 on superconductivity.

14.3 Surface Phonons

Real crystalline solids are finite in size with well-defined surface. At the surface, there is a transition from the bulk crystal to vacuum. Therefore, in a finite solid there is periodicity inside it which ends at the surface. Due to this transition, the atomic arrangement in the surface gets disturbed and the surface is reconstructed. The breakage in periodicity and surface reconstruction may change the force constants and lattice parameter of the solid which in turn changes the separation of a few outermost atomic layers. In other words, a few atomic layers below the surface also get strained. At finite temperatures, atoms in the surface vibrate about their equilibrium positions and the quantum of vibration is the surface phonon. Therefore, an elastic wave in a crystalline solid with finite surface propagates just like that in the bulk material but on reaching the surface it decays exponentially due to discontinuity in the periodicity. Therefore, phonon dispersion relations in finite crystalline solids can be studied using the lattice dynamical theory of bulk crystalline solids by incorporating the change, which allows the decay of elastic waves at the surface. The knowledge of surface phonon dispersion relations gives information about the surface relaxation, about the distance between the surface and an adsorbed atom and the quantity and type of defects on the surface.

The situation in semiconductors is more interesting as there appear dangling bonds which contain unpaired electrons and in the surface reconstruction the electrons in the dangling bonds get paired. In semiconductors, phonons interact with electrons thereby changing their optical and electrical properties. These effects are more important in two-dimensional electron systems and quantum dots. With the decrease in size of a crystal the surface to volume ratio increases. Therefore in a quantum dot the surface phonon frequencies increase which can couple electrons and affect their properties.

The experimental study of surface phonons is performed using two methods. First is the *electron energy loss* (EEL) method which is based on the fact that the electron energy decreases upon interaction with matter. The origin of this method lies in the Davisson and Germer experiment done in 1927 which demonstrated for the first time the wave nature of electrons. Since low energy electrons interact with only a few atomic layers below the surface, the energy loss due to surface phonon scattering is in the range 10^{-3} eV to 1 eV. The energy loss of electrons is measured as a function of scattering angle which yields information about the surface phonon dispersion relations. The EEL method is conceptually the same as Raman scattering or inelastic neutron scattering. Aldag and Stern (1965) performed the first theoretical study of surface vibrations using the inelastic scattering of low energy electrons from the surface. Li and coworkers (1980) developed detailed theory of electron scattering from surfaces. EEL spectroscopy acquired high resolution for an accurate determination of surface phonon curves mostly on the conducting materials and developed into a powerful experimental technique. Szeftel and coworkers (1983) succeeded for the first time in measuring phonon dispersion curves in Ni(100) surface using EEL spectroscopy. At approximately the same time, Lehwald and coworkers (1983) measured surface phonon dispersion curves on metallic surface Ni(100). Ibach and coworkers (1996) have given a detailed account of EEL spectroscopy of surface vibrations. Wallis and Maradudin (1966) developed a theory in the harmonic approximation for the effect of surface on the thermal diffuse scattering of X-rays or electrons from a crystal lattice. They also gave an alternate treatment based on thermodynamic Green's functions and applied it to monatomic simple lattice considering interactions up to the second nearest neighbors.

The second method for studying the surface phonons is *Helium (He) atom scattering* method. In 1930, Immanuel Estermann (1900–1973) and Otto Stern (1888–1969) performed the first experiment on He atom diffraction. Cabrera and coworkers (1969) and Manson and Celli (1971) formulated theory for inelastic atom scattering and pointed out that this method could be used for measuring surface phonon dispersion curves. Immediately after this several experimental groups were attracted to perform experiments for the measurement of surface phonon dispersion curves. Fisher and Bledsoe (1972) were the first to make precise measurements of the energy loss in He atom scattering from LiF(001) surface in the TOF method. At that time, there was a large spread in energy in the available He atom beam as a result of which the resolution of the results was poor. In 1977, Brusdeylins and coworkers produced helium nozzle beams with very high-speed ratios. Nozzle beam sources of He atoms were improved to a great extent up to about 1980 with energy spread less than 1 meV. The precise determination of surface phonon dispersion relations became possible with the use of these sources as one could measure very small changes in energy resulting from the inelastic scattering of He atoms from the vibrating lattice of a solid surface. Brusdeylins and coworkers (1981) performed the first successful measurement of surface phonon dispersion curves up to the boundary of first surface Brillouin zone (SBZ). This study renewed the interest in the He atom scattering applications, particularly the study of surface phonons. With the success of these initial experiments, many groups entered the field to study dispersion curves of surface phonons in different materials. There are two types of studies in the He scattering method, one is the TOF measurements and second involves the measurement of momentum of scattered He atoms by a LiF grating monochromator. The TOF measurements reveal up to sixth sharp maxima which can be attributed to the annihilation or creation of single surface phonons. From the measured energy loss and the scattering angle, the frequency and wave vector of the interacting surface phonons are determined. The phonon dispersion relations are evaluated in SBZ. For example, in the (100) surface of fcc crystal structure, the phonon dispersion relations are calculated along the ΓX and ΓM symmetry directions. The phonon frequencies $\omega(\mathbf{q})$ are measured as a function of wave vector $\mathbf{q}$ along the surface. It has been found that surface vibrations are localized at the outermost atomic layers. Using this technique, the Rayleigh mode dispersion curves of LiF, NaF, NaCl, and KCl etc. have been measured in the SBZ.

The two methods of measuring surface phonon dispersion relations have their own merits and demerits. He atoms are neutral, therefore, are non-destructive and insensitive to surface charges. He atom method has higher resolution of 0.5–1 meV than the EEL method having resolution of about 7 eV. Second, He atom scattering method is available for small energy differences (≤ 30 meV) while EEL method can be used for energy difference up to 500 meV. He atoms undergo single scattering from the surface layer only while electrons can go deep up to a few atomic layers (4 to 5 atomic layers) and can undergo more than one scattering processes. Lastly one can obtain He atom beam with higher flux than the electron beam but the detection of electrons is much easier than that of the He atoms. These two methods are nicely reviewed by Toennies (1991).

The theoretical calculations of surface phonons can be classified into two categories: *slab method* and *Green function method*. The usual method of studying lattice dynamics of surfaces is the slab method in which the crystalline solid is in the form of a thin slab with parallel faces containing a finite number of parallel atomic planes with their size extending to infinity. The infinite dimensions of the atomic planes allow us to apply periodic boundary conditions in the direction parallel to the surface thus simplifying the problem. Clark and coworkers (1965) introduced the slab method and calculated the mean-square displacements of atoms in an fcc crystal, in the high-temperature limit, as a function of distance from a free surface using the harmonic approximation. Allen and de Wette (1969) applied the slab method to noble gas solids and calculated various dynamical surface properties in the quasiharmonic approximation. They calculated surface specific heat for the (100) surface and found it to be positive and varying as T^2 at low temperatures. Allen and co-workers (1971) developed a general and systematic theory of slab method for the description of surface vibrations. It is applicable to unreconstructed clean surfaces, reconstructed surfaces and surfaces with adsorbed impurity particles, etc., provided the two-dimensional periodicity is preserved. The slab method describes the vibrational properties of the entire system that is phonons inside the slab (*bulk phonons*) and phonons on the surface of the slab (*surface phonons*). After these initial attempts, a number of scientists applied the slab method to study surface phonons in different materials. Chen and coworkers (1972) used the rigid-ion model in studying surface dynamics as the first order approach due to its simplicity. Later Chen and coworkers (1977) used shell model in the study of surface dynamics and gave excellent results in case of ionic crystals within the slab method. Various shell models like breathing shell model, quadrupole shell model, Weber's double shell model etc. have also been applied to study surface phonons. The multipole expansion method of Philip B. Allen proposed in (1977) has also been applied to study the surface phonon anomalies in noble metals. The multipole expansion method has the special merit of providing the phonon-induced surface charge density oscillations, which are directly related to the inelastic He atom scattering intensities. Second is the Green function method which is usually applied in conjunction with the density functional theory. This method gives detailed information about the lattice defects on the surface, vibrational response functions, and phonon density of states for semi-infinite crystals and particularly suited for the calculation of He atom scattering. Benedek (1973, 1976) have applied the Green function method to study the surface lattice dynamics of ionic crystals. Miglio and Benedek (1987) have given a detailed account of the study of surface phonons and surface structure by Green function method. Richard F. Wallis (1924–2015) have summarized the theoretical developments in studying the surface phonons while, in 2018 Giorgio Benedek & Peter J. Toennies have reviewed both theory and experimental studies, using helium atom scattering method, performed in studying surface phonons. Last but not the least it is remarked that almost all the methods described earlier in this chapter such as shell models, multipole expansion theory, density functional theory, Varma-Weber tight binding theory etc. have been applied to study surface phonons in ionic crystals and metals.

Bibliography

Adler SL (1962) Quantum theory of the dielectric constant in real solids. Phys Rev 126:413

Aldag J, Stern RM (1965) Surface thermal diffuse scattering for tungsten. Phys Rev Lett 14:857–860

Allen PB (1972) Neutron spectroscopy of superconductors. Phys Rev B 6:2577

Allen PB (1977) Charge density distortions and lattice dynamics: a general theory and application to Nb. Phys Rev B 15:5139

Allen RE, de Wette FW (1969) Calculation of dynamical surface properties of noble-gas crystals. I. The quasiharmonic approximation. Phys Rev 179:873

Allen RE, Alldredge GP, de Wette FW (1971) Studied of vibrational surface modes I. General formulation. Phys. Rev. B4:1648; (1971) Studied of vibrational surface modes II. Monatomic fcc crystals. Phys Rev B 4:1661

Allen PB (1980) In: Horton GK, Maradudin AA (eds) Dynamical properties of solids, vol 3, pp 95–196

Bardeen J (1937) Conductivity of monovalent metals. Phys Rev 52:688

Begbie GH, Born M (1947) Thermal scattering of X-rays by crystals—I. Dynamical foundation. Proc Roy Soc (London) A 188:179–188

Benedek G, Toennies JP (2018) The physics of surface phonons: theory and experimental studies with helium atom scattering. In: Atomic scale dynamics at surfaces (Springer series in surface sciences, vol 63). Springer, Berlin, pp 57–101

Benedek G (1973) Surface lattice dynamics of ionic crystals by the Green's function method. Phys Status Solidi 58:661–671; (1976) The Green's function approach to the surface lattice dynamics of ionic crystals. Surf Sci 61:603–634

Bhatia AB (1955) Vibration spectra and specific heats of cubic metals. I. Theory and application to sodium. Phys Rev 97:363

Bilz H, Gliss B, Hanke W (1974) In: Horton GK, Maradudin AA (eds) Dynamical properties of solids (North-Holland, Amsterdam), vol I, p 343

Bloch F (1936) On the magnetic scattering of neutrons. Phys Rev 50:259

Born M (1942) Theoretical investigations on the relation between crystal dynamics and **X-ray scattering**. Rep Prog Phys 9:294

Born M, Huang K (1954) Dynamical theory of crystal lattices. Oxford, London

Born M, Von Karman T (1912) Uber schwingungen im raumgittern. Physikalische Zeitschrift 13:297–309; (1913) Theory of specific heat. Physikalische Zeitschrift 14:15–71

Brillouin L (1953) Wave propagation in periodic structures: electric filters and crystal lattices, 2nd edn. Dover Publication, New York

Brockhouse BN (1955) Energy distribution of neutrons scattered by paramagnetic substances. Phys Rev 99:601

Brockhouse BN, Stewart AT (1955) Scattering of neutrons by phonons in an aluminum single crystal. Phys Rev 100:756; (1958) Normal modes of aluminum by neutron spectrometry. Rev Mod Phys 30:236

Brusdeylins G, Doak RB, Toennies JP (1981) Measurement of the dispersion relation for Rayleigh surface phonons of LiF(001) by inelastic scattering of He atoms. Phys Rev Lett 46:437

Cabrera N, Celli V, Manson R (1969) Theory of surface scattering and detection of surface phonons. Phys Rev Lett 22:346

Chen TS, Alldredge GP, deWette FW (1972) Surface phonons for the (100) face of lithium fluoride. Phys Lett A 40:401–402

Chen TS, deWette FW, Alldredge GP (1977) Studies of vibrational surface modes in ionic crystals: detailed shell model studies for the unrelaxed (001) face of seven crystals having the rocksalt structure. Phys Rev B 15:1167

Clark BC, Herman R, Wallis RF (1965) Theoretical mean square displacements of surface atoms in face-centred cubic lattices with applications to nickel. Phys Rev A 139:860

Cochran W (1959) Theory of the lattice vibrations of germanium. Proc Roy Soc (London) A253:260–276; (1959) Lattice dynamics of alkali halides. Phil Mag 4:1082–1086; (1963) Lattice vibrations. Rep Prog Phys 26:1; (1971) Lattice dynamics of ionic and covalent crystals. Crit Rev Solid State Sci 2:1–44

Cowley RL, Cochran W, Brockhouse BN, Woods ADB (1963) Lattice dynamics of alkali halide crystals. II. Experimental studies of KBr and NaI. Phys Rev 131:1025; (1963) Lattice dynamics of alkali halide crystals. III. Theoretical. Phys Rev 131:1030

Czachor A (1965) In: Inelastic scattering of neutrons, vol 1. International Atomic Energy Agency, Vienna, pp 181–204

Das SG (1973) Interpolation-scheme calculation of electron-phonon interaction in noble and transition metals: copper. Phys Rev B 7:2238

Davisson C, Germer LH (1927) The scattering of electrons by a single crystal of nickel. Nature 119:558; (1927) Diffraction of electrons by a single crystal of nickel. Phys Rev 30:705

De Launey J (1953) Lattice dynamics of body-centered and face-centered cubic metallic Elements-I. J Chem Phys 21:1975–1986; (1954) Lattice dynamics of body-centered and face-centered cubic metallic elements-II. J Chem Phys 22:1676–1677

Debye P (1912) Zur theorie der spezifischen waerme. Annalen der Physik (in German) 39:789–839

Debye P (1913) Interferenz von Röntgenstrahlen und wärmebewegung. Ann Phys 348:49

DeWames RE, Wolfram T, Lehman GW (1965) Lattice dynamics, heat capacities and Debye-Waller factors for Be and Zn using a modified axially symmetric model. Phys Rev 138A:717

Dolling G (1963) In inelastic neutron scattering in solids and liquids (IAEA, Vienna), vol II, p 37

Dunning JR, Pegram GB, Fink GA, Mitchell DP, Segre E (1935) Velocity of slow neutrons by mechanical velocity selector. Phys Rev 48:704

Dynes RC, Varma CM (1976) Superconductivity and electronic density of states. J Phys F 6:L215

Ehrenreich HE, Cohen MH (1959) Self-consistent field approach to the many electron problem. Phys Rev 115:786

Estermann I, Stern O (1930) Beugung von molekularstrahlen. Z Phys 61:95–125

Evenson WE, Fleming GS, Liu SH (1969) Generalized susceptibilities and magnetic properties of some bcc transition metals. Phys Rev 178:930

Evenson WE, Liu SH (1968) Generalized susceptibilities and magnetic ordering of heavy rare earths. Phys Rev Letts 21:432; (1969) Theory of magnetic ordering in the heavy rare earths. Phys Rev 178:783

Ewald PP (1921) Die berechnung optischer und elektrostatischer gitterpotentiale. Ann Phys 64:253

Fisher SS, Bledsoe JR (1972) One-phonon transitions observed in the scattering of He atoms from a LiF(001) surface. J Vac Sci Technol 9:814–818; (1974) Scattering of helium nozzle beams from LiF(001) and NaCl(001) crystal surfaces: I. Elastic and inelastic transitions. Surf Sci 46:129–156

Golibersuch CG (1964) Electron phonon interaction via the augmented plane wave method. Phys Rev 157:532

Gupta M, Freeman AJ (1976) Direct correlation of phonon anomalies in NbC with Fermi-surface-induced maxima in generalized susceptibilities. Phys Rev Letts 37:364

Halpern O, Johnson MH (1939) On the magnetic scattering of neutrons. Phys Rev 55:898

Hanke W (1978) Dielectric theory of elementary excitations in crystals. Adv Phys 27:278–341

Hanke W, Bilz H (1972) Phonons in metals. Inelastic scattering of neutrons. IAEA, Vienna, pp 3–28

Hanke W, Hafner WJ, Bilz H (1976) Phonon anomalies and superconductivity in transition-metal compounds. Phys Rev Lett 37:1560

Hanke W (1971) In: Nusimovici MA (ed) Proceedings of international conference on phonons. Flammarion, Paris, p 294

Hanke W (1973) Microscopic theory of dielectric screening and lattice dynamics in Wannier representation. I. Theory. Phys Rev B8:4585; (1973) Microscopic theory of dielectric screening and lattice dynamics in Wannier representation. II. Application to transition metals Pd and Ni. Phys Rev B8:4591

Hardy JR (1974) In: Horton GK, Maradudin AA (eds) Dynamical properties of solids, vol I. North-Holland, Amsterdam, p 157

Hayashi E, Shimizu M (1969) Dielectric screening in transition metals. II. Induced spin density in nickel. J Phys Soc Jpn 27:43–48

Ibach H, Balden M, Lehwald S (1996) Recent advances in electron energy loss spectroscopy of surface vibrations. J Chem Soc Faraday Trans 92:4771–4774

Johnson FA (1974) A bond charge model of lattice dynamics. I. Proc Roy Soc (London) 339A:73–83; Johnson FA, Moore K (1974) A bond charge model of lattice dynamics. I I. Proc Roy Soc (London) 339A:84–96

Joshi SK, Rajagopal AK (1968) Solid state physics, vol 22. Academic Press, New York, p 159

Karo AM, Hardy JR (1967) Low temperature second-order Raman spectra of CsCl, CsBr, CsI. Phys Rev 160:702; (1968) Lattice dynamics and second order Raman spectra of CsF. Phys Rev 168:1054; Hardy JR (1974) In: Horton GK, Maradudin AA (eds) Dynamical properties of solids, vol 1. North-Holland, Amsterdam, p 157

Keeton SC, Loucks TL (1968) Electronic structure of rare-earth metal. I. Relativistic augmented plane wave calculations. Phys Rev 168:672

Kellermann EW (1940) Theory of the vibrations of the sodium chloride lattice. Phil Trans Roy Soc (London) 238A:513

Klein BM, Papaconstantopoulos DA (1974) Electron-phonon interaction and superconductivity in transition metals and transition metal carbides. Phys Rev Lett 32:1193

Krebs K (1964) Dynamical model for lattice vibrations in metals. Phys Lett 10:12–13; (1965) Dispersion curves and lattice frequency distribution of metals. Phys Rev 138A:143

Laval J (1938) Crystallography. On the diffusion of X-rays by a crystal. C R Hebd Seances Acad Sci 207:169–170; (1939) Diffusing X-rays with crystals outside the bearing of selective reflection. C R Hebd Seances Acad Sci 208:1512–1514

Lax M (1965) In: Wallis RF (ed) Lattice dynamics. Pergamon Press, Oxford, p 179

Lehman GW, Wolfram T, DeWames RE (1962) Axially symmetric model for lattice dynamics of metals with application to Cu, Al, and ZrH_2. Phys Rev 128:1593

Lehwald S, Szeftel JM, Ibach H, Rahman TS, Mills DL (1983) Phys Rev Lett 50:518

Li CH, Tong SY, Mills DL (1980) Theory of large angle inelastic scattering of electrons by adsorbate vibrations. Phys Rev B 21:3057

Liu SH, Gupta RP, Sinha SK (1971) Generalized susceptibility function for rare earths and thorium and their alloys. Phys Rev B 4:1100

Manson R, Celli V (1971) Inelastic surface scattering of non-penetrating particles. Surf Sci 24:495–514

Maradudin AA, Montroll EW, Weiss GH, Ipatova IP (1971) Solid state physics, Suppl. 3. Academic Press, NewYork

McMillan WL (1968) Transition temperature of strong-coupled superconductors. Phys Rev 167:331

McMurray HL, Solbrig AW Jr, Boyter JK, Noble C (1967) The use of valence force potentials in calculating crystal vibrations. J Phys Chem Solids 28:2359–2368

Miglio L, Benedek G (1987) Study of surface phonons by means of the Green's function method. In: Schommers W, von Blanckenhagen P (eds) Structure and dynamics of surfaces II. Springer, Berlin, pp 35–71

Mitchell DP, Powers PN (1936) Bragg reflection of slow neutrons. Phys Rev 50:486

Moriarty JA (1970) Pseudopotential form factors for copper, silver, and gold. Phys Rev B 1:1363; (1972) Pseudo Green's functions and the pseudopotential theory of d-band metals. Phys Rev B 5:2066; (1972) Total energy of copper, silver, and gold. Phys Rev B 6:1239; (1972) Localized d states for pseudopotential calculations: application to the alkaline-earth metals. Phys Rev B 6:4445

Mueller FM (1967) Combined interpolation scheme for transition and noble metals. Phys Rev 153:659

Musgrave MJP, Pople JA (1962) A general valence force field for diamond. Proc Roy Soc (London) A 268:474–484

Newton I (1687) Philosophiae naturalis principia mathematica, IMPRIMATUR, S. Pepys, Reg Soc Praeses (Londini)

Nilsson G, Rolander S (1973) Lattice dynamics of copper at 80K. Phys Rev B 7:2393

Nozieres P, Pines D (1958) Electron interaction in solids. General formulation. Phys Rev 109:741

Olmer PP (1948) Dispersion des vitesses des ondes acoustiques dans l'aluminium. Acta Cryst 1:57–63

Papaconstantopoulos DA, Boyer LL, Klein BM, Williams AR, Morrouzi VL, Janak JF (1977) Calculations of the superconducting properties of 32 metals with Z≤49. Phys Rev B 15:4221

Pick RM, Cohen MH, Martin RM (1970) Microscopic theory of force constants in the adiabatic approximation. Phys Rev B1:910; (1968) Microscopic theory of force constants in solids. In: Neutron inelastic scattering, vol 1. IAEA, Vienna, pp 119–139

Schroder U (1966) A new model for lattice dynamics (breathing shell model) Solid State Commun 4:347; Nusslein V, Schroder U (1967) Calculations of dispersion curves and specific heat for Li and NaCl using the breathing shell model. Phys Stat Sol 21:309

Sham LJ (1969) Electronic contribution to lattice dynamics in insulating crystals. Phys Rev 188:1431

Sharma PK, Joshi SK (1963) Model for the lattice dynamics of metals and its application to sodium. J Chem Phys 39:2633–2638

Singh J, Singh N, Prakash S (1975a) Non-interacting band model for dielectric screening and local field corrections in bcc transition metals: application to paramagnetic chromium. Phys Rev B 12:3159–3165; (1975b) Phonon frequencies of paramagnetic chromium. Phys Rev B 12:3166–3173

Sinha SK (1968) Electron phonon interactions and phonon dispersion relations using the augmented plane wave method. Phys Rev 169:477

Sinha SK, Harmon BN (1975) Electronically driven instabilities. Phys Rev Lett 35:1515

Sinha SK, Harmon BN (1976) In: Douglass DH (ed) Superconductivity in d- and f-band metals. Springer, Boston, pp 269–296

Sinha SK, Gupta RP, Price DL (1971) Generalized screening model for lattice dynamics. Phys Rev Lett 26:1324; (1974) Dielectric function and phonon spectrum in grey tin. Phys Rev B 9:2564; (1974) Microscopic theory of dielectric screening and lattice dynamics. II. Phonon spectra and effective charges. Phys Rev B 9:2573

Sinha SK (1980) In: Horton GK, Maradudin AA (eds) Dynamical properties of solids, vol 3, pp 1–93

Svensson EC, Brockhouse BN, Rowe JM (1967) Crystal dynamics of copper. Phys Rev 155:619

Szeftel JM, Lehwald S, Ibach H, Rahman TS, Black JE, Mills DL (1983) Dispersion of adsorbate vibrational modes—the c(2×2) oxygen overlayer on Ni(100). Phys Rev Lett 51:268

Toennies JP (1991) Experimental determination of surface phonons by He atom and electron energy loss spectroscopy. Surface phonons. Springer, Berlin Heidelberg

Toya T (1958) Hartree-Fock method for the interaction of lattice vibrations with electrons in a metal. J Res Inst Catal 6:161; (1958) Normal vibrations of sodium. J Res Inst Catal 6:183; (1959) Normal vibrations of potassium, rubidium and cesium. J Res Inst Catal 7:60; (1958) Normal vibrations of copper. Prog Theo Phys (Kyoto) 20:974–976

Varma CM, Weber W (1977) Phonon dispersion in transition metals. Phys Rev Lett 39:1094

Varma CM, Weber W (1979) Phonon dispersion in transition metals. Phys Rev B 19:6142

Varma CM, Blount EI, Vashishta PD, Weber W (1979) Electron-phonon interactions in transition metals. Phys Rev B 19:6130

von Halban H, Preiswerk P (1936) Preuve expérimentale de la diffraction des neutrons. C R Acad Sci 203:73–75

Wakabayashi N (1977) Charge fluctuation model of lattice vibrations. Solid State Comm 23:737–740

Waller I (1923) Zur frage der einwirkung der wärmebewegung auf die interferenz von Röntgenstrahlen. Z Phys A 17:398–408

Wallis RF (1994) Surface phonons: theoretical developments. Surf Sci 299–300:612–627

Wallis RF, Maradudin AA (1966) Theory of surface effects on the thermal diffuse scattering of X rays or electrons from crystal lattices. Phys Rev 148:962

Warren JL, Yarnell JL, Dolling G, Cowley RL (1967) Lattice dynamics of diamond. Phys Rev 158:805

Weber W (1973) Lattice dynamics of transition-metal carbides. Phys Rev B8:5082

Weber W (1974) New bond-charge model for the lattice dynamics of diamond-type semiconductors. Phys Rev Lett 33:371

Weber W, Bilz H, Schroder U (1972) Resonant electronic polarization in the lattice dynamics of transition-metal compounds. Phys Rev Lett 28:600

Wiser N (1963) Dielectric constant with local field effects included. Phys Rev 129:62

Wollan EO, Borst LB (1944) Physics division semi-monthly report for the period ending May 31, 1944, p 4. Metallurgical project report no M-CP-1621, Clinton Laboratories, Oak Ridge, TN, USA

Wollan EO, Borst LB (1945) Physics section III monthly report for the period ending Dec 31, 1944, p 37. Metallurgical project report no M-CP-2222, Clinton Laboratories, Oak Ridge, TN, USA

Zachariasen WH (1940) A theoretical study of the diffuse scattering of X-rays by crystals. Phys Rev 57:597

Zachariasen WH (1945) Theory of X-ray diffraction in crystals. Wiley, New York

Ziman JM (1965) The T-matrix, the K-matrix, d-bands and l-dependent pseudopotentials in the theory of metals. Proc Phys Soc (London) 86:337

Zinn WH (1944) Report for month ending Aug 26, 1944, pp 7–9. Metallurgical project report no CP-2081, Argonne Laboratory, Chicago, IL, USA

Chapter 15
Superconductivity

Louis-Paul Cailletet (1832–1913) and Raoul-Pierre Pictet (1846–1929) worked independently and produced liquid oxygen in 1877. But they used completely different processes in liquefying oxygen: Cailletet liquefied oxygen by the expansion of a compressed gas while Pictet employed the principle of refrigeration. After this scientists started working to liquefy gases to achieve low temperatures. Zygmunt Florenty von Wroblewsky (1845–1888) and Karol Stanislaw Olszewsky (1846–1915) were the first to liquefy, in 1883, oxygen, nitrogen and carbon dioxide from the atmospheric air. In 1884, Olszewski liquefied hydrogen in a dynamic state achieving a record low temperature of 48 K and then in 1885, he liquefied argon. James Dewar (1842–1923), a British chemist and physicist, started work on the liquefaction of gases. In 1891, he designed and built a machine to produce liquid oxygen in industrial quantities. In 1892, he invented vacuum flask, known as Dewar flask, for the storage of liquefied gases and became very famous due to this invention. In 1898, he was able to liquefy hydrogen using the same machine. Dewar further initiated research into electrical resistance at low temperatures. James Dewar and John Ambrose Fleming (1849–1945) predicted that at absolute zero, pure metals would become perfect electromagnetic conductors. Later Dewar believed that there would be some resistance left at absolute zero. Walther Herman Nernst (1864–1941) proposed the third law of thermodynamics and stated that absolute zero was unattainable. Heike Kammerlingh Onnes (1853–1926), a Dutch physicist, set up a laboratory to liquefy Helium (He) and was able to liquefy it in 1908 thereby achieving temperature down to 4.2 K. Further, he achieved temperature near 1.5 K by reducing the pressure of liquid He. Onnes was studying the various properties of elements at low temperatures. In 1911, he found that the electrical resistance of mercury suddenly dropped down to zero at 4.2 K. In other words, one can say that electrical conductance of mercury suddenly became infinitely large at 4.2 K. Onnes repeated the experiment to confirm the sudden drop in the resistance of mercury. He reported that "mercury has passed into a new state"

J. S. Galsin, *History of Solid State Physics*,
https://doi.org/10.1007/978-981-95-0504-3_15

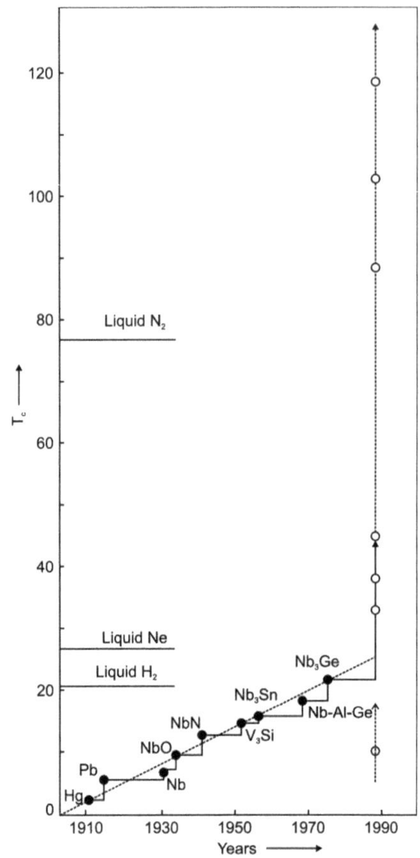

Fig. 15.1 The superconducting transition temperature T_C of various superconductors as a function of years in which they were discovered. The graph shows a sharp increase in the superconducting transition temperature at the right end. [modified from Bednorz and Muller (1988). Perovskite-type oxide: new approach to high T_C superconductivity. Rev Mod Phys 60:585–600]

and called it superconducting state and the phenomenon was named *superconductivity*. The temperature at and below which the material exhibits superconductivity is called superconducting transition temperature T_C. Later Onnes discovered superconductivity in tin (Sn) and lead (Pb). Kammerlingh Onnes also found that there is a critical current I_c above which the resistance rose to measurable value and the material goes to the normal state. It was found that the critical current increases with the decrease in temperature. Further, he found that the value of I_c depended on the shape of the material: for example, in a straight Pb wire $I_c = 8A$ while in a coiled Pb wire $I_c = 1A$. Onnes also found that the value of current flowing in a superconductor did not diminish rather persisted for a long time due to the superconducting state of the element (*persistent current*). In 1913, Kammerlingh Onnes and coworkers discovered that superconductivity is destroyed in the presence of applied magnetic field. There is a critical value of magnetic field H_c at and above which the superconductor makes a transition to the normal state. It was found that value of H_c increases with the decrease in temperature. The value of H_c ranges from 1.1 Oersted for tungsten to 830 Oersted for tantalum. Kammerlingh Onnes was awarded 1913 Nobel Prize in Physics for the discovery of superconductivity. With the onset of World War I, the work on superconductivity suffered immensely. During the World War I, in 1916, Francis Briggs Silsbee (1889–1967) gave a hypothesis on the relation between the critical current and critical magnetic field. According to him, "*The threshold value of current is equal to the value at which the magnetic field produced by the current is equal to the threshold value of magnetic field*". Later this hypothesis was verified and now it is known as Silsbee effect. The Kamerlingh laboratory became the center for the investigation of properties of solids at low temperatures and a number of scientists availed this facility for their research work. After the discovery of superconductivity in mercury, a number of scientists investigated the presence of superconductivity in various elements, alloys and compounds and the maximum critical temperature found, in 1973, was 23.2 K in Nb_3Sn: Fig. 15.1 shows the increase in T_c with time.

Portrait of Heike Kamerlingh Onnes (1853–1926), copyrighted free use, URL: https://en.wikipedia.org/wiki/Heike_Kamerlingh_Onnes#/media/File:Kamerlingh_portret.jpg

The discovery of superconductivity created great interest among theoreticians for its explanation. A number of scientists proposed different theories to explain superconductivity but failed to do so. In 1902, Lord Kelvin (1824–1907), using the law of corresponding states, argued that the resistivity of all substances diverges at sufficiently low temperatures- which we know today that it is incorrect. In 1915, Joseph John Thomson (1856–1940) tried to explain superconductivity in terms of fluctuating electric dipole moments but in vain. In 1922, Albert Einstein (1879–1955) proposed the theory of *molecular conduction closed chains* to explain superconductivity. He argued that supercurrents are carried by the closed molecular chains where the electrons undergo continuous cyclic exchanges between the molecular chains. But he did not give any justification for the existence of such conduction paths. According to Einstein no supercurrents can pass from one superconductor to another at the interface. But later it was experimentally proved that supercurrents can pass at the interface from lead to tin thereby disproving the Einstein theory.

Up to 1930, quantum theory in the one-electron approximation was formulated and used to solve many problems like the band theory of solids. There are many problems which could not be solved in the one-electron approximation like exchange and correlation effects, plasma, electrostatic screening, superconductivity, superfluidity in helium, and ferromagnetism etc. Actually, a solid contains interacting electrons of the order of 10^{23}. The above mentioned phenomena depend strongly on the interactions between the electrons therefore many-body theory is needed for their explanation. Such phenomena are called collective or cooperative phenomena.

In 1932, Willem Hendrik Keesom (1876–1956) and J. A. Kok studied for the first time specific heat C in superconducting tin (Sn) and found that there is a discontinuous jump in it at T_c. Below T_c the specific heat exhibits the exponential behavior as follows:

$$C = \text{constant} \times e^{-\Delta/k_B T} \tag{15.1}$$

where the parameter Δ is related to the band gap E_g $\left(E_g = 2\Delta\right)$ in a superconductor and k_B is the Boltzmann constant. Energy gap in a superconductor is an electronic property and is much smaller than the energy band gap due to the lattice. In 1938, they measured atomic heats of tin in both superconducting and nonsuperconducting states. In 1937, Keesom and van Laer evaluated the entropy of superconducting Sn and found that it is less than that of the normal state which shows that the superconducting state is a more ordered state. Therefore, from the study of specific heat and entropy two new features emerged: first is that the superconducting state is an ordered state and second, the band gap in superconductors is an electronic property and it is much smaller than the usual band gap due to the lattice.

In 1932, Ralph Kronig (1904–1995) proposed a two-lattice model: electron lattice and ion lattice. He said that the kinetic energy of electrons is sufficiently small and they crystallize in such a way as to minimize their Coulomb energy (electron lattice).

The vibrational frequencies of electron lattice were much higher than those of the ion lattice vibrations amounting to its strong rigidity. He assumed that the electron lattice as a whole move in the ion lattice without experiencing any resistance. Kronig had detailed discussions on this model with Felix Bloch (1905–1983) and Niels Bohr (1885–1962) but they could not come to consensus and decided not to proceed further with this model. Wolfgang Pauli (1900–1958) himself was involved in the theory of superconductivity and discussed about it with Niels Bohr in 1932 but could not arrive at any definite conclusion.

Both Felix Bloch and Lev Landau (1908–1968) were of the view that ferromagnetism and superconductivity are closely related to each other. Therefore, Felix Bloch tried to develop theory of superconductivity on the lines similar to that of ferromagnetism. He found that in the equilibrium state of a superconductor this theory yields no current. In 1933, Leon Brillouin (1889–1969) applying classical arguments showed that one can obtain metastable (nonequilibrium) state in a superconductor instead of equilibrium state which may give rise to a net current. In 1933, Lev Landau said the superconductors should not be considered as perfect conductors. His reasoning was based on the fact that resistivity just above T_c is finite that is electron mean-free path is finite and electrons must be undergoing scattering events. But these theories could not explain the phenomenon of superconductivity. Jacov Frenkel (1894–1952) also studied the occurrence of superconductivity in metals and in 1933 predicted that a metal in the superconducting state must behave as a diamagnetic material with large negative susceptibility. The interior of a superconducting metal will be screened from the external magnetic field by the surface currents induced by the lattice.

It is found that Ohm's law fails in superconductors as it yields two equilibrium states depending on the past history of the superconductor. But from thermodynamical considerations one must obtain a unique equilibrium state. Kammerlingh Onnes was also of the view that as the temperature is lowered in case of a superconductor in a magnetic field, the magnetic flux is frozen in the interior of the superconductor but this again showed nonuniqueness of the superconducting state. Due to this fact, for about 22 years it was thought that no theory is possible in superconductors. In 1933, Walther Meissner (1882–1974) and Robert Ochsenfeld (1901–1993) were working on the distribution of magnetic field in superconductors Sn and Pb in the presence of applied magnetic field. They found that magnetic field is expelled outside Sn and Pb when they were cooled below the transition temperature thus establishing the unique equilibrium state in a superconductor: it is called Meissner-Ochsenfeld effect. This experiment demonstrated for the first time that superconductors were more than just perfect conductors. After the discovery of Meissner-Ochsenfeld effect, scientists felt encouraged for the formulation of a physical theory in superconductors.

Walther Meissner (1882–1974), Public Domain, URL: https://en.wikipedia.org/wiki/Walther_Meissner#/media/File:Walther_Meissner.jpg

In order to study the thermodynamic properties of superconductors, Cornelis Jacobus Gorter (1907–1980) and Hendrik Casimir (1909–2000), in 1934, introduced the two-fluid model of superconductivity. In this theory electrons are classified into two categories: normal electrons and superelectrons. The normal electrons form a normal fluid possessing normal properties as usual. The superelectrons form superfluid and the superelectrons are endowed with several novel properties. For example the persistent current in a superconductor is due to the flow of superelectrons as they experience no resistance. They assumed that at absolute zero all the electrons behave as superelectrons. But the component of superelectrons decreases with the increase in temperature and goes to zero at T_c. In other words, the density of superelectrons depends on the temperature: decreases with the increase in temperature and goes to zero at critical temperature T_c. In 1935, Fritz London (1900–1954) and Heinz London (1907–1970), two brothers, put forward a theory, usually called *London theory*, to explain the electrodynamics of superconductors using the two-fluid model. They were able to explain the penetration of magnetic field into the superconductor near its surface and defined London penetration depth which is of the order of 10^{-6}–10^{-5} cm. This theory further accounted for the observation that not all the electrons become superconducting as soon as superconductivity is established.

Further, London brothers assumed that in the presence of a weak magnetic field the disturbance in the ground state wave function is negligible that is the ground state wave function is rigid. The rigidity of the wave function was assumed because of the existence of the energy band gap between ground state and the next excited state. In one of the research papers entitled "*Discussion of superconductivity and other low temperature phenomena*" presented at Royal Society of London, Fritz London suggested that electrons were coupled by some form of interaction in such a way that the lowest state may be separated by a finite energy gap from the lowest excited state. It is noteworthy that London theory is the first theory of superconductivity which gave explanation of the Meissner-Ochsenfeld effect.

John Bardeen (1908–991), Public Domain, URL: https://en.wikipedia.org/wiki/John_Bardeen#/media/File:Bardeen.jpg

In 1941, John Bardeen (1908–1991) said that electrons in a superconductor are coupled to small periodic distortions of the lattice. His theory of small lattice distortions gave rise to weak electron–phonon interactions which could not explain the superconducting behavior. Bardeen concluded that the electron-lattice coupling in a superconductor should be large, a conclusion that later followed from the Bardeen-Cooper-Schrieffer (BCS) theory yet for very different reasons. In 1948, Fritz London emphasized that superconductivity is a pure quantum phenomenon on macroscopic scale. This view was also one of the key elements of the elegant phenomenological theory by Ginzberg and Landau that was proposed soon after in 1950. There was another theory of superconductivity by Born and Cheng (1948).

In 1950, Emanuel Maxwell (1912–2000) and almost simultaneously Reynolds and coworkers (1950) found that the critical temperature of a superconductor depends on the isotopic mass of the constituent element. This fact indicated that the lattice vibrations and hence the electron–phonon interactions may play an important role in superconductivity. In 1950, Alfred Brian Pippard (1920–2008) found that the penetration depth changes only by 3% for the values of magnetic field between zero and the critical value which indicated the rigidity of ground wave function as suggested earlier by London brothers.

In 1950, Vitaly Ginzburg (1916–2009) and Lev Devidivich Landau (1908–1968) proposed a phenomenological theory for studying phase transition in superconductors. The free energy of a solid was expressed in terms of a series with terms having increasing powers of order parameter and variational technique was used to obtain differential equation. Near the transition temperature, they expressed the free energy F of a superconductor in terms of even powers of saturation current i_s as

$$F = F_0 + \frac{1}{2} a\, i_s^2 + \frac{1}{4} b\, i_s^4 + \ldots \tag{15.2}$$

Ginzberg and Landau combined Landau's theory of second order phase transition with Schrodinger-like wave equation and got success in explaining macroscopic properties of superconductors. This partially phenomenological theory of superconductivity called *Ginzburg–Landau theory* could give only qualitatively correct behavior of entropy and specific heat near the transition temperature. It proved that any internal magnetic flux in a superconductor exists in discrete amounts: an effect called *quantization of magnetic flux*. In 1961, Bascom S. Deaver (1930–present) and William M. Fairbank (1917–1989) and independently Doll and Nabauer (1961) observed experimentally the quantization of magnetic flux in hollow superconducting cylinders and the quantum of magnetic flux was found to be hc/2e. It confirmed London's prediction that superconductivity is a macroscopic quantum phenomenon and superconducting state is a single-valued quantum state. It also confirmed that electron pair-interaction of the BCS theory is important for the superconductors studied at that time.

In 1950, Herbert Frohlich (1905–1991) was working on isotopic effect in superconductivity. He supposed that an electron in a metal perturbs atoms in its neighborhood causing them to oscillate and the lattice perturbation then reacts on the electron: electron–phonon (e–p) interaction. If the electron–phonon interactions are important then the transition temperature T_c should vary as $M^{-1/2}$ where M represents the isotopic mass (isotopic effect). In 1952, Frohlich suggested that e–p interactions are important for the existence of superconductivity. He further remarked that it is not accidental that very good conductors do not become superconductors because strong e-p interactions give rise to large normal resistivity. In 1950, John Bardeen (1908–1991) was also working independently on superconductivity and both Frohlich and Bardeen came to the conclusion that lattice vibrations lead to a net attraction between electrons and are likely a cause of superconductivity. The ideas of Bardeen and Frohlich were incomplete, not wrong, therefore could not explain superconductivity but these ideas had a major impact on future developments. In 1953, John Bardeen gave formalism for T_c in terms of e–p interactions. But he obtained the value of T_c of the order of the Debye temperature θ_D which is an order of magnitude larger than the observed values. Richard Feynman (1918–1988) also struggled with the problem of superconductivity. He used the perturbation theory including lattice vibrations without changing the properties of electrons but was not able to obtain superconducting state. He concluded that the solution must be beyond the scope of perturbation theory which later turned out to be correct. In 1957, Feynman summarized the developments in superconductivity shortly before the BCS theory appeared. He said that Frohlich and Bardeen model was the correct approach but it needed to be solved adequately.

Ultimately Leon N. Cooper (1930–present), a postdoctoral from Princeton, and John Robert Schrieffer (1931–2019), graduate student from Illinois, joined John Bardeen in Urbana to work on superconductivity. In 1956, Cooper carried the work further and established theoretically that two electrons, with equal and opposite momenta and having opposite spins, interact with each other through the exchange of a phonon (e–p interaction), see Fig. 15.2. Such pairs of electrons are bound by a weak attractive force and are called *Cooper pairs* and these electron pairs yield ground state for a system of electrons. Schrieffer was given the task of writing down variational form of wave function of the ground state taking in to account the pairing of electrons. Schrieffer was excited to find that he could readily work out the calculation of the corresponding ground-state energy in terms of variational 'gap parameter' in the wave function. Schrieffer showed his mathematical equations for the ground state energy to Bardeen who immediately realized that this is the solution to the problem. Bardeen worked on driving the energy gap which was found to be the same as the gap parameter in the ground-state energy. In 1956, Michael Tinkham (1928–2010) and a post doctoral fellow Rolfe Eldridge Glover (1924–2004) presented first definitive measurement of energy gap at absolute zero using transmission and reflection of electromagnetic absorption in the far infrared region. On 15th February, a historic letter was sent to Physical Review by Bardeen about their theory of superconductivity usually called BCS theory. The BCS theory was officially announced at the American Physical Society meeting held from March 21–23, 1957 and in July, 1957, full paper on BCS theory was sent to Physical Review. In the BCS theory the concept of the Cooper pairs was used and it predicted the existence of small energy gap Δ in superconductors due to the weak attractive interaction. An electron pair could be broken easily by giving a small energy in the form of thermal or magnetic energy. Immediately after the publication many scientists made objections on the BCS theory but by early 1963, all rational objections to the BCS theory had been answered to the full satisfaction. The BCS theory explained satisfactorily all

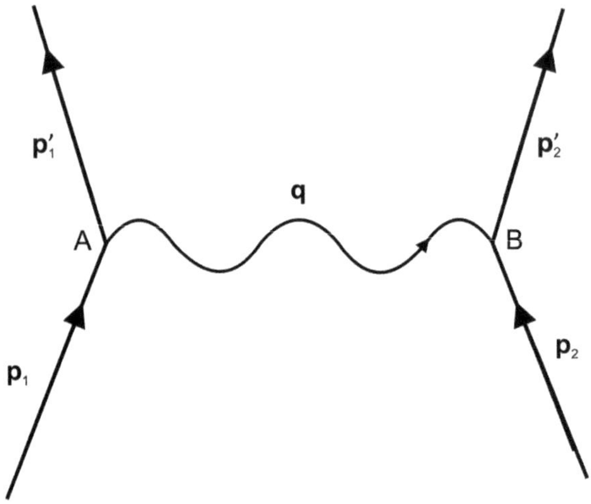

Fig. 15.2 The Feynman diagram for electron–phonon interaction. Here $\mathbf{p}_1$ and $\mathbf{p}_2$ are the momenta of two electrons before interaction while $\mathbf{p}'_1$ and $\mathbf{p}'_2$ are the corresponding momenta of two electrons after interaction. During the interaction a phonon with momentum $\hbar\mathbf{q}$ emitted at point A by the first electron and is instantaneously absorbed at point B by the second electron. The wavy line represents the phonon moving from point A to point B

aspects of superconductivity known at that point. It is noteworthy that for almost 50 years after Kammerlingh Onnes' discovery, theoretical physicists were unable to formulate a fundamental theory of superconductivity. It is because neither the conventional physics nor the elementary quantum mechanics could explain superconductivity because it treated the behavior of electrons and ions separately in a lattice. Bardeen, Cooper and Schrieffer won the 1972 Nobel Prize in Physics for the development of the BCS theory. From the above description one can conclude that the development of the BCS theory is among the most outstanding intellectual achievements in theoretical physics. Extensions of the BCS theory have been particularly successful in describing superfluidity of He^3. In 1958, Nikolay Nikolayevich Bogolyubov (1909–1992) showed that the BCS wave function can also be obtained using the canonical transformation of electronic Hamiltonian. In 1959, Lev Petrovich Gorkov (1929–2016) developed the microscopic foundations of the Ginzburg–Landau theory of superconductivity and showed that near the transition temperature T_c, the BCS theory is reduced to the Ginzburg–Landau theory.

On the other hand, in 1957, Alexi Abrikosov (1928–2017) using the Ginzburg–Landau theory classified superconductors into two categories: Type I and Type II superconductors. Type-I superconductors are those in which superconductivity vanishes abruptly when the magnetic field rises above a critical value H_c. In type-II superconductors, up to critical field H_{c_1}, the material is in the perfect diamagnetic state but above H_{c_1} it exhibits mixed state (known as vortex state). Such superconductors are characterized by negative surface energy. In the vortex state an increasing amount of magnetic field penetrates into the material but with no resistance to the flow of electric current as long as the current is not too large. Above a certain value H_{c_2} the superconductivity is completely destroyed. Abrikosov explained how magnetic flux penetrates in type-II superconductors. Depending on the geometry of the superconducting material, one may obtain Type I and Type II superconductors. Much before this, in 1934, Cornelis Jacobus Gorter (1907–1980) and Hendrik Casimir (1909–2000), for the first time, discussed about the inhomogeneous state in superconductors.

They proposed that a superconducting material is divided into small superconducting and normal regions which received greater attention of the scientific community. In 1934, de Haas and coworkers found the mixed state in alloys which are mostly the Type II superconductors. In 1936, Rudolf Peierls (1907–1995) studied about this inhomogeneous state in type I superconductors and named it as *intermediate state.* In 1937, Lev Landau developed a theory of intermediate state according to which a superconductor breaks up into laminar structure of alternating superconducting and normal layers. In 1947, Meshkovsky and Shalnikov gave experimental evidence of occurrence of intermediate state in superconductors. It is noteworthy that most of the pure elements, except niobium and carbon nanotubes, are type-I superconductors while almost all impure element and compound superconductors are of type-II. Type II superconductors are of industrial importance because they have high critical fields and therefore are used in making strong permanent magnets. Abrikosov, Ginzburg and Anthony J. Leggett (1938–present) shared 2003 Nobel Prize in Physics.

In 1958, Leo Esaki (1925–present) discovered *electron tunneling* in semiconductors and found that as the voltage is increased the current decreases (negative resistance). Tunneling in semiconductors gave birth to a new electron device called Esaki diode. Ivar Giaever (1929–present), a Norwegian mechanical engineer, had come up with an idea that energy gap could be observed directly by electron tunneling. He immediately started working on tunneling in superconductors and in 1960 observed tunneling in superconductors: he demonstrated electron tunneling through a thin layer of oxide surrounded on both sides by aluminum superconductors. It was found that, in general, electron tunneling can take place through a thin layer of oxide on both sides of which are either metals in the superconducting or normal state. Giaever demonstrated experimentally the existence of energy band gap in superconductors confirming the most important prediction of the BCS theory. Giaever's experimental demonstration of electron tunneling in superconductors stimulated the theoretical physicists in this direction. Brian Josephson (1940–present) started working on the phenomenon of superconductive tunneling. In 1962, he predicted theoretically that when the oxide layer between two superconductors is very thin, less than 10°A, the Cooper pairs as such can tunnel through the oxide layer from one superconductor to the superconductor on the other side of the oxide layer. This effect was named in his honor as *Josephson effect.* Josephson effect is of two types: dc Josephson effect and ac Josephson effect. This phenomenon is often called "weak superconductivity" because the critical parameters involved have much lower values. In 1963, Philip W. Anderson (1923–2020) and John Rowell (1935–present) confirmed experimentally the Josephson effect. The Josephson effect played the central role in the invention of superconducting quantum interference device (SQUID). SQUID is a very sensitive magnetometer used to measure very low magnetic fields, as low as 5×10^{-14} tesla, and therefore is a very sensitive detector of critical magnetic fields in superconductors. Again, SQUIDs are of two types: dc SQUID and ac SQUID depending on the type of Josephson effect used. In 1964, Jaklevic and coworkers (1964) invented dc SQUID while the ac SQUID was invented in 1965 by Silver and Zimmerman (1965) at Ford research laboratories. SQUIDs have many applications such as detectors in magnetic resonance imaging (MRI) and in the detection gravitational waves.

Ivar Giaever (1929–present), , URL: https://en.wikipedia.org/wiki/Ivar_Giaever#/media/File:Ivar_Gi%C3%A6ver_3.4jpg

In 1960, Gerasim Matveevich Eliashberg (1930–present) generalized the BCS theory by considering frequency dependent superconducting energy band gap $\Delta(\omega)$ resulting in a complex frequency dependent integral equation whose solution yielded T_c. In the theory, he introduced electron–phonon spectral function $\alpha^2 F(\omega)$ which can be extracted from the tunneling experiments. In 1968, William L. McMillan (1936–1984) was the first to solve numerically the Eliashberg integral equation to determine T_c and is given by

$$T_c = \frac{\omega_D}{1.45} \exp\left[\frac{-1.04(1+\lambda)}{\lambda - \mu - 0.62\lambda\mu}\right] \tag{15.3}$$

where

$$\lambda = 2 \int_0^\infty d\omega \, \frac{\alpha^2 F(\omega)}{\omega} \tag{15.4}$$

Here λ is a dimensionless parameter and is a measure of electron–phonon coupling strength and μ is the effective Coulomb repulsion. Equation (15.3) is called *McMillan*

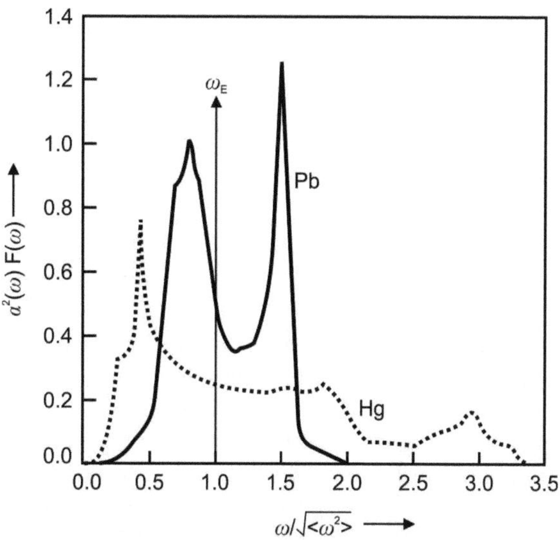

Fig. 15.3 The spectral function $\alpha^2(\omega)F(\omega)$ as a function of renormalized frequency for Pb and Hg metals obtained from the tunneling experiments [modified from McMillan and Rowell (1969) and Hubin and Ginsberg (1969)]. ω_E represents the Einstein frequency

equation and is given for value of λ up to 1.5. In 1969, William L. McMillan (1936–1984) and John Martin Rowell (1935–present) and Wilbert N. Hubin and Donald Maurice Ginsberg (1933–2007) obtained $\alpha^2F(\omega)$ for Pb and Hg from tunneling experiments and is shown in Fig. 15.3. Phillip B. Allen (1972) gave a method for calculating $\alpha^2F(\omega)$ which is related to the phonon density of states $F(\omega)$. They found that the theoretical results of $\alpha^2F(\omega)$ agree with the experimental results except that the peaks in the theoretical results are much sharp than predicted by the experiments.

In the strong coupling limit ($\lambda \geq 1.5$) McMillan equation underestimates the value of T_c. McMillan rigorously proved that in the strong coupling limit λ can be expressed in the following form

$$\lambda = \frac{\eta}{M\langle\omega^2\rangle} \tag{15.5}$$

where

$$\eta = n_\sigma(E_F)\langle I^2\rangle \tag{15.6}$$

$$\langle\omega^n\rangle = \frac{2}{\lambda}\int_0^\infty d\omega\, \alpha^2 F(\omega)\, \omega^{n-1} \tag{15.7}$$

Here M, $\langle\omega^2\rangle$, ω_D, $n_\sigma(E_F)$ are atomic mass, renormalized phonon frequency, Debye frequency, and density of electron states for spin σ. $\langle I^2\rangle$ is the average of the square of electron–phonon interaction matrix at the Fermi surface. McMillan evaluated η for most of the superconductors and found it to be nearly constant ($\approx 7\,\text{eV/A}^2$).

Therefore, λ and hence T_c depends upon the phonon frequencies $\langle \omega^2 \rangle$. Following John Joseph Hopfield (1933–present), Gaspari and Georffy (1972) developed an expression for η in terms of phase shifts using spherical bands. This theory was later extended to more realistic calculations by using nonspherical bands. Gomersall and Georffy (1974) and Rietschel (1975) generalized Eq. (15.5) for compounds and alloys and is given by

$$\lambda = \sum_{A} \frac{\eta_A}{M_A \langle \omega^2 \rangle} \tag{15.8}$$

where the subscript A denotes the type of atom in a compound. Later large number of ab initio calculations have been performed to evaluate λ and hence T_c.

The relation of T_c with enhancement factor λ is under serious debate. Different authors emphasize on different physical quantities involved in the expression for λ [Eq. (15.5) or Eq. (15.8)] and therefore leading to different criteria for high T_c. In the transition metal compounds the situation is still more complex as a number of expressions for λ have been proposed which are approximate and to some extent doubtful. Therefore, although in most of the superconductors a relation between phonon anomalies and high T_c is obtained but there is a little doubt about it as there are also a few counter examples in the literature [Varma and Dynes (1976)].

In 1986, Johannes Georg Bednorz (1950–present) and Karl Alex Muller (1927–present) added barium to the crystals of La_2CuO_4 (lanthanum-copper-oxide) and obtained the approximate composition $La_{1.85}Ba_{0.15}CuO_4$, which turned out to be the first successful high-temperature superconductor with $T_c \approx 35$ °K. Up to 1986, on the average there was an increase of 0.25 K per year in T_c of superconductors but Bednorz and Muller found a sudden jump of 12 K in $La_{1.85}Ba_{0.15}CuO_4$. This fact gave these materials the name high-T_c superconductors and the phenomenon was named high-T_c superconductivity. This discovery stimulated lot of activities in the field of high-T_c superconductivity. Worldwide eminent researchers shifted to the field of superconductivity in ceramics with the hope to discover room temperature superconductivity. Bednorz and Muller were awarded 1987 Nobel Prize in Physics for their important breakthrough in the discovery of high-temperature superconductivity in ceramic materials. Later it was found that most of the high-temperature superconductors are ceramics and most of them are type-II superconductors. All of the high-T_c superconductors can be classified into three categories:

1. The first group can be represented by the formula $La_{2-x}M_xCuO_4$ where M is a metal, which occupies some of the positions of La: a group of superconductors found by Bednorz and Muller. Another important member of this group is $La_{1.85}Sr_{0.15}CuO_{4-y}$ with $T_c = 35$ °K, where y represents the deficiency of oxygen atoms in the crystal structure.
2. The second group is represented by the formula $MBa_2Cu_3O_{7-y}$ where M can be one of the rare-earth elements. The structure of such high-temperature superconductors is usually called 123-structure. In 1987, Wu and coworkers (1987) discovered Yttrium compound with structure $YBa_2Cu_3O_{7-y}$ having $T_c = 93$ °K.

This was the first material found to exhibit superconductivity above 77 °K, the boiling temperature of liquid nitrogen. Y–Ba–Cu–O is a family of crystalline compounds famous for displaying high-temperature superconductivity.

3. The third group of high-T_c superconductors was discovered by Hazen and coworkers (1988) and can be represented by the chemical formula $Bi_2Sr_2Ca_nCu_{n+1}O_{n+6+y}$. For n = 0, one obtains superconductor $Bi_2Sr_2CuO_{6+y}$ with highest T_c less than 20 °K. For n = 1, one gets $Bi_2Sr_2CaCu_2O_{7+y}$ with highest T_c = 85 °K: these superconductors are called 2212 superconductors. In 1988, highest T_c = 110 °K was achieved for superconductor $Bi_2Sr_2Ca_2Cu_3O_{8+y}$ (with n = 2) and these are called 2223 superconductors. Sheng and Hermann (1988) discovered thallium-based superconductor $Tl_2Ba_2Ca_2Cu_3O_{10}$ having highest T_c = 127 °K. Schilling and coworkers (1993) discovered Hg-superconductors Hg–Ba–Ca–Cu–O which exhibit maximum T_c = 134 °K at ambient pressure. In the same year Chu and coworkers (1993) discovered Hg-superconductors with transition temperature above 150 °K.

Johannes Georg Bednorz (1950–present), CC BY-SA 4.0, URL: https://en.wikipedia.org/wiki/Georg_Bednorz#/media/File:8th_Silesian_Science_Festival,_Johannes_Georg_Bednorz_KsP_709,_09.12.2024.jpg

Karl Alexander Mueller (1927–present), CC BY-SA 3.0, URL: https://en.wikipedia.org/wiki/K._Alex_M%C3%BCller#/media/File:Karl_Alexander_Mueller_(cropped).jpg

It is noteworthy that most of the high temperature superconductors contain CuO_2 layers which are insulating in nature. When cuprate layers are doped with some impurities, generally ceramic impurities, these become metallic. The most common and simple example is that of La_2CuO_4 which consists of CuO_2 and LaO insulating layers. When a small amount of La is replaced by Sr, the latter act as dopant, it contributes holes to CuO_2 layers making the sample metallic. Here Sr impurities also act as electronic bridges enabling interlayer coupling. The structure of cuprate superconductors is closely related to the perovskite structure. Actually the structure of these compounds has been described as distorted, oxygen deficient multilayered perovskite structure. In cuprates the superconductivity resides in the CuO_2 layers and more is the number of CuO_2 layers higher in the value of T_c in these systems.

In late 1980s, high temperature SQUIDs were developed using high temperature superconductors particularly Y–Ba–Cu–O and are operated at liquid nitrogen temperature. These are used in making scanning SQUID microscope, which uses SQUID immersed in liquid helium as the probe.

As far as theoretical explanation high temperature superconductivity is concerned, two types of theories have been proposed: first is the *weak coupling theory* and second the *interlayer coupling theory*. In 1987, Anderson (1987) gave the first theoretical explanation of high temperature superconductivity based on *resonating valence bond* theory. Bickers and coworkers (1987) gave the first weak coupling theory in which high temperature superconductivity involves *d-wave pairing*. Inui and coworkers (1988) and Monthoux and coworkers (1992) gave the spin fluctuation theory and proposed that the high temperature cuprate superconductivity emerges from antiferromagnetic spin fluctuations. In these systems the mechanism is extremely similar to the conventional superconductors except that the role of phonons is exactly replaced by the spin density wave and phonons as such do not play any role. All the high temperature superconductors near the magnetic transition are strong spin density wave systems just as the conventional superconductors are strong phonon systems. In a high temperature superconductor, when an electron moves its spin creates a spin density wave around it. The spin density wave in turn causes a nearby electron to fall into the spin depression created, thus forming a Cooper pair. This mechanism is called d-wave pairing (having d-symmetry) where the pairing state has a node at the origin. As the temperature is lowered, more and more d-waves and hence more and more Cooper pairs are formed eventually leading to superconductivity. Gros and coworkers (1988) and Kotliar and Liu (1988) identified the d-wave pairing as a natural consequence of the resonating valence bond theory. The confirmation of the d-wave nature of cuprates was made by a variety of experiments such as angle resolved photoemission spectroscopy. Chakravarty and coworkers (1993) gave the explanation of high temperature superconductivity in terms of interlayer coupling model. According to this model, layered structure consists of BCS type superconductors having s-wave symmetry and these can enhance the superconductivity itself.

High temperature superconductivity has also been observed in nanomaterials. Hebard and coworkers (1991) found superconductivity in K_3C_{60}, carbon fullerenes doped with potassium, with $T_c = 18\ ^{o}K$. Later Darling and coworkers (2008) found $T_c = 38\ ^{o}K$ in Cs_3C_{60}. Savini and coworkers (2010) proposed that p-doped graphene is also capable of sustaining high temperature superconductivity.

A large number of high temperature superconductors with different compositions and having different structures were synthesized and as of 2021, the highest T_c reached at normal pressure is 133 ^{o}K in cuprate of mercury, barium, and calcium. High temperature superconductors exhibit zero resistance, strong diamagnetism, Meissner effect, magnetic flux quantization, penetration depth, energy gap, Josephson effect, and temperature dependence of specific heat and thermal conductivity etc. as found in normal superconductors. These facts indicate that conduction electrons form Cooper pairs as found in BCS theory. But it is difficult to explain high values of T_c because Cooper pairs are bound with weak attractive forces. It indicates that some other mechanisms must be operating in high temperature superconductors. At present there is no single theory, which explains all aspects of the high temperature superconductors.

The major advantage of high temperature superconductors is that they can be cooled using liquid nitrogen. The second advantage is that they can retain their superconductivity at magnetic fields higher than the previous superconducting materials. This property is important for constructing superconducting magnets, a primary application of high temperature superconductivity. One of the main drawbacks of ceramic superconductors is that they are brittle which makes the fabrication of wires from them very difficult. Overcoming these drawbacks is the subject of future research.

In the early years of high temperature superconductivity, there was a mad race in discovering materials which could exhibit high temperature superconductivity at room temperature. The researchers started reporting new high-T_c materials with year, date and time. But after working for a couple of years they came to the conclusion that no breakthrough was going to happen in the near future. So ultimately many of the researchers returned back to their original fields of research. At present the research in high-T_c superconductivity is going at the normal pace.

Bibliography

Abrikosov AA (1957a) The magnetic properties of superconducting alloys. J Phys Chem Solids 2:199–208

Abrikosov AA (1957b) On the magnetic properties of superconductors of the second group. Zh Eksp Teor Fiz 32:1442–1452; (1957) On the magnetic properties of superconductors of the second group. JEPT 5:1174–1182

Allen PB (1972) Neutron spectroscopy of superconductors. Phys Rev B 6:2577

Anderson PW (1987) The resonating valence bond state in La_2CuO_4 and superconductivity. Science 235:1196–1198

Anderson PW, Rowell JM (1963) Possible observation of the Josephson **superconducting tunneling effect**. Phys Rev Lett 10:230–232

Bardeen J (1941) Theory of superconductivity. Phys Rev 59:928

Bardeen J (1950) Zero-point vibrations and superconductivity. Phys Rev 79:167; (1950) Wave functions for superconducting electrons. Phys Rev 80:567; (1951) Relation between lattice vibration and London theories of superconductivity. Phys Rev 81:829; (1951) Electron-vibration interactions and superconductivity. Rev Mod Phys 23:261

Bardeen J, Pines D (1955) Electron-phonon interaction in metals. Phys Rev 99:1140

Bardeen J, Cooper LN, Schrieffer JR (1957a) Microscopic theory of superconductivity. Phys Rev 106:162; (1957b) Theory of superconductivity. Phys Rev 108:1175–1204

Bednorz JG, Muller KA (1986) Possible high T_c superconductivity in the Ba–Ca–Cu–O system. Z Phys B 64:189–193; (1988) Perovskite-type oxide: new approach to high T_c superconductivity. Rev Mod Phys 60:585–600

Bickers NE, Scalapino DJ, Scalettar RT (1987) CDW and SDW mediated pairing interactions. Int J Mod Phys B 1:687–695

Bogolyubov NN (1958) A new method in the theory of superconductivity. JEPT 7:41; (1958) A new method in the theory of superconductivity. Nuovo Cimento 7:794–805; (1958) Theory of superconductivity. Nauka I Zhizn (Russian periodical), Nr 8:17–20

Born M, Cheng KC (1948) Theory of superconductivity. Nature 161:968–969; (1948) Theory of superconductivity. Nature 161:1017–1018

Brillouin L (1933) Le champ self-consistent, pour des électrons liés; la supraconductibilité. J Phys Radium 4:333–361; (1933) Sur la stabilité du courant dans un supraconducteur. J Phys Radium 4:677–690; (1933) Comptes Rendus Hebdomadaires des Seances de L Academie des Sciences 96:1088

Cailletet LP (1877) De la condensation de l'oxygène et de l'oxyde de carbone. CR 85:1213–1214; (1878) Recherches sur la liquéfaction des gaz. Ann Chim Phys 15:131–144; (1878) Sur la liquéfaction des gaz. CR 86:97–98; (1879) Recherches sur la compressibilité des gaz. CR 88:61–65; (1885) The liquefaction of oxygen. Science 6:51–52

Chakravarty S, Sudbo A, Anderson PW, Strong S (1993) Interlayer tunneling and gap anisotropy in high-temperature superconductors. Science 261:337

Chu CW, Hor PH, Meng RL, Gao L, Huang ZJ, Wang YQ (1987) Evidence for superconductivity above 40 °K in the La–Ba–Cu–O compound system. Phys Rev Lett 58:405

Chu CW, Gao L, Chen F, Huang ZJ, Meng RL, Xue YY (1993) Superconductivity above 150 K in $HgBa_2Ca_2Cu_3O_{8+\delta}$ at high pressures. Nature 365:323–325

Cooper LN (1956) Bound electron pairs in a degenerate Fermi gas. Phys Rev 104:1189–1190

De Haas WJ, Voogd J, Jonker JM (1934) Quantitative untersuchung uber einen moglichen einfluss der achsenorientierung auf die magnetische ubergangsfigur. Physica 1:281–290

Deaver BS, Fairbank WM (1961) Experimental evidence for quantized flux in superconducting cylinders. Phys Rev Lett 7:43

Doll R, Nabauer M (1961) Experimental proof of magnetic flux quantization in a superconducting ring. Phys Rev Lett 7:51

Einstein A (1922) Theoretische bemerkungen zur supraleitung der metalle. In: Het Natuurkundig Laboratorium der Rijksuniversiteit te Leiden in de Jaren 1904–1922 (Eduard Ijdo, Leiden), p 429

Eliashberg GM (1960) Interaction of electrons with lattice vibrations in a superconductor. Zh Eksp Teor Fiz 38:966–976; (1960) Interactions between electrons and lattice vibrations in a superconductor. Sov Phys JEPT 11:696–702]; (1960) Temperature green's function for electrons in a superconductor. Zh Eksp Teor Fiz 39:1437–1441; (1961) Temperature green's function for electrons in asuperconductor. Sov Phys-JETP 12:1000–1002

Esaki L (1958) New phenomenon in narrow germanium p−n junctions. Phys Rev 109:603

Feynman RP (1957) Superfluidity and superconductivity. Rev Mod Phys 29:205

Fock VA (1930) A method for the solution of many-body problems in quantum mechanics. Z Phys 61:126

Frenkel J (1933) On a possible explanation of superconductivity. Phys Rev 43:907–912

Frohlich H (1950) Theory of the superconducting state. I. The ground state at the absolute zero of temperature. Phys Rev 79:845; (1952) Interaction of electrons with latticevibrations. Proc Roy Soc (London) A 215:291–298

Ganin AY, Takabayashi Y, Khimyak YZ, Margadonna S, Tamai A, Rosseinsky MJ, Prassides K (2008) Bulk superconductivity at 38 K in a molecular system. Nat Mater 7:367–371; Darling GR, Ganin AY, Rosseinsky MJ, Takabayashi Y, Prassides K (2008) Intermolecular overlap geometry gives two classes of fulleride superconductors: electronic structure of 38K T_c Cs3C60. Phys Rev Lett 101:136404

Gaspari GD, Georffy BL (1972) Electron-phonon interactions, d-resonances, and superconductivity in transition metals. Phys Rev Lett 28:801

Giaever I (1960) Energy gap in superconductors measured by electron tunnelling. Phys Rev Lett 5:147; (1974) Electron tunnelling and superconductivity. Rev Mod Phys 46:245–250

Ginzburg VL, Landau LD (1950) On the theory of superconductivity. Zh Eksperim i. Teor Fiz 20:1064–1082

Glover RE, Tinkham M (1956) Transmission of superconducting films at millimeter-microwave and far infrared frequencies. Phys Rev 104:844–845; (1957) Conductivity of superconducting films for photon energies between 0.3 and 40 $k_B T_c$, Phys Rev 108:243–256; Tinkham M (1956) Energy gap interpretation of experiments on infrared transmission through superconducting films. Phys Rev 104:845

Gomersall IR, Georffy BL (1974) A simple theory of electron mass enhancement in transition metal compounds. J Phys F 4:1204

Gorkov LP (1959) Microscopic derivation of the Ginzburg-Landau equations in the theory of superconductivity. J Exp Theor Phys 9:1364

Gorter CJ, Casimir HBG (1934) On superconductivity I. Physica 1:306; (1934) The thermodynamics of the superconducting state. Z Tech Phys 15:539; (1934) Two fluid model. Z Physik 35:963

Gros C, Poilblanc D, Rice TM, Zhang FC (1988) Superconductivity in correlated wavefunctions. Physica 153–155:543–548

Hartree DR (1928) The wave mechanics of an atom with a non-Coulomb central field. Proc Camb Philos Soc 24:89

Hazen R, Prewitt C, Angel R, Ross N, Finger L, Hadidiacos C, Veblen DR, Heaney PJ, Hor PH, Meng RL, Sun YY, Wang YQ, Xue YY, Huang ZJ, Gao L, Bechtold J, Chu CW (1988) Superconductivity in the high-T_c Bi–Ca–Sr–Cu–O system: phase identification. Phys Rev Lett 60:1174

Hebard AF, Rosseinsky MJ, Haddon RC, Murphy DW, Glarum SH, Palstra TTM, Ramirez AP, Kortan AR (1991) Superconductivity at 18 K in potassium-doped C_{60}. Nature 350:600–601

Hopfield JJ (1969) Angular momentum and transition-metal superconductivity. Phys Rev 186:443

Hubin WN, Ginsberg DM (1969) Electron tunneling into superconducting mercury films at temperatures below 0.4 °K. Phys Rev 188:716

Inui M, Doniach S, Hirschfeld PJ, Ruckenstein AE (1988) Coexistence of antiferromagnetism and superconductivity in a mean-field theory of high-T_c superconductors. Phys Rev B 37:2320

Jaklevic RC, Lambe J, Silver AH, Mercereau JE (1964) Quantum interference effects in Josephson tunnelling. Phys Rev Lett 12:159

Josephson B (1962) Possible new effects in superconducting tenneling. Phys Lett 1:251–253

Keesom WH, Kok JA (1932) Leiden commun. 221e [Proc Kon Akad Amsterdam 35:743]; (1934) Leiden Commun 230c, 230e. [Measurement of specific heat of thallium at liquid helium temperatures. Physica 1:175].

Keesom WH, van Laer PH (1937a) Relaxation phenomena in superconductivity. Physica 4:499–502; (1937b) Measurements of the latent heat of tin while passing from the superconductive to the non-superconductive state at constant temperature. Physica 4:487–493

Keeson WH, van Laer PH (1938) Measurements of the atomic heats of tin in the superconductive and in the non-superconductive state. Physica 5:193–201

Kotliar G, Liu J (1988) Superexchange mechanism and *d*-wave superconductivity. Phys Rev B 38:5142

Kronig R (1932) Zur theorie der supraleitfähigkeit. Z Phys 78:744–750; (1933) Zur theorie der supraleitfähigkeit II. Z Phys 80:203

Landau LD (1933) A possible explanation of the field dependence of the susceptibility at low temperatures. Physikalische Zeitschrift der Sowjetunion 4:675; (1933) Electron motion in crystal lattices. Physikalische Zeitschrift der Sowjetunion 3:664

Landau LD (1937) On the theory of superconductivity. JEPT 7:371; (1937) On the theory of superconductivity. Phys Z Sowjet 11:211

Landau LD (1943) On the theory of the intermediate state of superconductors. J Phys USSR 7:99; (1943) On the theory of the intermediate state of superconductors. JEPT 13:377

London F (1948) On the problem of the molecular theory of superconductivity. Phys Rev 74:562

London F, London H (1935) The electromagnetic equations of the supraconductor. Proc Roy Soc (London) A149:71–88; (1935) Supraleitung und diamagnetismus. Physica 2:341–354

Kelvin L (1902) XXIX. Aepinus atomized. Philos Mag 3:257–283

Maxwell E (1950) Isotope effect in the superconductivity of mercury. Phys Rev 78:477

McMillan WL (1968) Transition temperature of strong-coupled superconductors. Phys Rev 167:331

McMillan WL, Rowell JM (1969) In: Parks RD (ed) Superconductivity, vol 1. Marcell Dekker, New York, p 561

Meissner W, Ochsenfeld R (1933) Ein neuer effect bei eintritt der supraleitfahigkeit. Naturwissenschaften 21:787–788

Meshkovsky AG, Shalnikov AI (1947) The structure of the superconductors in the intermediate state. II, Zh Eksp Teor Fiz 17:851; (1947) J Phys (Moscow) 11:1–15

Monthoux P, Balatsky A, Pines D (1992) Weak-coupling theory of high temperature superconductivity in the antiferromagnetically correlated copper oxides. Phys Rev 46:14803

Olszewski C (1895) The liquification and solidification of argon. Nature 51:355; [Dewar J (1895) On liquifucation of gases—a claim for priority. Nature 51:245]

Onnes HK (1911) Further experiments with liquid helium. C. On the change of electric resistance of pure metals at very low temperatures, etc. IV. The resistance of pure mercury at helium temperatures. Comm Phys Lab Univ Leiden, no 120b; (1911) Further experiments with liquid helium. D. On the change of electric resistance of pure metals at very low temperatures, etc. V. The disappearance of the resistance of mercury. Comm Phys Lab Univ Leiden, no 122b; (1911). Further experiments with liquid helium. On the electrical resistance of pure metals, etc. VI. On the sudden change in the rate at which the resistance of mercury disappears. Comm Phys Lab Univ Leiden, no 124c; (1911) The resistance of platinum at helium temperature. Comm Phys Lab Univ Leiden, 119b:19–26

Peierls R (1936) Magnetic transition curves of supraconductors. Proc Roy Soc (London) A 155:613

Pictet R (1877) Expériences de M. Raoul Pictet sur la liquéfaction de l'oxygène. CR 85:1214–1216; (1878) Mémoire sur la liquéfaction de l'oxygène. Ann Chim Phys 13:145–229

Pippard AB (1953) An experimental and theoretical study of the relation between magnetic field and current in a superconductor. Proc Roy Soc (London) A 216:547

Reynolds CA, Serin B, Wright WH, Nesbitt LB (1950) Superconductivity of isotopes of mercury. Phys Rev 78:487; (1951) The isotope effect in superconductivity. I. mercury. Phys Rev 84:691

Rietschel H (1975) Electron-optical phonon coupling in superconductors. Z Phys B 22:133

Savini G, Ferrari AC, Giustino F (2010a) Doped graphene: a prototype high-T_c electron-phonon superconductor. Phys Rev Lett 105:037002; (2010) First principles prediction of doped graphene as a high-temperature electron-phonon superconductor. Phys Rev Lett 105:059902

Schilling A, Cantoni M, Guo JD, Ott HR (1993) Superconductivity above 130 K in the Hg–Ba–Ca–Cu–O system. Nature 363:56–58

Sheng ZZ, Hermann AM (1988) Superconductivity in the rare-earth-free Tl–Ba–Cu–O system above liquid-nitrogen temperature. Nature 332:55–58; (1988) Bulk superconductivity at 120K in the Tl–Ca/Ba–Cu–O system. Nature 332:138–139; Sheng ZZ, Hermann AM, El Ali A, Almasan C, Estrada J, Datta T, Matson RJ (1988) Superconductivity at 90 K in the Tl–Ba–Cu–O system. Phys Rev Lett 60:937

Silsbee FB (1916) A note on electrical conduction in metals at low temperatures. J Wash Acad Sci 6:597–602; (1918) Bulletin of bureau of standards 14:301

Silver AH, Zimmerman JE (1965) Quantum transitions and loss in multiply connected superconductors. Phys Rev Lett 15:888

Thomson JJ (1915) Conduction of electricity through metals. Philos Mag 30:192–202

Varma CM, Dynes RC (1976) In: Superconductivity in d- and f-band metals. In: Douglass DH (ed) Proceedings of 2nd Rochester conference. Plenum Press, New York, p 507; Dynes RC, Varma CM (1976) Superconductivity and electronic density of states. J Phys F 6:L215

Wroblewski Z, Olszewski K (1883) Sur la liquéfaction de l'oxygène et de l'azote, et sur la solidification du sulfure de carbone et de l'alcool. CR 96:1140–1142; (1883) Ueber die volumenanderungen von metallen und legirungen beim schmelzen. Annalen der Physik 20:228–243

Wu MK, Ashburn JR, Torng CJ, Hor PH, Meng RL, Gao L, Huang ZJ, Wang YQ, Chu CW (1987) Superconductivity at 93 °K in a new mixed phase Y–Ba–Cu–O compound system at ambient pressure. Phys Rev Lett 58:908–910

Chapter 16
Solid State Electronics

The subject of electronics emerged in the twentieth century from electricity and magnetism. Today electronics has made deep inroads in many areas. Electronic circuits and devices are used in the electronic gadgets, home appliances, computers, cell phones, cameras, televisions, automobiles, healthcare systems, medical diagnosis and industries etc. One is convinced that today it is really impossible to work without electronics. A large number of scientists had put in untiring efforts to bring about the present situation in electronics. In 1883, Thomas Alva Edison (1847–1931) was the first to observe the phenomenon of emission of some kind of charged particles during the passage of electricity from a heated filament to a metallic plate inside an incandescent lamp. Initially it was assumed that some new particles were emitted to produce current and it was given the name *Edison effect*. In 1897, Joseph John Thomson (1856–1940) demonstrated that the charged particles emitted from the heated filament are nothing but electrons. The British physicist Owen Willans Richardson (1879–1959) was aware of the Edison effect and started working on the emission of electrons from a hot filament and termed it "*thermionic emission*". In a metal, some electrons (valence electrons) are mobile inside it but cannot come out because of attractive forces. A definite amount of external energy, called work function W, is required to emit these electrons from the surface of a metal. When metals are heated to high temperatures of about 1000C, electrons can be emitted from their surface. In 1901, Richardson derived the first equation for thermionic emission using the classical theory of electronic conduction in metals According to Richardson the electronic current I at temperature T is given by

$$I = AT^2 e^{-\frac{W}{k_B T}} \tag{16.1}$$

where A is a constant and k_B is the Boltzmann constant. Later, in 1923, Saul Dushman (1883–1954) applied the quantum theory of free electrons and showed that the equation for thermal electric current is the same as Eq. (16.1) but the constant A is modified which becomes a universal constant. Therefore, Eq. (16.1) is usually called Richardson-Dushman equation. Owen Willans Richardson was awarded the 1928

J. S. Galsin, *History of Solid State Physics*,
https://doi.org/10.1007/978-981-95-0504-3_16

Nobel Prize in Physics for his work on thermionic emission and especially for the discovery of law after his name.

Joseph John Thomson (1856–1940), Public Domain, URL: https://en.wikipedia.org/wiki/J._J._Thomson#/media/File:J.J_Thomson.jpg

Thermionic emission became the base of early developments in electronics. In 1904, Johan Ambrose Fleming (1849–1945), a British engineer and physicist, invented two-electrode vacuum thermionic tube known as diode: two electrodes are cathode and anode. Diode is the first thermionic tube and Fleming got it patented in 1905. In a diode, some of the electrons emitted from the cathode reach the anode even when no electric field is applied thus constituting current. In actual practice, electric potential is applied between the two electrodes of a diode, anode being positively charged, to collect most of the emitted electrons at the cathode. Diode was capable of converting alternating electric currents and especially high-frequency alternating electric currents or electric oscillations, into continuous electric currents. The invention of diode was considered to be a great advancement in setting a stage for modern electronics. Fleming designed the radio transmitter with which the first transatlantic radio transmission was made. Fleming gave the *right hand rule* and *left hand rule* in physics. In 1914, Walter Hans Schottky (1886–1976) found that the application of electric field between the filament (negative) and plate (positive) lowers the surface barrier and hence increases the electron emission: it is known as *Schottky effect.*

Reginald Aubrey Fessenden (1866–1932) was interested in transmitting continuous sound particularly the human voice. He developed the concept of *amplitude modulation* (AM) in which an electrical signal with frequencies of sound waves was superimposed on a high constant frequency signal (radio wave) thereby modulating the amplitude of the radio wave. In 1900, he was able to transmit speech by radio over a distance of 1 mile (1.6 km) at Cobb Island, USA. His first transmitted words were "*Hello. One, two, three, four. Is it snowing where you are, Mr. Thiessen*". The transmitted signal was barely intelligible above the background noise. In 1901, Fessenden developed heterodyne principle in which two closely spaced radio signals are used to produce an audible tone that makes Morse code transmission much easier to hear but it remained nonpratical because of lack of producing local stable signal. He invented the word *heterodyne* from Greek word "heteros" means other and "dynamis" means force. Fessenden contributed significantly towards wireless communication. In his 1905 patent Fessenden stated that the frequency stability of his local oscillator was one part in one thousand. In 1906, Lee De Forest (1873–1961), an American engineer, invented vacuum triode with three *electrodes.* The third electrode, in the form of a wire gauze, is called *controlling grid* which is situated between the cathode and anode and controls current in the vacuum tube. He used triode to amplify radio frequency (RF) signals and also used it as electronic oscillator. Lee Forest filed a U. S. Patent in 1907 which he got in 1908. With the invention of triode stable oscillator became available and in 1906 Fessenden made the first two-way radiotelegraphic communication across the Atlantic Ocean. Lee Forest developed sound-on-film recording technique used for motion pictures. It made possible radio broadcasting, long distance telephone lines and talking motion pictures among other applications. Lee de Forest is called the "*Father of radio*". In 1910, Robert von Lieben (1878–1913), a self-taught Austrian physicist, independently invented a gas filled triode usually called *Lieben valve*, which is a distant ancestor of thyratron. Immediately after, it was improved and patented jointly by Lieben, Eugen Reisz (1879–1957) and Siegmund Strauss (1875–1942) in 1910. In 1915, Walter Hans Schottky (1886–1976) invented the screen grid tube and in 1919 he invented *tetrode*, the first multigrid vacuum tube. In tetrode an additional grid called *screening grid*, was introduced between the controlling grid and plate and was given positive voltage. The screened grid acts as an electrostatic shield which reduces the grid-to-plate capacitance thus minimizing feedback and improving amplifier stability and gain.

Edwin Howard Armstrong (1890–1954) invented an amplifier by giving positive feed-back to the grid which produced amplification of a signal hundred times larger than the earlier obtained amplification. In 1913, he applied for its patent and was issued as U. S. Patent 1,113,149 on Oct. 1914. In 1919, Edwin Armstrong developed a *superheterodyne* receiver circuit and applied for a US patent which he got in 1920. But this patent was challenged by Lucien Levy (1892–1965) and was canceled.

On appeal Levy's claim as inventor was accepted with some limitations. Actually, Lucien Levy invented the superheterodyne method of amplifying radio signals in radio receivers earlier and was awarded French patents in 1917 and 1918 that covered some of the basic ideas used in Armstrong's superheterodyne receiver. Armstrong developed a wide-band frequency modulation (FM) system and got it patented by the US in 1933. The first paper on FM radio was presented to the Institute of Radio Engineers, New York, on November 6, 1935, and published in 1936.

Albert Wallace Hull (1880–1966) made significant contributions to the development of vacuum tubes. In 1918, Wallace Hull developed dynatron vacuum tube which had three electrodes: a thermionic cathode, a perforated anode and a supplementary anode plate (negative resistance device). Wallace Hull also worked on gas filled tubes in 1920s and developed hot-cathode thyratrons (gas-filled triodes) and phanotrons (gas-filled diodes). In 1920, Wallace Hull invented magnetron and later John Randall (1905–1984) and Harry Boot (1917–1983) improved it in the form of modern cavity magnetron. James Sayers (1912–1993) played a crucial role in developing centimeter radar and in this process he further improved magnetron. In 1926, Gilles Holst (1886–1968) and Bernardus Dominicus Hubertus Tellegen (1900–1990) invented pentode with five electrodes: cathode, anode, controlling grid, screening grid and suppressor grid. Adding more grids within the vacuum tubes allows us to have finer control on current between cathode and anode by varying voltages on the grids. In 1948, Tellegen invented gyrator. All these tubes constituted standard equipment in radio receivers, radar sets, early television sets and other forms of communications for at least half a century, until it was replaced by solid state electronics in the mid twentieth century.

Printed circuit board (PCB) is an electrical device in which wiring and certain components consist of a thin coat of electrical conductor, in certain pattern, applied on an insulating board by any of the graphic-arts techniques. In 1925, Charles Ducas (1880–1966), an American inventor, patented the first circuit board designed with conductive material on a flat wooden board. In 1936, Paul Eisler (1907–1992), an Austrian inventor, developed the first PCB to operate radio system. Eisler called it "*Printing and etching of copper foil*" and it was the first method of producing PCB on mass scale and at economically affordable prices. Eisler is called "the father of Printed Circuit Boards". PCB mechanically supports and electrically connects the electrical or electronic components of a circuit using the conductive tracks: the components are generally soldered onto the PCB. PCB replaced the conventional wiring in most of the electronic equipments such as radio, television sets, computers, control systems and guided-missile electronic systems etc. They greatly reduced the size and weight of the equipment while improving the reliability and uniformity over the hand-soldered circuits formerly used. There are many types of PCBs including double-sided and multilayered versions.

The development of solid-state electronics started with the discovery of semiconductors. In 1833, Michael Faraday (1791–1867) observed the semiconductor effect for the first time. He observed that the electrical resistance of semiconducting silver sulphide decreased with the increase in temperature (negative temperature coefficient), an effect opposite to that in metals as described by Sir Humphry Davy (1778–1829). It is believed that in 1782, Allessandro Volta (1745–1827) used the term "semiconducting" for the first time. In 1839 Alexander Edmund Becquerel (1820–1891), father of Henri Becquerel, discovered the photovoltaic effect at a junction between a semiconductor and an electrolyte. In 1873, Willoughby Smith (1828–1891) discovered photoconductivity in selenium. He observed that light caused a dramatic decrease in the resistance of selenium resistor. In 1874, Karl Ferdinand Braun (1850–1918) observed that current flowed only in one direction at the contact between a metal and Galina crystal: the first semiconductor effect. In the same year, Arthur Schuster (1851–1934) discovered rectifying properties of copper and its oxide: copper-cupreous oxide system. From 1877–1883 Ferdinand Braun studied the rectifying action of semiconductors and found that it is most pronounced if one of the electrodes consists of a sharp wire tip pressed on to the material. In 1876, William Grylls Adam (1836–1915) and Richard Evans Day and independently by Charles Fritts (1850–1903) discovered photovoltaic effect in selenium. In 1883, Charles Fritts created the first solar cell by coating selenium with a thin layer of gold. The world's first rooftop solar array using Fritts' selenium cells was installed in 1884 on a New York city-rooftop. The Hall effect, discovered in 1879, gave information about the sign of the charge carriers and thus allowed a clear distinction between metals and semiconductors. Therefore, by the end of nineteenth century, the basic properties of semiconductors had been discovered.

Pure semiconductors generally called intrinsic semiconductors, are basically insulators at very low temperatures and have energy band gap much less than that in insulators. The most important intrinsic semiconductors are silicon (Si) and germanium (Ge). In 1824, Jons Jacob Berzelius (1779–1848) discovered Si and in 1886, Clemens Alexander Winkler (1838–1904) discovered Ge. At finite temperature some electrons from the full band (valence band) make transition to the next empty band (conduction band) resulting in a small intrinsic electric current in the presence of an applied electric field. With the increase in temperature more and more electrons jump to the conduction band resulting in the increase in current and hence decrease in resistance. In 1885, Shelford Bidwell (1848–1909) and in 1931, Bernhard Gudden (1892–1945) reported independently that the properties of semiconductors were due to the presence of impurities in them. Jagadish Chandra Bose (1858–1937) demonstrated the use of glena (lead sulphide) crystal with a point contact by a metal to detect millimeter electromagnetic waves. In 1901, he filed a U. S. patent as "*Cat whisker detector*" for a point contact rectifier for detecting radio waves. From 1902 to 1906, Greenleaf Whittier Pickard (1877–1956) tested a number of materials for making a point contact rectifier and got the best results with semiconductors. On Aug. 30, 1906, Pickard filed U. S. patent for a silicon point-contact detector and it was granted in Nov. 20, 1906. With his two associates Pickard set up a company "Wireless Specialty Apparatus Company" to make and sell silicon semiconductor devices and probably this was the first company to sell silicon semiconductor devices. In 1930s, William Bradford Shockley (1910–1989) worked for more than ten years and ultimately was able to develop theory successfully for solid-state devices but could not build a working model. In 1931, Alan Herries Wilson (1906–1995) realized that the presence impurities give rise to localized states within the forbidden energy gap. The impurities can become ionized at low enough temperatures thus giving free electrons in the conduction band. In semiconductors, one can have donor states near the conduction band (n-type semiconductor) or acceptor states near the valence band (p-type semiconductor) depending on the nature of the impurity. In 1932, Walter H. Schottky (1886–1976) worked on the theory of copper-cuprous oxide contact for the rectification process but he did not publish the results because it gave the wrong sign for the rectification effect. Schottky realized that at the interface between metal and semiconductor there is a layer of finite width around the contact in which electron density is less than that in the bulk: called *depletion layer*. He gave an exact calculation of the shape and width of the depletion layer in a semiconductor. He established that electrons will enter or leave the semiconductor until the potential difference between the two materials disappears. The depletion layer becomes wider when a reverse voltage is applied and narrower with the application of forward voltage. In 1938, Boris Davydov (1940–present), in 1939, Walter Schottky (1886–1976), and Nevill Francis Mott (1905–1996) worked independently and were successful in formulating the theory of rectifying contacts. In 1942, Hans Bethe (1906–2005) developed the theory of thermionic emission and of high frequency rectification by silicon crystals but it was applicable to very thin barrier layers.

William Bradford Shockley (1910–1989), CC BY 3.0, URL: https://en.wikipedia.org/wiki/William_Shockley#/media/File:William_Shockley,_Stanford_University.jpg

With the onset of World War II, the research in semiconductors got activated as these were of immense use as microwave radar detectors. A radar laboratory, usually called *Rad Lab*, was set up in MIT, USA, to work on microwave radar detectors. In most of the applications silicon detectors were used but Sperry Gyroscope Corporation, USA, for the first time introduced the germanium detector. In 1942, Karl Lark-Horovitz (1892–1958) said that both germanium and silicon were true intrinsic semiconductors. In the same year, Lark-Horovitz reported that both n- and p-type germanium had been produced. During the War-time it was found that the mobility of free electrons varied as $T^{3/2}$ with temperature. There were a number of other projects related to the military requirements. A lot of work was done on rectification from metal–semiconductor junction. Scientists also worked on the problem of detection of infrared radiation mainly for bombs and gunsights. Another burning problem during the World War II was the development of solid state amplifiers but before making these a method to insert a third electrode (grid) in the cold rectifier was needed.

Russel Shoemaker Ohl (1898–1987) and Jack Hall Scaff had understood the effect of impurities in producing n- and p-type silicon and they were the first to grow p–n junction in silicon in 1940: it was a silicon point contact detector. Ohl was also probably the first to find the photovoltaic effect in silicon. In 1948, Ralph Bray (1921–2000) had observed minority carrier injection, an effect that helped in making the transistor a reality.

After the World War II, in 1945, Bell set up the Solid-State Group led by William Bradford Shockley (1910–1989) with the aim of producing solid state devices. John Bardeen (1908–1991) and Walter Houser Brattain (1902–1987) were asked to join his group and after working for two years, in 1947, they could invent point-contact transistor which has three terminals just like a triode. Actually John Robinson Pierce (1910–2002) gave the device the name 'transistor' because of the concept of transresistance in analogy with transconductance in vacuum tubes. In 1949, John Shive (1913–1984) made point-contact transistor with the emitter and collector placed on the opposite sides of a very thin slice of germanium (0.01 cm). He showed that the conduction in such a point-contact transistor was indeed taking place in the bulk. In 1948, Shockley developed his theory of the P–N junction that led to the development of the junction transistor two years later. In 1950, Shockley designed a new type of transistor called bipolar junction transistor, which was superior to the point-contact transistor and replaced it. Thus the invention of transistor was, in large part, Shockley's creation. Actually this work was kept secret but later on July 4, 1951 the announcement of the invention of bipolar junction transistor was made. Shockley explained the action of transistor in analogy with the triode vacuum tube. Perhaps for the first time in the history of semiconductors the theoretical framework preceded a practical device. The invention of a transistor triggered an explosive growth in the theoretical and experimental studies of semiconductors. With the help of transistor, scientists were able to produce solid-state amplifiers, another important invention in solid state electronics. Bardeen, Brattain and Shockley were honored with the 1956 Nobel Prize in Physics for their researches on semiconductors and their invention of a transistor.

After the invention of a transistor the main problem was its mass production that is transistor technology which includes synthesis of pure single crystal of semiconductors, the precise introduction of impurity concentration and its temperature dependence. Theoretical understanding is equally important in developing the transistor technology. In 1953, Frank Herman (1926–2021) and Joseph Callaway (1931–1994) studied the energy band structure of silicon and found it to be anisotropic in nature and exhibited multivalleys in it. Gerald L. Pearson (1905–1987) observed experimentally the anisotropy in the magnetoresistance of germanium and silicon which confirmed anisotropy in band structure of silicon. In 1954, Gerald L. Pearson (1905–1987) and Conyers Herring (1914–2009) showed multivalley structure from the magnetoresistance measurements.

It was found that semiconductors could be more useful if their conductivity is increased to a significant level. In this regard, the findings of Shelford Bidwell (1848–1909) and Bernhard von Gudden (1824–1886) were considered to be of immense use. John Robert Woodyard (1904–1981) invented the process of doping of semiconductors and got it patented in 1950. Doping is the process of introducing small amounts of impurities in semiconductors. With the process of doping n- and p-type extrinsic semiconductors were synthesized. In n-type semiconductors the majority charge carriers are negatively charges electrons and the minority charge carriers are positively charged holes. Opposite is the case in p-type semiconductors as for as the charge carriers are concerned. Doping of magnetic impurities had been done to produce ferromagnetic semiconducting alloys possessing different properties. In 1950, Gordon Kidd Teal (1907–2003) and J. B. Little invented a method by applying Czochralski method to produce extremely pure single crystals of germanium and silicon used in making greatly improved transistors. In 1951, Gordon Kidd Teal and coworkers invented a modification of the process that produced configuration necessary for the fabrication of bipolar junction transistors. Morgan Sparks (1916–2008) is most remembered for developing the first silicon transistor in 1951. In 1954, Morris Tanenbaum (1928–2023) and coworkers were the first to develop commercial silicon transistor on January 26, 1954. A few months later in 1954, Teal and his team working independently at Texas Instruments created a similar working silicon transistor. Both the devices were made by the same method of controlled doping of silicon single crystals. Transistor replaced vacuum tube triode and has very small size as compared to the vacuum tube triode. Further, vacuum tubes were operated at 220 V while the transistors had low operating voltage (6–9 V) and were more efficient. Transistor being a solid-state device has much longer life and ideally speaking infinite life. The invention of transistor was the first step toward the miniaturization of electronic components. In 1954 itself, Daryl Muscott Chapin (1906–1995) and coworkers created first silicon p–n junction photocell for converting sun light into electric power.

After the invention of transistor attempts were made to further miniaturize the electronic components. For many years efforts were made to combine several components in one device. Jack Kilby (1923–2005) gave the idea of integrated circuits (ICs) and in 1958 made the first IC of germanium and demonstrated its working successfully. Kilby described his new device as a body of semiconductor in which were embedded and integrated the various electronic circuits. Kilby filed for the U. S. patent for his integrated circuit in 1959. Kilby's IC was a hybrid IC and had external wire connections which made it unsuitable for mass production. After six months, in 1959, Robert Norton Noyce (1927–1990) independently invented the first monolithic IC chip made of silicon: a new variety of IC which is more practical than Kilby's hybrid IC. He had put all the components on a silicon chip connecting with copper lines: modern IC chips are the Noyce monolithic planar chips. Actually, there was a conflict between Kilby and Noyce about the invention of ICs and later there happened a compromise between the Kilby and Noyce and both Patents were considered necessary to manufacture ICs. Julius Edgar Lilienfeld (1882–1963) gave the basic principle of field-effect transistor (FET) and got it patented in 1925. In 1959, Mohamed Martin Atalla (1924–2009) invented metal–oxide–semiconductor field effect transistor (MOSFET) at Bell laboratories. In 1960, Mohamed Atalla and his student Dawon Kahng (1931–1992) proposed the concept of metal–oxide–semiconductor (MOS) integrated circuit chip and materialized the same. The fabrication of MOSFETs is comparatively easy which made them useful for integrated circuits. In 1963, Steven Hofstein (1938–present) and Fredric Heiman (1939–2020) published a paper on silicon MOSFET. In the same year 1963, Frank Wanlass (1933–2010) proposed the first complementary metal-oxide semiconductor (CMOS) circuit. MOSFETs are usually fabricated on one side of the chip in the form of flat *two-dimensional integrated circuits*. With the invention of MOSFET it became possible to build high density integrated chips. Nearly all modern IC chips are MOS integrated circuits built from MOSFETs. With the advancement of IC technology, in the late 1960s, MOS chips led to large scale integration (LSI) with hundreds of transistors on a single MOS chip. Robert E. Kerwin (1932–present) and coworkers developed self-aligned silicon gate MOSFET and got it patented in 1969, a key element in the fabrication of very large-scale integration (VLSI). Actually, Robert W. Bower (1936–2024) and Hans G. Dill invented the self-aligned silicon gate MOSFET and presented their work in the 1966 IEEE International Electron Devices Meeting held in Washington D. C. They applied for its Patent on Oct. 27, 1966 but Us Patent 34,722,712 was issued on Oct. 14, 1969. In a legal battle between Bower and Kerwin, the court decided the invention of self-aligned silicon gate technology in favor of Kerwin and coworkers. In 1968, Federico Faggin (1941–present) developed CMOS integrated circuits. While working at Fairchild semiconductors, USA, in 1968, he created self-aligned MOS single-gate technology (SGT). In 1971, using SGT, Faggin designed the world's first single-chip commercial microprocessor Intel-4004 (4-bit CPU). After that he designed microprocessors Intel 8008 (8-bit CPU) and 8080 (8-bit CPU), improved versions of Intel-4004 and ZilogZ80. In 1978, Faggin developed single-chip microcontroller ZilogZ8. The development of IC technology is considered to be the second step in the miniaturization of electronic components. Therefore, transistor is viewed as an ancestor of microelectronics.

In 1965, Gordon Moore (1929–2023) was asked by Electronics Magazine what will happen in the semiconductor components industry over the next ten years. He gave a law, called Moore's law, regarding the miniaturization of electronic components according to which the number of transistors in a chip would double every year. But it was found that the number of transistors increased to double in every two years. So, in 1975, Moore modified his law accordingly: that the number of transistors would become double in every two years. It is worth mentioning here that *three-dimensional ICs* were synthesized by stacking several layers of transistors. While working at Texas instruments in USA, Robert W. Haisty and coworkers proposed the concept of three-dimensional MOS integrated circuits and got it patented in 1964. Katsuhiro Onoda and coworkers proposed the concept of three-dimensional MOS integrated circuit memory chip at NEC Corporation, Japan and got it patented in 1969. By 2017, the transistor size reduced to a few microns to 10 nm. Integrated circuits are usually classified into three categories depending on their functions: analog, digital and mixed signal IC.

1. Digital ICs contain logic gates, flip flops, multiplexers and other electronic circuits. Microprocessors and microcontrollers are formed from them.
2. Analog ICs contain linear integrated circuits and radiofrequency circuits. These perform analog functions such as amplification, active filtering, demodulation and mixing.
3. Mixed signal ICs combine analog and digital circuits on a chip such as analog-to-digital and digital-to-analog converters.

It is noteworthy that the invention of integrated circuits and MOSFETs revolutionized drastically the technology in different spheres. Today we have synthesized a large number of basic electronic components for different applications. Different types of diodes such as Zener diode, Schottky diode, tunnel diode, light emitting diode, photodiode, laser diode, current regulating diode etc. have been invented. A variety of triodes such as basic triodes, light house tubes, Nievistor tubes, amplifying triodes and oscillating triodes were invented. A variety of solid state transistors such as bipolar junction transistors (PNP and NPN), field effect transistors (N-channel and P-channel), insulated-gate bipolar transistors (voltage-driven MOSFET etc.) have been devised for different purposes.

Another aspect of solid-state electronics in which the immense progress has been made is computer. In 1614, John Napier (1550–1617), a Scottish mathematician, published his discovery of logarithm. Shortly after it around 1622, William Oughtred (1574–1660) invented the earliest form of slide rule but its description did not appear in print. In 1630, one of his pupil Richard Delamain published at London as a small pamphlet entitled Grammelogia. Around 1850, Amedee Mannheim (1831–1906), a French artillery officer, developed a new version of slide rule that became popular among the engineering students. In 1642, Blaise Pascal (1623–1662) invented the first mechanical calculator called Pascaline which was able to perform addition and subtraction. Pascal made this mechanical calculator when he was not even 19 and later it was improved producing a number of models of Pascaline. In 1673, Gottfried Wilhelm Leibniz (1646–1716) designed the first calculating machine that could perform all the four basic arithmetic operations: addition, subtraction, multification and division. Early in the industrial revolution (duration of industrial revolution was from 1760 to 1840), some mechanical computing devices were made. In 1679, Leibniz described the very principle of binary computer. Further, he gave precise description how electronic computers function. Computer is a machine that can carry a number of arithmetic and logical operations automatically via computer programming. Leibniz has been called the founder of computer science. Then the electrical calculators for analog calculations came and ultimately the electronic computers during World War II. Charles Babbage (1791–1871) originated the concept of digital programmable computer. In 1823, Babbage developed prototype difference engine which used the method of differences to compute the value of polynomial functions. During mid 1830s, Babbage developed plans for Analytical Engine and its design provided the basic framework for the development of today's computers. It is because of this reason that Charles Babbage is considered to be the "*Father of computers*".

With time a number of companies came into existences, which were dealing with data tabulating machines and time clocks. In 1911, a number of these companies merged to form Computing-Tabulating-Recording (CTR) Company. In 1924, its name was changed to International Business machines (IBM). In 1938, Konrad Zuse (1910–1995), a German civil engineer, developed world's first programmable computer named Z1. Despite some mechanical engineering problems, it had all the basic ingredients of modern machines, using binary system and having today's standard separation of storage and control. Later he improved upon the computer and made Z2 in 1940, Z3 in 1941 and Z4 model in 1949: Z4 became the world's first commercial computer. Due to World War II, Zuse's work went largely unnoticed in the United Kingdom and United States. In the beginning, IBM made computers with card-based systems using vacuum tubes. In 1954, IBM made the first calculating machine IBM 608 that used solid-state transistors instead of vacuum tubes.

Along with the development of computers, an equally important field of computer software was also developed. Software consists of programming instructions; commands that tell the computer how to execute various tasks. Software is an interface between the computer systems and humans who use them. In 1948, Tom Kilburn (1921–2001), a computer scientist, wrote the world's very first piece of software at the Manchester University, England. The computer programming was written in machine language or assembly language which required writing instructions in binary or hexadecimal arithmetic. In 1957, John Backus (1924–2007) created computer programming language FORTRAN (Formula Translation) and was one of the very first high-level programming languages. It shortened the process of programming and made computer programming more accessible. Later a number of updates of FORTRAN like FORTRAN2, FORTRAN3, FORTRAN4, FORTRAN77 and FORTRAN90 etc. appeared. FORTRAN became the language of first choice in late fifties and sixties. FORTRAN set the stage for the development of other high-level computer programming languages like COBOL, BASIC, Pascal, Java and C, C+, C++ which appeared over the next two decades. The first solid-state electronic calculator was created in 1961 and the pocket-sized devices became available in the 1970s. In 1971, Busicom Co., Ltd produced the first pocket-sized electronic calculator using light emitting diode (LED) display. Various companies produced a large number of calculators, programmable or nonprogrammable.

Generally, computers are classified into three generations. In 1938, John Vincent Atanasoff (1903–1995) and Clifford Berry (1918–1963) built first prototype electronic digital computer named Atanasoff-Berry computer. Further improvements in the design of the digital computer continued up to 1946. This phase of computers is called First generation. These computers used vacuum tubes making them very heavy and were capable of performing a variety of Boolean logical operations on its data. In the Second generation of computers spanned from 1947 to 1962, transistors were used in place of vacuum tubes in computers. The first computer built and used was Universal Automatic Computer (UNIVAC1). In 1953, IBM made the first 700 series computers and then made a number of series with improved versions. These computers were having memory and operating systems, mechanical input systems (card readers) and output devices (printers). In mid 1960s, the computers were quite heavy and large in size and were installed in separate buildings called computer centers. In the Third generation (1963 onward) integrated circuits were used which reduced the size of computers immensely and at the same time these were more powerful and reliable. The third generation computers could run many computer programs simultaneously. Pier Giorgio Perotto (1930–2002) led the design team that built Programma 101 in Olivetti Company: it is usually called Olivetti Programma101 or Perottina or P101. It is all in one computer and can be considered the first commercial programmable desktop calculator. In this computer no ICs were used since they were in their infancy. It was launched at the 1964 New York's World Fair and its mass production began in 1965. In 1970s, Perotto designed computer named the Olivetti P6060, the first personal computer with integrated floppy-disk drive, the other models were Olivetti P6040 (the first microprocessor based computer) and the Olivetti P6066 (improved version of P6060). Other desktop computers came in mid 1970s and the first home computers, TRS-80, Apple II and Commodore Personal Electronic Transact, appeared in 1977. In 1980, Microsoft Disc Operating System (MS-DOS) was launched and in 1981 IBM introduced the personal computer (PC) for home and office use. Three years later, Apple gave us Macintosh computer and in 1990s Window Operating System was introduced. In 1990, the all-in-one style desktop computers were invented with computer towers that user could store on the floor under the desk. Full-sized desktops came by moving the processor of the desk itself and working with larger monitors. Cathode ray tubes (CRTs) had been used for video display for several decades until being replaced by plasma, liquid crystal display (LCD), solid state devices such as thin film transistors (TFTs), light emitting diodes (LEDs) and organic light emitting diodes (OLEDs).

Very recently in December 2024, Hartmut Neven (1964–present), founder of Google Quantum Artificial Intelligence, announced that Google has made a new quantum chip named *Willow* which acts as the most convincing 'prototype quantum computer (the details of it appeared in Nature under the name Google quantum AI and collaborators). Unlike traditional bits (0 and 1), Willow used qubits which can exist in multiple states simultaneously enabling exponential computational power. One of the main feature of Willow is that it can reduce errors exponentially as we scale up using more qubits This breakthrough confirms quantum computing potential to tackle problems beyond the reach of classical computers. Neven emphasized that quantum computing paired with artificial intelligence will drive humanity closer to addressing its greatest challenges.

The development of television is another interesting field in solid-state electronics. In 1897, Karl Ferdinand Braun (1850–1918), German physicist, invented cathode ray tube (CRT) and the earliest version of CTR was called the "*Braun tube*". The Braun tube was a cold-cathode diode, a modification of the Crooke's tube with phosphorous-coated screen. Braun was the first to conceive the use of CRT as a display device and is called "the father of television". In 1922, John Bertrand Johnson (1887–1970) and Harry Weinhart developed the first CRT with hot cathode which became the commercial product. The main advantage of the hot cathode CRT was to reduce the accelerating voltage of the anode. Vladimir Kosma Zworykin (1888–1982), Russian-born American inventor, worked on television experiments at Westinghouse Laboratories in Pittsburgh. He applied for two television patents one on Dec. 23, 1923 and second in 1925 in the Unites States. Second patent had essentially the same contents but with minor changes. Zworykin described cathode ray tube as both the transmitter and receiver. On 26 January 1926, John Logie Baird (1888–1946) gave the world's first demonstration of true television before 50 scientists in an attic room in central London. In 1927, he demonstrated television between London and Glasgow, over a distance of 438 miles and formed the Baird Television Development Company, Ltd. Baird Television Ltd filed an application for patent on May 18, 1929 and Patent US1925554A was issued on Sept. 5, 1933. In the end of 1930 Zworykin joined Radio Corporation of America (RCA) and started developing a new type of cathode ray transmitter named iconoscope. In 1932, he presented the first iconoscope and in 1934 presented the image iconoscope. In 1934, Telefunken in Germany manufactured the first commercial electronic TV sets with CRT. So, CRT became one of the main parts in a television, except the electronic circuit, and it was called picture tube. On the other hand, in 1926, Kenjiro Takayanagi (1899–1990) demonstrated a CRT television (TV) in Japan that received images with 40-line resolution and later he improved it to 100-line resolution. In 1933, he developed the main component of a television camera only months after Zworykin did so in the United States. In 1935, Takayanagi had invented an early all-electronic CRT television. Kenjiro Takayanagi is referred to as "*the father of Japanese television*". In 1944, Baird produced world's first color CRT and demonstrated world's first live working television system. In 1954, RCA used color CRTs in first mass-produced color TV sets. The size of CRTs increased from 19 inches in 1938 to 43 inches in 1989. In 1987, Zenith Company developed flat screen CRTs for computer monitors and these were later used in TV sets. Quantum computing is the most recent area in computer technology. The field of optoelectronics also occupies a central place in electronics. Optoelectronics is the communication between optics and electronics and includes the study, design and manufacture of hardware that convert electrical energy into light and vice versa through semiconductors. Optoelectronic device is basically an electronic device involving light. This academic field covers a wide range of optoelectronic devices such as telecommunication lasers, blue laser, laser diode, optical fiber, LED traffic lights, photodiodes, solar cells, image pick-up devices, information displays, optical storage and remote sensing systems etc.

A project called PLATO (Programmed Logic for Automatic Teaching Operation) was started at the University of Illinois to conduct research on the use of computers in education. During the PLATO programme, in 1964, Donald Bitzer (1934–present) and Hiram Slottow (1921–1989) developed the first plasma screen and Bitzer applied for its patent in 1966 which was granted in 1971. Plasma display panel (PDP) was the first flat panel display that comprises of very small cells containing plasma of ionized gas which responds to electric fields. The Plasma TV sets were the first ones with flat picture screen which were produced commercially. In 1967, E. Stredde developed the first colored plasma panel. The plasma TVs remained very popular up to 2007. With the invention of LCD display panels, the Plasma TVs lost their market after 2007. In digital imaging, a pixel is a smallest addressable and controllable element of a picture represented on the screen. Word pixel is a combination of two words, pix (from pictures shorten to pics) and el (for element). This word was first introduced in 1965, by Frederic Crockett Billingsley (1921–2002) to describe picture element of scanned images in support of American space probes to the Moon and Mars. Some authors explain pixel as "picture cell" as early as 1972. Pixels can be used as a unit of measurement for the picture: 500 pixels per inch, 3000 pixels per inch etc. More are the pixels used to represent an image more closer is the result to the original image. The number of pixels in an image is sometimes called the resolution although resolution has a more specific definition.

LCD is a flat panel display that uses light modulating properties of liquid crystals combined with polarizers. Liquid crystals do not emit light directly instead a backlight or reflector is used to produce images in color or monochrome. In 1964, George Harry Heilmeier (1936–2014) invented LCD display at RCA laboratories which marked the origin of a whole new generation of technical devices. In 1968, Heilmeier and coworkers discovered a new class of nematic crystals for LCD display. In 1968, RCA held a conference in New York city on the discovery of the flat electronic display. In 1970, Wolfgang Helfrich (1932–present) invented twisted nematic (TN) mode of operation and then Martin Schadt (1938–present) and Wolfgang Helfrich constructed such a device which gave LCD the first commercial success and the results were published in 1971. Initially the LCD manufacturers supplied small-size displays to portable products such as digital watches and pocket calculators: seven segment displays were used in these products. In 1988, Sharp Corporation demonstrated a 14-inch active-matrix full-color-full-motion display using TFT array. Observing this Japan launched LCD industry and large size displays were supplied to personal computers and then to TV receivers. LCD was a new technology, which replaced heavy and bulky CRTs displays by LCD screens in nearly all applications. Today LCDs are used in a wide range of applications including televisions, computer monitors, instrument panels, aircraft cockpit displays, indoor and outdoor signals, LCD projectors, digital cameras, watches, digital clocks, calculators, mobile phones and smart phones etc.

The comparison shows that plasma displays are much costlier than LCD displays and consume more power, which is the main reason why plasma TVs lost its market. LCDs give better resolution than plasma screens. But plasma screens have edge in terms of picture quality (bright picture): these have a wide color gamut and can be produced in fairly large sizes: up to about 3.8 m. Moreover, the pixels in a plasma screen can switch on and off thousand times faster than the pixels in an LCD screen, so one gets clear pictures with less blur in plasma screens. One can view plasma screens from a wider angle without seeing distortion in color.

These days the LCD displays are being replaced by light emitting diode (LED) displays. In 1961, James Robert Biard (1931–present) and Garry E. Pittman invented the first infrared LED diode from Texas Instruments and they filed a patent on Aug. 8, 1962 which was issued on Dec. 20, 1966. Infrared LED is commonly used in remote controls for TVs and other electronic devices. Nick Holonyak 1928–2022) was inspired by the work of Biard and Pittman on the infrared LED diode, which emits light that is invisible to humans. In the very next year 1962, Nick Holonyak (1928–2022) and S. F. Bevacqua produced the first LED, which was primarily red in color. Other inventors created LEDs that produced other colors, like yellow, orange, and green. In 1968, Howard C. Borden and Gerald P. Pighini developed the first LED display while working at Hewlett-Packard. In the next year that is 1969, Borden and Pighini made the first alphanumeric LED device using integrated circuits and it created another revolution in digital display technology. These LEDs were single color LEDs so was the LED display. In 1977, James P. Mitchell demonstrated the first monochromatic all-LED flat panel television screen. In the early 1990s, Isamu Akasaki (1929–2021) and coworkers invented the first bright blue light emitting LED. The blue light emitting LED was a key to white light emitting LED that made a breakthrough in illumination.

In the recent developments, the LCDs are being replaced by organic LEDs (OLEDs), which can be easily made into different shapes. These days the latest display screens are made of OLEDs. In 1987, Ching Wan Tang (1947–present) and Steven Van Slyke (1956–present), chemists, built the first practical OLED, which used a two-layer structure: one transporting electrons and other transporting holes. These two layers were such that electron–hole recombination takes place in the middle of the organic layer and hence the emission of light. Later multilayer polymers were developed marking a start of plastic electronics. In 1999, Kodak and Sanyo entered into partnership and developed the world's first 2.4 inches active-matrix, full color OLED display. Thereafter the plastic electronic industry grew rapidly and today we have OLED TV sets. LED and OLED being solid state devices have longer life and cannot be easily damaged with external shock unlike fluorescent or incandescent bulbs. In addition OLEDs have lower response time, wider color gamut, virtually infinite color contrast and viewing angles, thinner and having lower weight for a given display size and a similar profile.

The latest development is of a small and portable computer called *laptop*: a computer which fits in the user's lap. In a laptop all the input/output components, display screen, key board, data storage device, an optical disc drive, small speakers, touchpad/trackpad, operating system, processor and memory are combined in a single unit. Laptop can be powered from an internal battery or an external power supply using an AC adapter. In 1981, Adam Osborne (1939–2003) developed the first truly portable computer called Osborne-I and is recognized as the first true laptop computer. It weighed 24.5 pounds and had a 5 inch display. In 1981, Epson released the Epson HX-20, the first portable computer with a built-in printer. In 1983, Radio Shack released the TRS-80 model-100 portable computer with LCD display in the USA. Actually TRS-80 model-100 was originally manufactured by Kyocera Electronic Company, USA, and sold in Japan, but later the rights were sold to Radio Shack. In 1986, Commodore International released the first portable computer named Commodore SX-64 with a full-color display screen. In the same year 1986, IBM released a laptop, the PC Convertible, weighing 12 pounds: the first computer with weight under 15 pounds. In 1992, Olivetti developed and released the first laptop featuring a touchpad. After this a number of manufacturers released laptops with improved versions and with lightweight.

Another very important and useful solid-state electronic device is *mobile phone*, which caused revolution in the world of communication. A mobile phone is a portable telephone that can make and receive calls over a radio frequency link while the user is moving within a telephone service area. The development of MOS and LSI technology, information theory and cellular networking led to the development of affordable mobile communication. In 1973, John Francis Mitchell (1928–2009) and Martin Cooper (1928–present) of Motorola made the first handheld mobile phone weighing 2 kg (4.4 lbs). Martin Cooper is called "Father of the cell phone. In 1979, Nippon Telegraph and Telephone (NTT) launched the world's first cellular network in Japan. The advancements in the development of mobile phones are classified into three generations. The first generation (1G) systems used analog cellular technology and could support more than one simultaneous calls. In 1983, Mitchell and his team from Motorola were the first to introduce the truly portable commercial cell phone to the public. In 1990s, the adoption of MOSFET-based RF power amplifiers and RF circuits led to the introduction of digital signal processing in wireless communication. In 1991, the second generation (2G) digital cellular technology was launched in Finland based on Global System for Mobile (GSM) standard: the first digital mobile telephone standard. In 2001, the third generation (3G) was launched in Japan on the Wide-band Code Division Multiple Access (WCDMA) standard. Then came two 4G technologies based on the Worldwide Interoperability to Microwave Access (WiMAX) standard. The 5G technology has been launched which is a major phase in mobile communication standards beyond the 4G/IMT-Advanced standards where IMT stands for International Mobile Telecommunications. Further, 6G network technology is expected to be developed and released by 2030.

Most recent are the *smart mobile phones* that combine cellular and mobile computing functions into one unit. Smart phones use MOS integrated circuit chips along with various sensors and support wireless communication protocols such as blue tooth, WiFi (Wireless Fidelity) or satellite navigation. Smart phones support wider software, internet (including browsing over mobile broadband) and multimedia features (including music, cameras, video calling and gaming etc.) alongside core phone functions such as voice calls and text messaging etc. Today smart phones are so advanced that whole world is in your hands.

There were many early attempts made for the development of *Internet* for sending information from one computer to another. Robert Kahn (1938–present) came up with some initial ideas of transmission for what later became the Transmission Control

Protocol (TCP). In 1973, Vinton Gray Cerf (1943–present) joined Kahn and they together developed TCP and Internet Protocol (IP). After this the protocol was divided into two separate layers: first the host to host communication would be handled by TCP and the internetwork communication with IP. The two together are usually referred to as TCP/IP and form an important part of the basis for the modern Internet. Vinton Cerf (1943–present) is the architect of the Internet and is known as *the father of Internet*. The first workable prototype of the internet came in late 1960s with the creation of the Advanced Research Projects Agency Network (ARPANET) originally funded by the U. S. Department of Defense. Throughout the 1980s, researchers and scientists used to send files and data from one computer to another. However in 1991, Tim Berners-Lee (1955–present), a computer programmer of Switzerland, introduced the world-wide web (www or simply the web) of information that anyone on the internet could retrieve in addition to sending files from one place to another: the internet that we know today. The web helped popularize the internet among people.

In 1991, Percival joined the Commonwealth Scientific and Industrial Research Organisation (CSIRO) in the team of O'Sullivan as a key member and during this time, they developed the highly successful Wireless LAN technology which was patented by the CSIRO in 1996. This technology is a part of all recent WiFi implementations. WiFi is used for local area networking of devices and internet access, allowing nearby digital devices to share data by radio waves. WiFi technology is most widely used for computer networks in homes, small office networks to link desktops, laptops, smart phones, smart speakers, printers, tablets, kindles, and game consoles to a wireless router to connect them to internet. Victor Hayes (1941–present) and Bruce Tuch were involved in designing the first WiFi protocol standards; they released its first version 802.11a in 1997 and its updated version 802.11b, in 1999. Victor Hayes chaired the IEEE committee that created the 802.11 standards in 1997, therefore, he has been called the *father of Wi-Fi*. In 1999, Apple used the WiFi system for the first time in iBook series of laptops, a major commercial breakthrough. When WiFi first started, from 1998–2005, it was used for emailing and browsing the internet enterprise. From 2005–2010, WiFi was used for emailing, and browsing the internet enterprise, playing and downloading videos, music and applications including social media and voice services. Now it has become a household necessity because all the electronic equipments can be operated with WiFi system.

In 1954, Harold Hopkins (1918–1994) and Narinder Singh Kapany (1926–2020) at Imperial College, London made image-transmitting bundles with over 10,000 fibers, and subsequently achieved image transmission through a 75 cm long bundle. In 1960, Kapany coined the term *fiber optics* after writing an article in *Scientific American* that introduced the topic to a wide audience. He subsequently wrote the first book entitled, "*Fiber Optics-Principles and Applications*" published by Academic Press in 1967. In 1965, Charles Kuen Kao (1933–2018) and George Alfred Hockham (1938–2013) were the first to mention that in optical fiber the attenuation of optical signal could be reduced below 20 decibels per kilometer making fibers a practical communication medium. In 1966, Charles Kao made a discovery that led to breakthrough in fiber optics. Kao realized that by carefully purifying silica glass, bundles of thin glass fibers could be synthesized that would be capable of carrying huge amount of information in the form of optical signal over long distances with minimal signal attenuation. In 1970, they could achieve attenuation of about 20 decibels per kilometer by doping silica with titanium and patented it in the same year. A few years later they were able to reduce it to 4 decibels per kilometer after refinements in the synthesis and design of optical fibers. Ultimately after several refinements the optical fiber became much cheaper than the traditional copper wires. Optical fiber can also act as a laser medium when doped with certain rare-earth elements. For the discovery of optical fiber Charles Kao was awarded 2009 Nobel Prize in Physics. Today optical fibers are used in signal communication which permits transmission over longer distances with less distortion and at higher bandwidths than electrical cables.

Nanoelectronics is the most recent field in electronics. Organic light emitting diodes have organic material as their light emitting layer. Organic materials are classified into two categories: small molecule LED (SMOLED) and polymeric LED (PLED). Examples of nanophotonic applications that involve the emission of light include quantum dots, OLEDs, sensor applications and the next generation silicon-based emitting devices. The wavelength of light emitted by a quantum dot depends on its size which generally ranges from 10–100 nm in diameter. Some of the advantages of nanoelectronic devices are lower noise, less dissipation of energy, occupy less space and require lower input signal etc. It will yield efficient and cheaper solar cells. Nanophotonics will provide highly dense storage photonic systems.

In 1916, Albert Einstein (1879–1955) proposed that three processes are responsible for the formation of a spectral line: spontaneous emission, stimulated emission and absorption of energy. To illustrate these processes consider an atom with two states having energy E_1 and E_2 with $E_2 \rangle E_1$. Let the atom be in the ground state with energy E_1. If an energy $E_2 - E_1 = h\nu$ is supplied to the atom from an external source, it will absorb energy and will go to the excited state with energy E_2. The atom can come back to the ground state by emitting energy $h\nu$ spontaneously as the excited state is unstable. The transition to the ground state and hence the emission of radiation can also be simulated with the help of external radiation. Einstein associated coefficients, called Einstein coefficients, with each of absorption, simultaneous and stimulated emissions and developed a theory which ultimately proved to be the quantum theory of lasers. Einstein coefficients are the measure of probability of absorption, simultaneous and stimulated emissions. After the proposal of Einstein coefficients, the work on stimulated emission of radiations got an impetus. One has to wait for about forty years until, in 1953, James Power Gordon (1928–2013), Herbert J. Zeiger (1925–2011) and Charles Hard Townes (1915–2015) invented, for the first time, the amplification by stimulated emission of microwave radiations in ammonia and Townes named this device ammonia *maser* (microwave amplification by stimulated emission of radiations). James Gordon also worked on the measurement in quantum physics and introduced the concept of "measurement operator". Nikolay Basov (1922–2001) and Alexander Prokhorov (1916–2002) described independently, the theoretical principle governing the operation of a maser. They also proposed a method for the production of population inversion using inhomogeneous electric and magnetic fields. In 1952, they presented their results at a national conference and these were published in 1954. Modern masers can be designed to produce coherent waves in microwave, radio wave and infrared wave regions. For this reason Charles Townes suggested replacing microwave by word molecular in maser. Charles Townes, Nikolay Basov, and Alexander Prokhorov were awarded 1964 Nobel Prize in Physics for their theoretical work leading to maser.

Charles Hard Townes (1915–2015), Public Domain, URL: https://en.wikipedia.org/wiki/Charles_H._Townes#/media/File:Charles_Townes.jpg

Later Charles Townes and Nikolay Basov described the principle of optical maser which was later called *laser* (light amplification by stimulated emission of radiations): the name laser was given by Gordon Gould (1920–2005) in 1957. In 1958, Arthur Leonard Schawlow (1921–1999) and Charles Townes gave a theoretical interpretation of stimulated emission, which led to the discovery of laser. In 1960, Theodore Harold Maiman (1927–2007) invented the first laser which he got patented in his name: it used synthetic ruby crystal which emitted pink colored coherent light. In 1960, Ali Javan (1926–2016), William Ralph Bennett (1930–2008) and Donald Herriott (1928–2007) constructed the first gas laser using a mixture of He and Ne gases. It is noteworthy that the He–Ne laser is the first laser capable of continuous operation in the infrared region. Lasers emit coherent light which can be focused in a small spot and can travel great distances. Nicolaas Bloembergen (1920–2017) is best known for developing a new field of science now called nonlinear optics. He showed that by mixing two or more beams of laser light in the visible region one can produce a single laser source with different frequency in the infrared and ultraviolet regions: a field of laser spectroscopy which allows precise observations of atomic structure using lasers. In 1957, Kai Manne Borje Siegbahn 1918–2007) and co-workers were able to measure the first photoelectron spectrum of MgO with extremely sharp lines and with expected intensities. With this work Siegbahn developed the method of Electron Spectroscopy for chemical analysis now usually called X-ray Photoelectron Spectroscopy (XPS). In 1958, Siegbahn and coworkers also obtained the first evidence of chemical shifts for a metal and its oxides and for Auger electron lines. Both Bloembergen and Siegbahn were awarded 1981 Nobel Prize in Physics.

Jun-ichi Nishizawa (1926–2018), a Japanese engineer, made a number of inventions in electronics. He invented PIN diode in 1950, static induction transistor and static induction thyristor in 1955, and proposed the concept of a semiconductor maser and semiconductor laser in 1957. In 1963, he also proposed fiber optic communications: the use of optical fiber for optical communication. In 1962, Robert Noel Hall (1919–2016) demonstrated the first semiconducting laser diode device, which was made of gallium arsenide and emitted near-infrared band of spectrum. Hall also developed magnetron for radar jamming that led to the development of microwave oven. Again in 1962, Nick Holonyak Jr. (1928–2022) demonstrated the first semiconductor laser with a visible emission. This semiconductor laser could only be used in pulsed-beam operation when cooled to liquid nitrogen temperature (77 K). Zhores Ivanovich Alferov (1930–2019) developed semiconductor hetrojunctions and contributed significantly towards the creation of hetrostructure physics and electronics. In 1966, Alferov's lab created the first lasers based on heterostructures, again with pulsed-beam operation. In 1968, Alferov and coworkers produced the first continuous-wave semiconductor heterojunction laser operating at room temperature. This achievement came a month ahead of Izuo Hayashi (1922–2005) and coworkers as they also produced a continuous-wave room temperature heterojunction laser in 1969–1970. A number of other lasers were invented which are used for different purposes and some of them are listed below:

Zhores Ivanovich Alferov (1930–2019), CC BY 4.0, URL: https://en.wikipedia.org/wiki/Zhores_Alferov#/media/File:Zhores_Alfyorov_2012.jpg

In 1964, Chandra Kumar Naranbhai Patel (1938–present) invented carbon dioxide laser and was most powerful continuously operating laser of its times: also capable of pulsed-power output operation. It is being used world-wide as a cutting tool in surgery and industry. There are number of advanced diode-pumped solid state (DPSS) lasers and Nd:YAG laser is one of them (Nd:YAG means Neodymium-doped Yttirium, Aluminium Garnet crystal). In 1964, J. E. Geusic and coworkers invented the Nd:YAG laser and demonstrated its action. YAG laser can be operated both in pulsed and continuous modes. This laser has a number of applications in medicine and surgery. It is used for cutting and welding semiconductors, steel and various alloys. It is also used in military and laser designators and laser rangefinders. In 1982, Peter Moulton (1946–present) invented Titanium-sapphire laser (sapphire crystal Al_2O_3 doped with Ti) in its continuous-wave version. He has also made important advances in high-power diode-pumped solid state lasers, parametric oscillators, and long-wavelength bulk and fiber lasers. Later these lasers were shown to generate ultrashort pulses through Kerr lens mode-locking. These lasers are used mainly in scientific research because of their tenability and their ability to generate ultra-short pulses.

In 1971, John Madey (1943–2016) developed the first free-electron laser (FEL). To create an FEL, a beam of electrons is accelerated to almost the speed of light with the help of electron accelerators. These are tunable lasers and can be built for a wider frequency range than any other type of lasers: frequency ranges from microwaves, infrared, visible spectrum and ultraviolet to X-rays. As it causes minimal collateral damage to adjacent tissues therefore it is used in performing very precise surgeries on brain tumors, and optic nerve etc. It is used in melting fats without harming the overlying skin. FEL is used for military purposes as one can achieve very high power peaks with it. It is a hot candidate for antiaircraft and antimissile directed energy weapons. The apparatus of FEL is bulky and expensive (needs accelerator, and many vacuum pumps to create vacuum for electron beams).

Bibliography

Adams WG, Day RE (1877) The action of light on selenium. Proc Roy Soc 25:113–117

Akasaki I, Amano H, Itoh K, Koide N, Manabe K (1992) GaN-based UV/blue light emitting devices. Inst Phys Conf Ser 129:851–85; Akasaki I, Amano H (2006) Breakthroughs in improving crystal quality of GaN and invention of the p–n junction blue-light-emitting diode. Jpn J Appl Phys 45:9001–9010

Amano H, Akasaki I (1990) Fabrication and properties of GaN p–n junction LED. Mater Res Soc. Extended abstract (EA-21), pp 165–168 (Fall Meeting 1989)

Armstrong EH (1913) Patent was applied in 1913 which was issued as U. S. patent 1113149 on Oct 1914

Armstrong EH (1914) Operating features of audion. Electrical world, pp 1149–1152; (1917) Operating features of audion. Ann NY Acad Sci 27:215–243

Armstrong EH (1936) A method of reducing disturbances in radio signaling by a system of frequency modulation. Proceedings of the IRE, vol 24 [presented first, on Nov 6, 1935, to the Institute of Radio Engineers, New York], pp 689–740

Atalla MM (1960) Semiconductor triode, U. S. patent. 3056888 filed on Aug 17, 1960 and granted on Oct 2, 1962
Atanasoff JV (1973) Computing machine for the solution of large systems of linear algebraic equations. In: Brian R (ed) The origins of digital computer, selected papers. Chapter 7.2. Springer, New York
Babbage HP (1982) Babbage's calculating engines. Tomash Publishers and MIT Press, Los Angeles and Cambridge, Mass. [The best edition of the contemporary writings of Babbage and others concerning his machines, collected and published by his son after Babbage's death]
Backus JW (1954) The IBM speedcoding system. J Assoc Comput Mach 1:4–6; Backus JW (1979) The history of Fortran I, II, and III. Ann Hist Comput 1(1). (July–September 1979)
Backus JW, Herrick H (1954) IBM 701 speedcoding and other automatic programming systems. In: Symposium on automatic programming for digital computers, office of technical services. US Department of commerce, Washington DC, pp 106–113
Bardeen J (1947) Surface states and rectification at a metal-semiconductor contact. Phys Rev 71:717–727; Brattain WH (1947) Evidence for the surface states on semiconductors from change in contact potential. Phys Rev 72:345; Brattain WH, Shockley (1947) Density of surface states on silicon deduced from contact potential measurements. Phys Rev 72:345
Bardeen J, Brattain WH (1948) The transistor, a semiconductor tiode. Phys Rev 74:230; (1948) Nature of the forward current in germanium point contacts. Phys Rev 74:231; (1949) Physical principles involved in transistor action. Phys Rev 75:1208
Basov NG, Belenov EM, Letokhov VS (1965) Diffraction synchronization of lasers. Sov Phys Tech Phys 10:845
Basov NG, Bogdankevich OV, Grasyuk AZ (1966) 9B4-Semiconducting lasers with radiating mirrors. IEEE J Quantum Electron 2:594–597
Becquerel AE (1839) Recherches sur les effets de la radiation chimique de la lumiere solaire au moyen des courants electriques. Comptes Rendus de L'Academie des Sciences 9:145–149; (1839) Memoire sur les Effects d'Electriques Produits Sous l'Influence des Rayons Solaires. Comptes Rendus de l'Academie des Sciences 9:561–567
Berry CE, Doyle EW, Sibyl MR, Washburn HW (1946) A computer for solving linear simultaneous equations. J Appl Phys 17:262–272
Berzelius JJ (1825) On the mode of obtaining silicium, and on the characters and properties of that substance. Philos Mag J 65:254–267 [Discovered in 1824 but published in 1825]
Bethe HA (1942) Theory of high frequency rectification by silicon crystals. MIT radiation laboratory, United States Department of Atomic Energy, USA
Biard JR (1963) Photosensitive transistor chopper using light emissive diode, US patent 3304431 filed on Nov 29, 1963 and issued on Feb 14, 1967; (2002) The invention of the LED: a presentation delivered at the University of Texas at Dallas (UTD)
Biard JR, Pittman G (1966) Semiconductor radiant diode, patent: US3293513 filed on Aug 8, 1962 and issued on Dec 20, 1966
Bidwell S (1885) On the sensitiveness of selenium to light, and the development of a similar property in Sulphur. Proc Phys Soc (London) 7:129
Bitzer DL, Slottow HG (1966) The plasma display panel-a digitally addressable display with inherent memory. Proceeding fall Joint computer conference vol 29, AFIPS conference proceeding, Washington, DC, pp 541–547; Bitzer DL (1999) Inventing the ac plasma panel. Inf Display 15:22–27
Bitzer DL, Slottow HG, Wilson RH (1971) Gaseous display and memory apparatus. U.S. patent 3,559,190, Jan 26, 1971
Borden HC, Pighini GP (1969) Solid state displays. Hewlett-Packard J 2–12
Bose JC (1901) Detector for electrical disturbances, U. S. patent 755,840 (Filed on Sept 30, 1901 and issued on March 29, 1904)
Bower RW, Dill HG (1966) Insulated gate field effect transistors fabricated using the gate as source-drain mask. IEEE Int Electron Devices Meet 12:102–104

Braun F (1874) Uber die stromleitung durch schwefelmetalle [On current conduction through metal sulfides]. Ann Phys Chem 153:556–563; translated by Sze SM (1991) WorldScientific Publishing, p 377

Braun F (1897) Ueber ein verfahren zur demonstration und zum studium des zeitlichen verlaufs variabler strÿme (On a process for the display and study of the course in time of variable currents). Annalen der Physik und Chemie 60:552–559

Braun F (1906) Elektrotechnische Z 52:1199

Chapin DM, Fuller CS, Pearson GL (1954) A new silicon p-n junction photocell for converting solar radiation into electric power. J Appl Phys 25:676–677

Cooper M, Dronsuth RW, Leitich AJ, Lynk CN, Mikulski JJ, Mitchell JF, Richardson RA, Sangster JH (1973) All signalling, Patent US403725A, filed on Oct 17, 1973 by Motorola Inc, and granted on Sept 16, 1975

Davydov B (1938) The rectifying action of semiconductors. J Tech Phys (USSR) 5:87–95

De Forest L (1907) Application for the patent of triode was filed on Jan 29, 1907 which was granted in Feb 18, 1908 as US patent 879,532

Edison TA (1883) US patent 307,031, the first US patent for an electronic device was filed on Nov 15, 1883

Einstein A (1916) Strahlungs-emission und -absorption nach der quantentheorie. Verh Dtsch Phys Ges 18:318–323

Einstein A (1917) Zur quantentheorie der strahlung (On the quantum theory of radiation). Phys Z 18:121–128

Faggin F, Klein T, Vadasz L (1968) Insulated gate field effect transistor integrated circuits with silicon gates. IEEE International Electron Devices Meeting, Washington, D.C

Faraday M (1833) On a new law of electric conduction. Philos Trans Roy Soc (London) 123:507–515

Fessenden RA (1907) Recent progress in wireless telephony. Sci Am 68–69

Fessenden RA (1908) Wireless telephony, transactions of the American institute of electrical engineers, vol XXVII, Part 1, pp 553–629

Fleming JA (1905) Instrument for converting alternating electric currents into continuous currents, US patent no. 803,684 patented on Nov 7, 1905

Fritts CE (1883) On a new form of selenium photocell. Am J Sci 26:465

Geusic JE, Marcos HM, Van Uitert LG (1964) Laser oscillations in Nd-doped yttrium aluminium, yttrium gallium and gadolinium garnets. Appl Phys Lett 4:182–184

Google quantum AI and collaborators (2024) Quantum error correction below the surface code threshold. Nature. https://doi.org/10.1038/s41586-024-08449-y

Gordon JP, Zeiger HJ, Townes CH (1955) The maser—new type of microwave amplifier, frequency standard, and spectrometer. Phys Rev 99:1264–1274

Gudden B (1931) Conduction electrons and photoelectrons in insulators and semiconductors. Z Phys 32:825–833

Haisty RW, Johnson RE, Mehal EW (1964) Filed patent. 390,298 on Aug 18, 1964 for three-dimensional integrated circuits and methods of making same and was instantly granted

Hall EH (1879) On a new action of the magnet on electric currents. Am J Math 2:287–292

Hall RN, Fenner GE, Kingsley JD, Soltys TJ, Carlson RO (1962) Coherent light emission from GaAs junctions. Phys Rev Lett 9:366

Hayashi I, Panish M, Foy P (1969) A low-threshold room-temperature injection laser. IEEE J Quantum Electron 5:211–212

Hayashi I, Panish M, Foy P, Sumski S (1970) Junction lasers which operate continuously at room temperature. Appl Phys Lett 17:109–111; Panish MB, Hayashi I, Sumski S (1970) Double-heterostructure injection laser with room temperature thresholds as low as 2300 A/cm^2. Appl Phys Lett 16:326–327

Heilmeier GH (1970) Liquid crystal display devices. Sci Am 222:100–106

Heilmeier GH, Zanoni LA, Barton LA (1968) Anew electro-optic effect in certain classes of nematic liquid crystals. Proc IEEE 56:1162–1171; Helmeier GH, Castellano J, Zanoni LA (1969) Guest-host interactions in nematic liquid crystals. Mol Cryst Liq Cryst 8:293–304

Herman F, Callaway J (1953) Electronic band structure of germanium crystals. Phys Rev 89:518
Hofstein SR, Heiman FP (1963) The silicon insulated-gate field-effect transistor. Proc IEEE 51:1190–1202
Holonyak N, Bevacqua SF (1962) Coherent (visible) light emission from $Ga(As_{1-x}P_x)$ junctions. Appl Phys Lett 1:82–83
Holst G, Tellegen BDH (1934) Means for amplifying electrical oscillations. US patent 1945040 issued on January 1934 [Invention of pentode]
Hopkins HH, Kapany NS (1954) A flexible fibrescope, using static scanning. Nature 173:39–41
Hull AW (1921) The effect of a uniform magnetic field on the motion of electrons between coaxial cylinders. Phys Rev 18:31
Javan A, Bennett WR, Herriott DR (1961) Population inversion and continuous optical maser oscillation in a gas discharge containing a He–Ne mixture. Phys Rev Lett 6:106
Johnson JB (1922) A low voltage cathode ray oscillograph. JOSA 6:701–712
Kahng D (1960) Electric field controlled semiconductor device. US patent. 3102230 applied on May 31, 1960 and granted on Aug 27, 1963
Kahng D, Atalla MM (1960) Silicon-silicon dioxide field induced surface devices. IRE-AIEE solid state device research conference at Carnegie institute of technology, Pittsburg, PA
Kao KC, Hockham GA (1966) Dielectric-fibre surface waveguides for optical frequencies. Proc IEE 113:1151–1158
Kerwin RE, Klein DL, Sarace JC (1969) Method for making MIS structures, patent: US3475234 issued on Oct 28, 1969 and assigned to bell telephone laboratories Inc
Kilburn T (1948) A storage system for use with binary digital computing machines, manchester.ac.uk (Ph.D. thesis). University of Manchester
Kilby J (1959) US patent 3138743 for miniaturized electronic circuits the first integrated circuit was filed on Feb 6, 1959, and then patent. U.S. 3138744 refiled on May 6, 1959
Lilienfeld JE (1925) US patent 1,745,175 for method and apparatus for controlling electric currents, was first filed in Canada on Oct 22, 1925 describing a field-effect transistor
Madey JMJ (1971) Stimulated emission of bremsstrahlung in a periodic magnetic field. J Appl Phys 42:1906–1913
Maiman TH (1960) Stimulated optical radiation in ruby. Nature 187:493–494
Mitchell JP (1978a) Mitchell's abstract title light emitting diode television screen was published in the honors group of the 37th annual science talent search for the Westinghouse science scholarship and awarded in 1978, this qualified Mitchell for the UNI paper presentation that followed
Mitchell JP (1978) Mitchell's modular LED x-y (horizontally and vertically digitally scanned array system) was cited in the 29th international science and engineering exposition book of abstracts, published by the science service, Washington D.C., p 97
Moore G (1965) Cramming more components onto integrated circuits. Electron Mag 38:114–117
Mott NF (1939) The theory of crystal rectifiers. Proc Roy Soc (London) 171:27–144
Napier J (1614) The description of the wonderful canon of logarithms (PDF). Translated by Wright, Edward; Bruce, I., 17 century maths.com. Retrieved 14 March 2022
Nishizawa J (2009) Extension of frequencies from maser to laser. Proc Jpn Acad Ser B Phys Biol Sci 85:454–465
Noyce RN (1959) US patent 2,981,877 was filed on July 30, 1959 and issued on April 25, 1961 for monolithic ICs
Ohl RS. Light-sensitive electric device, US patent 2402662 (Filed on May 27, 1941, and issued on June 25, 1946)
Onoda K, Igarashi R, Wada T, Nakanuma S, Tsujide T. Filed US patent 3665423A on March 15, 1969 for Memory matrix using MOS semiconductor element but is pending
O'Sullivan JD, Daniels GR, Percival T, Ostry D, Deane J, (1996) Wireless LAN, patent US 5487069 published 23 Jan 1996
Patel CKN (1964) Continuous-wave laser action on vibrational-rotational transitions of CO_2. Phys Rev 136A:1187

Pearson GL, Herring C (1954) Magneto-resistance effect and the band structure of single crystal silicon. Physica 20:975–978
Pearson GL, Suhl H (1951) Magneto-resistance effect in oriented single crystals of germanium. Phys Rev 83:768–776
Pickard GW (1906) Means for receiving intelligence communicated by electric waves. US patent 836,531 (Filed on August 20, 1906 and Issued on Nov 20, 1906)
Richardson OW (1901) On the negative radiation from hot platinum. Philos Camb Philos Soc 11:286–295
Schadt M, Helfrich W (1971) Voltage-dependent opticalactivity of atwisted nematic liquid crystal. Appl Phys Lett 18:127–128
Schawlow AL, Townes CH (1958) Infrared and optical masers. Phys Rev 112:1940–1949
Schottky W (1939) Zer halbleitertheorie der sperrschicht und spitzengleichrichter. Z Phys 113:367–414
Schuster A (1874) On unilateral conductivity. Phil Mag 48:251–257
Shaff JH, Ohl RS (1947) Development of silicon crystal rectifiers for microwave radar receivers. Bell Syst Tech J 26:1
Shockley W (1951). US patent 2,569,347 for circuit element utilizing semiconductive material (filed on June 26, 1948)
Shockley W (1949) The theory of p-n junctions in semiconductors and p-n junction transistors. J Bell Sys Tech 28:435–489 [reprinted by Sze SM (1991) Semiconductor devices: pioneering papers, World Scientific Publishing]
Shockley W, Sparks M, Teal G (1951) P–n junction transistors. Phys Rev 83:151
Siegbahn K, Nordling C, Sokolowski E (1957) Precision method for obtaining absolute values of atomic binding energies. Phys Rev 105:1676; (1957) Proceeding Rehovoth conference nuclear structure, North-Holland publication Co., Amsterdam, p 291
Siegbahn K, Nordling C, Sokolowski E (1958) Chemical shifts of photo and auger electron lines, p 291. In: Rehovoth conference nuclear structure, proceeding, Rehovoth, Israel 8–14 Sept 1957. North–Holland Publication, Amsterdam; (1958) Chemical shift effect in inner electronic levels of Cu due to oxidation. Phys Rev 110:776
Sive JN (1949) The double surface transistor. Phys Rev 75:689
Smith W (1873) Effect of light on selenium during the passage of electric current. Nature 7:303. [In Feb 1873, Smith presented the paper entitled, Electrical properties of engineers]
Stredde E (1967) The development of a multicolor plasma display panel, University of Illinois, Coordinated science laboratory report R-370, Nov 1967. [colored display panel]
Tanenbaum M (1954) Bell labs notebook no. 25505, p 30 [Jan 26, 1954]
Tanenbaum M, Valdes LB, Buehler E, Hannay NB (1955) Silicon n–p–n grown junction transistor. J Appl Phys 26:686–692
Tang CW (2018) Organic light-emitting diode (OLED) US patent no 4,356,429, inducted in 2018
Tang CW, Van Slyke SA (1987) Organic electroluminescent diodes. Appl Phys Lett 51:913–915; Tang CW, Van Slyke SA, Chen CH (1989) Electroluminescence of dopped organic thin films. J Appl Phys 65:3610–3616
Teal GK (1954) Some recent developments in silicon and germanium materials and devices, Presented at the national conference on airborne electronics Dayton, Ohio, USA
Teal GK, Little JB (1950) Growth of germanium single crystals. Phys Rev 78:647
Teal GK, Sparks M, Buehler E (1951) Growth of germanium single crystals containing p–n junctions. Phys Rev 81:637
Thomson JJ (1897) Cathode rays. The Electrician 39:104; (1897) On cathode rays. Phil Mag 44:223–316
Wanlass FM, Sah CT (1963) Nanowatt logic using field-effect metal-oxide semiconductor triodes. International solid-state circuits, conference digest of technical papers, Philadelphia, USA, pp 32–33; (1963) Nanowatt logic using field-effect metal-oxide semiconductor triodes in proceeding technological digital. IEEE

Wanlass FM (1963) US patent 3,356,858 was filed on June 18, 1963 for low stand-by power complementary field effect circuitry and was issued Dec 5, 1967
Williams FC, Kilburn T (1949) A storage system for use with binary-digital computing machines. Proc IEE—Part II: Power Eng 97:183–200
Williams R, Heilmeier GH (1966) Possible ferroelectric effects in liquid crystals and related liquids. J Chem Phys 44:638
Wilson AH (1931) The theory of electronic semiconductors. Proc Roy Soc (London) 133:458–491; (1931) The theory of electronic semiconductors-II. Proc Roy Soc (London) 134:277–287
Winkler C (1886) Mittheilungen uber das germanium. J Prakt Chem 142:177–229
Woodyard JR (1944) Doping of semiconductors. US patent no. 2530110, filed in 1944 and granted in 1950
Zworykin VK (1928) Television system, US patent: 1691324 (filed in 1925 and issued on Nov 13, 1928); (1935) Television system, US patent: 2022450 (filed on Dec 29, 1923 and issued on Nov 26, 1935); (1938) Television systems US patent: 2141059 (filed on Dec 29, 1923 and issued on Dec 20, 1938)

Chapter 17
Smart Materials

There are some materials which have the capacity to change composition, structure and other properties in response to some external (environmental) stimulus and then regain their original state upon the removal of the stimulus. The external stimulus may be temperature, mechanical pressure, electric and magnetic fields etc. Because of their responsiveness, such materials are called "*smart materials*". The term smart materials emerged in late 1980s and 1990s particularly with the development of shape memory alloys. There is not a single person who is credited for coining this term. But it is believed that in 1990, Toshinori Takagi formalized the concept of smart materials to the responsive materials. In 2013, Pablo Zavattieri gave the name active materials to the responsive materials. These are also called intelligent materials although it would be more accurate to say "reactive materials" to all the responsive materials. Sometimes it is argued that it is not possible to come up with an exact definition of smart materials. Some scientists draw a distinction between smart and truly intelligent materials. Some scientists are of the opinion that no material by itself is truly smart but simply responsive. At present there are no materials, which are able to make their own decisions or repair on their own. Smartness and intelligence existed in nature even before the discovery of smart materials. Plants, animals and birds have the ability to adapt to their changing environment. For example sunflowers turn towards sun, lotus flowers bloom with the sunrise and fade away with the sunset, leaves of Mimosa Pudica collapse suddenly when touched. Animals also are intelligent enough but their intelligence is a combination of skills and abilities that allows them to live and adapt to their specific environments. Octopuses, dolphins and chimpanzees are very intelligent animals and can perform some complex tasks. Birds like parrots have the ability to recognize human faces and have a high aptitude for communication. There are different types of smart materials and commonly used are briefly discussed here.

J. S. Galsin, *History of Solid State Physics*,
https://doi.org/10.1007/978-981-95-0504-3_17

Basically smart materials are materials designed to have the specific response to an external stimulus and then reverse back upon the removal of the stimulus. Most of these are synthesized in the laboratory with unusual but desirable properties not found in nature. Smart materials can act as actuators or sensors depending on the nature of their property. Although the name smart material is new but these have been discovered much earlier. A brief historical account of some of the smart materials is given below.

Pyroelectric Materials: There are certain polar dielectrics, in which change in temperature causes the movement of atoms from their neutral positions. As a result, these materials exhibit finite spontaneous polarization, which changes with temperature, and thus produce finite voltage across the material. In 1707, Johann Georg Schmidt found, for the first time, that hot tourmaline crystal attract the ashes from the warm or burning coals, as the magnet attracts iron filings, but repels them after the contact. In 1717, Louis Lemery (1677–1743) found, that small scraps of non-conducting material were first attracted to tourmaline crystal, but then repelled by it after they come in contact. Carl von Linne (1707–1778) was the first to relate this phenomenon to electricity in 1747 although this was not proven. In 1756, Franz Ulrich Theodor Aepinus (1724–1802) established its relation with electricity. In 1824, Sir David Brewster (1781–1868) gave this effect the name pyroelectricity and the materials which exhibit this effect are called pyroelectric materials. The name pyroelectricity is derived from a Greek letter '*pyro*' which means fire. Pyroelectrics are hard and crystalline materials which are classified into two categories: ferroelectrics and nonferroelectrics. In 1878, William Thomson (1824–1907), and in 1897, Woldemar Voigt (1850–1919) developed a theory for the physical processes responsible for pyroelectricity.

Piezoelectric Materials: There are some materials in which when pressure is applied, negative and positive charges appear on the opposite faces of it. This conversion of mechanical energy into electrical energy is called piezoelectric effect: piezo is a Greek word which means 'to press'. The phenomenon of piezoelectricity was discovered in 1880 by French physicists Pierre Curie (1859–1906) and Paul-Jacques Curie (1855–1941) (brothers) leading to some of the mechanisms behind piezoelectricity. It is noteworthy that piezoelectricity and ferroelectricity are closely related to pyroelectricity. In 1881, Jonas Ferdinand Gabriel Lippmann (1845–1921) deduced inverse piezoelectric effect through mathematical calculations from fundamental thermodynamical principles. According to the inverse piezoelectric effect the application of electric field produces internal mechanical strain in a piezoelectric material. Curie brothers quickly demonstrated experimentally the inverse piezoelectric effect. During the World War I, Paul Langevin (1872–1946) began working for the detection of submarines using ultrasonic sound. In 1916 and 1917, Langevin and Constantin Chilowsky (1880–1958) invented an ultrasonic submarine detector which was the first real application work on piezoelectric devices and got it patented. However, the war was over by the time it became operational.

Jonas Ferdinand Gabriel Lippmann (1845–1921), Public Domain, URL: https://en.wikipedia.org/wiki/Gabriel_Lippmann#/media/File:Gabriel_Lippmann2.jpg

Barium titanate is a very strong piezoelectric ceramic material and was discovered independently during the World War II by E. Wainer and A. N. Solomon in 1942, T. Ogawa and Y. Waku in 1944 and by Wul and Goldman in 1945 in three different countries. Barium titanate has maximum permittivity, possesses perovskite structure and is used as a sensor of mechanical vibrations in sonar devices. Polycrystalline barium titanate has positive temperature coefficient of resistance making it a useful material for thermostats and self-regulating electric heating systems. In 2016, Genchi and coworkers have pointed out that due to immense biocompatibility, barium titanate nanoparticles have recently been employed as nanocarriers for drug delivery.

Invar Alloys: It is well-known that most of the materials expand on heating. In 1896, Charles Edouard Guillaume (1861–1938) discovered an alloy of nickel and steel which showed no expansion with the increase in temperature. Alloys showing no thermal expansion were named invar alloys and these are a good example of smart materials. A great deal of effort has gone into finding more invar alloys and these are used in watches so that there is no risk of jamming when they warm up. Many materials get more malleable when they are heated: some become better electrical conductors and others become worse.

Shape-Memory alloys: In 1932, Gustav Arne Olander (1902–1984), at a meeting of Swedish Metallurgical Society, reported that gold-cadmium alloy has two unusual properties that is pseudoelastic behavior and memory effect: the first recorded observation of smart materials. He observed that gold-cadmium alloy can be bent while cooling but returns to its original shape after heating: a shape memory alloy. In 1938, Greninger and Mooradian observed martensic phase in brass (Cu–Zn) at low temperatures but it disappeared with the increase in temperature (martensic phase transition). After about a decade in 1949, Kurdjumow and Khandros and then in 1951, Chang and Read reported that thermoelastic behavior in alloys governs the memory effect: the original idea of thermoelastic behavior is due to Kurdjumow and Khandros. In 1959, William J. Buehler (1896–1987) prepared an alloy of nickel and titanium, usually called Nitinol, after the names of its components. At the laboratory management meeting held in 1961, which Buehler could not attend, David Saville Muzzey, one of the Associate Technical Directors, bent the sample of NiTi alloy and then placed his pipe lighter under the bent sample to see the effect of increase in temperature. To his surprise, the bent sample came back to its original shape. It showed that NiTi alloy remembers its original shape. This effect was called *shape-memory effect* and the alloy which showed this effect was called shape-memory alloy as it remembered its original shape. This remarkable property of NiTi alloy was discovered accidentally. Therefore NiTi alloy is considered to be the first shape-memory alloy. In 1962, the United States Naval Ordnance Laboratory developed an alloy of nickel and titanium (NiTi). In 1962, Fredericks Wang (1932–2022) joined the group of Buehler and in 1965 jointly studied the properties of nitinol particularly the martensite phase transition and established its occurrence. After this researchers synthesized a number of shape-memory alloys which are metallic in nature. Depending on the temperature and internal stress, shape-memory alloy exhibits two phases: martensite phase exists at low temperatures and austenite phase exists at higher temperatures. The concept of thermoelastic martensite also assumed increased importance because of interest in shape-memory alloys and the related effects. The most effective and widely used alloys include NiTi, CuZnAl and CuAlNi. Such type of materials can be used for coffee-pots, thermostats, superelastic spectacle frames and stents for veins. In 1941, Vernon and Vernon discovered polymers which possessed the same properties as exhibited by shape-memory alloys and these are called shape memory polymers. These can be deformed when cold but returns to their original shape when heated. In 1955, Hitov and coworkers and in 1967, Perrone, discovered heat shrinkable covalently cross-linked polyethylenes which became another important milestone in the development of shape memory polymers. Compared with shape memory alloys, shape memory polymers have the advantages of high elastic deformation, low cost, low density, and potential biocompatibility and biodegradability.

Magnetostrictive Materials: The application of external magnetic field changes magnetization of magnetic materials. Magnetostriction is the property of magnetic materials that causes change in their shape or dimensions during the variation of magnetization when an external magnetic field is applied. In other words magnetostriction is the property of magnetic materials that produces magnetostrictive strain in the presence of external magnetic field. This effect was first discovered in 1842 by James Prescott Joules (1818–1889) and is called Joule effect. The property of magnetostriction originated from the presence of domains. When magnetic field is applied, the boundaries between the domains shift and in addition the domains rotate; both of these effects cause the change in the dimensions of a material. The change in the magnetic domains of material results in a change in the material's dimensions is a consequence of magnetocrystalline anisotropy that is it takes more energy to magnetize a crystalline material in one direction than in another. The materials which exhibit the property of magnetostriction are called magnetostrictive materials and these are similar to piezoelectric materials. These materials are nonlinear, and exhibit some hysteresis (less than piezoelectric materials). Some of the mostly used magnetostrictive materials are $Tb_xDy_{1-x}Fe_2$ (Terfenol, here Ter means Terbium, Fe means iron, NOL means Naval Ordinance Laboratory, and D means Disprosium), Fe_xGa_{1-x} (Galfenol), Fe_xAl_{1-x} (Alfer) and cobalt ferrite $CoFe_2O_4$. Another very common magnetostrictive material is amorphous alloy $Fe_{0.81}Si_{0.035}B_{0.135}C_{0.02}$ with its trade name Metglas 2605SC. Emilio Villari (1836–1904) discovered the inverse magnetostrictive effect that is magnetization changes when subjected to mechanical stress. Magnetostrictive materials are used both as sensors and actuators.

Electrostrictive Materials: Electrostriction is a property of all dielectrics which causes change in their shape when an external electric field is applied. It is just analogous to magnetostriction. In ionic solids, the application of electric field causes the displacement of positive ions in the direction of the field and that of negative ions in the opposite direction due to electrostriction. This displacement produces strain (elongation) in the direction of the field (of the order of 0.1% at very high voltages). The strain so produced is proportional to the square of polarization (nonlinear property). It should be noted that the reversal of the electric field does not reverse the direction of deformation.

Although all dielectrics exhibit some electrostriction, there is a special class of ferroelectrics called relaxor ferroelectrics, which exhibit extraordinarily high electrostrictive properties as compared to others. They are nonlinear materials (varying quadratically with field) but exhibit negligible hysteresis. The most commonly used are lead magnesium niobate (PMN), lead magnesium niobate-lead titanate (PMN-PT) and lead lanthanum zirconate titanate (PLZT). The electrostriction produced in lead magnesium niobate is of the order of 0.1% for quite high voltages. In 2013, Hiroyuki Takenaka and coworkers and in 2014, Daniel Phelan and coworkers tried to study the mechanism behind the electrostrictive property but it is still not completely understood. Relaxor Ferroelectric materials are used for energy storage and its conversion as they have high dielectric constants, which is orders-of-magnitude higher than those of conventional ferroelectric materials.

In addition we have some polymers which exhibit the property of electrostriction usually called electrostrictive graft polymers. These comprise of flexible backbone chains with branching side chains. The side chains form cross links with neighboring backbone chains and form crystal units. In 1995, Dagani coined the term "smart polymers" for such polymers that could adapt to their environment. In 2004, Wang and coworkers put forward mechanism of electrostrictive graft polymers according to which when an electric field is applied, it causes the rotation of whole of the polymer unit thereby producing electrostrictive strain and deformation of the polymer.

Photovoltic Materials: The phenomenon of converting light into voltage or electric current is called photovoltaic effect and the materials that convert light into voltage are called photovoltaic materials. In 1839, Alexander Edmond Becquerel (1820–1891), a French physicist, discovered the principle of photovoltaic effect. While performing experiment with metal electrode and electrolyte he observed that conductance rises with illumination. In 1873, Willoughby Smith (1828–1891) discovered that the conductivity of selenium increased significantly when exposed to strong light (photovoltaic effect). He presented a paper on Feb. 12, 1873 at the meeting of the Society of Telegraph Engineers entitled, "Electrical properties of selenium and the effect of light there on". A brief summary of this paper was published on Feb. 20, 1873 in the journal Nature. In 1876, William Grylls Adams (1836–1915) and Richard E. Day discovered that illuminating a junction between selenium and platinum also exhibits photovoltaic effect. They further showed that the intensity of the photovoltaic action depends on the square root of the illuminating power of light. This fact demonstrated that electricity could be produced from light without moving any parts and led to the modern solar cell. In 1883, Charles Fritts (1850–1903) constructed the first solar cell by coating selenium with a thin layer of gold. Fritt's cell, with high cost, had a conversion electrical efficiency of only about 1%. Fritts was hopeful that his solar cells might compete with Edison's coal-fired power plants. Only a few years later in 1888, inventor Edward Weston (1850–1936) received two patents for solar cells—U.S. Patent 389,124 and U.S. Patent 389,425. For both patents, Weston said that sun energy is transformed into electrical energy, or electrical energy into mechanical energy. In 1887, Heinrich Rudolf Hertz (1857–1894) discovered that ultraviolet light changes the voltage at which sparks between two metal electrodes would be initiated. In 1941, Russel Shoemaker Ohl (1898–1987) developed a silicon PN junction solar cell which is the modern solar cell and patented it with US Patent 2,402,662. Ohl's cell reached efficiency above 5% by the 1960s. A lot of improvements have been made in constructing solar cells and by 2022 the efficiency gradually reached reached to about 21%. These days solar energy has become very popular for day to day household needs and even governments are promoting the production of solar energy.

Alexander Edmond Becquerel (1820–1891), Public Domain, URL: https://en.wikipedia.org/wiki/Edmond_Becquerel#/media/File:Edmond_Becquerel-by_Nadar.jpg

Magnetic shape-memory alloys: Magnetic shape-memory alloys are a particular class of shape-memory alloys and these produce forces and change their shape in response to a magnetic field. In 1996, Kari Ullakko and coworkers suggested a new class of actuator materials in which the magnetic control of shape-memory effect is quite fast and precise and can develop magnetically-induced strains of several percent. Magnetic shape-memory alloys are ferromagnetic alloys of nickel, manganese and gallium (Ni-Mn-Ga). In 2007, Sebastian Faehler argued that the mechanism of magnetic shape-memory alloys is due to magnetically-induced reorientation of the elementary cells comprising the alloy and not by rotation of the magnetic moments within the cells. In 2013, Sozinov and coworkers obtained a deformation of about 12% in single crystalline Ni–Mn–Ga alloys and with further improvements in the synthesis of magnetic shape-memory alloys a deformation of about 20% was achieved. In magnetic shape-memory alloys also there occurs a martensitic phase transition: at low temperatures it exhibits martensitic phase and at and above the transition temperature it goes to the austenite phase. The magnetic shape-memory effect occurs at low temperatures in the martensite phase of the alloy. Other magnetic shape-memory alloys are Fe–Pd alloys, Ni–Fe–Ga alloys, and several derivatives of the basic Ni–Mn–Ga alloy which further contain Fe, Co or Cu. Magnetic shape-memory alloys are used as actuators where fast and precise motion is required: they are much faster than the heating/cooling cycles required for conventional shape-memory alloys. Therefore they are used in robotics, medical surgery, valves, dampers, and sorting etc.

Magnetorheological (MR) Fluids: There are some fluids, usually a type of oil, whose viscosity is low. But when magnetic field is applied their viscosity increases immensely approaching the point of becoming viscoelastic solid. These constitute smart fluids and are called magnetorheological (MR) fluids. When in the activated state, the yield stress of MR fluids can be controlled very accurately by varying the magnetic field intensity. In 1948, Jacob Rabinow (1910–1999) discovered MR fluid for the first time and it was used in MR clutch at the U.S. National Bureau of Standards. In a MR fluid the magnetic particles are of micrometer size. It has been found that in the normal state, the magnetic particles in MR fluid are distributed randomly in the carrier oil. When a magnetic field is applied the magnetic particles arrange themselves forming chains in the direction of magnetic flux. Practically the MR fluids acquire properties comparable to a solid when in the active state, until the point of yield. A number of advances have been made in the twenty-first century. Bell and coworkers (2007) investigated the effect of shape of particles on the viscosity of MR fluids. It was observed that the fluids containing wires have substantially greater viscosity than those fluids containing only spherical particles. Zimmerman and coworkers (2009) synthesized nanowire-based MR fluids and observed that the presence of nanowires affect the flow of MR fluids. It has been found that these MR fluids show no sedimentation for months together and this property has been attributed to their lower close-packing density. Further, Spaggiari and Dragoni (2012) investigated experimentally the flow of MR fluid as a function of pressure and magnetic field. They found that the internal pressure enhances many times the performance of MR fluids.

The MR fluids have a number of applications in the automobile industry: for example these are used as MR dampers (brakes) in automobiles. MR fluid disc brake is a device that transmits a torque by the shear force of MR fluid. The fluid is inserted between the rotating and fixed discs and a magnetic field is applied on the fluid.

Electrorheological (ER) fluids: There are some fluids in which extremely fine but electrically active non-conducting particles (up to 50 μm in diameter) are suspended in an electrically insulating fluid. In 1947, Willis M. Winslow found that the apparent viscosity of these fluids changes reversibly by an order of 10^5 in response to an electric field and obtained the US patent. In these fluids the variation in viscosity is not related by a simple relation to electric field, therefore are called ER fluids. As an example, a typical ER fluid can go from the consistency of a liquid to that of a gel, and back, with response time on the order of milliseconds. This effect is sometimes called electrorheological effect or the *Winslow effect* after the name of its discoverer. In 1949, Winslow wrote and published an article on ER fluids. The ER fluids are smart fluids and their behavior is quite similar to the MR fluids but here magnetic field is replaced by electric field. In 1997, Wing Yim Tam and coworkers observed that in some ER fluids the particles are either ferroelectric in nature or made from conducting material coated with an insulator. They have fabricated doubly coated particles with a conducting inner layer and insulating outer layer. In doubly coated particles one obtains enhanced yield stress by an order of magnitude as compared with the bare particles or singly coated particles. These dielectric particles form chains in the presence of an electric field and align just in the same way as in MR fluids. In 2003, Weijia Wen and coworkers discovered giant electrorheological (GER) fluid consisting of urea coated nanoparticles of Barium Titanium Oxalate suspended in silicone oil. Small size of particles (nanosize) and their urea coating give rise to large dielectric constant which is responsible for high yielding strength as compared to the simple RE fluids. One of the main advantages of GER fluid is that there is a near-linear relation between electric field intensity and yield strength in contrast to quadratic variation usually observed. Further, the GER suspension shows low current density over a wide temperature range of 10–120 °C with a reversible response time less than 10 ms.

There are some smart materials which are sensitive to their "pH value" and change their color or opacity as a result of change in acidity. The common examples of such materials are liquid crystal display and chromatic eye lenses. There are certain paints known as monochromatic or photochromatic paints which change their color with the increase in temperature and exposure to light. Polymer gels such as hydrogels have the ability to swell or shrink (up to 1000 times in volume) due to small changes in temperature. In addition to these, many more smart materials have been designed in the laboratory. It is noteworthy that smart materials have nonlinear properties and are likely to exhibit hysteresis. The nonlinear nature is one of the major obstacles in the applications of smart materials.

Smart materials can be classified in to two categories: *active* and *passive materials*. In 1998, J. A. Fairweather defined active smart materials as those, which have the ability to modify their structural or material properties with the application of light and electric, magnetic or thermal fields etc. As has been explained above pyroelectric and piezoelectric materials, shape-memory alloys, magnetostrictive and electrostrictive materials, ER and MR fluids fall under the category of active smart materials. The active materials can be used as *actuators* which are nothing but devices which can perform output function. For example, an atomic force microscope (AFM) uses piezoelectric actuators to realize scanning along three directions. On the other hand, smart materials which are not active are called passive smart materials. Passive smart materials act as *sensors* which can sense a change in some physical characteristics of a material and perform an electrical input function. Fiber optic is the most common example of passive smart materials. There are smart materials which can be used as both actuator and sensor simultaneously and the common example is piezoelectric crystals. It is referred to as *self-sensing actuation* and can be quite advantageous for active control applications. Both sensors and actuators when used collectively are known as *transducers* because they are used to convert one form of energy into another form. Transducers can be used to sense a wide range of different forms of energy such as movement, electrical signals, thermal or magnetic energy, etc. One can define transducer as a device that converts one physical quantity into another.

Table 17.1 The table lists some of the common smart materials used as actuators and sensors with their stimulus and response

Material	Stimulus	Response
Sensors		
Pyroelectrics	Temperature change	Electric polarization
Piezoelectrics	Mechanical strain	Electric polarization
Electrostrictors	Mechanical strain	Electric polarization
Magnetostrictors	Mechanical strain	Change in magnetization
Electroactive polymers	Mechanical strain	Electric polarization
Electroluminescent	Electric field	Light emission
Photoluminescent	Incident light	Light emission
Electrochromic	Electric field	Color change
Actuators		
Piezoelectrics	Electric current	Mechanical strain
Electrostrictors	Electric current	Mechanical strain
Magnetostrictors	Magnetic field	Mechanical strain
Shape memory alloys	Temperature change	Mechanical strain
Electroactive polymers	Electric field	Mechanical strain
Electrorheological fluids	Electric field	Viscosity change
Magnetorheological fluids	Magnetic field	Viscosity change

Some of the smart materials which are usually used as actuators and sensors are listed in Table 17.1 with their stimulus and response.

Sometimes smart materials are classified into materials which exhibit direct coupling and those which exhibit indirect coupling. Piezoelectric crystals, shape memory alloys, magnetostrictive ceramics and magnetic shape-memory alloys are examples of active materials that exhibit a direct coupling. On the other hand, a change in the electric and magnetic fields in ER and MR fluids can couple with the mechanical behavior through a change in the viscosity of the fluid, indirect coupling.

The ultimate goal in the field of smart materials is the development of *smart structures*. In 1968, Henry R. Clauser (1941–2022) conceptualized the basic idea of the synthesis of smart structures. Seven years later in 1975, Henry Clauser himself demonstrated the concept of smart structures. After this work, activity in this area started increasing and grew rapidly in the 1990s. In 1989, Junyi Takaki proposed the idea of smart structures which have actuators, sensors, power conditioning, control logics, and computers as the key elements. From these elements, different types of smart or intelligent structures can be constructed. In 1993, Craig Rogers gave the concept of adaptive and controlled structures using smart materials. If the structures have distributed actuators to alter characteristics in a desired manner these are called *Adaptive Structures*: these structures may not have sensors. Conventional aircraft wings with flaps and ailerons, and rotor blades with servoflaps are examples of adaptive structures. If the structures have distributed sensors only to monitor the characteristics of the structures, these are called *Sensory Structures*. These structures may detect strain, displacement, acceleration, temperature, electromagnetic properties etc. If the smart structures have the overlap of both the adaptive and sensory structures these are called *Controlled Structures.* Generally these comprise of actuators, sensors, and a feedback control system to actively control the characteristics of the structure. There is a subset of controlled structures called *Active Structures.* In active structures, integrated actuators and sensors have load-carrying capability. *Intelligent or Smart structures* further form a subset of the active structures and they have highly integrated control logic and power electronics in addition to the distribution of actuators and sensors. The development of smart structures is possible through one of the three approaches. Nanotechnology, the first approach, is rapidly entering the world of smart materials and taking them to the next level. These fundamentally new smart materials may incorporate nanosensors, nanocomputers, and nanomachines (nanomotors, nanorobots) into their structure (see Chap. 18). The *nanostructured materials* constitute new materials with smart functionality which can be synthesized at the atomic and molecular levels. But till date scientists could not materialize this method and is in the conceptual state. In the second approach, actuators and sensors are attached to a conventional structure that adaptively responds to external disturbances. The actuators and sensors normally do not constitute the load-carrying structure. Even though this is a relatively mature methodology, it is not expected to be a structurally efficient scheme. In the third approach, active plies representing actuators and sensors are synthesized with non-active plies to form a laminated structure. A major drawback of this method is that once the structure is cured, it is not possible to replace nonfunctional plies. Although this approach appears attractive in terms of structural efficiency, there are issues related to the integrity of the system.

In today's world smart materials are playing a vital role in every walk of life. The technology of smart materials by nature is a highly interdisciplinary field. Starting from basic sciences such as physics and chemistry it covers applied sciences and various engineering fields. The major role of these materials is to enhance the system performance at low cost and increase the comfort level in everyday life. There is a tendency to look for ways of making buildings adaptive so that they do not fight the change but accommodate such as in earthquake zones. We want buildings that can respond to earth tremors instead of trying to withstand like trees that bend rather than snap in the breeze. Smart materials have a wide range of applications in the fields of defense, aerospace, robotics, marine systems, machine tools, automotive engineering, health monitoring, nuclear industries, structural engineering, biomedical applications and electronics especially in computers and mobiles etc.

Bibliography

Adams WG, Day RE (1877) The action of light on selenium. Proc Roy Soc (London) 25(113–117):171–178

Aepinus FUT (1756) Memoire concernant quelques nouvelles experiences électriques remarquables [Memoir concerning some remarkable new electrical experiments]. Histoire de l'Académie royale des sciences et des belles lettres (Berlin) 12:105–121

Becquerel AE (1839) Le spectre solaire et la constitution de la lumière électrique. Comptes Rendus 9:561; (1940) Comptes Rendus 11:702
Bell RC, Miller ED, Karli JO, Vavereck AN, Zimmerman DT (2007) Influence of particle shape on the properties of magnetorheological fluids. J Mod Phys B 21:5018–5025
Brewster D (1824) Observation on the pyro-electricity in minerals. Edinbg J Sci 1:208–218
Buehler WJ, Gilfrich JW, Wiley RC (1963) Effects of low-temperature phase changes on the mechanical properties of alloys near composition TiNi. J Appl Phys 34:1475–1477
Buehler WJ, Wang FE (1968) A summary of recent research in the nitinol alloys and their potential application in ocean engineering. Ocean Eng 1:105–120
Chang LC, Read TA (1951) Behavior of the elastic properties of AuCd, transactions of the AIME. J Metals 191:47
Chilowsky CM, Langévin MP (1916) Procédés et appareils pour la production de signaux sous-marins dirigés et pour la localisation à distance d'obstacles sous-marins. French patent #502913
Chilowsky CM, Langévin MP (1917) Production of submarine signals and the location of submarine objects, US patent #1471547
Clauser HR (1968) Modern materials concepts make structure key to progress. Mater Eng 68:38–42
Clauser HR (1975) From static to dynamic materials in design. ASME J Mech Eng 97:20–26
Curie J, Curie P (1880) Développement par compression de l'électricité polaire dans les cristaux hémièdres à faces inclinées. Bulletin de la Société Minéralogique de France 3:90–93
Faehler S (2007) An Introduction to actuation mechanisms of magnetic shape memory alloys. ECS Trans 3:155–163
Fairweather JA (1998) Designing with active materials: an impedance based approach, 1st edn. UMI, New York, p 196
Fritts CE (1883) On a new form of selenium photocell. Am J Sci 26:465
Genchi GG, Marino A, Rocca A, Mattoli V, Ciofani G (2016) Barium titanate nanoparticles: promising multitasking vectors in nanomedicine. Nanotechnology 27:232001
Greninger AB, Mooradian VG (1938) Strain transformation in metastable beta copper–zinc and beta copper–titanium alloys. AIME TRANS 128:337–369
Guillaume CÉ (1919) The anomaly of the nickel-steels. Proc Phys Soc (London) 32:374; (1896) La Nature 24:234
Hitov JJ, Rainer WC, Redding EM, Sloan AW, Stewart WD (1955) Polyethylene product and process, US patent. 2877500 (Google Patents) application filed on June, 17, 1955 and granted on, March 17, 1959
Joule JP (1842) On a new class of magnetic forces. Ann Electr, Magn, Chem 8:219–224; (1847) On the effects of magnetism upon the dimensions of iron and steel bars. The London, Edinb, Dublin Philos Mag J Sci 30:225–241
Kurdjumov GV, Khandros LG (1949) First reports of the thermoelastic behaviour of the martensitic phase of Au–Cd alloys. Dokl Akad Nauk SSSR 66:211–213
Linnaeus C (1747) Flora Zeylanica: Sistens Plantas Indicas Zeylonae Insulae [The Flora of Ceylon: consisting of Indian plants of the island of Ceylon]. Laurentii Salvii, Stockholm, Sweden
Lippmann G (1881) Principe de la conservation de l'electricite. Annales de chimie et de physique (in French) 24:145
Ogawa T, Waku S (discovered in 1944); Ogawa T (1947) Busseiron, Kenkyu No. 6, p 1
Orlander A (1932) An electrochemical investigation of solid cadmium-gold alloys. Am Chem Soc 54:3819–3833
Perrone RJ (1967) Silicone-rubber, polyethylene composition; heat shrinkable articles made there from and process there for, US patent 3326869 granted in 1967 (Google patents)
Phelan D, Stock C, Rodriguez-Rivera JA, Chi S, Leão J, Long X, Xie Y, Bokov A, Ye Z-G, Ganesh P, Gehring PM (2014) Role of random electric fields in relaxors. Proc Natl Acad Sci US 111:1754–1759
Rabinow J (1948) Magnetic fluid clutch. Natl Bur Stan Tech News Bull 32:54–60
Rogers CA (1993) Intelligent material systems—the dawn of a new materials age. J Intell Mater Syst Struct 4:4–12

Schmidt JG (1707) Curiöse Speculationes bey Schalflosen Nächten (Conrad Stossen, Chemnitz and Leipzig), p 269 [An English translation of the relevant passage appears in Lang SB (1974) Sourcebook of pyroelectricity, vol 2, p 96, Gordon and Breach, New York]
Smith W (1873) Effect of light on selenium during the passage of electric current. Nature 7:303. [In Feb. 1873, Smith presented the paper entitled, electrical properties of selenium and the effect of light thereon, at the meeting of the society of telegraph engineers]
Sozinov A, Lanska N, Soroka A, Zou W (2013) 12% magnetic field-induced strain in Ni–Mn–Ga-based non-modulated martensite. Appl Phys Lett 102:021902; Pagounis E, Szczerba MJ, Chulist R, Laufenberg M (2015) Large magnetic field-induced work output in a NiMnGa seven-layered modulated martensite. Appl Phys Lett 107:152407
Spaggiari A, Dragoni E (2012) Effect of pressure on the flow properties of magnetorheological fluids. J Fluid Eng 134:091103
Takagi T (1990) A concept of intelligent materials. J Intell Mater Syst Struct 1:149; (1989) A concept of intelligent materials. Proceedinds of the international workshop on intelligent systems. Tokyo, Japan, pp 1–10
Takenaka H, Grinberg I, Rappe AM (2013) Anisotropic local correlations and dynamics in a relaxor ferroelectric. Phys Rev Lett 110:147602
Tam WY, Yi GH, Wen W, Ma H, Sheng P (1997) New electrorheological fluid: theory and experiment. Phys Rev Lett 78:2987–2990
Thomson W (1878) On the thermoelastic, thermomagnetic and pyroelectric properties of matter. Philos Mag 5:4–26
Ullakko K (1996) Magnetically controlled shape memory alloys: a new class of actuator materials. J Mater Eng Perform 5:405–409
Ullakko K, Huang JK, Kokorin VV, O'Handley RC (1997) Magnetically controlled shape memory effect in Ni_2MnGa intermetallics. Scripta Mater 36:1133–1138
Vernon LB, Vernon HM (1941) Process of manufacturing articles of thermoplastic synthetic resins. Google patents, vol 18, p 348; Behl M, Lendlein A (2007) Shape-memory polymers. Mater Today 10:20–28
Voigt W (1897) Versuch zur bestimmung des wahren specifischen electrischen momentes eines turmalins (Experiment to determine the true specific electric moment of a tourmaline). Ann Phys 60:368–375
von Linné C (Linnaeus) (1747) Flora zeylanica: Sistens plantas indicas zeylonae insulae (Stockholm and Sweden), p 8. [A translation of the relevant passage appears in Lang (1974) p 103]
Wainer E, Solomon AN (1942) Titanium alloy manufacturing Co. Report No 8
Wang FE, Buehler WJ (1972) Additional unique property changes observed during TiNi transition. Appl Phys Lett 21:105–106
Wang FE, Buehler WJ, Pickart SJ (1965) Crystal structure and a unique martensitic transition of TiNi. J Appl Phys 36:3232–3239
Wang FE, De Savage BF, Buehler WJ, Holster WR (1968) J Appl Phys 39:2166–2179
Wang FE, Pickart SJ, Alperin HA (1972) Mechanism of the TiNi martensitic transformation and the crystal structures of TiNi-II and TiNi-III phases. J Appl Phys 43:97–112
Wang Y, Changjie S, Eric Z, Ji S (2004) Deformation mechanisms of electrostrictive graft elastomers, smart materials and structures. Inst Phys Publishing 13:1407–1413
Wen W, Huang X, Yang S, Lu K, Sheng P (2003) The giant electrorheological effect in suspensions of nanoparticles. Nat Mater 2:727–730
Winslow WM (1947) Method and means for translating electrical impulses into mechanical force. U. S. patent 2417850 dated 25 March 1947
Winslow WM (1949) Induced fibration of suspensions. J Appl Phys 20:1137–1140
Wul B, Goldman JM (1945) Dielectric constants of titanates of metals of the second group. CA Acad Sci URSS 46:139–142
Zimmerman DT, Bell RC, Karli JO, Filer JA, Wereley NM (2009) Elastic percolation transition in nanowire-based magnetorheological fluids. Appl Phys Lett 95:014102

Chapter 18
Nanophysics and Nanotechnology

Over the past few decades, scientists had been working in the miniaturization of machines, particularly electron devices, and ultimately they discovered nanomaterials. Nano is a Greek word meaning dwarf or extremely small and one nanometer (nm) is equal to 10^{-9} m. Nanomaterials are solids having dimensions in the range of 1–100 nm. Nanomaterials were used by humans as far back as the ninth century and some examples are given below. Pottery from middle ages often retained a distinct gold or copper colored metallic glitter. This luster was created by applying a thin metallic film on the transparent surface of a glazing. The beautiful red color of glass window panes in some old cathedrals is due to gold particles dispersed in the glass matrix. Small colloidal particles of silver are used in image formation in photography. In 1857, Michael Faraday (1791–1867) studied the preparation and properties of colloidal suspensions of gold. He discovered that their optical properties differed from those of the corresponding bulk metal. Faraday demonstrated how gold particles produced different colored solutions under certain lighting conditions. Later these particles came to be called nanoparticles and it can be considered as the birth of nanoscience.

In general one can define a nanoparticle as a cluster of atoms (molecules) bonded together and having size within the range 1–100 nm. In 1914, Richard Adolf Zsigmondy (1865–1929) made a detailed study on gold sols and other nanoparticles with sizes down to 10 nm. He observed nanoparticles for the first time and measured their size with the help of ultramicroscope developed by him. The methods of synthesis of nanoparticles and in general nanomaterials fall under two categories: top-down and bottom-up approaches. Breaking down a bulk material to get nanosized particles is called a top-down approach such as ball milling, sputtering, chemical etching etc. The bottom-up approach refers to the build-up of nanostructures from the individual atoms or molecules by physical and chemical methods such as self-assembly method, chemical deposition method, solgel method, molecular beam epitaxy method etc.

J. S. Galsin, *History of Solid State Physics*,
https://doi.org/10.1007/978-981-95-0504-3_18

Kim Eric Drexler (1955–present), CC BY-SA 3.0, URL: https://en.wikipedia.org/wiki/K._Eric_Drexler#/media/File:Drexler763x1000.jpg

Nanotechnology is a relatively recent multidisciplinary subject although the development of its central concepts happened over a long period of time. In 1959, Richard Phillips Feynman (1918–1988) delivered a talk entitled "*There is Plenty of Room at the Bottom*" to American Physical Society in which he said that there are no fundamental physical reasons why materials cannot be fabricated by maneuvering individual atoms. It was an indication towards the synthesis of very small clusters of atoms of the order of nm size. But nobody paid attention to the talk of Feynman for about 20 years. In 1974, Norio Taniguchi (1912–1999), a Japanese scientist, was the first to use the term "Nanotechnology" in a conference to describe semiconductor processes such as thin film deposition and ion beam milling exhibiting characteristic control on the order of nm. However, this term was not used again until 1981 when Kim Eric Drexler (1955–present) published his first paper on nanotechnology. The term nanotechnology was independently used by Drexler in his book entitled, "*Engines of Creation: The Coming Era of Nanotechnology*". In this book he proposed the idea of nanoscale "assembler" which would be able to build a copy of itself and of other items of arbitrary complexity. Drexler popularized the concept of nanotechnology and founded the field of "Molecular Nanotechnology". Drexler has been called the "*Godfather of nanotechnology*". On the other hand, in 1943,

Joseph Mattauch (1895–1976) and coworkers originated the concept of clusters. But the science of clusters came into existence in 1980s: a cluster contains atoms up to about 3×10^7. In 1981, Gerd Binning (1947–present) and Heinrich Rohrer (1933–2013) invented scanning tunneling microscope which provided a new ground for the synthesis of very small particles. Nanotechnology came into existence with these two inventions. One should distinguish between nanoscience and nanotechnology. Nanoscience is the study of structures and molecules on the scale of nanometers ranging between 1–100 nm and nanotechnology utilizes it in practical applications such as devices.

Harold Walter Kroto (1939–2016), CC BY-SA 3.0, URL: https://en.wikipedia.org/wiki/Harry_Kroto#/media/File:Harry_Kroto.jpg

In 1985, Harold Walter Kroto (1939–2016) and coworkers were studying the structure of carbon-rich giant stars when they observed some special spectral lines corresponding probably to some type of long chain molecules of carbon or nitrogen. Later in the same year, they were able to produce clusters of carbon atoms in the laboratory by the vaporization of solid carbon. The analysis of carbon clusters by the mass spectrometry exhibited a number of peaks in the spectrum and out of these the strongest peak was observed at 60 carbon atoms and the next strongest peak at 70 carbon atoms. The molecules of carbon corresponding to these strong peaks are represented as C_{60} and C_{70}. The C_{60} molecules are also called buckyballs or fullerenes after the name of American architect and futurist Buckminster Fuller who designed a geodesic spherical dome using sticks at the Montreal World Exhibition, held in 1967, which looked quite similar to the structure of C_{60} molecule. C_{60} molecules are nanoparticles of carbon atoms and form the third allotropic form of carbon. In 1990, Wolfgang Kraetschmer (1942–present) and Donald R. Huffman (1935–present) discovered a method for synthesizing and purifying large quantities of fullerenes in the form of solid C_{60}. This opened the door for their characterization and functionalization by large number of investigators.

Magnetic nanoparticles form a class of nanoparticles which can be manipulated using magnetic field. Magnetic nanoparticles comprises of two elements, a magnetic material like Fe, Ni, Co and a chemical component that has functionality. In 1930, Frenkel (1894–1952) and Doefman first predicted nanoparticles of a ferromagnetic material with size less than 15 nm which would have single domain. Such nanoparticles are in a state of uniform magnetization in any field. In a magnetic nanoparticle all the atoms have finite magnetic moments. When magnetic moments of all the atoms are aligned parallel to one another, nanoparticle exhibits maximum magnetic moment usually called "*supermoment*". There are two types of magnetic nanoparticles: ferrite (or iron oxide) and metallic nanoparticles. Ferrite nanoparticles are the most studied magnetic nanoparticles. Metallic magnetic nanoparticles exhibit higher magnetic moment than ferrite magnetic nanoparticles having same size but metallic magnetic nanoparticles suffer from a major disadvantage being pyrophoric and are easily oxidized in the presence of air or oxidizing agent. Magnetic nanoparticles have a wide range of applications like magnetic memory and magnetic storage media, magnetic recording heads, magnetic resonance imaging and in biomedicine and health science. In 1978, Andrew E. Senyei and coworkers and Kenneth J. Widder and coworkers proposed the concept of drug delivery using micro- and nanoparticles. They said that therapeutic agents can either be attached to or encapsulated within the magnetic micro- or nanoparticles. The magnetic nanoparticles can be manipulated by the external magnetic field which can drive them to the target organ for gene or drug delivery.

In conventional solid state physics, the experimental value of a particular property of a solid in the bulk form is some sort of an average value of the property which may be quite different from the local behavior inside the solid. The 4d-transition metal rhodium (Rh) is nonmagnetic in nature. But in 1994 Cox and coworkers showed that small clusters of Rh metal exhibit finite magnetic moment which increases with the decrease in size of the cluster. In 1998, Marc Respaud and coworkers found that in case of cobalt nanoparticles the average magnetic moment of each atom increases with the decrease in its size. At the same time there is also an increase in magnetic anisotropy with the decreasing size. They attributed both the effects to the increasing surface area, more specifically with the increase in surface to volume ratio, of the nanoparticle. In 2003, Charles P. Poole and Frank J. Owens have also pointed out that the nanoparticles of paramagnetic Re metal exhibit significant increase in the net magnetic moment when the number of atoms becomes less than 20. On the other hand, in 1999, Richard Kodama found that the magnetic moment of some oxide nanoparticles decreases with the decrease in their size. It shows that surface effects can lead to an increase or a decrease in the magnetic moment of nanoparticles depending on their nature. The above results give an indication that there is a possibility to produce magnetic materials from nanoparticles of paramagnetic materials.

Scientists are trying to synthesize materials with nanoparticles as the basic unit of structure in place of an atom or molecule in conventional solids. Such materials in the bulk form are called *nanostructured materials* and these are of two types: crystalline and amorphous nanostructured materials. Scientists are able to synthesize amorphous nanostructured materials but the crystalline nanostructured materials have not become a reality yet. Today we have 103 stable elements although the total number of elements in the periodic table is more. On the left side of the periodic table there are metals, on the right side there are insulators/inert gasses but in the middle there are semiconductors and semimetals. When the crystalline nanostructured materials will become a reality as expected in the near future, one will be able to synthesize a few tens of crystalline nanostructured materials with entirely new properties from a single element of the periodic table. Therefore, one will be able to produce thousands of new crystalline materials with new properties which will bring a revolution in solid state physics. It shows that there is immense scope of nanoscience and nanotechnology in the near future, probably in the twenty-first century. From crystalline nanostructured materials of magnetic nanoparticles possessing supermoments one may be able to obtain super ferromagnets in which magnetic nanoparticles will form domains. A super ferromagnetic domain can be defined like a ferromagnetic domain, the only difference being that the atomic magnetic moments are replaced by the supermoments of the individual nanoparticles.

The concept of nanoscience can be well understood if the dimensions of the bulk material are reduced one by one to the nm range yielding different types of nanosolids having different properties. As regards the dimensionality, nanomaterials are characterized as quantum wells, quantum wires, and quantum dots etc. Figure 18.1 presents the schematic representation of quantum well, quantum wire, quantum dot and quantum ring. In 1963, Zhores Ivanovich Alferov (1930–2019) and Rudolf F. Kazarinov proposed the concept of a quantum well. In the same year, Alferov filed a patent for an application proposing a double-heterojunction laser. Herbert Kroemer (1928–2024) was also working independently on double heterojunction and in 1963 proposed the concept of double-heterojunction laser and filed a US Patent in 1963 a few months later. In 1966, Alferov's laboratory created the first laser based on heterojunction although it could not lase continuously. They continued the work in this direction and ultimately in 1968, Alferov and coworkers produced the first continuous-wave semiconductor heterojunction laser operating at room temperature. Alferov and Kroemer were awarded the 2000 Physics Nobel Prize for the development of semiconductor heterojunctions. In 1970, Leo Esaki (1925–present) and Raphael Tsu (1931–present) invented synthetic superlattices. They suggested that a heterostructure made up of alternating thin layers of semiconductors with different band gaps should exhibit interesting and useful properties. The development of quantum well devices is a great advancement in crystal growth techniques. It is because the quantum-well devices required structures of high purity with few defects. Therefore, great control over the growth of these heterostructures allowed the development of semiconductor devices that can have very fine-tuned properties.

Since then, much effort and research have gone into the study of the physics of quantum well systems as well as in the development of quantum-well devices. The theory surrounding the quantum well devices has led to significant advancements in the production and efficiency of many modern components such as light-emitting diodes and transistors etc. In 1972, Charles Howard Henry (1937–2016) suddenly got an idea that a double heterostructure may act as wave guide for electrons and its central thin active layer would have electron modes. Such a heterojunction can act as a quantum well laser with modes as electron states of the quantum well. In 1973, he suggested Raymond Dingle to work on this idea and in 1974 Henry and coworkers performed experiments and found the idea to be a reality. In 1975, they invented a quantum laser in which the wavelength of emitted light depends on the thickness of the thin quantum well layers. On the other hand in conventional lasers the wavelength of emitted light depends on the composition of the quantum layer. In 1975, they applied for the patent of a quantum laser entitled, "Quantum effects in Heterojunction Lasers" which was granted in 1976.

Nanotubes have single dimension (1D) and are usually called quantum wires. In 1991, Sumio Iijima (1939–present) discovered carbon nanotubes (CNTs) while synthesizing fullerenes using electric arc discharge technique. High resolution transmission electron microscope (TEM) was employed for the observation of CNTs. CNTs that Iijima observed were called multi-walled nanotubes (MWNTs): MWNT is a nanotube containing a number of nanotubes inside it. Two years later in 1993, Sumio Iijima and Toshinari Ichihashi and Donald Bethune and coworkers independently synthesized single-walled nanotubes (SWNTs) by adding transition metal catalysts to carbon electrodes in an electric discharge. The single nanotube has a very small diameter, of the order of 1 nm, and is not straight but curled or looped. Later both experiments and calculations showed that SWNTs can be either metallic or semiconducting in nature depending on chirality and diameter. It is further found that the energy band gap of a nanotube depends on its chiral angle and diameter. Because of the strong bonding in the sp^2—configuration, carbon nanotubes are chemically inert but are able to transport large electric current. According to theory, carbon nanotubes are also able to conduct heat nearly as well as diamond or sapphire. In 2004, Konstantin Sergeevich Novoselov (1974–present) and coworkers discovered graphene which can carry electrons and holes up to 1013 per square centimeter with a high mobility of about 10,000 square centimeters per volts- sec. After the discovery of graphene carbon-based materials became the backbone of almost every field of science.

Nanoring is a cyclic one-dimensional nanostructure and is one of the recent developments in nanoscience. Nanorings are generally synthesized using bottom up approach. In 2001, Li and coworkers were the first to synthesize nanorings of gallium nitride (GaN). In 2004, Kong and coworkers synthesized single crystal nanorings of zinc oxide (ZnO) using epitaxial self-cooling of polar nanobelts. Later several other common nanostructure compounds have been synthesized into nanorings.

Quantum dots (QDs) are tiny nanocrystals of semiconducting material with a diameter in the range of 2–10 nm and are usually called zero-dimensional material. In 1980, Alexei Ekimov (1945–present) and Aleksei A. Onushchenko (1952–present)

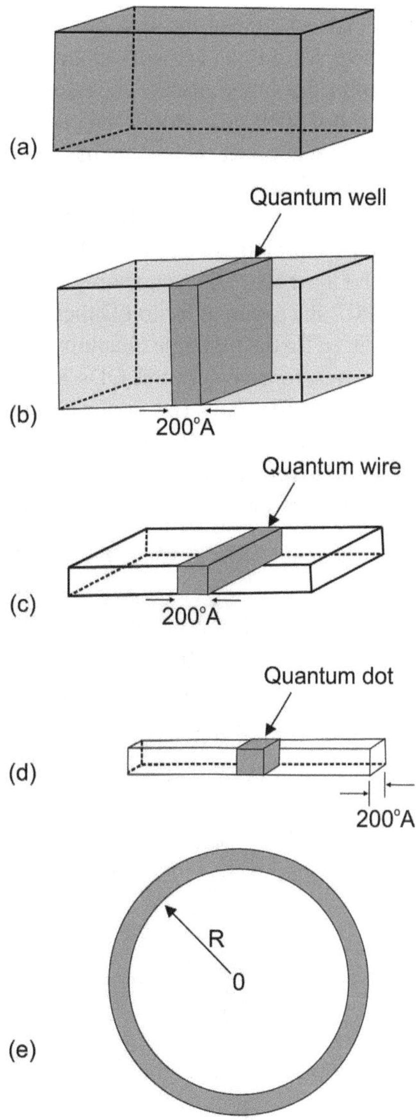

Fig. 18.1 Schematic representation of (**a**) bulk material, (**b**) quantum well with width 200 Å, (**c**) quantum wire with width and height of 200 Å, (**d**) quantum dot with all the three dimensions of 200 Å, and (**e**) quantum ring

produced QDs in a molten glass matrix for the first time and published their results in 1981. In 1982, Alexander L. Efros and Alexei L. Efros published the first theory of the confinement of electrons in QDs aiming at explaining the behavior of these very small crystals. In 1982, Rossetti and coworkers discovered QDs of cadmium sulfide in a colloidal suspension and published their results in 1983. Their TEM investigation showed a typical particle diameter of 35A which corresponds to six unit cells. In 1993, Christopher Bruce Murray and coworkers produced high quality QDs in the colloidal suspension that had less than 5% size variation. This enabled scientists to control the size of QDs and fine-tune the color of their fluorescence. In 1995, Murray and coworkers invented a technique for creating three-dimensional quantum dot superlattices. The size and spacing of the dots within the superlattice are controlled with near atomic precision. This technology provides a general method for the creation of three-dimensional superlattices for nanocrystallites of different materials. In 2004, Xiaoyou Xu and coworkers accidentally discovered a new class of carbon materials called carbon quantum dots (CQDs) with size below 10 nm during the purification of SWNTs. CQDs have excellent optical and electronic properties and can be used for catalysis, energy conversion, photovoltaic devices and nanoprobes. In 2007, the group of Benoit Dubertret (1970–present) [Allen and coworkers (2007)] produced for the first time quantum plates. It was an extremely important step forward compared to conventional QDs as the quantum plates are brighter and easier to implement for industrial applications. In a semiconducting QD the electrons and holes are strongly confined and these possess discrete electron energy states just like atoms. Therefore, QDs are sometimes referred to as *artificial atoms*. It was shown that the electron wave functions in QDs resemble those in real atoms.

Alexei Ekimov (1945–present), CC BY 2.0, URL: https://en.wikipedia.org/wiki/Alexey_Ekimov#/media/File:US_Embassy_Sweden_2023_Nobel_Reception_(53390424929)_(cropped).jpg

The optical and electronic properties of QDs differ from the bulk material due to quantum confinement. When a suitable electromagnetic radiation is allowed to fall on a QD, electrons are excited from the valence to the conduction state. With the fall of the electron back to the valence state, light is emitted whose color depends on the size and shape of the QD. Larger QDs (5–6 nm) emit longer wavelengths (orange or red color) while smaller QDs (2–3 nm) emit shorter wavelengths (blue or green color). This property makes QDs suitable for laser action.

The early 2000s saw the beginning of commercial applications of nanotechnology. Most of the applications are limited to the bulk use of passive amorphous nanomaterials. Titanium oxide and zinc oxide nanoparticles are used in sunscreen, cosmetics and in some food products. Silver nanoparticles are used for packaging, clothing and disinfectants. Carbon nanotubes are used for stain-resistant textiles, and cerium oxide as a fuel catalyst. Nanowires and nanostructured materials may be used to create cheaper and more efficient solar cells. Nanosensors and nanorobots could be used to detect cancerous cells in the human body.

In order to study the structure of nanomaterials one needs special type of microscopes. Among these are electron microscopes and probe-type microscopes. Hans Walter Hugo Busch (1884–1973) suggested that magnetic fields could be used to direct a beam of electrons in just the way light is refracted by optical lenses. In 1926, Busch developed an electromagnetic lens which indicated the possibility of electron microscope using these lenses and laid the theoretical basis for electron microscope. In 1931, Max Knoll (1897–1969), a German engineer, and Ernst August Friedrich Ruska (1906–1988), a German physicist, demonstrated that a magnetic coil could act as an electron lens and in 1932 developed prototype electron microscope which was the first practical demonstration of the principle of electron microscopy. In 1933, they improved the design of electron microscope and obtained resolution greater than that of an optical microscope. In 1938, Ernst Ruska at Siemens produced the first commercial electron microscope and in the same year Siemens obtained the patent for the same. The basic principle of the electron microscope is that an electron beam is passed through the specimen and the electrons interact with the atoms near the surface of the specimen to produce image of the surface. Basically three types of electron microscopes came into existence.

1. Transmission electron microscope (TEM), which was discovered by Knoll and Ruska (see Fig. 18.2). In a high resolution TEM, the highest resolution is 50 pm and magnification above 50 million times. The major disadvantage of TEM is that it needs extremely thin sections of specimens ($\approx$10 nm).
2. In 1937, Manfred von Ardenne (1907–1997) invented the earliest scanning electron microscope (SEM) (see Fig. 18.3). In a SEM, a very narrow electron beam scans the rectangular area of surface of the specimen (raster scanning) which produces image of the surface. As the electrons do not travel through the specimen one can use the specimen in the bulk form. We want to point out that in a SEM the resolution of the image obtained is less than that obtained in a TEM.
3. Shortly after the discovery of TEM, in 1933, Ernst Ruska developed the reflection electron imaging technique to surpass the resolution limit of optical microscopy. Reflection electron microscopy is very similar to transmission electron microscopy, but the main difference being that the high energy electron

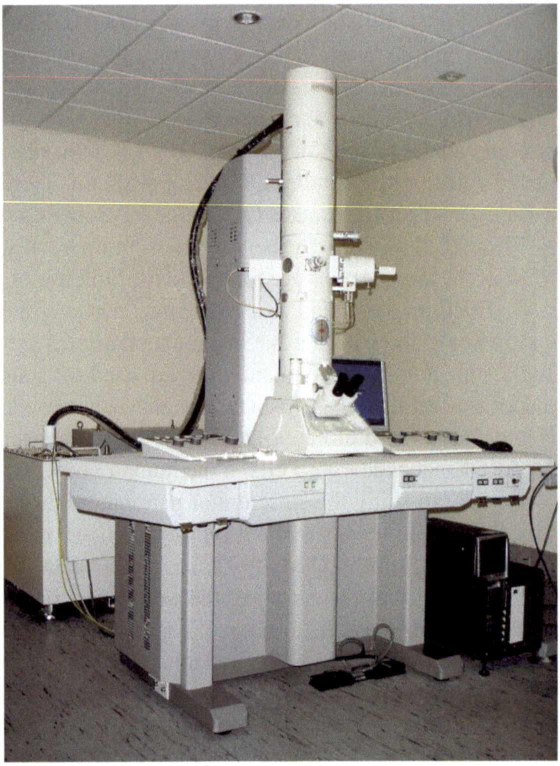

Fig. 18.2 Image of transmission electron microscope, Hitachi (*Courtesy EMN laboratory, Punjab Agricultural University, Ludhiana*)

beam (100–200 keV) interacts with the sample surface at a glancing angle of incidence of 5 to 10°. Therefore, the sample no longer needs to be thin and electron transparent. In a reflection electron microscope the elastically scattered electron beam is detected. With reflection electron microscope one can perform spin-polarized low energy electron microscopy which yields information about the microstructure of magnetic domains. There is another electron microscope named scanning transmission electron microscope (STEM) which was invented in 1938 by Manfred von Ardenne. With the advancement of technology modern models of STEM were made having a resolution of 2A. Therefore, modern STEM has the ability to produce images of single heavy atoms and is suitable for the study of atomic structure of some elements. There are some disadvantages of TEM. The specimens have to be placed in ultrahigh vacuum conditions. The electron beam may damage some delicate samples like living cells. One major disadvantage of TEM is the need of extremely thin sections of the specimens, typically about 100 nm.

After the invention of electron microscopes, the scientists made continuous efforts to achieve resolution of the order of the size of an atom. In 1936, while working at Siemens, Erwin Wilhelm Muller (1911–1977) invented a field emission electron microscope which gave a resolution of 2 nm. In 1951, he invented the field ion

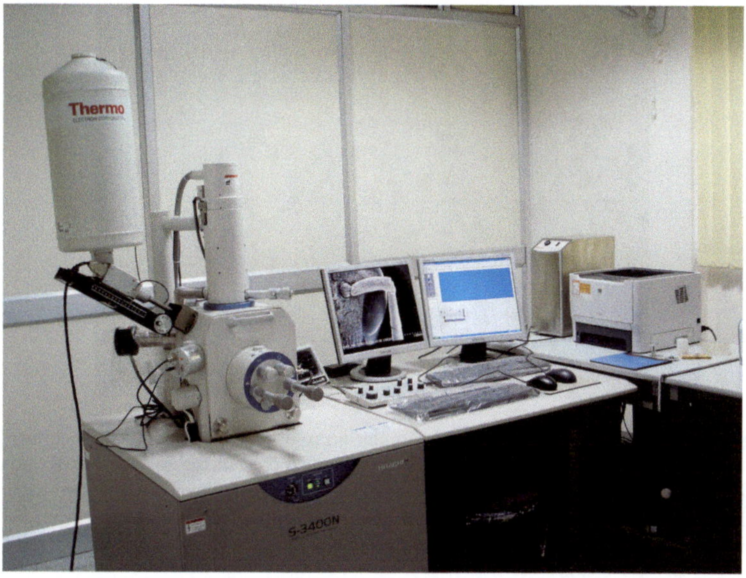

Fig. 18.3 Scanning electron microscope, Hatachi (Courtesy EMN laboratory, Punjab Agricultural University, Ludhiana)

microscope and in 1955, made improvements in its design to achieve a resolution of 0.25 nm. It was the first instrument in human history with which atoms were seen. Muller further developed an atom-probe field ion microscope that has important applications in the microanalysis of nanomaterials, especially alloys. The field ion microscopy is of immense importance in surface science and nanoscience especially in the surface analysis of solids.

In 1981, Gerd Binnig (1947–present) and Heinrich Rohrer (1933–present) developed the scanning tunneling microscope (STM) which is a probe-type microscope and its patent was granted in Aug., 1982. It works on the principle of quantum electron tunneling through very thin potential barriers. With STM one is able to obtain a resolution of the order of 10 pm. It means the image of individual atoms can be obtained with STM. In 1983, Binnig and coworkers published the first STM image of Si(111) $-$ (7×7) reconstructed surface. Further, with the help of the stylus of STM one can move individual atoms from one position to another on a substrate, thus *quantum corrals* can be obtained. In 1990, Donald M. Eigler (1953–present) and Erhard K. Schweizer were the first to manipulate 35 individual xenon atoms on the nickel surface with the help of STM and formed the letters of IBM. A quantum corral made by Donald M. Eigler by placing Fe atoms in a circle on a copper surface using STM is shown in Fig. 18.4. The main disadvantage of STM is that it can be used to study only the conducting and semiconducting surfaces and thus has important applications in semiconductor physics and microelectronics. It is noteworthy that STM works under ultra high vacuum conditions. Gerd Binning and Heinrich Rohrer were awarded 1986 Nobel Prize in Physics for their discovery of STM.

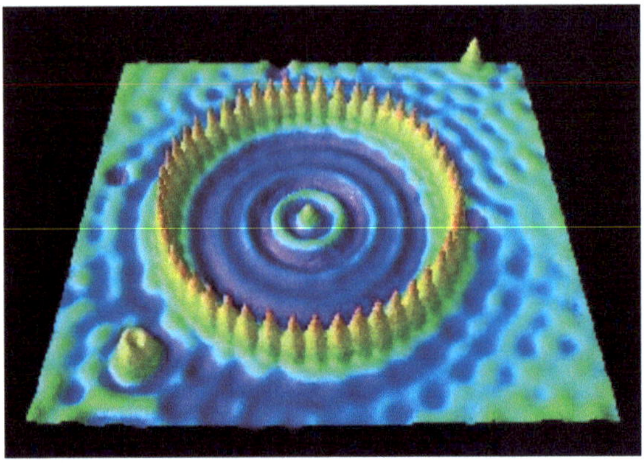

Fig. 18.4 Quantum corral in the form of circular array of Fe atoms placed on a copper surface using the tip of an STM. The ripples inside the corral represent the surface distribution of the electron density that arises from the quantum mechanical energy levels of a circular two-dimensional potential well (created by Don Eigler, IBM Almeden Research Center and taken from Wikipedia with link http://nisenet.org/.../scientificimage-quantumcorral)

Gerd Binnig (1947–present), CC BY 2.0, URL: https://en.wikipedia.org/wiki/Gerd_Binnig#/media/File:Gerd_Binnig_at_the_Memorial_Symposium_for_Heinrich_Rohrer_(cropped)_2.jpg

In 1986, Gerd Binnig (1947–present), Cavin Quate (1923–2019) and Christoph Gerber (1942–present) invented atomic force microscope (AFM) for studying surfaces of insulating materials (see Fig. 18.5a and b). AFM is a very powerful probe-type microscope that can image almost any type of surface including ceramics, polymers, glass, composite materials, and biological samples. In 1987, Martin and coworkers introduced a modified version of AFM which enables the precise measurement of the force between the tip and a sample over a tip- sample distance of 30-50A and this distance is kept constant. The AFM can be operated in two modes: *contact mode* and *dynamic mode.* In AFM the lateral resolution is of the order of 30 nm (less than that of STM) due to convolution and the vertical resolution can be up to 0.1 nm. So STM gives better resolution because of the exponential dependence of tunneling current on distance. One of the main advantages of AFM over STM is that it works with a sample placed in air or with a sample immersed in a liquid. It was realized by Binnig and coworkers that the concept of force microscopy is a general one and is no way limited to atomic forces. Magnetic force microscope was made which is just similar to AFM but the tip of the stylus is made of magnetic material. In 1987, Martin and Wickramasinghe obtained the first magnetic force image by measuring magnetic force by means of a magnetic recording head. Shortly after this in 1987, Juan Jose Saenz (1960–2020) and coworkers and Rolf Allenspach and coworkers independently studied the structure of magnetic domains walls in magnetic materials using magnetic force microscopy.

In 1986, Dmitri Averin and Konstantin Likharev proposed the idea of a new three-terminal device called single-electron tunneling transistor. In 1987, Theodore Alan Fulton (1941–present) and Gerald J. Dolan (1945–2008) fabricated single electron tunneling transistor and demonstrated its working. The single electron transistor works on the principle of quantum tunneling in which the transfer of charge through the barrier is quantized. It can be viewed as a field-effect transistor in which the channel is replaced by two tunnel junctions forming a metallic island. The single-electron tunneling transistor comes in two versions: semiconducting and metallic. Fulton and Dolan invented the metallic single-electron tunneling transistor.

The next stage and perhaps the ultimate stage of miniaturization in electronics is the development of nanodevices. In late 1990s, silicon MOSFET with nanoscale dimensions was synthesized which made the beginning of nanoelectronics: Complementary MOS (CMOS) having dimensions of 22 nm was synthesized. In 1989, Digh Hisamoto and coworkers fabricated the first FinFET transistor which was initially called "Depleted Lean-Channel Transistor" or DELTA Transistor and it laid the foundation of modern electronics. A FinFET is a type of field-effect transistor (FET) that has a thin vertical fin instead of being completely planar. The gate is fully "wrapped" around the channel on three sides formed between the source and the drain. Therefore, FinFET is a three-dimensional, nonplanar double-gate MOSFET. The dimension of FinFET in 2013 was 16 nm which was reduced to 5 nm in 2018.

Single molecule devices are another possibility. In 1959, Arthur Von Hippel (1898–2003) mentioned about molecular electronics for the first time who suggested

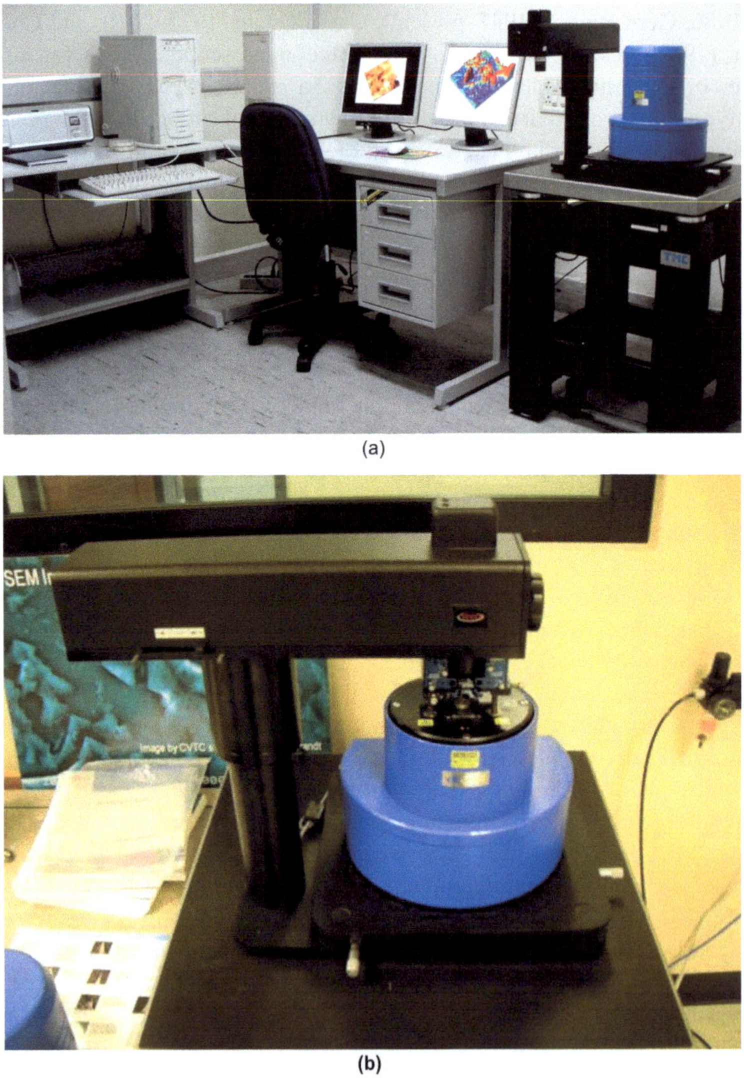

Fig. 18.5 (**a**) Image of atomic force microscope, blue in color, Hatachi (**b**) Open head image of atomic force microscope, Hitachi (*Courtesy EMN laboratory, Punjab agricultural university, Ludhiana*)

a bottom-up procedure of developing electronics from atoms and molecules, an idea he named molecular engineering. In 1974, Arieh Aviram and Mark Ratner put forward the theory of transport through a modified charge-transfer molecule and constructed a rectifier based on the use of a single organic molecule. It consists of a donor pi system and an acceptor pi system separated by a sigma- bonded (methylene) tunneling bridge that would allow transport only in one direction and named it

Molecular Rectifier. It was a breakthrough that inspired many years of research in the field of molecular electronics. The molecular electronics technology is still in its infancy but brings hope for truly atomic scale electronic systems in the future. This is one of the many possible ways in which the molecular diode/transistor might be synthesized by organic chemistry.

Based on nanoscience and nanotechnology a number of new fields such as DNA nanotechnology, nano-oncology, nanoelectronics, nanoinformatics, nanobiosensors, nanomedicines, nanopharmaceutics etc. have emerged. Scientists are working for new breakthroughs in nanoscience and nanotechnology in order to make human life easier and more comfortable. Memory storage devices can be made using the concept of high magnetoresistance. Giant magnetoresistance can be obtained by separating two ferromagnetic layers by a nonmagnetic layer having several nm thickness. Similarly one can have tunneling magnetoresistance, just like giant magnetoresistance, which is based on spin-dependent tunneling of electrons. Both giant magnetoresistance and tunneling magnetoresistance effects can be used to increase the density of memory chips (density of terabyte per square inch) and can act as main memory for computers. Nanotransistors will immensely reduce the size of integrated circuits. Nanosized electronic components can improve the quality of display screens in addition to reducing the power consumption and decreasing the weight and thickness of screens.

In the early 2000s, the field of nanotechnology was the subject of growing public awareness and controversies with prominent debates about the potential applications as well as feasibility of applications. Potential applications of QDs include single-electron transistors, solar cells, LEDs, diode lasers, single photon sources, second harmonic generation, and quantum computing. They operate like single-electron transistor and show the Coulomb blockade. QDs have potential use in making diode lasers and amplifiers. The QD lasers are cheap and offer a higher beam quality than the conventional laser diodes. QDs are most suitable for photovoltaic devices such as solar cells, LEDs, and quantum dot displays. If the band gap and chiral angle of CNTs are controlled, these would also be a promising candidate for future nanoscale devices.

Nanoradio is a technology in which nanotubes act as radio transmitters and receivers. In 1995, Nasreen Chopra and coworkers were the first to synthesize boron nitride nanotubes with diameters in the range of 1–3 nm. In 2003, Adam Fennimore and coworkers created and successfully operated carbon nanotube chemical sensors. In 2007, Kenneth Jensen and coworkers built the world's first fully functional and fully integrated radio receiver from a single carbon nanotube. The nanotube performs all the functions of a radio like antenna, amplifier, demodulator and tunable band-pass filter. Only voltage is supplied from outside by a battery. In 2008, Jensen and coworkers constructed a carbon nanotube-based nanomechanical resonator. This device is essentially a mass spectrometer with resolution at the atomic scale. Whole of this work was done in the group of Alex Zettl (1956–present) at the University of California, Berkeley. Other large numbers of applications of nanotechnology are under development such as transistors, integrated circuits from carbon nanotubes, and forming PN junctions in graphene etc. Scientists are trying to fabricate computers with nanodevices. FETs have been made using both semiconducting carbon nanowires and heterostructured semiconducting nanowires. Attempts are being made to use nanowires or small molecules in place of CMOS. The laws of quantum mechanics can be used to make quantum computers which will enable us to use the fast quantum algorithms. The quantum computer has a quantum bit memory space termed "*Qubit*" and can perform several computations at the same time.

Attempts to fabricate mechanical devices on the molecular level have yielded analogues of rotors, gears, switches, shuttles etc. In 1999, Nagatoshi Koumura and coworkers invented the world's first unidirectional light driven molecular rotary motor with controlled conversion of energy. The repetitive unidirectional rotation was found around a central carbon–carbon double bond in a chiral helical alkene in which the full rotation of 360° is performed in four steps. In 2005, Yasuhiro Shirai and coworkers were able to synthesize thermally-driven single-molecule nanocars having directional control. These nanocars have a new type of fullerene-based wheel-like rolling motion which was later replaced by a better system of motion. In 2006, Rieske Eelkema and coworkers designed a synthetic light-driven molecular motor embedded in a liquid crystal film. It can rotate microscale objects on the surface of the film. In 2006, Wesley Browne and Ben Feringa presented a review of the work done on the control of motion at the molecular level including solid state and surface mounted rotors and its use in the development of synthetic molecular machines. In 2009, Guillaume Vives and James Tour designed a single molecule nanocar which comprises a chassis with wheels at the ends of the axles: C_{60} fullerenes are used for the wheels. These nanocars convert heat, light and electric field into controlled motion on the surface. They demonstrated the directional rolling of the nanocar on the gold surface by the scanning tunneling microscope. Vives and Tour further synthesized nanotrains and nanotrucks for the transport of atoms and molecules. Later, in 2011, Tibor Kudernac and coworkers synthesized a motorized nanocar which had molecular motors attached to the chassis as rotating wheels. They synthesized a molecule with four functional units acting as motors. These four motors rotate in the same direction thereby producing linear translational motion of the molecule across the Cu(111) surface through a paddle-wheel type movement. Their design provides a starting point for the exploration of more sophisticated molecular mechanical systems with directionally controlled motion. Most of the work on molecular machines, especially nanocars was done under the supervision of Bernard Lucas Feringa (1951–present) and has laid the foundation for a key component of future molecular nanotechnology that is nanomachines and nanorobots powered by molecular motors.

Bernard Lucas Feringa (1951–present), , URL: https://en.wikipedia.org/wiki/Ben_Feringa#/media/File:Ben_Feringa_-_april_2017_05.png

Richard Feynman was the first who mentioned the word nanobot (or nanorobot) in 1959 in his famous talk "There is plenty of room at the bottom". In today's scientific world nanorobot is a tiny nanomachine programmed to carry out a particular task with great precision at nanoscale. Nano robotics is an emerging technology for creating nanorobots. In 1998, Robert Freitas performed the first study on nanorobots which was related to medical nanorobots called" respirocytes" resembling red blood cells. In 1998, Desai and coworkers created one of the earliest therapeutically useful F nanodevices. Actually, Adriano Cavalcanti is given the credit of being the inventor of medical nanorobots and is considered the world's leading expert on nanorobotics and nanorobot circuit manufacturing. In 2017, scientists created nanorobots that targeted cancer cells and killed them leaving healthy cells unharmed. In 2020, an article was published in "Nature Nanotechnology" stating that scientists at Michigan State University and Stanford University created nanoparticles that ate away plaque inside arteries that can cause heart attack. The question arises whether such nanoparticles and nanorobots are real? At present one can say that many real examples of nanotechnology do exist but nanorobots are imaginary. A number of other molecular machines were synthesized such as nanorobotic arms, DNA nanomachines and self-assembly machines etc. There is a range of artificial molecular- and nano-structures that show promise for making next-generation nanomachines. Artificial nanosystems are capable of performing useful functions, very often by means of doing mechanical work at the nanoscale. Jean-Pierre Sauvage (1944–present), Sir James Fraser Stoddart (1942–present) and Bernard Lucas Feringa (1951–present) were awarded 2016 Nobel Prize in Chemistry for designing and synthesizing molecular machines.

QDs have a number of potential applications in biology especially biological sensors. Nanomaterials are thousand times smaller than the cell of human body therefore these can slip into a cell without being recognized by the immune system. Pharmacological agents can be put into the buckyballs to deliver the medicine more effectively inside the cell. Radioactive elements can also be put into a buckyball, which can travel through the blood stream and emit radiation. Because buckyballs are excreted intact, the radiation source is removed from the body, reducing the complication of radioactive toxicity. QDs can act as much brighter dye which is more stable than the organic dyes. The use of QDs for highly sensitive cellular imaging has seen major advances. Nanomachines might be possible, which could be directed to correct defects in the cell, to kill cancerous cells without using radiotherapy and chemotherapy or to repair cell damage. In 2010, a team of scientists at the University of California, San Diego, created nanorobots that could move through the bloodstream and deliver drugs directly to tumors. This marked a significant milestone in the field of nanomedicine. In 2018, researchers at the Max Planck Institute for Intelligent Systems, Germany, developed a swarm of tiny, magnetically controlled nanorobots that could navigate through the bloodstream and remove harmful substances.

Nanoparticles are so small that they can directly enter the body of living animals through the pores of their skin. There is a possibility that nanoparticles of some materials may be toxic in nature therefore they must be tested for their toxicity. The toxic nanoparticles need to be produced under controlled environments. In 2002, Michael Crichton (1942–2008), writer and film-maker, published an article, "Prey", in Seed magazine about nanotechnology, genetic engineering and artificial intelligence. In this article he expressed his fear that the artificial swarm of nanorobots which develop intelligence may be a threat to their human inventors. In general, there is a need to investigate the toxicity of the various nanomaterials. If the nanoparticles of a particular material happen to be really toxic, they can be a threat to the life of all living animals and human beings and can create a severe health hazard.

The future of nanoscience and nanotechnology seems to be bright. The nanoparticles are quantum systems and there is a need to understand theoretically the various properties exhibited by them. The main thrust in nanoscience and nanotechnology is to devise methods to synthesize crystalline nanostructured bulk materials on one side and produce nanomachines on the other. It would not be wrong to say that twenty-first century belongs to nanoscience and nanotechnology.

Bibliography

Alferov ZI, Kazarinov RF (1963) Authors certificate 28448 (U.S.S.R)

Allen CN, Lequeux N, Chassenieux C, Tessier G, Dubertret B (2007) Optical analysis of beads encoded with quantum dots coated with a cationic polymer, Adv Mater 19:4420; (2007) Nanocrystal coated beads for optical coding and whispering gallery modes. Art no 67962Z, In: Armitage J (ed) Photonics, vol. 6796. Spie Int. Soc. optical engineering, Bellingham, pp Z7962–Z7962; Nasilowski M, Mahler B, Lhuillier E, Ithurria S, Dubertret B (2016) Two-dimensional colloidal nanocrystals. Chem Rev 116:10934–10982

Allenspach R, Salemink H, Bischof A, Weibel E (1987) Tunneling experiments involving magnetic tip and magnetic sample. Z Phys B67:125–128

Averin DV, Likharev KK (1986) Coulomb blockade of single-electron tunneling, and coherent oscillations in small tunnel junctions. J Low Temp Phys 62:345–373

Aviram A, Ratner MA (1974) Molecular rectifiers. Chem Phys Lett 29:277–283

Bethune DS, Klang CH, De Vries MS, Gorman G, Savoy R, Vazquez J, Beyers R (1993) Cobalt-catalysed growth of carbon nanotubes with single-atomic-layer walls. Nat (London) 363:605–607

Binnig G, Quate CF, Gerber C (1986) Atomic force microscope. Phys Rev Lett 56:930–933

Binnig G, Rohrer H (1981) Scanning tunneling microscope. U. S. patent no. 4343993A issued on Aug 10, 1982

Binnig G, Rohrer H, Gerber C, Weibel E (1983) 7×7 reconstruction on Si(111) resolved in real space. Phys Rev Lett 50:120–123; (1982) Tunneling through controllable vacuum gap. Appl Phys Lett 40:178

Browne WR, Feringa BL (2006) Making molecular machine work. Nat Nanotechnol 1:25–35

Busch H (1926) Berechnung der bahn von kathodenstrahlen im axialsymmetrischen elektromagnetischen felde. Ann Phys 386:974–993

Cavalcanti A (2009) Nanorobot invention and linux: the open technology factor—an open letter to UNO general secretary, CANNXS project, vol 1, pp 1–4; Cavalcanti A, Shirinzadeh B, Zhang M, Kretly LC (2008) Nanorobot hardware architecture of medical defense. Sensors 8:2932–2958; Cavalcanti A, Shirinzadeh B, Fukuda T, Ikeda S (2009) Nanorobot for brain aneurysm. Int J Robot Res 28:558–570

Chopra NG, Luyken RJ, Cherrey KC, Vincent HC, Cohen ML, Louie S, Zettl A (1995) Boron nitride nanotubes. Science 269:966–967

Cox AJ, Louderback JG, Apsel SE, Bloomfield LA (1994) Magnetism in 4d- transition metal clusters. Phys Rev B 49:12295

Desai TA, Chu WH, Tu JK, Beattie GM, Hayek A, Ferrari M (1998) Biotechnol Bioeng 57:118

Dingle R, Wiegmann W, Henry CH (1974) Quantum states of confined carriers in very thin $Al_x Ga_{1-x}As$–GaAs–$Al_x Ga_{1-x}As$ heterostructures. Phys Rev Lett 33:827

Dingle R, Henry CH (1975) U. S. patent no. 3982207 filed on March, 7, 1975 and issued on Sept 21, 1976

Drexler KE (1981) Molecular engineering: an approach to the development of general capabilities for molecular manipulation. Proce Natl Acad Sci (USA) 78:5275–5278

Drexler KE (1986) Engines of creation: the coming era of nanotechnology. Knopf Doubleday Publishing Group

Eelkema R, Pollard MM, Vicario J, Katsonis N, Ramon BS, Bastiaansen CWM, Boer DJ, Feringa BL (2006) Nanomotor rotates microscale objects. Nature 440:163–163

Efros AL, Efros AL (1982) Interband absorption of light in semiconductor sphere. Sov Phys Semicond 16:72–775; (1982) Interband absorption of light in semiconductor sphere. Fizika Tverdogo Tela 16:1209–1214

Eigler DM, Schweizer EK (1990) Positioning single atoms with a scanning tunneling microscope. Nature 344:524–526

Ekimov AI, Onushchenko AA (1981) Quantum size effect in three-dimensional microscopic semiconductor crystals. JETP Lett 34:345; (1982) Quantum size effect in the optical-spectra of semiconductor micro-crystals. Sov Phys Semicond 16:775–778
Ekimov AI, Efros AL, Onushchenko AA (1985) Quantum size effect in semiconductor microcrystals. Solid State Commun 56:921–924; (1996) Growth and optical properties of semiconductor nanocrystals in a glass matrix. J Lumin 70:1–20
Esaki L, Tsu R (1970) Superlattice and negative differential conductivity in semiconductors. IBM J Res Dev 14:61–65; (1969) IBM research report RC2418
Faraday M (1857) The Bakerian lecture. Experimental relations of gold (and other metals) to light Philos. Trans R Soc (London) 147:145
Fennimore AM, Yuzvinsky TD, Han W, Fuhrer MS, Cumings J, Zettl A (2003) Rotational actuators based on carbon nanotubes. Nature 424:408–410
Feynman RP (1960) There's plenty of room at the bottom. Eng Sci 23:22
Freitas R (1998) Exporatory design in medical nanotechnology: a mechanical artificial red cell. Artif Cells Blood Subsit Immobil Biotechnol 26:411–430
Frenkel J, Doefman J (1930) Spontaneous and induced magnetisation in ferromagnetic bodies. Nature 126:274–275
Fulton TA, Dolan GJ (1987) Observation of single-electron charging effects in small tunnel junctions. Phys Rev Lett 59:109
Hisamoto D, Kaga T, Kawamoto Y, Takeda E (1989) A fully depleted lean-channel transistor (DELTA)-a novel vertical ultra thin SOI MOSFET. International technical digest on electron devices meeting, pp 833–836
Iijima S (1991) Helical microtubules of graphitic carbon. Nat (London) 354:56–58
Iijima S, Ichihashi T (1993) Single-shell carbon nanotubes of 1-nm diameter. Nat (London) 363:603–605
Jensen K, Kim K, Zettl A (2008) An atomic-resolution nanomechanical mass sensor. Nat Nanotechnol 3:533–537
Jensen K, Weldon J, Garcia H, Zettl A (2007) Nanotube radio. Nano Lett 7:3508–3511
Knoll M, Ruska E (1932) Beitrag zur geometrischen elektronenoptik I. Ann Phys 404:607–640; (1932) Beitrag zur geometrischen elektronenoptik II. Ann Phys 404:641–661; (1932) Das elektronenmikroskop. Z Phys 78:318–339
Kodama RH (1999) Magnetic nanoparticles. J Magn Magn Mater 200:359–372
Kong XY, Ding Y, Yang R, Wang ZL (2004) Single-crystal nanorings formed by epitaxial self-coiling of polar nanobelts. Science 303:1348–1351
Koumura N, Zijlstra RWJ, van Delden RA, Harada N, Feringa BL (1999) Light-driven monodirectional molecular rotor. Nature 401:152–155
Krätschmer W, Lamb LD, Fostiropoulos K, Huffman DR (1990) Solid C_{60}: a new form of carbon. Nature 347:354–358; Huffman DR (1991) Solid C_{60}. Phys Today 44:22–29
Kroemer H (1963) A proposed class of heterojunction injection lasers. Proc IEEE 51:1782–1783
Kroto HW, Heath JR, O'Brien SC, Curl RF, Smalley RE (1985) C_{60}: Buckminsterfullerene. Nat (London) 318:162–163; Kroto H (1988) Space, stars, C_{60}, and soot. Science 242:1139–1145
Kudernac T, Ruangsupapichat N, Parschau M, Marcia BM, Katsonis N, Harutyunyan SR, Ernst K, Feringa BL (2011) Electrically driven directional motion of a four-wheeled molecule on a metal surface. Nature 479:208–211
Li ZJ, Chen XL, Li HJ, Tu QY, Yang Z, Xu YP, Hu BQ (2001) Synthesis and Raman scattering of GaN nanorings, nanoribbons and nanowires. Appl Phys A 72:629–632
Martin Y, Wickramasinghe HK (1987) Magnetic imaging by force microscopy with 1000 Å resolution. Appl Phys Lett 50:1455–1457
Mattauch J, Ewald H, Hahn O, Strassmann F (1943) Hat ein caesium-isotop langer halbwertszeit existiert? Ein beitrag zur deutung ungewöhnlicher linien in der massenspektrographie. Z Phys 120:598–617
Müller EW (1936) Experimente zur theorie der elektronenemission unter dem einfluß starker felder. Z Phys 37:838–841

Müller EW, Panitz JA, Brooks MS (1968) The atom-probe field ion microscope. Rev Sci Instrum 39:83–86; (1970) The atom-probe field-ion microscope. Naturwissenschaften 5:222–230
Müller EW (1951) Das feldionenmikroskop. Z Phys 131:136–142; (1965) Field ion microscopy. Science 149:591–601
Murray CB, Norris DJ, Bawendi MG (1993) Synthesis and characterization of nearly monodisperse CdE (E = S, Se, Te) semiconductor nanocrystallites. J Amer Chem Soc 115:8706–8715
Murray CB, Kagan CR, Bawendi MG (1995) Self-organization of CdSe nanocrystallites into three-dimensional quantum dot superlattices. Science 270:1335–1338
Novoselov KS, Geim AK, Morozov SV, Jiang D, Zhang Y, Dubonos SV, Grigorieva IV, Firsov AA (2004) Electric field effect in atomically thin carbon film. Science 306:666–669
Poole CP, Owens FJ (2003) Introduction to nanotechnology. Wiley, New York, pp 88–89
Respaud M, Broto JM, Rakoto H, Fert AR, Thomas L, Barbara B, Verelst M, Snoeck E, Lecante P, Mosset A, Osuna J, Ely TO, Amiens C, Chaudret BB (1998) Surface effects on the magnetic properties of ultrafine cobalt particles. Phys Rev B 57:2925–2935
Rossetti R, Nakahara S, Brus LE (1983) Quantum size effects in the redox potentials, resonance Raman spectra, and electronic spectra of CdS crystallites in aqueous solution. J Chem Phys 79:1086–1088
Saenz JJ, Garcia N, Grütter P, Meyer E, Heinzelmann H, Wiesendanger RL, Rosenthaler L, Hidber HR, Güntherodt HJ (1987) Observation of magnetic forces by the atomic force microscope. J Appl Phys 62:4293–4295
Senyei AE, Widder KJ, Czerlinski C (1978) Magnetic guidance of drug carrying microspheres. J Appl Phys 49:3578–3583
Shirai Y, Osgood AJ, Zhao Y, Kelley KF, Tour JM (2005) Directional control in thermally driven single-molecule nanocars. Nano Lett 5:2330–2334
Taniguchi N (1974) On the basic concept of nano-technology. Proceeding International conference products engineering Tokyo, Part II, Japan society of precision engineering
Vives G, Tour JM (2009) Synthesis of single-molecule nanocars. Acc Chem Res 42:473–487
Von Ardenne M (1938) Das elektronen-rastermikroskop, theoretische grundlagen. Z Phys 109:553–572; (1938) Das elektronen-rastermikroskop, praktische ausführung. Zeitschrift für technische Physik 19:407–416
Von Hippel AR (1956) Molecular engineering. Science 123:315–317
Von Hippel AR (1959) Molecular science and molecular engineering, the technology. Press of MIT and Wiley, Inc, New York
Von Hippel AR, Landshoff R (1959) Molecular science and molecular engineering. Phys Today 12(10):48
Widder KJ, Senyei AE, Scarpelli GD (1978) Magnetic microspheres: a model system of site specific drug delivery in vivo. Proc Soc Exp Biol Med 158:141–146
Xu X, Ray R, Gu Y, Ploehn HJ, Gearheart L, Raber K, Scrivens WA (2004) Electrophoretic analysis and purification of fluorescent single-walled carbon nanotube fragments. J Amer Chem Soc 126:12736–12737
Zsigmondy ER (1914) Colloids and ultramicroscope. Wiley, New York

Epilogue

The history of a scientific discipline is crucial, as it illuminates its origins, the trajectory of intellectual inquiry, the challenges encountered, and how those challenges were overcome. Human rationality evolved alongside the rise of civilizations across different parts of the world. The early thinkers—keen observers and serious thinkers—were called philosophers. Their insights emerged from speculation, observation, and lived experience, rather than systematic experimentation. In antiquity, humans began exploring their immediate environment: the Earth, Sun, stars, planets, light from celestial bodies, and the motion of objects; both living and non-living—on the Earth's surface. These observations gave rise to the earliest sciences: astronomy, optics, and mechanics.

In the pre-Christian era, philosophers like Thales of Miletus (626/623BC–548/545BC), Pythagoras (570BC–495BC), Socrates (470BC–399BC), Plato (428/427BC–348/347BC), Aristotle (384BC–322BC), and Archimedes (287BC–212BC) gave foundational ideas and hypotheses. Aristotle, often credited as the first scientist, introduced logical reasoning into scientific inquiry. He proposed the geocentric model of the universe, which was later refined by Claudius Ptolemy (100–170) and remained dominant for nearly 1500 years. When I was a student in the 1950s and 1960s, we often slept on rooftops, watching stars, moon, and planets. At night, they appeared to move from east to west, just as the Sun did during the day. I believe the geocentric model of universe may have originated from such simple, intuitive observations.

In 1543, Nicolaus Copernicus (1473–1543) proposed the heliocentric model of the universe. Johannes Kepler (1571–1630) later expanded it into a detailed theory. He is best known for his laws of planetary motion, which revolutionized scientific understanding. Meanwhile, optics and mechanics were also progressing. By the sixteenth century, speculative theories began to face failure such as the geocentric model and the notion that vision occurred through light emitted from the eyes.

Galileo Galilei (1564–1642) asserted that scientific conclusions must be based on experimental evidence. He is credited with establishing the role of experimentation in science. Along with Isaac Newton (1643–1727), he laid the foundation for classical mechanics. Newton's laws of motion and gravitation transformed the field.

J. S. Galsin, *History of Solid State Physics*, https://doi.org/10.1007/978-981-95-0504-3

The science of heat evolved in the sixteenth and seventeenth centuries, while thermodynamics took shape in the nineteenth century. Optics advanced steadily from the 17th to nineteenth centuries. James Clerk Maxwell (1831–1879) proposed the electromagnetic theory of light, which became well-established by the end of the nineteenth century. The subjects of electricity and magnetism also matured during this period.

Solids have been known to humankind since the Stone Age, around 2.6 million years ago. Initially, they were seen as structureless matter. In 1591, Thomas Harriot (1560–1621) first observed hexagonal symmetry in snowflakes. Johannes Kepler later noted that snowflakes consist of spherical droplets arranged in hexagonal patterns—marking the first recognition of symmetry in solids. By the late seventeenth century, scientists had discovered cleavage planes in solids, giving rise to the science of macroscopic crystallography. Originally part of mineralogy, crystallography evolved through the work of scientists like René Just Haüy (1743–1822), William Hallowes Miller (1801–1880), and Ludwig Seeber (1793–1855). In 1803, John Dalton (1766–1844) formally introduced the concept of atoms, although the idea dates back to Democritus (460BC–370BC). Dalton's model laid the groundwork for classifying solids based on atomic arrangements. Major contributions to crystallography came from August Bravais (1811–1863), Evgraf Fedorov (1853–1919), and Arthur Schoenflies (1853–1928).

By the end of the nineteenth century, many physicists believed that science had reached completion and only more precise measurements were needed. However, classical physics failed to explain phenomena such as atomic line spectra, blackbody radiation, and the photoelectric effect. In the last decade of nineteenth century, the discoveries of electrons, X-rays, and radioactivity signaled the onset of a new scientific era.

Max Planck (1858–1947), in 1900, proposed the concept of energy quanta to explain blackbody radiation—a revolutionary idea that unknowingly laid the foundation for quantum mechanics. In 1905, Albert Einstein published three landmark papers: the first proved the existence of atoms through Brownian motion in liquids; the second explained the photoelectric effect via quantized light (photons); the third introduced the theory of special relativity, challenging Newtonian assumptions and revolutionizing mechanics. Therefore, 1905, is assumed to be the real birth of quantum mechanics.

As science transitioned from macroscopic to microscopic understanding, quantum mechanics in conjunction with relativity theory emerged as essential for explaining atomic and subatomic systems. Under specific conditions, quantum effects also manifest at the macroscopic level—for instance, in superconductivity and Bose–Einstein condensation at very low temperatures. The atomic model's refinements enabled deeper development of quantum theory. Early quantum mechanics evolved in Europe through major contributions from Niels Bohr, Arnold Sommerfeld, Werner Heisenberg, Max Born, Wolfgang Pauli, and Paul Dirac. Scientists soon realized that quantum mechanics could be used to study the physics of solids at atomic level. Later it was found that solid-state phenomena like magnetism, superconductivity, band theory, lattice vibrations, magnetic resonance phenomena, and quantum Hall effect cannot be explained without quantum mechanics.

With the advent of nanoscience and nanotechnology in the late twentieth century, researchers discovered that material properties change drastically at the nanoscale. Insulators may become conductors or semiconductors, and non-magnetic substances can exhibit magnetism. At this scale, the properties of materials may differ from their bulk characteristics. Nanosolids are inherently quantum systems, though comprehensive theoretical interpretations are still evolving. There are some quantum mechanical concepts in solids which have no classical analogue such as intrinsic electron spin, electron–electron exchange interactions, and quantum tunneling.

Richard Phillips Feynman (1918–1988) is considered the most influential physicist of the second half of the twentieth century. In the beginning of "*Feynman Lectures in Physics, Vol. I*", he wrote "*I believe that the atomic hypothesis (that is all things are made of atoms) contained most of the information of scientific knowledge of the world*". It has been found that both the internal atomic structure and special arrangement of atoms are crucial in determining the properties of solids. For example, line spectrum of elements, lasers, masers, magnetic resonance etc. are explained by the internal atomic structure. Whether a solid is a metal, insulator or semiconductors is determined by the different arrangements of atoms in it. In magnetic materials the different arrangements of atoms with finite magnetic moment yield different types of magnetic materials such as ferromagnetic, antiferromagnetic and ferrimagnetic materials etc. Further, it has been observed that same atoms may exhibit different order under different conditions. The best example is that of carbon atoms which yield dramatically different materials based on their arrangement: diamond (an insulator), graphite (a conductor), carbon nanotubes (conductors/semiconductors), and coal.

Naturally occurring solids are largely crystalline, likely due to Earth's gradual cooling from a molten state. Amorphous solids, with only short-range atomic order, are generally man-made and formed through rapid solidification.

Since electrons in atoms move at speeds approaching that of light, quantum mechanics combined with relativity (relativistic quantum mechanics) is essential for understanding their behavior. Matter, as per relativity, is a form of energy and quantum theory quantizes it. Thus, the universe consists of different forms of quantized energy fields. The concept of Anaximander (610BC–546BC) that all matter comprises of *apeiron* seems to be correct. According to conservation of energy, the energy present at the time of the Big Bang still exists today, though transformed into various forms. Following the Big Bang, hydrogen formed first, then helium and lithium. Heavier elements were synthesized later inside the stars. The quantization of energy is encapsulated by nonrelativistic Schrödinger equation, and field quantization by Dirac and Schrödinger relativistic equations. Every elementary particle has a quantized field associated with it. Particle interactions are governed by the exchange of quanta of field. For example, in proton-proton interaction the fields of two protons interact with the exchange of mesons.

The twentieth century is usually called the century of technologies as it witnessed technological leaps—from solid-state electronics and transistors to integrated circuits, computers, televisions, cell phones, airplanes, nuclear power, and solar energy. These developments marked the birth of the information age reducing the world to a village. The ongoing miniaturization of components has led to nanoscience and nanotechnology, and emerging fields like nano-optics, nanoelectronics, nanomedicines, and nanocosmetics etc. Nanomachines have also been made although these are still in its infancy. Although bulk crystalline nanostructured materials are still under development, future breakthroughs are expected to build upon nanoparticles as foundational units. It is believed that twenty-first century belongs to nanoscience and nanotechnology.

In conclusion, quantum mechanics explains most solid-state properties. In quantum theory, one begins with known initial conditions, solves complex equations with approximations, and validates the outcomes through experiment. But the question is what actually the quantum mechanics is? The physicists have different opinion about it. Einstein, skeptical of its probabilistic interpretation, famously said, "*God does not play dice.*" Richard Feynman in his book, "*Lectures on Physics, Vol. 3*" described Young's double slit experiment using bullets, electrons and light and it illustrates quantum uncertainty and the observer effect. If we measure which slit an electron passes through, interference disappears—not because electrons change their nature, but because their wave functions collapse. Further, what actually happens between the holes and screen is not known and it is just like a black box. Feynman described the double-slit experiment as the central mystery of quantum mechanics and said "*it is impossible to explain it in any classical way*".

Even as late as the nineteenth century, many scientists believed that the work of science was nearly at an end. But it has been seen how new subjects of quantum mechanics and relativity emerged in the beginning of the twentieth century and revolutionized physics at the atomic and subatomic level: a part of history. Simultaneously the miniaturization of material components gave birth to solid state electronics, and nanoscience and nanotechnology. Many scientists are of the opinion that nanotechnology is the ultimate technology which I believe may not be strictly true because science never truly concludes. At present we do not know what type of new physics is in store of us and what type of new technologies may emerge in future. The theories which are assumed to be correct today may undergo modifications or may prove to be false in future. This is how the science progresses. Max Planck (1858–1947) had rightly said, "*We have no right to assume that any physical laws exist or if they have existed up to now, that they will continue to exist in a similar manner in future*".

MIX
Papier aus verantwortungsvollen Quellen
Paper from responsible sources
FSC® C105338

If you have any concerns about our products,
you can contact us on
ProductSafety@springernature.com

In case Publisher is established outside the EU,
the EU authorized representative is:
Springer Nature Customer Service Center GmbH
Europaplatz 3, 69115 Heidelberg, Germany

Printed by Libri Plureos GmbH
in Hamburg, Germany